Likelihood and Its Extensions

Likelihood serves as a unifying concept in both the theory and practice of statistical science. This is, in a sense, inevitable when probability models are used as a basis for inference. While the key ideas were set out in Fisher in 1922, and further developed throughout the 1930s and 40s, it was the ubiquity of the personal computer and the development of general-purpose software that made likelihood-based inference the method of choice in a wide variety of applications.

This book provides an overview of the many "adjective"-likelihood functions that have been developed in various contexts, aiming to include a wide array of inference functions used in the current literature, while recognizing that a comprehensive treatment is not possible, as research on likelihood-based inference continues.

This book is intended for readers with diverse backgrounds who have an interest in, or a need for, statistical methods in complex models. Some familiarity with likelihood-based inference and the main principles of estimation and hypothesis testing are assumed. The authors have used this text for senior undergraduate and graduate courses in inference.

Nancy Reid is University Professor Emeritus of Statistical Sciences at the University of Toronto.

Cristiano Varin is Professor of Statistics at Ca' Foscari University of Venice.

Grace Y. Yi is Professor and Canada Research Chair in Data Science at the University of Western Ontario.

MONOGRAPHS ON STATISTICS AND APPLIED PROBABILITY

Recently Published Titles

Dynamic Treatment Regimes
Statistical Methods for Precision Medicine
Anastasios A. Tsiatis, Marie Davidian, Shannon T. Holloway, and Eric B. Laber 164

Sequential Change Detection and Hypothesis Testing
General Non-i.i.d. Stochastic Models and Asymptotically Optimal Rules
Alexander Tartakovsky 165

Introduction to Time Series Modeling
Genshiro Kitigawa 166

Replication and Evidence Factors in Observational Studies
Paul R. Rosenbaum 167

Introduction to High-Dimensional Statistics, Second Edition
Christophe Giraud 168

Object Oriented Data Analysis
J.S. Marron and Ian L. Dryden 169

Martingale Methods in Statistics
Yoichi Nishiyama 170

The Energy of Data and Distance Correlation
Gabor J. Szekely and Maria L. Rizzo 171

Sparse Graphical Modeling for High Dimensional Data: Sparse Graphical Modeling for High Dimensional Data
Faming Liang and Bochao Jia 172

Bayesian Nonparametric Methods for Missing Data and Causal Inference
Michael J. Daniels, Antonio Linero, and Jason Roy 173

Functional Data Analysis with R
Ciprian M. Crainiceanu, Jeff Goldsmith, Andrew Leroux, and Erjia Cui 174

Mixture Models: Parametric, Semiparametric, and New Directions
Weixin Yao and Sijia Xiang 175

Seminal Ideas and Controversies in Statistics
Roderick J.A. Little 176

Theory of Non-Regular Factorial Designs
Ching-Shui Cheng and Boxin Tang 177

Robust Small Area Estimation: Methods, Theory, Applications, and Open Problems
Jiming Jiang and J. Sunil Rao 178

Hierarchical Modeling and Analysis for Spatial Data, Third Edition
Sudipto Banerjee, Alan E. Gelfand, and Bradley P. Carlin 179

Likelihood and Its Extensions
Nancy Reid, Cristiano Varin, and Grace Y. Yi 180

For more information about this series, please visit: https://www.crcpress.com/Chapman--HallCRC-Monographs-on-Statistics--Applied-Probability/book-series/CHMONSTAAPP

Likelihood and Its Extensions

Nancy Reid, Cristiano Varin, and Grace Y. Yi

CRC Press
Taylor & Francis Group
Boca Raton London New York

CRC Press is an imprint of the
Taylor & Francis Group, an **informa** business
A CHAPMAN & HALL BOOK

Designed Cover Image: Nancy Reid, Cristiano Varin, and Grace Y. Yi

First edition published 2026
by CRC Press
2385 NW Executive Center Drive, Suite 320, Boca Raton, FL 33431

and by CRC Press
4 Park Square, Milton Park, Abingdon, Oxon, OX14 4RN

CRC Press is an imprint of Taylor & Francis Group, LLC

ISBN: 978-1-498-71974-2 (hbk)
ISBN: 978-1-041-34357-8 (pbk)
ISBN: 978-0-429-09093-6 (ebk)

DOI: 10.1201/9780429090936

Typeset in Latin Modern font
by KnowledgeWorks Global Ltd.

Publisher's Note: This book has been prepared from camera-ready copy provided by the author.

To our families

Contents

Preface

Likelihood serves as a unifying concept in both the theory and practice of statistical science. This is in a sense inevitable once we have decided to use probability models as a basis for inference. While the key ideas were set out in Fisher in 1922, and further developed by Cramér, Wald, Wilks, Neyman, and others throughout the 1930s and 40s, it was the ubiquity of the personal computer and the development of general-purpose software that made likelihood-based inference the method of choice in a wide variety of applications.

As data and models increased in complexity, inference based on classical likelihood summaries such as the maximum likelihood estimator or the likelihood ratio test were not always feasible or useful: the asymptotic theory might be found to lead to poor approximations in finite samples; computation of the likelihood function might be too difficult; in some cases the likelihood function could not even be given mathematical expression.

As a result, many variations on the likelihood theme started to emerge in the literature. The adjectives marginal, conditional, pseudo, quasi, composite, empirical and others were placed in front of "likelihood" and the related inference functions were proposed as alternatives to, or extensions of, Fisher's original program.

This book was conceived as an effort to provide an overview of these many adjective-likelihoods that have arisen in various contexts. The literature was somewhat scattered, and the same adjectives were sometimes used for somewhat different constructs. As the work progressed, we were guided by the complexities of modern data to expand from likelihood-based inference to related topics. A comprehensive treatment is impossible; new papers on various aspects of likelihood-like inference appear regularly in the statistical literature. We have attempted to provide a reasonably thorough overview of the subject as we see it now, recognizing that many topics of current interest could not be included.

The first chapter introduces the classical theory of parametric inference based on the likelihood function, and some of the statistical models used as examples throughout the book. Some, such as linear and generalized linear models, are foundational building blocks for any course in statistics; others, such as the Ising model, or models for extremes, are somewhat more specialized, and introduced to motivate the developments in later chapters. Chapter 2 continues with classical approaches to inference in the presence of nuisance parameters: marginal and conditional, profile, and partial likelihood functions are discussed. Chapter 3 expands on the theme of data complexity by introducing a range of models for time series, spatial data, longitudinal data, missing and mis-measured data, and survival data. Chapter 4 presents inferential theory under model mis-specification, with particular emphasis on composite likelihood. Various versions of composite likelihood for particular data structures have been discussed in the literature under the name pseudo-likelihood, as indicated in the references in that chapter. Chapter 5 discusses quasi-likelihood and links its development to the theory of estimating equations. Chapter 6 considers nonparametric and semiparametric models, of increasing importance for modern applications. The theory of likelihood inference becomes more difficult in this setting, but the motivation is the same; to have a principled set of methods for moving from a probability model to valid inference. Chapter 7 discusses various versions of regularization which have been devised to accommodate models with high-dimensional parameters.

We have aimed to make the work accessible to students and researchers who may be familiar with the classical theory, or use it in their work, but have an interest in expanding their knowledge of statistical theory, or in using a likelihood-based method in applications where the structure of their data requires going beyond the classical methods. Some familiarity with likelihood-based inference and the main principles of estimation and hypothesis testing are assumed. We have used drafts of this text for graduate-level courses in inference and in specialized courses in likelihood-based inference.

We owe a great debt to many colleagues and students who have commented on earlier drafts. In particular, we would like to highlight the very helpful contributions from Alan Agresti, Heather Battey, Ruggero Bellio, Anthony Davison, Wenqing He, Nils Lid Hjort, Art Owen, Nigel Petersen, Alessandra Salvan, Nicola Sartori, Mary Thompson, and Ruoyong Xu.

List of Symbols

$Y \in \mathbb{R}^d$	random variable ($d = 1$) or random vector ($d > 1$)
$y \in \mathbb{R}^d$	observed value
$Y = \{Y_1, \ldots, Y_n\}$	sample of n random variables
$y = \{y_1, \ldots, y_n\}$	sample of n observations
$1(Y \in A)$	indicator function of the event $\{Y \in A\}$
1_d	column vector of d ones
$\mathcal{I}_{d\times d}$	identity matrix of dimension $d \times d$
$\mathrm{pr}(Y \in A)$	probability of the event $\{Y \in A\}$
$f(y)$	density function
$F(y)$	distribution function
$\widehat{F}_n(y)$	empirical distribution function
$S(y)$	survival function
$\{f(y;\theta);\ \theta \in \mathbb{R}^p\}$	parametric model
θ_0	true parameter value
$h(y)$	density function for true model (if needed)
$H(y)$	distribution function for true model (if needed)
θ_*	least-false parameter value
$L(\theta) \propto f(y;\theta)$	likelihood function
$L(\theta;y)$, $L(\theta;Y)$	if needed to specify dependence of $L(\theta)$ on y or Y
$\ell(\theta) = \log L(\theta)$	log-likelihood function
$\ell(\theta;y)$, $\ell(\theta;Y)$	if needed to specify dependence of $\ell(\theta)$ on y or Y
$\pi(\theta), \pi(\theta \mid y)$	prior and posterior functions
$u(\theta) = \partial\ell(\theta;y)/\partial\theta$	score function (observed)
$U(\theta) = \partial\ell(\theta;Y)/\partial\theta$	score function (random vector)
$\widehat{\theta} = \widehat{\theta}(y) = \arg\sup_\theta \ell(\theta;y)$	maximum likelihood estimate
$\widehat{\theta} = \widehat{\theta}(Y) = \arg\sup_\theta \ell(\theta;Y)$	maximum likelihood estimator
$I(\theta) = \mathrm{E}_\theta\{U(\theta)U^{\mathsf{T}}(\theta)\}$	expected Fisher information function (n observations)
$J(\theta) = -\partial^2\ell(\theta;y)/\partial\theta^2$	observed Fisher information function (n observations)
$J(\widehat{\theta})$	observed Fisher information
$K(\theta) = \mathrm{E}_\theta\{J(\theta)\}$	sensitivity matrix
$i(\theta)$, $j(\theta)$, $k(\theta)$	information functions for a single observation
$w(\theta) = 2\{\ell(\widehat{\theta}) - \ell(\theta)\}$	likelihood ratio statistic
$s(\theta) = J^{-1/2}(\widehat{\theta})U(\theta)$	Rao's score statistic
$q(\theta) = J^{1/2}(\widehat{\theta})(\widehat{\theta} - \theta)$	Wald statistic
$r(\theta) = \mathrm{sign}(\widehat{\theta} - \theta)w^{1/2}(\theta)$	likelihood root statistic
$\theta = (\psi, \lambda)^{\mathsf{T}}$, $\psi \in \mathbb{R}^q, \lambda \in \mathbb{R}^{p-q}$	parameter of interest and nuisance parameter
$\widehat{\lambda}_\psi = \arg\sup_\lambda \ell(\psi,\lambda;y)$	constrained maximum likelihood estimate
$U_\psi(\theta), U_\lambda(\theta)$	partitioned score vector
$\begin{bmatrix} J_{\psi\psi} & J_{\psi\lambda} \\ J_{\lambda\psi} & J_{\lambda\lambda} \end{bmatrix}$	partitioned observed information matrix

$\begin{bmatrix} J^{\psi\psi} & J^{\psi\lambda} \\ J^{\lambda\psi} & J^{\lambda\lambda} \end{bmatrix}$	partitioned inverse observed information matrix
$L_p(\psi), \ell_p(\psi)$	profile likelihood function
$L_m(\psi), \ell_m(\psi)$	marginal likelihood function
$L_c(\psi), \ell_c(\psi)$	conditional likelihood function
$\Psi(\theta; y)$	estimating function
$\Gamma_\Psi(\theta) = \mathrm{E}\{-\partial\Psi(\theta; Y)/\partial\theta^\top\}$	sensitivity matrix
$\Sigma_\Psi(\theta) = \mathrm{E}\{\Psi(\theta; Y)\Psi^\top(\theta; Y)\}$	variability matrix
$G_\Psi(\theta) = \Gamma_\Psi(\theta)\Sigma_\Psi(\theta)\Gamma_\Psi^\top(\theta)$	Godambe information
$L_{c\ell}(\theta), \ell_{c\ell}(\theta)$	composite likelihood function
$U_{c\ell}(\theta), u_{c\ell}(\theta)$	composite likelihood score function

1

Likelihood

1.1 Models and likelihood

1.1.1 Statistical models

Statistical inference is broadly concerned with the problem of learning from data; in particular, data obtained from a system subject to random fluctuations. This problem is made somewhat more concrete by considering a family of probability models, which, it is hoped, describe at least some of the essential features of the system. These probability models are often specified by a relatively small class of probability distributions, parametrized by one or more unknown parameters, and this class is used to define a likelihood function. Inference based on the likelihood function is very widely used in applications and plays a key role in statistical theory.

As scientific problems and technological advances provide data of ever-increasing size and complexity of structure, probability models also become increasingly complex. This has meant in practice that it is often not feasible to base inference on the likelihood function, and a variety of simplifications, both computational and conceptual, have been proposed. Many of these are treated at length in subsequent chapters.

In this chapter we introduce models and likelihood and outline the key concepts in the theory of likelihood inference.

We assume that there is a notional random variable Y, assumed to follow a probability distribution with cumulative distribution function $F(y)$, for y taking values in $\mathcal{Y}$, a set of possible values usually called the sample space. Typically Y will be a scalar random variable or a vector random variable of length d, and the distribution function will have an associated density function $f(y)$, with respect to counting measure or Lebesgue measure. The distribution will nearly always depend on some unknown parameters θ, modelled as taking values in a parameter space Θ. The parameter space is often a subset of $\mathbb{R}^p$, but may be more general, even infinite-dimensional. The family of distributions is denoted by $\mathcal{F} = \{f(\cdot;\theta);\theta \in \Theta\}$.

The problem of statistical inference is to use observed values of Y to make statements about the data-generating model, or its parameters, and to associate measures of uncertainty with these statements. In some areas of application, this inference may be a first step in an analysis where the final goal is to predict new observations from the same stochastic system.

The observations are not necessarily independent realizations of Y, although this may be a suitable starting point. It will very often be the case that additional information in the form of covariates x is observed. With a slight abuse of notation we will write $Y = \{Y_1, \dots, Y_n\}$ for a sample of random variables, and $y = \{y_1, \dots, y_n\}$ for the sample of observations. It is sometimes convenient to consider Y or y as vectors in $\mathbb{R}^n$, in which case $Y = (Y_1, \dots, Y_n)^\mathsf{T}$; by convention, all vectors are column vectors. When each Y_i is a random vector in $\mathbb{R}^d$, then Y is an $n \times d$ matrix.

DOI: 10.1201/9780429090936-1

Example 1.1 (Regression illustrations) We illustrate these ideas with the following set of regression models, where for $i = 1, \ldots, n$, we assume

$$Y_i = \mu_i + \epsilon_i = \beta_0 + \beta_1 x_i + \epsilon_i,$$

with x_i known constants.

(a) The errors $\{\epsilon_1, \ldots, \epsilon_n\}$ are assumed to be independent, identically distributed standard normal variables. The joint density for the sample y is

$$f(y; \beta_0, \beta_1) = \left(\frac{1}{\sqrt{2\pi}}\right)^n \exp\left\{-\frac{1}{2}\sum_{i=1}^{n}(y_i - \beta_0 - \beta_1 x_i)^2\right\}. \tag{1.1}$$

(b) The errors $\{\epsilon_1, \ldots, \epsilon_n\}$ are assumed to be independent, identically distributed with density $f_0(\epsilon)$ or $f_0(\epsilon; \nu)$ where the form of f_0 is known, possibly up to a number of unknown parameters. For example, $f_0(\cdot)$ might be the density of Student's t distribution with ν degrees of freedom.

(c) The errors $\{\epsilon_1, \ldots, \epsilon_n\}$ are assumed to be independent, identically distributed, each having the same, unknown, density $f(\cdot)$, possibly satisfying some smoothness assumption.

(d) The error vector $\epsilon = (\epsilon_1, \ldots, \epsilon_n)^{\mathsf{T}}$ is assumed to follow a known joint distribution $f_0(\epsilon; \nu)$. This could be appropriate, for example, if the vector $y = (y_1, \ldots, y_n)^{\mathsf{T}}$ is measurements on members of a family, cluster, or spatial area, or measurements of a process evolving in time.

(e) We could replace the linear regression model by a non-linear regression model with $\mu_i = \mu(x_i; \beta)$, of some known form, the errors having any of the distributions outlined above.

(f) We could replace the linear regression model by a nonparametric regression model where $\mu_i = \mu(x_i)$ is an unknown function, assumed to have some smoothness, and the errors having any of the distributions outlined above.

(g) In any of the models above, we might have a vector, rather than a scalar, explanatory variable for each observation, which we typically write $x_i^{\mathsf{T}} = (x_1, \ldots, x_p)$.

□

Often in regression settings, particularly in observational studies, the x_i are themselves observations of a random variable or vector X_i with some probability distribution. In Example 1.1, the model for Y_i is then conditional on the observed values x_i; there may be in some applications interest in modelling the joint distribution of Y_i and X_i.

In general, we would expect to require that the dimension of the data, n, be greater than the dimension of the parameter space in order to construct inference about the parameter. This is not true in cases (c) and (f) above, as the parameter space is a space of unknown functions, and hence effectively of infinite dimension. In practice, the dimension will be reduced by, for example, assuming that $\mu(x_i; \beta)$ is a linear combination of a given set of basis functions. Model (c) is an example of a semi-parametric model, and Model (f) is often described as a nonparametric regression, although in that setting the distribution of $\epsilon = \{\epsilon_1, \ldots, \epsilon_n\}$ is often assumed to be normal. Models for dependent data are discussed at length in Chapter 3. In Chapter 4 the assumption that the model is correctly specified is relaxed. In Chapter 5 models are considered in which only some aspects of the probability distribution are specified; typically the expected value and variance. Likelihood methods for nonparametric models are discussed in Chapter 6. In regression models generalizing those above when the dimension p of the regression parameter is larger than the sample size, some form of regularization or assumption of sparsity is required, and this is the focus of Chapter 7.

Models considered in this chapter will henceforth be parametric, with a parameter $\theta = (\theta_1, \ldots, \theta_p)^{\mathsf{T}}$ taking values in a p-dimensional space Θ, typically $\mathbb{R}^p$ or a subset of $\mathbb{R}^p$. By our convention that all vectors are column vectors, a partition of θ into two sub-vectors will be written, for example, $\theta = (\theta_1^{\mathsf{T}}, \theta_2^{\mathsf{T}})^{\mathsf{T}}$.

Some additional comments on modelling are provided in Appendix 1.A.

1.1.2 The likelihood function

For a single random variable Y with density function $f(y;\theta)$, the likelihood function is defined to be

$$L(\theta; y) = c(y) f(y;\theta), \quad \theta \in \Theta. \tag{1.2}$$

It is a function of θ, proportional to the density function evaluated at the observed value of the random variable. The form of (1.2) emphasizes that there is an equivalence class of likelihood functions, all differing by arbitrary multiplicative constants that may depend on y but not on θ.

The likelihood function is invariant under one-to-one differentiable transformations of continuous random variables Y that do not involve θ, as the density of $Z = g(Y)$ is

$$f_Z(z;\theta) = f_Y\{g^{-1}(z);\theta\}|dg^{-1}(z)/dz|, \tag{1.3}$$

so the likelihood function based on Z is in the same equivalence class as the likelihood function based on Y. A similar equivalence holds for discrete distributions.

Similarly, for a sample of random variables $Y = \{Y_1, \ldots, Y_n\}$, with observations $y = \{y_1, \ldots, y_n\}$, the likelihood function for θ is proportional to the joint density of y; we will use the notation in (1.2) again for this setting, writing

$$L(\theta; y) = c(y) f(y;\theta), \quad \theta \in \Theta. \tag{1.4}$$

The equivalence class of likelihood functions defined by (1.4) depends on the sample y only through the minimal sufficient statistic for θ. Recall that a statistic $S = s(y)$ is sufficient for θ if the conditional distribution of Y, given S, is free of θ, and is minimal sufficient if it is a function of every sufficient statistic; see Appendix 1.B.

The likelihood function is invariant to re-parametrization, as it can be expressed in a new parametrization $\eta = \eta(\theta)$ without materially changing its interpretation, as long as the transformation is one-to-one. Often, $L(\theta; y)$ is shortened to $L(\theta)$, with the observed value of the relevant random variable or sample hidden. When we are interested in the probability distribution of various quantities derived from the likelihood function, we write $L(\theta; Y)$.

Example 1.2 (Bernoulli distribution) Suppose $\{Y_1, \ldots, Y_n\}$ is a sample of independent and identically distributed random variables, each taking the value 1 with probability θ and 0 with probability $1 - \theta$. The likelihood function for θ based on a sample of observations $y = \{y_1, \ldots, y_n\}$ is

$$L(\theta; y) = \prod_{i=1}^{n} \theta^{y_i}(1-\theta)^{1-y_i} = \theta^{\Sigma_i y_i}(1-\theta)^{n-\Sigma_i y_i}, \quad 0 \le \theta \le 1. \tag{1.5}$$

From this we see that $\Sigma_i y_i$, or more precisely the random variable $\Sigma_i Y_i$, is the minimal sufficient statistic for θ. □

Example 1.3 (Binomial distribution) If S is the total number of successes in n independent Bernoulli trials with constant probability of success θ, the likelihood function for θ from an observed value s is

$$L(\theta; s) = \binom{n}{s}\theta^s(1-\theta)^{n-s} \propto \theta^s(1-\theta)^{n-s}, \quad 0 \le \theta \le 1, \tag{1.6}$$

which is the same as (1.5), as $s = \Sigma_i y_i$. □

Example 1.4 (Negative binomial distribution) If instead N is the random number of independent Bernoulli trials required to observe s successes, then the likelihood function for θ based on an observed value n is given by the negative binomial probability function

$$L(\theta; n) = \binom{n-1}{s-1}\theta^s(1-\theta)^{n-s}, \quad 0 \le \theta \le 1. \tag{1.7}$$

□

As functions of θ, the likelihood functions in these three examples are in the same equivalence class, showing in particular that the likelihood function is unaffected by the sampling rule; this is true in general. However, the distribution of a quantity derived from $L(\theta; Y)$ does depend on the sampling rule.

Example 1.5 (Normal distribution) Suppose $\{Y_1, \dots, Y_n\}$ are independent and identically distributed normal random variables with expected value μ and variance σ^2. The likelihood function for $\theta = (\mu, \sigma^2)^\mathsf{T}$ is

$$\begin{aligned} L(\mu, \sigma^2; y) &\propto \sigma^{-n} \exp\left\{-\frac{1}{2\sigma^2}\sum_{i=1}^n (y_i - \mu)^2\right\}, \\ &= \sigma^{-n} \exp\left(-\frac{\Sigma_i y_i^2}{2\sigma^2} + \frac{\mu\Sigma_i y_i}{\sigma^2} - \frac{n\mu^2}{2\sigma^2}\right), \\ &= \sigma^{-n} \exp\left\{-\frac{\Sigma_i (y_i - \overline{y})^2}{2\sigma^2} - \frac{n(\overline{y}-\mu)^2}{2\sigma^2}\right\}, \quad \mu \in \mathbb{R}, \sigma^2 > 0. \end{aligned} \tag{1.8}$$

The likelihood function depends only on y only through the sufficient statistic $S = (\Sigma_i y_i^2, \Sigma_i y_i)$, or equivalently $(\Sigma_i (y_i - \overline{y})^2, \overline{y})$, and in particular if σ^2 is assumed to be known, the likelihood function for μ depends only on $\overline{y}$. □

Example 1.6 (Stationary Gaussian autoregression) As a first example with dependent data, suppose for $i = 1, \dots, n$ that the conditional density of Y_i, given $Y_{i-1} = y_{i-1}$, is normally distributed with expected value $(1-\rho)\mu + \rho y_{i-1}$ and variance σ^2. The likelihood function for $\theta = (\mu, \rho, \sigma^2)^\mathsf{T}$ based on $y = (y_1, \dots, y_n)^\mathsf{T}$ is

$$L(\theta; y) \propto f_1(y_1; \theta)\prod_{i=2}^n \frac{1}{\sigma}\exp\left[-\frac{1}{2\sigma^2}\{y_i - \mu - \rho(y_{i-1} - \mu)\}^2\right], \quad \mu \in \mathbb{R}, \sigma^2 > 0, |\rho| \le 1. \tag{1.9}$$

Various assumptions on the density of the first observation could be considered. If we assume that the marginal distributions of all the variables are the same, then we would use for Y_1 a normal density with expected value μ and variance $\sigma^2/(1-\rho^2)$. Autoregressive models are discussed in Section 3.3.1. □

Example 1.7 (Nonhomogeneous Poisson process) Consider observation of event times over an interval $[0, t_0)$, $t_0 > 0$, and let $N(t)$ be the number of events observed in $[0, t)$. In continuous time this determines a process $\{N(t)\} = \{N(t) : 0 \le t < t_0\}$. A Poisson process is defined by the following properties:

(a) $N(0) = 0$;

(b) $\text{pr}[N(t+h) = N(t) + 1 \mid \{N(t)\}] = \lambda(t)h + o(h)$;

(c) $\text{pr}[N(t+h) = N(t) \mid \{N(t)\}] = 1 - \lambda(t)h + o(h)$;

The notation $o(h)$ indicates a function of h that approaches 0 as $h \to 0$. The order notation $o(\cdot)$ and $O(\cdot)$ and probabilistic versions are described in Exercise 1.12. The rate function $\lambda(t)$ of the Poisson process is the instantaneous probability of an event at time t, given the history of the process up to time t. It is assumed to be non-negative, with finite integral on $[0, t_0)$. The process can equivalently be defined by requiring $\{N(t)\}$ to have independent increments; i.e., the evolution of the process from time t forward is independent of the history before time t.

If we observe $\{N(t)\}$ to increase by 1 at times $y_1 < \cdots < y_n$ in the interval $[0, t_0)$, the joint probability of this is

$$
\begin{aligned}
L\{\lambda(\cdot); y\} &= \prod_{i=1}^{n} \lambda(y_i)h \prod_{j \in \mathcal{M}} \{1 - \lambda(a_j)h\}\{1 + o(1)\}, \\
&= h^n \prod_{i=1}^{n} \lambda(y_i) \exp\left[\sum_{j \in \mathcal{M}} \log\{1 - \lambda(a_j)h\}\right] \{1 + o(1)\}, \\
&\propto \prod_{i=1}^{n} \lambda(y_i) \exp\left[\sum_{j \in \mathcal{M}} \{-\lambda(a_j)h\}\right] \{1 + o(1)\}, && (1.10) \\
&\approx \prod_{i=1}^{n} \lambda(y_i) \exp\left\{-\int_0^{t_0} \lambda(y)\, dy\right\}, && (1.11)
\end{aligned}
$$

where the interval $[0, t_0)$ has been divided into a large number of small intervals of length h, with left endpoints a_j, $j = 0, 1, \ldots, \lceil t_0/h \rceil$ and $a_0 = 0$. There are n intervals that each contain a single event; the set $\mathcal{M}$ denotes the collection of intervals that do not contain any event. In (1.10) we drop the factor h^n without changing the equivalence class, and the approximation in (1.11) comes from taking the limit as $h \to 0$. □

Example 1.8 (Homogeneous Poisson process) The rate function $\lambda(t)$ in (1.11) could be parametrized by a relatively small number of parameters. For example, if $\lambda(t) = \theta$, then

$$L(\theta; y) = \theta^n \exp(-\theta t_0), \quad \theta > 0, \tag{1.12}$$

which is the same likelihood function as that for a sample of n independent observations from the exponential distribution, each with density

$$f(y_i; \theta) = \theta \exp(-\theta y_i), \quad y_i, \theta > 0, \tag{1.13}$$

with observed total failure time $t_0 = \Sigma_i y_i$. The homogeneous Poisson process is often used as a starting point for time series or spatial processes, as it corresponds to complete randomness in the distribution of the events in time or space. Poisson processes for spatial data are discussed in Section 3.4.3. □

Example 1.9 (Exponential regression) Suppose Y_i follows an exponential distribution with expected value μ_i, with $\log \mu_i = \beta_0 + \beta_1 x_i$, and we have n independent observations from this model. The log-likelihood function is

$$\ell(\beta_0, \beta_1; y) = -n\beta_0 - \beta_1 \sum_{i=1}^{n} x_i - \sum_{i=1}^{n} y_i \exp\{-(\beta_0 + \beta_1 x_i)\}, \quad \beta_0, \beta_1 \in \mathbb{R}, \tag{1.14}$$

which is easily generalized to more complex regression models for μ_i, as in Example 1.1. The model in (1.14) is one of a class of generalized linear models, which are built on exponential families. □

Example 1.10 (Exponential families) The exponential family distribution for a random variable Y has a density of the form

$$\begin{aligned} f(y;\theta) &= \exp\{\varphi(\theta)^{\mathsf{T}} s(y) - c(\theta) - h(y)\}, \\ &= \exp\{\Sigma_{j=1}^{p} \varphi_j(\theta) s_j(y) - c(\theta) - h(y)\}, \quad \theta \in \Theta. \end{aligned} \tag{1.15}$$

The parameter $\varphi(\theta)$ is called the canonical parameter, and the canonical form of the density is

$$f(y;\varphi) = \exp\{\varphi^{\mathsf{T}} s(y) - c(\varphi) - h(y)\}, \quad \varphi \in \mathcal{P}, \tag{1.16}$$

where with a slight abuse of notation $c(\varphi) = c\{\theta(\varphi)\}$.The density $f(y;\varphi)$ in (1.16) is often called the exponential tilt of the base density $f_0(y) = \exp\{-h(y)\}$. If Θ is such that the parameter space for φ is

$$\mathcal{P} = \left\{\varphi : \int \exp\{\varphi^{\mathsf{T}} s(y) - h(y)\}\,\mathrm{d}y < \infty\right\}, \tag{1.17}$$

then the exponential family has full rank. If the set $\mathcal{P}$ is also open, then the full rank exponential family is called regular. We assume that the model is presented in reduced form, i.e., with no linear constraints on φ; otherwise, it is possible to rewrite the family of densities with fewer than p parameters.

Suppose $Y_1, \ldots, Y_n$ are independent, identically distributed, each with a distribution of the form (1.16). Then the joint density of $Y = (Y_1, \ldots, Y_n)$ is

$$f(y;\varphi) = \exp\left\{\varphi_1 \sum_{i=1}^{n} s_1(y_i) + \cdots + \varphi_p \sum_{i=1}^{n} s_p(y_i) - nc(\varphi) - \sum_{i=1}^{n} h(y_i)\right\}, \tag{1.18}$$

showing that the exponential family model is closed under sampling. The statistic $s(y) = (\Sigma_i s_1(y_i), \ldots, \Sigma_i s_p(y_i))$ is minimal sufficient for φ in a full exponential family, assuming there is no linear constraint on φ. Examples 1.2 through 1.6 are exponential family models.

If in (1.15) the canonical parameter vector φ is a non-linear function of θ, the model is a curved exponential family. In curved exponential families the dimension of the minimal sufficient statistic exceeds that of the parameter, and as a result the maximum likelihood estimator is not minimal sufficient. □

Exponential families are discussed further in Section 2.3.

Example 1.11 (A curved exponential family) Suppose $(Y_1, Y_2)^{\mathsf{T}}$ follows a bivariate normal distribution with expected value $(0, 0)^{\mathsf{T}}$ and variance-covariance matrix

$$\begin{bmatrix} 1 & \theta \\ \theta & 1 \end{bmatrix}. \tag{1.19}$$

The likelihood function based on a single observation is

$$\begin{aligned} L(\theta; y_1, y_2) &= \frac{1}{2\pi} \frac{1}{(1-\theta^2)^{1/2}} \exp\left\{-\frac{1}{2(1-\theta^2)}(y_1^2 - 2\theta y_1 y_2 + y_2^2)\right\}, \\ &= \exp\left\{-\frac{y_1^2 + y_2^2}{2(1-\theta^2)} + \frac{\theta y_1 y_2}{1-\theta^2} - \frac{1}{2}\log(1-\theta^2) - \log(2\pi)\right\}, \quad -1 \le \theta \le 1, \end{aligned}$$

which is a curved exponential family, as $\varphi_2(\theta) = -2\theta\varphi_1(\theta)$. In independent sampling from this model $(\Sigma_i(Y_{i1}^2 + Y_{i2}^2), \Sigma_i(Y_{i1}Y_{i2}))$ is minimal sufficient for θ. □

Example 1.12 (Generalized linear models) Generalized linear models are a class of regression models based on the exponential family that have found wide application. For a sample of response variables $\{Y_1, \ldots, Y_n\}$ each with a vector of associated covariates x_i, the form of the density for the ith response is

$$f(y_i; \beta, \phi) = \exp[\{y_i\theta_i - b(\theta_i)\}/a(\phi) + h(y_i, \phi)], \tag{1.20}$$

where ϕ is a dispersion parameter, and $a(\cdot)$ and $b(\cdot)$ are known functions. A regression model is defined using a given link function $g(\mu_i)$ of the form

$$g(\mu_i) = x_i^{\mathsf{T}}\beta, \tag{1.21}$$

where $\mu_i = \mathrm{E}(Y_i)$. If ϕ is known, then this density is a linear exponential family, and the canonical parameters θ_i are functions of β determined by the link function. Differentiating $\int f(y_i; \beta, \phi)\, \mathrm{d}y_i = 1$ under the integral sign gives the results

$$\mu_i = \mathrm{E}(Y_i) = b'(\theta_i), \quad \mathrm{var}(Y_i) = a(\phi)b''(\theta_i). \tag{1.22}$$

As in Example 1.1, if the covariates are random, then the model describes the conditional distribution of Y_i, given $X_i = x_i$. The variance is conventionally written $a(\phi)V(\mu_i)$, with $V(\cdot)$ called the variance function. The logistic regression model for binary data, treated below in Example 1.23, is a generalized linear model, with $a(\phi) = 1$ and $V(\mu_i) = \mu_i(1 - \mu_i)$. Generalized linear models are discussed in Examples 3.8, 3.24, 3.20, and in Section 5.2, where the model is used to motivate generalized estimating equations.

As noted in McCullagh and Nelder (1989, Section 2.2), $a(\phi)$ is very often of the form ϕ/w_i where w_i is a known weight. For example, if $Y_i \sim \mathrm{Binom}(m_i, p_i)$, the weight is $1/m_i$. □

In many applications for which regression models of various forms are used, the covariate values are sampled, or measured, along with the responses, and then inference is based on the conditional density of the responses, given the observed covariate values. So in (1.22), we could write $\mathrm{E}(Y_i \mid x_i)$ and $\mathrm{var}(Y_i \mid x_i)$ instead. This notation will be used in Chapter 3, where most of the envisioned applications are to observational data. In experimental settings the covariate values may be under the control of the investigator, and hence are fixed values. It is sometimes theoretically convenient to postulate a distribution $F_X(\cdot)$ for the covariates and average over their values, particularly in high-dimensional settings. This is called "model-X" in Candès et al. (2018), for example. In the analysis of data, regression models are nearly always conditioned on the observed values of the covariates ("fixed-X"), on grounds of interpretability. Formally, functions of the covariates are ancillary for the regression parameters.

Example 1.13 (Generalized linear models with random effects) It is a common modelling strategy to introduce dependence among responses in generalized linear models, and other regression models, through an assumed latent variable that is unobserved. Suppose the distribution of Y_i, given $X_i = x_i$ and $Z_i = z_i$ follows a generalized linear model with link function

$$g(\mu_i) = x_i^{\mathsf{T}}\beta + z_i^{\mathsf{T}}b, \tag{1.23}$$

where b is a random vector with some assumed distribution, typically $N(0, \sigma_b^2 I)$, and z_i is a known covariate vector. The likelihood function based on the observed sample is

$$L(\beta, \sigma_b^2, \phi; y) = \prod_{i=1}^{n} \int f\{y_i \mid b; \beta, \phi\} f(b; \sigma_b^2)\, \mathrm{d}b, \tag{1.24}$$

where the first density in the integrand has the form (1.20). This likelihood function will rarely be available in closed form, which motivates some of the alternatives to likelihood discussed in Chapters 3, 4, and 5. An exception is if the distribution of Y_i is normal and $\mu_i = x_i^{\mathsf{T}}\beta$; see Exercise 1.9 and Examples 3.3 and 3.6. □

Example 1.14 (Multivariate normal distribution) Suppose Y is a d-dimensional vector following a normal distribution with expected value 0 and covariance matrix Σ. The density is

$$f(y;\Sigma) = \frac{1}{(2\pi)^{d/2}}\frac{1}{|\Sigma|^{1/2}}\exp\left\{-\frac{1}{2}(y^{\mathsf{T}}\Sigma^{-1}y)\right\}, \quad y \in \mathbb{R}^d. \tag{1.25}$$

The sufficient statistic for Σ is the $d \times d$ matrix yy^{T}. The (a, b) element of Σ, σ_{ab}, is the covariance between the components y_a and y_b. The concentration or precision matrix is $\Omega = \Sigma^{-1}$, with (a, b) element ω_{ab}, say. The partial correlation between y_a and y_b, conditional on the other elements of y, is $-\omega_{ab}/\surd(\omega_{aa}\omega_{bb})$. Sometimes it is helpful for interpretation to express the dependence between components of y in a graph, with nodes for each component y_a, and edges between the nodes representing partial correlation. In particular, the absence of an edge between nodes y_a and y_b is equivalent to $\omega_{ab} = 0$. □

Example 1.15 (Transformation models) Transformation models are models for which a transformation on the sample space and a corresponding transformation on the parameter space leave the density function unchanged. Location, scale, location-scale, and location-regression models are all examples of transformation models.

A one-parameter location family on $\mathbb{R}$ has a density function $f(y;\theta)$ of the form $f_0(y-\theta)$, for $-\infty < \theta < \infty$, where $f_0(y)$ is the standard form of the density and θ is the location parameter. The normal distribution with expected value θ and known variance is a location family. The t_ν distribution with density function

$$f(y;\theta) = \frac{\Gamma\{(\nu+1)/2\}}{\Gamma(\nu/2)\sqrt{\pi\nu}}\left\{1+\frac{(y-\theta)^2}{\nu}\right\}^{-(\nu+1)/2}, \quad y, \theta \in \mathbb{R}, \tag{1.26}$$

and known degrees of freedom ν is also a location family; the Cauchy distribution is a special case with $\nu = 1$. The exponential location density function is

$$f(y;\theta) = \exp\{-(y-\theta)\}, \quad y-\theta \geq 0,\ \theta \in \mathbb{R}; \tag{1.27}$$

the support of the density is the interval $[\theta, \infty)$. Members of the same location family are simply shifted along the axis, relative to each other, and all have the same shape.

The location-scale family of distributions on $\mathbb{R}$ has density

$$f(y;\theta) = \theta_2^{-1} f_0\{(y-\theta_1)/\theta_2\}, \quad \theta_1 \in \mathbb{R}, \theta_2 \in \mathbb{R}^+. \tag{1.28}$$

The density $f_0(y)$ is the standard form of the density, θ_1 is the location parameter, and θ_2 is the scale parameter. The normal distribution with expected value θ_1 and variance θ_2^2 is a location-scale family, and the standard normal is the standard form. Other examples include the $t_\nu(\theta_1, \theta_2)$ density,

$$f(y;\theta) = c_\nu(1/\theta_2)\{1+(y-\theta_1)^2/(\nu\theta_2^2)\}^{-(\nu+1)/2}, \tag{1.29}$$

and the logistic(θ_1, θ_2) density,

$$f(y;\theta) = \frac{\exp\{-(y-\theta_1)/\theta_2\}}{\theta_2[1+\exp\{-(y-\theta_1)/\theta_2\}]^2}. \tag{1.30}$$

Many texts use μ and σ for the location and scale parameter although they do not always correspond to the expected value and standard deviation of the distribution. Any continuous density on $\mathbb{R}$ can be embedded in a location-scale family, and in fact most location-scale families are constructed this way. If the distribution of the random variable Y is a member of the location-scale family, then it can be expressed as

$$Y = \theta_2 Z + \theta_1, \tag{1.31}$$

where Z has the standard distribution with density function $f_0(z)$.

Discrete distributions are not members of the location-scale family, essentially because the parameter and the variable do not take values on the same space. □

Example 1.16 (Regression-scale model) A generalization of the location-scale model is the regression-scale model

$$Y = X\beta + \sigma\epsilon, \tag{1.32}$$

where Y is a vector of length n, X is a known $n \times p$ matrix, β is a vector of unknown parameters of length p, and ϵ is a random vector of length n that follows a known distribution $f_0(\cdot)$. If we let $Y^* = Xb + cY$, where $b \in \mathbb{R}^p$ and $c > 0$, then we can write

$$Y^* = X(b + c\beta) + c\sigma\epsilon, \tag{1.33}$$

which is a member of the same family, as long as the parameter space is $\mathbb{R}^p \times \mathbb{R}^+$. Transformation models are also discussed in Section 2.3. □

A sample of size n from a transformation model typically permits a reduction by conditioning on an ancillary statistic. An ancillary statistic is a function of Y whose distribution does not depend on θ; see Example 1.25 and Appendix 1.B.

1.2 Inference based on the likelihood function

The maximum likelihood estimate of θ is conventionally denoted by $\widehat{\theta}$. It is a function of the observed sample,

$$\widehat{\theta} = \widehat{\theta}(y) = \arg\sup_{\theta} L(\theta; y). \tag{1.34}$$

The maximum likelihood estimator is the corresponding random variable $\widehat{\theta}(Y)$.

If the parameter space is one- or two-dimensional, a form of direct inference can be obtained by plotting the likelihood function, more usefully the likelihood ratio $L(\widehat{\theta}; y)/L(\theta; y)$. Fisher (1956) suggested that values of θ for which the ratio exceeds 15 be declared "implausible", and other cut-off values have been suggested in the literature; see the bibliographic notes.

In general study of statistical theory and methods, we are usually interested in properties of our statistical methods under the model $f(y; \theta_0)$, where θ_0 is the notional true value of θ that generated the data $y = \{y_1, \ldots, y_n\}$. This requires considering the distribution of $L(\theta; Y)$, or relative ratios such as $L\{\widehat{\theta}(Y); Y\}/L(\theta; Y)$, where $Y = \{Y_1, \ldots, Y_n\}$ is a sample of random variables. For this, a number of derived quantities are routinely used in parametric inference.

The score function $u(\theta; y) = \partial\ell(\theta; y)/\partial\theta$ is typically used to obtain the maximum likelihood estimate, which in regular models satisfies

$$u(\widehat{\theta}; y) = 0. \tag{1.35}$$

If the support of the density does not depend on θ,

$$\mathrm{E}_\theta\{u(\theta;Y)\} = \int \frac{\partial \ell(\theta;y)}{\partial\theta} f(y;\theta)dy = \frac{\partial}{\partial\theta}\int f(y;\theta)\,\mathrm{d}y = 0, \tag{1.36}$$

assuming the interchange of derivative and integral is valid. The random variable corresponding to the score function $U(\theta) = u(\theta;Y)$ has expected value 0 and is an example of an unbiased estimating function; see Section 5.3.

The expected Fisher information is the variance of $U(\theta)$:

$$I(\theta) = \mathrm{var}_\theta\{U(\theta)\} = \mathrm{E}_\theta\left\{\frac{\partial \ell(\theta;Y)}{\partial\theta}\frac{\partial \ell(\theta;Y)}{\partial\theta^{\mathsf{T}}}\right\}. \tag{1.37}$$

Differentiating (1.36) again,

$$0 = \int \frac{\partial^2 f(y;\theta)}{\partial\theta\partial\theta^{\mathsf{T}}}\,\mathrm{d}y = \int \left(\frac{\partial \ell(\theta;y)}{\partial\theta}\right)\left(\frac{\partial \ell(\theta;y)}{\partial\theta}\right)^{\mathsf{T}} f(y;\theta)\,\mathrm{d}y + \int \frac{\partial^2 \ell(\theta;y)}{\partial\theta\partial\theta^{\mathsf{T}}} f(y;\theta)\,\mathrm{d}y, \tag{1.38}$$

$$= \mathrm{var}_\theta\{U(\theta)\} + \mathrm{E}_\theta\{\partial^2\ell(\theta;Y)/\partial\theta\partial\theta^{\mathsf{T}}\} = I(\theta) + \mathrm{E}_\theta\{-J(\theta)\}. \tag{1.39}$$

The negative second derivative of the log-likelihood function is called the observed Fisher information function, and when evaluated at the maximum likelihood estimate is called the observed Fisher information:

$$J(\widehat{\theta}) = -\left.\frac{\partial^2 \ell(\theta;y)}{\partial\theta\partial\theta^{\mathsf{T}}}\right|_{\theta=\widehat{\theta}}. \tag{1.40}$$

The identities in (1.36) and (1.38) are called Bartlett identities, and further identities can be obtained for higher-order derivatives of the log-likelihood function, assuming these exist; see Exercise 1.11. Vector and matrix derivatives are reviewed in Appendix 1.C.

Example 1.17 (Exponential regression, cont'd) The log-likelihood function is given in (1.14). Writing $\theta = (\beta_0, \beta_1)^{\mathsf{T}}$, we have

$$u(\theta;y) = \begin{bmatrix} \sum_i y_i \exp\{-(\beta_0+\beta_1 x_i)\} - n \\ \sum_i y_i x_i \exp\{-(\beta_0+\beta_1 x_i)\} - \sum_i x_i \end{bmatrix},$$

$$J(\theta) = \begin{bmatrix} \sum_i y_i \exp\{-(\beta_0+\beta_1 x_i)\} & \sum_i y_i x_i \exp\{-(\beta_0+\beta_1 x_i)\} \\ \sum_i y_i x_i \exp\{-(\beta_0+\beta_1 x_i)\} & \sum_i y_i x_i^2 \exp\{-(\beta_0+\beta_1 x_i)\} \end{bmatrix},$$

$$I(\theta) = \begin{bmatrix} n & \sum_i x_i \\ \sum_i x_i & \sum_i x_i^2 \end{bmatrix}.$$

If $\Sigma_i x_i = 0$, then the expected information matrix is diagonal, and parameters β_0, β are said to be orthogonal; see Section 2.1. □

Example 1.18 (Exponential family, cont'd) Suppose the density for Y is an exponential family in the canonical form (1.16). Then $-\partial^2\ell(\varphi)/\partial\varphi\partial\varphi^{\mathsf{T}} = \partial^2 c(\varphi)/\partial\varphi\partial\varphi^{\mathsf{T}}$, so observed and expected Fisher information are identical. The score equation defining $\widehat{\varphi}$ is

$$s = \left.\frac{\partial c(\varphi)}{\partial\varphi}\right|_{\varphi=\widehat{\varphi}}, \tag{1.41}$$

and as $\mathrm{E}_\varphi(S) = \partial c(\varphi)/\partial\varphi$, the maximum likelihood estimate of φ is the parameter value for which the expected value of the sufficient statistic is equal to its observed value.

One reason for the popularity of exponential family models, and by extension generalized linear models, is that under mild regularity conditions, the corresponding log-likelihood function is concave, so the maximum likelihood estimator is indeed the unique root of the score equation, and numerical methods are relatively easily implemented. □

Example 1.19 (Multivariate normal distribution, cont'd) Suppose we have an independent, identically distributed sample of d-dimensional vectors y_i, each from the d-dimensional multivariate normal distribution of Example 1.14, with expected value 0 and covariance matrix Σ. The log-likelihood function is

$$\ell(\Sigma; y) = -\frac{n}{2}\log|\Sigma| - \frac{1}{2}\sum_{i=1}^{n} y_i^{\mathsf{T}}\Sigma^{-1}y_i, \tag{1.42}$$

where in the left-hand side $y = (y_1^{\mathsf{T}}, \ldots, y_n^{\mathsf{T}})^{\mathsf{T}}$. The derivatives needed to calculate the score function and Fisher information are easier to obtain for $\Omega = \Sigma^{-1}$, so we write

$$\ell(\Omega; y) = \frac{n}{2}\log|\Omega| - \frac{1}{2}\sum_{i=1}^{n} \mathrm{tr}(y_i y_i^{\mathsf{T}}\Omega),$$

from which we have the score function

$$u(\Omega; y) = \frac{n}{2}\Omega^{-1} - \frac{1}{2}\sum_{i=1}^{n} y_i y_i^{\mathsf{T}}, \tag{1.43}$$

showing that $\widehat{\Sigma} = \widehat{\Omega}^{-1} = (1/n)\sum_{i=1}^{n} y_i y_i^{\mathsf{T}}$, assuming $n > d$. The observed Fisher information function is $(n/2)d\Omega^{-1}/d\Omega$, and as the score function is a matrix, this is a four-way array, equal to $(n/2)(\Sigma \otimes \Sigma)$, where $\otimes$ is the Kronecker product. It can be useful to vectorize the elements of Ω: let $\omega = \mathrm{vech}(\Omega)$, a vector of length $d(d+1)/2$ of the lower triangle of Ω, and similarly $w = \mathrm{vech}(\widehat{\Omega})$. The log-likelihood function is $\ell(\omega; w) = (n/2)\log|\Omega| - (n/2)\omega^{\mathsf{T}}Dw$, where D is a diagonal matrix of a $d(d+1)/2$-vector with each element of the vector either 2 or 1, depending on whether the element corresponds to a covariance or a variance. An expression for the observed Fisher information for the vectorized parameter is

$$J\{\mathrm{vech}(\Omega)\} = \frac{n}{4}DI_{\mathrm{ss}}(\Sigma)D; \tag{1.44}$$

the matrix $I_{\mathrm{ss}}(\Sigma)$ is called the Isserlis matrix of Σ, with entries $\mathrm{cov}(w_{ij}, w_{rs}) = \sigma_{ir}\sigma_{js} + \sigma_{is}\sigma_{jr}$ (DiCaterina et al., 2025). If $\Omega_{rs} = 0$, the components Y_r and Y_s of Y are conditionally independent, given the other components of Y. This is often of interest in applications of Gaussian graphical networks; see Example 7.8. □

If the random variables $Y_1, \ldots, Y_n$ are independent, then $\ell(\theta; Y)$ is a sum of independent random variables, as is $U(\theta) = u(\theta; Y)$, and under suitable conditions a central limit theorem for U will be available:

$$\frac{1}{\sqrt{n}}U(\theta) = \frac{1}{\sqrt{n}}\sum_{i=1}^{n} U_i(\theta) \xrightarrow{d} N\{0, i(\theta)\}, \tag{1.45}$$

where $U_i(\theta) = u(\theta; Y_i)$, and $i(\theta) = I(\theta)/n$ is assumed to be finite and positive. If the Y_i are identically distributed, $i(\theta) = \mathrm{var}\{u(\theta; Y_1)\}$, the Fisher information in one observation. This central limit theorem leads to the following limiting distributions of related likelihood quantities, under the model $f(y;\theta_0)$, presented here for a scalar parameter θ:

$$I^{-1/2}(\theta_0)U(\theta_0) = i^{-1/2}(\theta_0)U(\theta_0)/\sqrt{n} \xrightarrow{d} N(0,1), \tag{1.46}$$

$$I^{1/2}(\theta_0)(\widehat{\theta} - \theta_0) = i^{1/2}(\theta_0)\sqrt{n}(\widehat{\theta} - \theta_0) \xrightarrow{d} N(0,1), \tag{1.47}$$

$$2\{\ell(\widehat{\theta}) - \ell(\theta_0)\} \xrightarrow{d} \chi^2_1. \tag{1.48}$$

We have suppressed the dependence of each derived quantity on Y (and on n) for notational convenience. Result (1.47) is often described as asymptotic consistency and normality of the maximum likelihood estimator; this is made more precise in Section 1.7.

When θ is a vector in $\mathbb{R}^p$, the Fisher information function is a $p \times p$ matrix, and it is more convenient to express the limit theorems in their quadratic forms. Under the model $f(y;\theta_0)$, and under suitable regularity conditions,

$$w_u(\theta_0) = U(\theta_0)^{\mathsf{T}} I^{-1}(\theta_0)U(\theta_0) \xrightarrow{d} \chi^2_p, \tag{1.49}$$

$$w_e(\theta_0) = (\widehat{\theta} - \theta_0)^{\mathsf{T}} I(\theta_0)(\widehat{\theta} - \theta_0) \xrightarrow{d} \chi^2_p, \tag{1.50}$$

$$w(\theta_0) = 2\{\ell(\widehat{\theta}) - \ell(\theta_0)\} \xrightarrow{d} \chi^2_p. \tag{1.51}$$

The argument leading to these results, and the required regularity conditions, are discussed in Section 1.7.

1.3 Significance functions

In the case that θ is a scalar parameter, the limiting results (1.46)–(1.48) give the following finite-sample distributional approximations under $f(y;\theta)$:

$$s(\theta) = J^{-1/2}(\widehat{\theta})U(\theta) \mathrel{\dot{\sim}} N(0,1), \tag{1.52}$$

$$q(\theta) = J^{1/2}(\widehat{\theta})(\widehat{\theta} - \theta) \mathrel{\dot{\sim}} N(0,1), \tag{1.53}$$

$$w(\theta) = 2\{\ell(\widehat{\theta}) - \ell(\theta)\} \mathrel{\dot{\sim}} \chi^2_1, \tag{1.54}$$

$$r(\theta) = \text{sign}(\widehat{\theta} - \theta)w^{1/2}(\theta) \mathrel{\dot{\sim}} N(0,1). \tag{1.55}$$

These approximations rely on the asymptotic equivalence of observed and expected Fisher information and the consistency of $\widehat{\theta}$; see Section 1.7.2.

The notation emphasizes the quantities $s(\theta)$, $q(\theta)$, $r(\theta)$ and $w(\theta)$ as functions of θ, but to study their distribution under the model we consider them as functions of Y as well. A pivotal quantity is a function of Y and θ with a known distribution: thus s, q, r, and w are all approximate pivotal quantities. The approximate pivotal quantity $r(\theta)$ in (1.55) is known as the likelihood root.

For $\theta \in \mathbb{R}$ we can plot $\Phi\{q(\theta)\}$, $\Phi\{s(\theta)\}$, or $\Phi\{r(\theta)\}$ as functions of θ for fixed y and obtain approximate p-values for testing any value of θ; here $\Phi(\cdot)$ is the standard normal distribution function. These approximate significance functions are denoted generically by $p(\theta)$. The approach to inference based on significance functions avoids the somewhat artificial distinction between point estimation and hypothesis testing. Significance functions are also sometimes called p-value functions or confidence distribution functions; see the bibliographic notes. The conventional two-sided p-value for testing $H_0 : \theta = \theta_0$ is $2\min\{p(\theta_0), 1 - p(\theta_0)\}$.

In (1.52) and (1.53) we can standardize by $I(\theta)$ or $I(\widehat{\theta})$ and the approximation is still valid. Standardization of the maximum likelihood estimator by the observed Fisher information

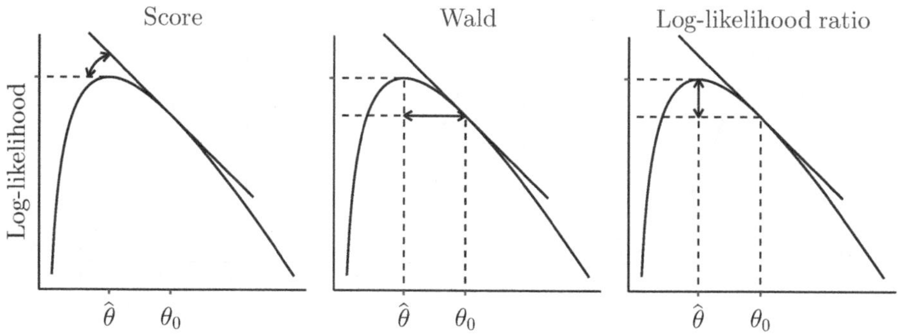

FIGURE 1.1
Different measures of the discrepancy between the maximum likelihood estimate $\widehat{\theta}$ and the hypothesized value θ_0. Rao's score statistic measures the standardize slope of the log-likelihood curve at θ_0. The Wald statistic measures the standardized horizontal distance between $\widehat{\theta}$ and θ_0. The likelihood ratio statistic measures the vertical distance between the peak of the log-likelihood function and the point on the curve that corresponds to θ_0.

$J(\widehat{\theta})$ is typically implemented in statistical packages, but also arises naturally in higher-order approximation to the distribution and is recommended on general grounds. Sometimes the score function is standardized by $I^{-1/2}(\theta)$ on grounds of simplicity, especially if $s(\theta)$ is only to be evaluated at a single value θ_0.

While there is nothing to distinguish the quality of the four approximations in (1.53)–(1.55), numerical work indicates that the normal approximation to the distribution of $r(\theta)$ is generally the most accurate in finite samples, and has the advantage of preserving sign information in the pivotal quantity. The distribution of the standardized maximum likelihood estimator $q(\theta)$ is usually better approximated by the standard normal distribution than is that of the standardized score function $s(\theta)$. However, confidence intervals based on q are necessarily symmetric about $\widehat{\theta}$, which may be inappropriate if the log-likelihood function is far from symmetric about its maximum. When evaluated at a fixed θ_0, $q(\theta_0)$ is often called the Wald statistic and $s(\theta_0)$ is often called Rao's score statistic. One peculiarity of the Wald statistic is that it is not guaranteed to be increasing in $|\widehat{\theta} - \theta_0|$ for fixed sample size n; this is called the Hauck–Donner effect.

Figure 1.1 illustrates how Rao's score, Wald and likelihood ratio statistics measure the distance between the maximum likelihood estimate $\widehat{\theta}$ and some hypothesized value θ_0.

The score statistic and the log-likelihood ratio statistic are invariant under reparametrization, while the Wald statistic is not. As a point estimator, however, the maximum likelihood estimator is equivariant, i.e., if $\eta = g(\theta)$ for some one-to-one function $g(\cdot)$, then $\widehat{\eta} = g(\widehat{\theta})$. The maximum likelihood estimator also has the smallest asymptotic variance, among estimators that are consistent and asymptotically normal. Such an estimator is called asymptotically (fully) efficient. For a competing estimator $\widetilde{\theta}$, say, that also has an asymptotically normal distribution,

$$\sqrt{n}(\widetilde{\theta} - \theta_0) \xrightarrow{d} N(0, \widetilde{\sigma}^2), \tag{1.56}$$

the ratio of $\widetilde{\sigma}^2$ to the asymptotic variance of the maximum likelihood estimator is called the relative efficiency. Estimators satisfying (1.56) are $\sqrt{n}$-consistent.

Example 1.20 (Gamma mean parameter) Suppose $Y_1, \ldots, Y_n$ are independent and identically distributed from a gamma distribution with known shape parameter β and

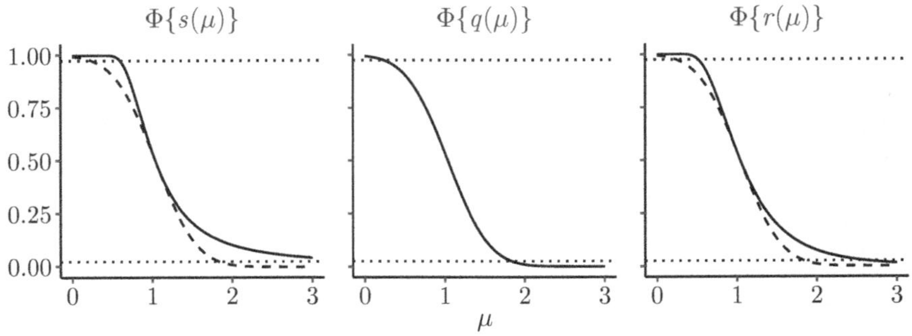

FIGURE 1.2
Plot of significance functions $p(\mu)$ based on $\Phi\{s(\mu)\}$, $\Phi\{q(\mu)\}$, and $\Phi\{r(\mu)\}$, for a sample of size 2 from a gamma distribution with shape 3 and expected value 1. The Wald significance function $\Phi\{q(\mu)\}$ (middle panel) is shown as a reference dashed curved in the plots of $\Phi\{s(\mu)\}$ (left panel) and $\Phi\{r(\mu)\}$ (right panel). The horizontal dotted lines intersect the significance function at values of μ that are the endpoints for an approximate equi-tailed 95% confidence interval.

unknown expected value μ. The density of Y_i is

$$f(y_i;\theta) = \frac{1}{\Gamma(\beta)}\left(\frac{\beta}{\mu}\right)^{\beta} y_i^{\beta-1}\exp\left(-\frac{\beta}{\mu}y_i\right),$$

and the log-likelihood function for the sample $y = (y_1, \ldots, y_n)$ is

$$\ell(\mu; y) = -n\beta\log\mu - (\beta/\mu)\Sigma_i y_i,$$

where we have omitted any terms not depending on μ. The pivotal quantities are

$$\begin{aligned}
s(\mu) &= \sqrt{n\beta}(\overline{y}-\mu)(\overline{y}/\mu^2),\\
q(\mu) &= \sqrt{n\beta}(\overline{y}-\mu)/\overline{y},\\
r(\mu) &= \operatorname{sign}(\overline{y}-\mu)[2n\beta\{\overline{y}/\mu - 1 - \log(\overline{y}/\mu)\}]^{1/2}.
\end{aligned}$$

As expected, the results are identical to those that would be obtained for a sample of size $n\beta$ from an exponential distribution with expected value μ.

The significance functions $\Phi\{s(\mu)\}$, $\Phi\{q(\mu)\}$, and $\Phi\{r(\mu)\}$ are plotted in Figure 1.2, for the sample $(0.240, 1.834)$ from a gamma distribution with shape parameter $\beta = 3$ and true $\mu = 1$. Also shown are the $(0.975, 0.025)$ values of the significance function, which provide the limits for approximate equi-tailed 95% confidence intervals. The confidence interval computed with the Wald significance function $q(\mu)$ is $(0.207, 1.867)$ and is symmetric around $\widehat{\mu} = \overline{y} = 1.037$, as it must be. In contrast, the confidence intervals from the score $(0.595, 4.052)$ and likelihood ratio $(0.512, 2.609)$ significance functions are not symmetric about $\widehat{\mu}$. The large upper bound of the score confidence interval illustrates the inaccuracy of the asymptotic approximation for the distribution of the score statistic, which can arise with a small sample size or when the shape of the log-likelihood function is far from quadratic.

□

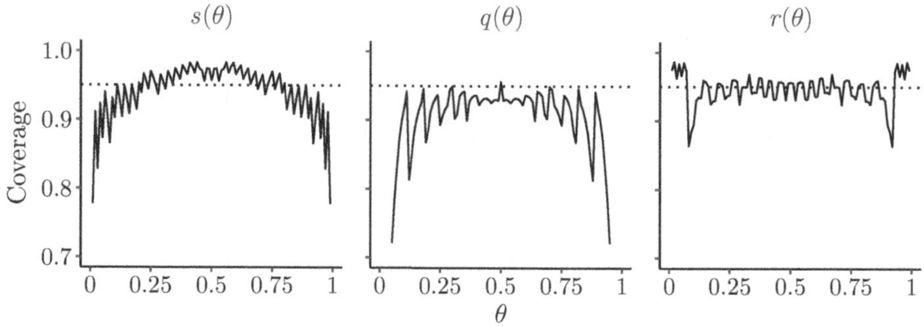

FIGURE 1.3
Coverage of 95% confidence intervals computed with pivotal quantities $s(\theta)$, $q(\theta)$ and $r(\theta)$ for a binomial distribution with $n = 25$ and probability θ. The coverage of $q(\theta)$ is truncated at 0.7. Adapted from Agresti (2002, Figure 1.3).

Example 1.21 (Binomial, cont'd) The log-likelihood function for the binomial probability, from (1.6) is $\ell(\theta) = y\log\{\theta/(1-\theta)\} + n\log(1-\theta)$, leading to

$$\begin{aligned} s(\theta) &= \{\widehat{\theta}(1-\widehat{\theta})\}^{1/2}\left\{\frac{\sqrt{n}(\widehat{\theta}-\theta)}{\theta(1-\theta)}\right\}, \\ q(\theta) &= \frac{\sqrt{n}(\widehat{\theta}-\theta)}{\{\widehat{\theta}(1-\widehat{\theta})\}^{1/2}}, \\ r(\theta) &= \left[2n\left\{\widehat{\theta}\log\left(\frac{\widehat{\theta}}{\theta}\right) + (1-\widehat{\theta})\log\left(\frac{1-\widehat{\theta}}{1-\theta}\right)\right\}\right]^{1/2}. \end{aligned} \tag{1.57}$$

The approximations to the significance functions, given by (1.52)–(1.54) are continuous, whereas the exact distributions are discrete. As a result there is oscillation of the coverage of confidence intervals constructed from the significance function, as illustrated in Figure 1.3. □

Example 1.22 (Normal distribution, cont'd) The log-likelihood function for $\theta = (\mu, \sigma^2)^{\mathsf{T}}$ is

$$\ell(\theta; y) = -n\log\sigma - \frac{1}{2\sigma^2}\sum_{i=1}^{n}(y_i-\mu)^2. \tag{1.58}$$

If σ^2 is known, then it is easily verified that $s(\mu) = q(\mu) = r(\mu) = \sqrt{n}(\overline{y}-\mu)/\sigma$, and the approximations in (1.52)–(1.55) are exact. If instead μ is assumed known, then $\widehat{\sigma}^2 = \Sigma_i(y_i-\mu)^2/n$, and

$$\begin{aligned} s(\sigma^2) &= \sqrt{\frac{n}{2}}\left(\frac{\widehat{\sigma}^2-\sigma^2}{\sigma^4}\right)\widehat{\sigma}^2, \\ q(\sigma^2) &= \sqrt{\frac{n}{2}}\left(\frac{\widehat{\sigma}^2-\sigma^2}{\widehat{\sigma}^2}\right), \\ r(\sigma^2) &= \sqrt{n}\left\{\log\left(\frac{\sigma}{\widehat{\sigma}}\right) - 1 + \left(\frac{\widehat{\sigma}^2}{\sigma^2}\right)\right\}^{1/2}. \end{aligned} \tag{1.59}$$

If both parameters are of interest, then the quadratic forms in (1.49) and (1.50) are based on

$$u(\theta) = \begin{bmatrix} \sum_i (y_i - \mu)/\sigma^2 \\ \sum_i (y_i - \mu)^2/(2\sigma^4) - n/(2\sigma^2) \end{bmatrix} \tag{1.60}$$

and

$$I(\theta) = \begin{bmatrix} n/\sigma^2 & 0 \\ 0 & n/(2\sigma^4) \end{bmatrix}. \tag{1.61}$$

□

Example 1.23 (Logistic regression) Suppose Y_i follows a binomial distribution with m_i independent Bernoulli trials, each with probability of success $p_i(\beta)$, where

$$\text{logit}(p_i) = \log\left(\frac{p_i}{1-p_i}\right) = x_i^{\mathsf{T}}\beta,$$

and x_i is a vector of constants. For a sample of independent observations from this model $y = \{y_1, \ldots, y_n\}$, the log-likelihood function is

$$\ell(\beta; y) = \sum_{i=1}^{n} [y_i x_i^{\mathsf{T}}\beta - m_i \log\{1 + \exp(x_i^{\mathsf{T}}\beta)\}], \tag{1.62}$$

and the associated score function and information matrix are readily computed. The maximum likelihood estimator satisfies the p equations

$$\sum_{i=1}^{n} y_i x_{ij} = \sum_{i=1}^{n} m_i p_i(\widehat{\beta}) x_{ij}, \quad j = 1, \ldots, p; \tag{1.63}$$

i.e., the maximum likelihood estimate is obtained by equating the sufficient statistics to their expected values. See also Exercise 1.2. □

1.4 Bayesian inference

In some settings it is natural to add to the model specification for Y, a model specification for θ, implicitly thereby assuming that the parameters also follow a probability distribution, with density $\pi(\theta)$, which in turn might depend on further parameters. The density function $f(y; \theta)$ is now interpreted as the conditional density of Y, given θ, and combined with the prior density defines the posterior density of θ, given y:

$$\pi(\theta \mid y) = \frac{L(\theta; y)\pi(\theta)}{\int L(\theta; y)\pi(\theta)\,\mathrm{d}\theta}. \tag{1.64}$$

The denominator is the marginal density of y, denoted by $f(y)$.

With this introduction of a new source of randomness, it can be argued that all the information about θ provided by the data is given by the posterior density $\pi(\theta \mid y)$. This posterior can be used to provide direct probability statements about θ, such as "the posterior probability that θ lies between a and b is 0.75", and so on. If θ is one- or two-dimensional, the posterior density can be plotted. In higher dimensions some summaries could be constructed instead, although in dimensions higher than about four or five, exact numerical integration to normalize the posterior may not be feasible. In some cases analytic calculation of the posterior is possible; see Exercise 1.19. There are a number of sampling-based approaches to approximating the integral as well: the most well-studied is the Markov chain Monte Carlo method.

The meaning to be ascribed to the prior probability distribution is not always clear. The subjective approach to Bayesian inference regards the prior as a distribution that expresses an analyst's personal uncertainty about the values of the parameter. With such a prior, the conclusions from the posterior are also subjective in the same sense. In contrast to this there may in some settings be a well-established distribution available either from previous experiments or from a precise physical theory, in which case the Bayesian posterior is simply an application of the rules of conditional probability. In many applied settings, priors are chosen simply to make analytical or numerical calculations easier, with the hope or presumption that the form used will not affect the conclusions very much. There are several proposals for noninformative priors, which are designed to give credibility intervals that also have an interpretation as confidence intervals to some order of approximation.

We now describe a limit theorem for the posterior distribution, analogous to those in (1.46)–(1.48). For $a < b$, define

$$a_n = \widehat{\theta} + a\{J(\widehat{\theta})\}^{-1/2}, \quad b_n = \widehat{\theta} + b\{J(\widehat{\theta})\}^{-1/2},$$

where the maximum likelihood estimate $\widehat{\theta}$ and the observed Fisher information $J(\widehat{\theta})$ also depend on n. The limit result is

$$\int_{a_n}^{b_n} \pi(\theta \mid y)\, \mathrm{d}\theta \xrightarrow{p} \int_a^b \phi(t)\, \mathrm{d}t, \tag{1.65}$$

where $\phi(\cdot)$ is the density of the standard normal distribution. The associated approximation is expressed as

$$\pi(\theta \mid y) \mathrel{\dot\sim} N\{\widehat{\theta}, J^{-1}(\widehat{\theta})\}, \tag{1.66}$$

showing that pivotal quantities from Bayesian posteriors are asymptotically equivalent to those based on (1.53), the standardized maximum likelihood estimator. This is sometimes expressed by saying "the data swamps the prior".

Calculations similar to the Taylor series expansions of Section 1.7.2 show that an equivalent approximation is

$$\pi(\theta \mid y) \mathrel{\dot\sim} N\{\widehat{\theta}_\pi, J_\pi^{-1}(\widehat{\theta}_\pi)\}, \tag{1.67}$$

where the posterior mode $\widehat{\theta}_\pi$ maximizes $\pi(\theta \mid y)$ in (1.64), and

$$J_\pi(\theta) = -\partial^2 \log \pi(\theta \mid y)/\partial\theta\partial\theta^{\mathsf{T}}.$$

Asymptotic theory cannot tell us which of the two approximations (1.66) or (1.67) is more accurate in finite samples, although only the second depends on the prior.

The result that the posterior density is approximately normal is often called the Bernstein–von Mises theorem in the Bayesian literature.

An approximation to the posterior density that is more accurate than the normal approximation, and does depend on the prior, can be obtained from a Taylor series expansion of the log-likelihood function and the prior density around the maximum likelihood value $\widehat{\theta}$.

Assuming here that the model is sufficiently regular that these calculations can be formally verified, we have

$$\begin{aligned} f(y) &\doteq \int \exp\left\{\ell(\widehat{\theta};y) + (\theta-\widehat{\theta})\ell'(\widehat{\theta};y) + \frac{1}{2}(\theta-\widehat{\theta})^2\ell''(\widehat{\theta};y)\right\}\pi(\theta)\,\mathrm{d}\theta, \\ &\doteq \exp\{\ell(\widehat{\theta};y)\}\pi(\widehat{\theta})\int \exp\left\{-\frac{1}{2}(\theta-\widehat{\theta})^2 J(\widehat{\theta})\right\}\mathrm{d}\theta, \\ &= \exp\{\ell(\widehat{\theta};y)\}\pi(\widehat{\theta})\sqrt{2\pi}\{J(\widehat{\theta})\}^{-1/2}, \end{aligned} \tag{1.68}$$

as $\ell'(\widehat{\theta};y)=0$ and the integrand in the second line is the kernel of a normal density with variance $J^{-1}(\widehat{\theta})$. The resulting approximation to the posterior is

$$\pi(\theta \mid y) \doteq \frac{1}{\sqrt{2\pi}}\{J(\widehat{\theta})\}^{1/2}\frac{L(\theta;y)}{L(\widehat{\theta};y)}\cdot\frac{\pi(\theta)}{\pi(\widehat{\theta})}. \tag{1.69}$$

Approximation (1.69) is called the Laplace approximation to the posterior density, and can be shown to have relative error $O(n^{-1})$. The Laplace approximation for the posterior density of a vector parameter is

$$\pi(\theta \mid y) \doteq \left(\frac{1}{\sqrt{2\pi}}\right)^p |J(\widehat{\theta})|^{1/2}\frac{L(\theta;y)}{L(\widehat{\theta};y)}\frac{\pi(\theta)}{\pi(\widehat{\theta})}. \tag{1.70}$$

The integral defining $f(y)$ could alternatively be written

$$f(y) = \int \exp\{\ell(\theta;y) + \log\pi(\theta)\}\,\mathrm{d}\theta = \int \exp\{\ell_\pi(\theta;y)\}\,\mathrm{d}\theta,$$

and expanded about the posterior mode $\widehat{\theta}_\pi = \arg\sup_\theta \ell_\pi(\theta;y)$, which leads to

$$\begin{aligned} f(y) &\doteq \int \exp\left\{\ell_\pi(\widehat{\theta}_\pi;y) - \frac{1}{2}(\theta-\widehat{\theta}_\pi)^2 J_\pi(\widehat{\theta}_\pi)\right\}\mathrm{d}\theta \\ &= L_\pi(\widehat{\theta}_\pi;y)\sqrt{2\pi}\{J_\pi(\widehat{\theta}_\pi)\}^{-1/2}, \end{aligned}$$

and the alternative, but asymptotically equivalent, Laplace approximation

$$\pi(\theta \mid y) \doteq \frac{1}{\sqrt{2\pi}}\{J_\pi(\widehat{\theta}_\pi)\}^{1/2}\frac{L_\pi(\theta;y)}{L_\pi(\widehat{\theta}_\pi;y)}, \tag{1.71}$$

where $\mathrm{L}_\pi(\theta;y)=\exp\{\ell_\pi(\theta;y)\}$.

1.5 Model assessment and prediction

There are two aspects of model assessment that are relevant for many practical settings. The first is to see whether or not the model class assumed is plausible in light of the data. The second, which is closely related but more specific, is to assess whether a model in the same class, but with fewer parameters, fits the data as well as a model with more parameters. This second aspect arises most often in the context of regression models with large numbers of covariates; the simpler model in the same class would typically have some of the components of β in the regression equal to zero, to reflect the possibility that a regression model with fewer covariates is adequate for the observed data.

The first assessment can make use of the reduction in dimension by sufficiency or ancillarity, if either of these is available. For example, suppose the model for Y factorizes as

$$f(y;\theta) \propto f_1\{s(y);\theta\}f_2\{t(y) \mid s(y)\}, \tag{1.72}$$

where f_1 and f_2 are the marginal and conditional densities of their respective arguments; this is a version of the factorization theorem of Appendix 1.B. An assessment of the model $f_2\{t(y) \mid s(y)\}$ is in effect an assessment of the model class $\{f(y;\theta);\theta \in \Theta\}$, being free of the value of θ.

Example 1.24 (Poisson distribution) Suppose we have n independent observations from the density

$$f(y_i;\theta) = \theta^{y_i}\exp(-\theta)/y_i!, \quad y_i = 0,1,2,\ldots;\theta > 0. \tag{1.73}$$

The conditional distribution of $y = (y_1,\ldots,y_n)$ given $s(y) = \Sigma_i y_i = n\overline{y}$ is multinomial, with density

$$f_2(y \mid s) = \frac{s!}{y_1!\ldots y_n!}\left(\frac{1}{n}\right)^{y_1}\cdots\left(\frac{1}{n}\right)^{y_n}, \quad \sum_{i=1}^{n} y_i = s, \tag{1.74}$$

Various functions of y could be proposed as test statistics of the Poisson assumption, to be assessed using this multinomial distribution. In Cox and Hinkley (1974, Ex 3.4), $t(y) = \Sigma_i(y_i - \overline{y})^2/\overline{y}$ is suggested as a natural statistic for assessing the fit of the Poisson; see also Exercise 3.11 in the same reference, and Example 5.13 in Davison (2003). □

A similar analysis applies if there is an identified ancillary statistic for a model, so that, for example,

$$f(y;\theta) = f\{s(y) \mid a(y);\theta\}f_2\{a(y)\}, \tag{1.75}$$

in this case the marginal distribution of the ancillary statistic provides a basis for assessing the model free of the unknown parameter θ.

Example 1.25 (Transformation models, cont'd) If $y_1,\ldots,y_n$ are independent and identically distributed with density

$$f(y_i;\theta) = \theta_2^{-1}f_0\{(y_i - \theta_1)/\theta_2\}, \tag{1.76}$$

the vector of residuals with ith entry $a_i = (y_i - \widehat{\theta}_1)/\widehat{\theta}_2$ has a distribution free of θ; see Exercise 1.10. Often these residuals are plotted, and a visual assessment of their consistency with the density f_0 is made. For example, in normal-theory linear regression the residuals are plotted on a normal probability scale. As with Example 1.24, a statistic can be based on the distribution of these residuals, such as the standardized skewness, $\widehat{\gamma}_1 = \Sigma_i a_i^3/(\Sigma_i a_i^2)^{3/2}$, which should be approximately zero if the normal model is supported by the data. □

In treatments of generalized linear models, twice the difference of log-likelihoods comparing the regression model that specifies $g(\mu_i) = x_i^{\mathsf{T}}\beta$ with a fully unconstrained model where $\mathrm{E}(y_i) = \mu_i$ is called the deviance. The fully unconstrained model is sometimes called saturated. Choosing among nested generalized linear models is often done by analysis of deviance, repeatedly using result (1.48) or (1.51). The contribution from the ith of n independent observations from the model is called the deviance contribution from y_i; these deviance contributions play the role of residuals in diagnostic methods.

Example 1.26 (Logistic regression, cont'd) Suppose $Y_i \sim \text{Bin}(m_i, p_i), i = 1, \ldots, n$ are independently distributed. The saturated model does not assume any regression relation between expected values of Y_i, so in this model the maximum likelihood estimate is $\widehat{p}_i = y_i/m_i, i = 1, \ldots, n$. The deviance comparing this model to the logistic regression model of Example 1.17 is

$$\begin{aligned} D &= 2\sum_{i=1}^{n}[\ell(\widehat{p}_i; y_i) - \ell\{p_i(\widehat{\beta}); y_i\}] \\ &= 2\sum_{i=1}^{n} y_i \log\{y_i/m_i p_i(\widehat{\beta})\} + 2\sum_{i=1}^{n}(m_i - y_i)\log\left\{(m_i - y_i)/[m_i\{1 - p_i(\widehat{\beta})]\right\} \\ &= \sum_{i=1}^{n} d_i\{y_i, p_i(\widehat{\beta})\}, \end{aligned}$$

where the elements d_i are the contributions to the deviance. Residuals and diagnostics for logistic regression are discussed in Agresti (2002, Section 6.2). □

In a normal-theory linear regression, a model which allows the expected value to be arbitrary has maximum likelihood estimates $\widehat{\mu}_i = y_i$, and $\ell(\widehat{\mu}; y)$ is identically zero, so the deviance comparing this to the regression model $\mu_i = x_i^{\mathsf{T}}\beta$ is simply the sum of the squared residuals $\Sigma_i(y_i - x_i^{\mathsf{T}}\widehat{\beta})^2$.

We might hope to choose the best model among a set of competing models by finding the one that has the largest value of the maximized log-likelihood function, or the smallest deviance. However, this will always choose the most complex model, as we can always do at least slightly better on the data set we are fitting by making the model more complex. This is a familiar problem in linear regression, where adding additional covariates is guaranteed to reduce the residual sum of squares. A commonly used approach to this problem is to add to the log-likelihood function a penalty for model complexity. The simplest of these is Akaike's Information Criterion,

$$\text{AIC} = -2\ell(\widehat{\theta}; y) + 2p, \tag{1.77}$$

where p is the number of parameters estimated in $\ell(\theta; \cdot)$. Models with smaller values of AIC are preferred over models with larger values; the term $2p$ is a penalty for fitting models with larger number of parameters. Model selection is taken up in detail in Chapter 4.

A related approach to model selection is derived from a Bayesian approach, comparing different models using the marginal density $f(y)$ or, equivalently, its logarithm. Using a vector parameter version of (1.68),

$$\log\{f(y)\} \doteq \ell(\widehat{\theta}; y) + \log\pi(\widehat{\theta}) + \frac{p}{2}\log(2\pi) - \frac{1}{2}\log|J(\widehat{\theta})|. \tag{1.78}$$

Under the assumption that the prior $\pi(\theta)$ is $O(1)$, $J(\widehat{\theta})$ is $O_p(n)$, and $\widehat{\theta}$ is consistent, $\log|J(\widehat{\theta})| \doteq p\log n + C$. As a function of p and n, the approximation to the marginal density is

$$\log\{f(y)\} \doteq \ell(\widehat{\theta}; y) - \frac{p}{2}\log n.$$

The Bayesian information criterion based on comparing models using the marginal density is formulated by analogy with (1.77) as

$$\text{BIC} = -2\ell(\widehat{\theta}; y) + p\log n. \tag{1.79}$$

As with AIC, models with smaller values of BIC are preferred over models with larger values. Although derived from a Bayesian point of view, the BIC is commonly used in a non-Bayesian context. Since $\log n$ exceeds 2 for $n \geq 8$, the BIC criterion imposes a stronger penalty on increasing model complexity. As a result, it tends to favour simpler models compared to AIC.

Consider two nested models $M_1 \subset M_2$ with q and p parameters, respectively, and $q < p$, and suppose that M_1 is the correct model. The AIC criterion will select M_1 with probability $\text{pr}\{\text{AIC}(M_1) < \text{AIC}(M_2)\}$ which is approximately $\text{pr}\{\chi^2_{p-q} < 2(p-q)\}$ for large n. For fixed $p-q$, this probability does not converge to 1. Conversely, BIC is a consistent model selection criterion because the probability of selecting M_1 is approximately $\text{pr}\{\chi^2_{p-q} < (p-q)\log n\}$ which converges to 1 as $n \to \infty$.

However, in finite samples, BIC may result in underfitting, as it generally favours models that are too parsimonious. In regression models, underfitting based on BIC is related to the so-called omitted variable bias.

In Section 4.4 the model selection criterion AIC is derived as the optimal solution to a prediction problem. For a likelihood-based approach to prediction, suppose for illustration that $y = (y_1, \ldots, y_n)^{\mathsf{T}}$ is a vector of independent observations each with density $f(y_i; \theta)$, and that we are interested in predicting a new observation, $\widetilde{y}$, say, from the same model. One simple solution is to use the model $\widetilde{y} \sim f(\cdot; \widehat{\theta})$, where $\widehat{\theta}$ is estimated from the original sample y. Prediction intervals, or a most-probable value, or some other summary statistics can all be calculated from this estimated density, usually numerically.

This plug-in approach has the drawback that it ignores the variability in the estimation of θ. An approach closer to profile likelihood, treating θ as a nuisance parameter, is the profile predictive likelihood function

$$L(\widetilde{y}; y) \propto \sup_\theta f(\widetilde{y}, y; \theta) = \sup_\theta f(\widetilde{y}; \theta) \prod_{i=1}^n f(y_i; \theta), \tag{1.80}$$

the final expression only valid under our assumption that $(\widetilde{y}, y_1, \ldots, y_n)$ are independent and identically distributed. The maximizing value $\widetilde{\theta}_y$, say is a function of $(\widetilde{y}, y)$. The function $L(\widetilde{y}; y)$ is proportional to the conditional density of $\widetilde{y}$ given y, and in this context it is natural to normalize it to integrate to 1, and use the density $f(\widetilde{y}; \widetilde{\theta}_y)$ for prediction. From a non-Bayesian point of view, prediction of a random variable is qualitatively different from estimation of a parameter, as the latter is assumed to have as a target a fixed value, usually denoted by θ_0, whereas $\widetilde{y}$ is random. To reflect this difference in perspective, the function $L(\theta, \widetilde{y}; y) \propto f(\widetilde{y}, y; \theta)$ is called an extended likelihood function in Pawitan (2026, Chapter 16).

A Bayesian approach to prediction is more straightforward, at least conceptually; $\widetilde{y}$ and θ are both considered random variables and θ can be integrated out of their joint posterior density, leading to the posterior predictive density,

$$\pi(\widetilde{y} \mid y) \propto f(\widetilde{y} \mid \theta)\pi(\theta \mid y) d\theta. \tag{1.81}$$

The prior predictive density replaces $\pi(\theta \mid y)$ in (1.81) by $\pi(\theta)$. The prior predictive distribution is closely related to the Bayes factor for model selection; see the bibliographic notes.

Example 1.27 (Normal distribution) Suppose an observed sample $\{y_1, \ldots, y_n\}$ is to be used to predict a new observation $\widetilde{y}$, and each of y_i and $\widetilde{y}$ are assumed independent and normally distributed with expected value θ and known variance σ_0^2. Then (1.80) is

$$\begin{aligned} L(\widetilde{y}; y) &\propto \sup_\theta \exp\left[-\frac{1}{2\sigma_0^2}\{n(\bar{y}-\theta)^2 + (\widetilde{y}-\theta)^2\right] \\ &\propto \exp\left\{-\frac{n}{2(n+1)\sigma_0^2}(\widetilde{y}-\bar{y})^2\right\}, \end{aligned}$$

showing that the conditional distribution of $\widetilde{y}$, given $\bar{y}$, is normal with expected value $\bar{y}$ and variance $\sigma_0^2(1+1/n)$, as expected and as could be calculated directly from the model. A prediction interval of size $(1-2\alpha)$, for example, is $\bar{y} \pm z_\alpha \sigma_0 \surd(1+1/n)$. In Cox (2006, Section 8.1) this interval is motivated by a slightly different argument: if we assume $\widetilde{y}$ follows a normal distribution with expected value $\widetilde{\theta}$, then this prediction interval consists of all values $\widetilde{y}$ that are not rejected by a test of the hypothesis that $\theta = \widetilde{\theta}$, based on the test statistic $\widetilde{y} - \bar{y}$. □

If σ^2 is unknown, the predictive likelihood approach leads to prediction intervals for $\widetilde{y}$ based on Student's t_n distribution. Some variations of predictive likelihood that lead to prediction intervals based on the t-distribution, but with a degrees-of-freedom adjustment, are described in Bjørnstadt (1990). Using the testing approach, the predictive distribution of $\widetilde{y} - \bar{y}$, standardized by its estimated standard error, is Student's t with $n-1$ degrees of freedom.

In Example 1.27 the same result could be obtained with a flat prior for θ. This is not always the case, as the following example shows.

Example 1.28 (Binomial prediction) Suppose y is an observation from a binomial(n, θ) distribution and it is desired to predict the number of successes, $\widetilde{y}$ in a further m trials with the same parameter θ. The predictive likelihood for $\widetilde{y}$ is

$$L(\widetilde{y}; y) \propto \binom{n}{y}\binom{m}{\widetilde{y}}\left(\frac{y+\widetilde{y}}{n+m}\right)^{y+\widetilde{y}}\left(1-\frac{y+\widetilde{y}}{n+m}\right)^{n+m-(y+\widetilde{y})}, \quad \widetilde{y} = 0, \ldots, m,$$

based on the profile value $\widetilde{\theta}_y = (y+\widetilde{y})/(n+m)$. The posterior predictive density $f(\widetilde{y} \mid y) = \int_0^1 f(\widetilde{y} \mid y, \theta)\pi(\theta \mid y)\, \mathrm{d}\theta$, with the flat prior $\pi(\theta) = 1\{0 \le \theta \le 1\}$, is

$$\binom{n}{y}\binom{m}{\widetilde{y}}(n+1)\Big/\left\{\binom{m+n}{y+\widetilde{y}}(n+m+1)\right\}, \quad \widetilde{y} = 0, \ldots, m.$$

□

More relevant models for prediction would typically involve regression models with covariates, in which case the observation or observations to be predicted are often assumed to follow the same conditional distribution, given the covariates, as the observed responses; see Exercise 1.20.

Prediction of random variables arises more naturally in applications of models that include unobserved latent variables, or random effects, such as the generalized linear mixed model of Example 1.13 above, or the linear mixed-effects model (1.99) below. The extended likelihood function $L(\theta, b; y)$, say, based on the joint distribution of y and the vector b of random effects, is the h-likelihood of Lee and Nelder (1996).

Accuracy of prediction is very important in many machine-learning applications, especially in regression models with a large number of potential covariates, where inference about the regression parameters may not be very interpretable. Predictive accuracy is then an assessment of a fitted model on data that has not been used to tune the model and is very useful for model comparison. Estimating predictive accuracy for high-dimensional regression is discussed in Section 7.6.

In nonparametric models, in which the model parameter is effectively infinite-dimensional, overfitting to the observed data can easily occur, and models and methods are nearly always compared on the quality of the predictions on new data. Prediction intervals, or more generally prediction sets, are evaluated by evaluating that the probability that $\widetilde{y}$, say, is included in the prediction set with sufficiently high probability. One method of construction of nonparametric prediction sets is conformal prediction; see Section 6.9 for a very brief introduction.

1.6 Some more complex models

This book is mainly focussed on some of the many extensions to likelihood functions and the associated inference. In several cases these extensions have been developed because the model is very complex, and the likelihood function cannot be expressed in closed form, or possibly even computed. In this section we illustrate some of the issues with examples from the literature that are taken up in more detail in later chapters.

Example 1.29 (Correlated Poisson observations) Consider a model for n observations from a Poisson distribution, with density function for the ith observation

$$f(y_i \mid \alpha_i; \theta) = \exp(y_i \log \mu_i - \mu_i)/y_i!, \quad \log \mu_i = \beta + \alpha_i. \tag{1.82}$$

A time series of counts can be defined by modelling α_i as unobserved latent variables, for example,

$$\alpha_i = \rho\alpha_{i-1} + \epsilon_i, \quad |\rho| < 1,$$

where ϵ_i are uncorrelated zero mean normal variables with variance $\sigma^2 > 0$. The likelihood function for $\theta = (\beta, \rho, \sigma^2)^{\mathsf{T}}$ is

$$L(\theta; y_1, \ldots, y_n) = \int_{\mathbb{R}^n} \prod_{i=1}^{n} f(y_i \mid \alpha_i; \theta) f(\alpha_i; \theta) \, \mathrm{d}\alpha_i. \tag{1.83}$$

This is an example of a parameter-driven model in the terminology of Cox (1981); these models are addressed in Section 3.3, and the Poisson model here is a special case of Example 3.20. Davis and Yau (2011) studied various versions of composite likelihood (Chapter 4) for this and other time series models. □

Example 1.30 (Multivariate extremes) Suppose we have n observations of, for example, windspeed, at each of d locations in space. In a study of extreme values, we may be interested in the joint distribution of the maximum windspeed at these d locations. Davison and Huser (2015) show that after suitable centering and scaling, these component-wise maxima $Y_1, \ldots, Y_d$, say, follow the multivariate extreme value distribution:

$$\mathrm{pr}(Y_1 \le y_1, \ldots, Y_d \le y_d) = \exp\{-V(y_1, \ldots, y_d)\}, \tag{1.84}$$

where $V(\cdot)$ is called the spectral measure. For $d = 2$, parametrized versions of $V(\cdot)$ have been proposed in the literature; for $d > 2$ the situation is much more complex. Computing the likelihood function is conceptually straightforward, as the density of Y is given by the d-dimensional derivative of (1.84), but for moderately large d this is not computable, and the lack of families of parametrized models for $d > 2$ complicates the situation. Davison and Huser (2015) use pairwise likelihood as an approximation; see Example 4.27 and the bibliographic notes. □

Example 1.31 ($M/G/1$ queue) Suppose we observe the times of departure y_i of customers $i = 1, \ldots, n$, from a queue with a single server. The arrival times t_i are assumed to follow a dependent exponential process: $t_i = t_{i-1} + e_i$, where e_i are independent exponential random variables with expected value θ_1, and $t_0 = 0$. The distribution of y_i depends on the arrival time and the number of customers in the queue. Assuming the t_i are unobserved, the likelihood function is

$$L(\theta; y) = \int \cdots \int f(t_1 \mid \theta) \prod_{i=2}^{n} f(t_i \mid t_{i-1}; \theta) \prod_{i=1}^{n} f(y_i \mid t_i, s_{i-1}; \theta) \mathrm{d}t_1 \ldots \mathrm{d}t_n,$$

where $s_i = \Sigma_{j=1}^{i} y_j$. This model, with the distribution of Y_i, given (S_i, T_i) assumed to be uniformly distributed over the interval $(\theta_2 + \max(0, T_i - S_{i-1}), \theta_3 + \max(0, T_i - S_{i-1}))$, was used in Heggland and Frigessi (2004) in a discussion of indirect inference (Section 4.6), and in Fearnhead and Prangle (2012) to illustrate the use of approximate Bayesian computation. Shestopaloff and Neal (2014) consider calculation of the likelihood function and posterior distributions using Markov chain Monte Carlo methods. □

Example 1.32 (Ising model) The Ising model is a Markov random field for a vector $y = (y_1, \ldots, y_n)^{\mathsf{T}}$, with $y_i = \pm 1$, where y_i records a binary value for the ith node in a graph with vertex set $V = \{1, \ldots, n\}$ and edge set $E \subset V \times V$. The density is

$$f(y; \theta) = \frac{1}{Z(\theta)} \exp\left(\sum_{(i,j) \in E} \theta_{ij} y_i y_j \right),$$

and the parameter θ_{ij} measures the strength of the interaction between nodes i and j. The partition function $Z(\theta)$ is the normalizing constant for this density, in principle determined by summing over all binary vectors y, but in practice typically not computable for large graphs. Estimation of sparse Ising models using an approximate likelihood and regularization is treated in Ravikumar et al. (2010) and Xue et al. (2012). See also Example 3.27. Existence of the maximum likelihood estimator in the Ising model is addressed in Skalski and Stroiński (2025). □

1.7 Asymptotic theory

1.7.1 Consistency of maximum likelihood estimators

Suppose $Y_1, \ldots, Y_n$ are independent and identically distributed random variables, each with density function $f(y; \theta_0)$, for some notional true value $\theta_0 \in \Theta \subset \mathbb{R}^p$. We assume throughout that the parameter is identifiable, i.e., that $f(y; \theta) = f(y; \theta')$ if and only if $\theta = \theta'$. The likelihood function based on $Y = \{Y_1, \ldots, Y_n\}$ is, in the limit as $n \to \infty$, larger at the true value θ_0 than it is at any other value:

$$\lim_{n \to \infty} \mathrm{pr}_{\theta_0}\{L(\theta_0; Y)/L(\theta; Y) > 1\} = 1. \tag{1.85}$$

Here and below, the dependence of the vector Y on n is implicit. To prove (1.85), an application of Jensen's inequality gives

$$\mathrm{E}_{\theta_0}\{\ell(\theta; Y)\} \leq \mathrm{E}_{\theta_0}\{\ell(\theta_0; Y)\}. \tag{1.86}$$

Under the assumption that $E_{\theta_0}\{|\ell(\theta; Y_1)|\} < \infty$, $\ell(\theta; Y)/n$ converges in probability to $E_{\theta_0}\ell(\theta; Y_1)$, and (1.85) follows.

This suggests that the value $\widehat{\theta}(Y)$ that maximizes $\ell(\theta; Y)$ converges to the true value θ_0. However, since $\ell(\theta; \cdot)$ is a function on the parameter space Θ, its behaviour in small regions or at isolated points in Θ could be unusual enough to cause problems. One solution is to make more assumptions: for example the original argument due to Wald (1949) assumed Θ is compact and $f(\cdot; \theta)$ is continuous in θ.

The more usual presentation relies on the Cramér–Rao conditions to show that there is a solution of the score equation $u(\theta; Y) = 0$ that is consistent for θ_0. This solution, if unique, is the maximum likelihood estimator if the likelihood function is maximized at an interior point of Θ. If there is more than one solution to the score equation, then the approach does not say which of the roots is the consistent for θ_0.

The Cramér–Rao conditions for consistency are:

A1: The support of the density for Y_1, i.e., the set $\{y : f(y; \theta) > 0\}$ does not depend on θ.

A2: The true value θ_0 is contained in an open subset of Θ, and for all θ in this open subset, the density $f(y; \theta)$ is differentiable with respect to θ for all y in the support of the density.

Under the identifiability assumption and conditions A1 and A2, there exists a solution $\widehat{\theta} = \widehat{\theta}(Y)$ of the score equation $U(\theta) = 0$ that maximizes the likelihood function and is consistent, i.e.,

$$\widehat{\theta} \xrightarrow{p} \theta_0, \quad n \to \infty.$$

1.7.2 Asymptotic distributions of likelihood quantities

Further conditions are needed to establish the asymptotic distribution of the maximum likelihood estimator and related quantities. We follow Knight (2000, Section 5.4), and assume that $Y_1, \ldots, Y_n$ are independent and identically distributed replicates of a random variable Y with density $f(y; \theta)$ and log-likelihood function $\ell(\theta; y)$, which satisfies

A2′: $f(y; \theta)$ is three times continuously differentiable with respect to θ for all y in the support of the density.

A3: $E_\theta\{u(\theta; Y)\} = 0$, and $E_\theta\{u(\theta; Y)u^\top(\theta; Y)\} = E_\theta\{-\partial u(\theta; Y)/\partial\theta^\top\}$, and these expectations exist and are finite for all θ in the open subset defined in A2.

A4: The matrix $i(\theta) = E_\theta\{u(\theta; Y)u^\top(\theta; Y)\}$ is positive definite for all θ in the open subset defined in A2.

A5: There exist functions $M_{abc}(\cdot)$ such that

$$\left|\frac{\partial^3}{\partial\theta_a\partial\theta_b\partial\theta_c}\ell(\theta; y)\right| \le M_{abc}(y), \quad ||\theta - \theta_0|| \le \delta$$

and $E_{\theta_0}\{M_{abc}(Y)\} < \infty$.

Theorem 1.33 Under the identifiability assumption, and conditions A1, A2′, A3–A5,

$$\sqrt{n}(\widehat{\theta} - \theta_0) \xrightarrow{d} N\{0, i^{-1}(\theta_0)\}. \tag{1.87}$$

□

Proof We sketch the argument for the case that θ is a scalar parameter. A Taylor series expansion of the equation $\ell'(\widehat{\theta}; y) = 0$ about θ_0 gives

$$0 = \ell'(\widehat{\theta}) = \ell'(\theta_0) + (\widehat{\theta} - \theta_0)\ell''(\theta_0) + \frac{1}{2}(\widehat{\theta} - \theta_0)^2 \ell'''(\theta_n^*), \tag{1.88}$$

where $|\theta_n^* - \theta_0| \leq |\widehat{\theta} - \theta_0|$, and the dependence of $\ell(\theta)$ on y is suppressed for convenience. Re-arranging terms,

$$-\frac{\ell'(\theta_0)}{\ell''(\theta_0)} = (\widehat{\theta} - \theta_0)\left\{1 + \frac{1}{2}(\widehat{\theta} - \theta_0)\frac{\ell'''(\theta_n^*)}{\ell''(\theta_0)}\right\}, \tag{1.89}$$

and introducing dependence on n explicitly,

$$\frac{\ell'(\theta_0)/\sqrt{n}}{-\ell''(\theta_0)/n} = \sqrt{n}(\widehat{\theta} - \theta_0)\left\{1 - \frac{1}{2}(\widehat{\theta} - \theta_0)\frac{\ell'''(\theta_n^*)/n}{-\ell''(\theta_0)/n}\right\}.$$

This can be re-expressed as

$$\frac{\ell'(\theta_0)}{\{ni(\theta_0)\}^{1/2}}\left\{\frac{i(\theta_0)}{-\ell''(\theta_0)/n}\right\} = \{ni(\theta_0)\}^{1/2}(\widehat{\theta} - \theta_0)(1 + Z_n), \tag{1.90}$$

where Z_n is shorthand for the remainder term.

The first factor on the left hand side of (1.90) converges in distribution to a standard normal random variable, by the central limit theorem for $U(\theta_0) = \ell'(\theta_0)$, as at (1.45). The second factor converges in probability to 1, by the weak law of large numbers for $-\ell''(\theta_0)$. To see that the remainder term Z_n converges in probability to 0, we assume that $\widehat{\theta}$ converges in probability to θ_0, which implies that θ_n^* does as well. The limit in probability of $\ell'''(\theta_n^*)/n$ is $\mathrm{E}_{\theta_0}\{\ell'''(\theta_0; Y)\}$, which is assumed to be finite. Thus

$$(\widehat{\theta} - \theta_0)I^{1/2}(\theta_0) = \frac{U(\theta_0)}{I^{1/2}(\theta_0)}\{1 + o_p(1)\}, \tag{1.91}$$

showing that the limiting distribution of the standardized maximum likelihood estimator is a standard normal distribution. □

A similar argument gives the result (1.48) for the likelihood ratio statistic. The Taylor series expansion

$$\ell(\theta_0) = \ell(\widehat{\theta}) + (\theta_0 - \widehat{\theta})\ell'(\widehat{\theta}) + \frac{1}{2}(\theta_0 - \widehat{\theta})^2\ell''(\widehat{\theta}) + \frac{1}{6}(\theta_0 - \widehat{\theta})^3\ell'''(\theta_n^*),$$

can be re-arranged to give

$$\ell(\widehat{\theta}) - \ell(\theta_0) = \frac{1}{2}(\widehat{\theta} - \theta_0)^2\{-\ell''(\widehat{\theta})\} + \frac{1}{6}(\widehat{\theta} - \theta_0)^3\ell'''(\theta_n^*),$$

from which we have

$$2\{\ell(\widehat{\theta}) - \ell(\theta_0)\} = (\widehat{\theta} - \theta_0)^2 I(\theta_0)\frac{-\ell''(\widehat{\theta})}{I(\theta_0)}\left\{1 + \frac{1}{3}(\widehat{\theta} - \theta_0)\frac{\ell'''(\theta_n^*)}{-\ell''(\widehat{\theta})}\right\}, \tag{1.92}$$

$$= (\widehat{\theta} - \theta_0)^2 I(\theta_0)(1 + \widetilde{Z}_n). \tag{1.93}$$

The consistency of $\widehat{\theta}$ ensures that $-\ell''(\widehat{\theta})/n \xrightarrow{p} i(\theta_0)$, and $\widetilde{Z}_n = o_p(1)$ follows as above.

The result for $\theta \in \mathbb{R}^k$ follows by the same argument, once the consistency of $\widehat{\theta}$ has been established.

The conditions above are stronger than necessary; in particular Assumption A5 can be weakened to a condition on second derivatives. As we will see in Chapters 4 and 5, the techniques for establishing limiting distributions of more general estimators are very similar to those outlined above; van der Vaart (1998, Chapter 5) treats the general case first and then maximum likelihood estimation as a special case.

Another approach to establishing these limit theory results, called local asymptotic normality, is to show that the log-likelihood function as a function of θ converges to the log-likelihood function of θ corresponding to a normal random variable with expected value θ. The results for the score, Wald and likelihood ratio statistics then follow, under some regularity assumptions. This approach, originally due to Le Cam (1960), is presented in van der Vaart (1998, Chapter 7).

1.7.3 Bias of maximum likelihood estimators

The maximum likelihood estimator is typically biased in finite samples, even when consistent. This bias is a direct consequence of the invariance property of maximum likelihood estimation. Methods that reduce the bias of the maximum likelihood estimator summarized in this section have the side effect of violating parametrization invariance and may increase bias under other parametrizations. However, in situations where interest is focussed on a specific parametrization, bias reduction may improve the finite-sample performance of estimation and inferential procedures.

The bias of the maximum likelihood estimator $\widehat{\theta}$ is denoted by $B(\theta_0) = \mathrm{E}_{\theta_0}(\widehat{\theta} - \theta_0)$. If θ_0 and $B(\cdot)$ were known, the modified estimator $\widetilde{\theta} = \widehat{\theta} - B(\theta_0)$ would have zero bias. Bias-reduction methods aim to approximate $\widetilde{\theta}$. The bias of the maximum likelihood estimator is asymptotically $O(n^{-1})$ and typically has the form

$$B(\theta_0) = \frac{b_1(\theta_0)}{n} + \frac{b_2(\theta_0)}{n^2} + \dots, \tag{1.94}$$

with $b_j(\theta_0)$ free of n. The methods described below are designed to reduce the bias to $O(n^{-2})$. Bias-reduction methods can be explicit or implicit. In explicit methods, also known as bias correction, the bias term $B(\theta_0)$ is estimated and then subtracted from $\widehat{\theta}$.

The jackknife is a classical approach to explicit bias correction, and may be applied to any estimator with bias of the form (1.94). The maximum likelihood estimator with the jth observation removed, $\widehat{\theta}_{(-j)}$, has the same bias as $\widehat{\theta}$ with n replaced by $n-1$, which suggests the bias estimator $\widehat{B}_{\text{jack}} = (n-1)(\bar{\theta} - \widehat{\theta})$, where $\bar{\theta} = n^{-1}\Sigma_j \widehat{\theta}_{(-j)}$ is the average of n leave-one-out estimators. The jackknifed maximum likelihood estimator is

$$\widetilde{\theta}_{\text{jack}} = n\widehat{\theta} - (n-1)\bar{\theta},$$

and has bias $O(n^{-2})$.

Another explicit method for bias correction is bootstrap simulation. The bootstrap estimate of the bias is $\widehat{B}_{\text{boot}} = \bar{\theta}^* - \widehat{\theta}$, where $\bar{\theta}^* = B^{-1}\Sigma_b \widehat{\theta}^{*b}$ is the average of the maximum likelihood estimators $\widehat{\theta}^{*b}$ computed at each of the $b = 1, \dots, B$ bootstrap samples. Bootstrap samples are created either by drawing observations with replacement from the original data or by simulating from the fitted model. The former is called the nonparametric bootstrap and the latter the parametric bootstrap. The bootstrap bias-corrected maximum likelihood estimator is

$$\widetilde{\theta}_{\text{boot}} = 2\widehat{\theta} - \bar{\theta}^*.$$

A third explicit bias correction method is based on the result that for $\theta \in \mathbb{R}^p$,

$$b_1(\theta_0)/n = -I^{-1}(\theta_0)A(\theta_0), \tag{1.95}$$

where the p-vector $A(\theta)$ has components

$$A_j(\theta) = \frac{1}{2}\mathrm{tr}[I^{-1}(\theta)\{(P_j(\theta) + Q_j(\theta)\}], \quad j = 1, \ldots, p,$$

where $P_j(\theta) = \mathrm{E}_0\{U(\theta)U^{\mathsf{T}}(\theta)U_j(\theta)\}$, $Q_j(\theta) = -\mathrm{E}_0\{J(\theta)U_j(\theta)\}$, and $U_j(\theta)$ is the jth component of $u(\theta; Y_1, \ldots, Y_n)$. The corresponding first-order bias-corrected maximum likelihood estimator is

$$\widetilde{\theta} = \widehat{\theta} - b_1(\widehat{\theta})/n = \widehat{\theta} + I^{-1}(\widehat{\theta})A(\widehat{\theta}).$$

Implicit bias-reduction methods avoid computation of the maximum likelihood estimator, effectively preventing bias rather than subtracting it. The best-known class of implicit methods is due to Firth (1993a). In the simplest case, Firth's method consists in computing the estimator $\widetilde{\theta}$ by solving the modified score equations

$$U(\widetilde{\theta}) + A(\widetilde{\theta}) = 0. \tag{1.96}$$

Applying this to (1.95) shows that $b_1(\widetilde{\theta})/n = I^{-1}(\widetilde{\theta})U(\widetilde{\theta})$, which is asymptotically equivalent to $\widehat{\theta} - \widetilde{\theta} = b_1(\widetilde{\theta})/n$, showing that $\widetilde{\theta}$ has, in general, bias $O(n^{-2})$ and asymptotic distribution the same as that of the maximum likelihood estimator.

If θ is the canonical parameter of an exponential family model (Example 1.10), solving (1.96) is equivalent to maximizing the penalized log-likelihood function

$$\ell(\theta) + \frac{1}{2}\log|I(\theta)|, \tag{1.97}$$

which corresponds to computing the maximum-a-posteriori estimator of θ using Jeffreys' invariant prior; see Section 2.8. Kosmidis and Firth (2021) show in binomial regression models that the maximizer of this penalized log-likelihood function exists. Maximizing this penalized log-likelihood function is a form of regularized estimation that is widely used in logistic regression, where it is sometimes termed "Firth's logistic regression'; see Example 2.9. It is especially helpful for fitting models to categorical data, where the maximum likelihood estimator may not exist due to separability or quasi-separability. Other forms of penalized likelihood estimation are discussed in Chapter 7.

1.8 Non-regular models

The regularity conditions needed to establish the asymptotic distributions summarized in (1.49), (1.50) and (1.51) are outlined in Section 1.7. Informally, once consistency of the maximum likelihood estimator is established, the conditions needed on the model are those that enable the use of Taylor series expansions, convergence in probability of various log-likelihood derivatives, and a central limit theorem for the score function. Establishing consistency of the maximum likelihood estimator can be difficult, although it may be enough to show that there is a consistent root of the score equation, which can be easier.

Models that do not support these assumptions are broadly classified as "non-regular", but this somewhat vague term covers a wide variety of potential issues. In this section we describe a few of the more common forms of non-regularity; see the bibliographic notes for references to more complete accounts.

1.8.1 Accumulation of information

The results outlined in Sections 1.7.1 and 1.7.2 are presented for n independent and identically distributed variables $Y_1, \ldots, Y_n$ with density $f(y; \theta_0)$. But as can be seen there, the key results used are that the standardized score function converges in distribution to a normal distribution, and the average observed information converges in probability to its expected value. As long as these two results hold, the asymptotic theory described in Section 1.7 can be extended to non-i.i.d. settings. With independent observations, this is often relatively easily established. For example, for generalized linear models, discussed in Section 1.12, we obtain the classical limiting results for likelihood inference under an assumption that the expected Fisher information matrix is $O(n)$, which in turn entails an assumption on X, the $n \times p$ matrix of covariates. Knight (2000, Chapter 8), for example, requires $\max_i x_i^\top (X^\top X)^{-1} x_i \to 0$ as $n \to \infty$.

Models with some types of dependence among the Y_i can also be accommodated, under constraints on the dependence that ensure appropriate accumulation of information. A central limit theorem may still be available if the dependence decays sufficiently quickly, as, for example, in an autoregressive model of order 1, or the ARMA model in Example 3.19. However if there is long-range dependence in the sequence of observations a different asymptotic analysis will be needed; see, for example, Beran (1994).

Example 1.34 (Change-point problems) Suppose the model is

$$y_i = \begin{cases} \mu + \beta_1 x_i + \epsilon_i, & i = 1, \ldots, \tau, \\ \mu + \beta_2 x_i + \epsilon_i, & i = \tau + 1, \ldots, n, \end{cases} \tag{1.98}$$

where τ is an unknown change-point of the slope coefficient of the regression model. Often the hypothesis of most interest is $H_0 : \beta_1 = \beta_2$; under H_0, τ is not identifiable. A likelihood-based statistic for testing H_0 is

$$W_{\max} = \max_{\tau} L(\widehat{\beta}_{1\tau}, \widehat{\beta}_{2\tau}, \tau) / L(\widehat{\beta}, \widehat{\beta}),$$

where $\widehat{\beta}_1$ and $\widehat{\beta}_2$ are estimated from the first τ and the last $(n - \tau)$ observations, respectively. The asymptotic distribution of this test statistic is determined by the distribution of the maximum of a stochastic process.

If interest focusses on estimation of τ, rather than detection, then different arguments are needed. Intuitively the information about τ in the data is contained only in the observations near τ, and depending on the sampling model this information may not increase, or increase very slowly, as the sample size increases. The value $\widehat{\tau}$ at which $W_{\max}$ is achieved is often used as an estimator, but convergence of $\widehat{\tau}$ or $W_{\max}$ to their limiting distributions can be very slow. Examples of change-point inference are discussed in Davison (2003, Example 4.40) and Barndorff-Nielsen and Cox (1994, Section 3.8). □

1.8.2 Parameter space topology

A standard regularity condition is that the true value of the parameter is contained in an open subset of the parameter space. This rules out discrete parameter spaces, for example. Another commonly encountered non-regularity is one or more of the components of θ having true values on the boundary of the parameter space.

Example 1.35 (Boundary values) Consider a linear mixed-effects model,

$$Y_{ij} = \mu + b_i + \epsilon_{ij}, \quad j = 1, \ldots, m, \ i = 1, \ldots, n, \tag{1.99}$$

say, where $b_i \sim N(0, \sigma_b^2)$ and $\epsilon_{ij} \sim N(0, \sigma^2)$ are assumed to be independent over their subscripts and with respect to each other. If there is much less variation between groups than within groups, the likelihood function could be decreasing in $\sigma_b^2 \in [0, \infty)$, in which case the maximum likelihood estimate of σ_b^2 would be 0. As there is positive probability that $\widehat{\sigma}_b^2 = 0$, the limiting distribution of the maximum likelihood estimator described in Section 1.2 will not be valid. Similarly, the likelihood ratio statistic for $H_0 : \sigma_b^2 = 0$ will not follow a χ^2 distribution under H_0, but rather a mixture of a point mass at 0 and a χ^2 distribution (Chernoff, 1954; Self and Liang, 1987). □

Example 1.36 (Generalized extreme value distribution) The generalized extreme-value distribution has distribution function

$$F(y; \mu, \tau, \xi) = \exp\left[-\left\{1 + \xi\left(\frac{y-\mu}{\tau}\right)\right\}^{-1/\xi}\right], \quad -\infty < \xi, \mu < \infty, \tau > 0, \tag{1.100}$$

where μ and τ are location and scale parameters, respectively, ξ is a shape parameter, and the range of y is constrained by $1 + \xi(y - \mu)/\tau > 0$. The Gumbel distribution is the limit of F as $\xi \to 0$,

$$F(y; \mu, \tau) = \exp\left\{-\exp\left(-\frac{y-\mu}{\tau}\right)\right\}, \quad -\infty < y < \infty.$$

For $\xi > -1/2$, maximum likelihood inference is regular, but for $\xi < 0$, the support of the density depends on the parameter and a different asymptotic analysis is needed. A simpler version of this is illustrated in Exercise 1.23. □

The regularity conditions also assume that θ is identifiable from the data; i.e., if $f(y; \theta_1) = f(y; \theta_0)$, then $\theta_1 = \theta_0$. This is a particular problem in testing natural null hypotheses if these are compatible with two or more expressions of the model.

Example 1.37 (Indeterminate parameters) A two-component mixture model has density

$$f(y; \theta, p) = pf_1(y; \theta_1) + (1-p)f_2(y; \theta_2), \quad 0 \le p \le 1, \tag{1.101}$$

where θ_1 and θ_2 may be vector or scalar. The hypothesis $H_0 : p = 0$ indicates the absence of a second class. For a simple example, if f_1 is the density for a $N(0, \sigma^2)$ distribution, and f_2 that for a $N(\mu, \sigma^2)$ distribution, the parameters of the model are not identifiable under H_0 (Exercise 1.24). There are many useful models that have the property that one or more parameters are not identifiable for some subset of the parameter space.

Shi and Drton (2025, Section 1.1) provide a helpful review of the literature on likelihood ratio tests for mixture models when the component densities are normal. Their focus is on the split likelihood ratio test proposed in Wasserman et al. (2020); see Exercise 2.10. □

1.8.3 Singular information matrix

A key regularity condition for establishing the results of Section 1.2 is that the expected Fisher information matrix $I(\theta)$ is non-singular for all $\theta \in \Theta$; if $\theta \in \mathbb{R}$ this requires that $I(\theta) > 0$. When testing a boundary value of the parameter space, as in the random effects model in Example 1.35 above, it may be that $I(\theta_0) = 0$, where θ_0 is the boundary value of interest (often 0). When θ_0 is not on the boundary of the parameter space, $I(\theta_0) = 0$ is equivalent to $\ell'(\theta_0) = 0$ with probability 1. An example is inference for θ when $Y \sim N(\theta^q, 1)$ for an odd integer $q > 0$; the information function is $I(\theta) = nq^2\theta^{2(q-1)}$ and at $\theta_0 = 0$ is identically 0 (Brazzale and Mameli, 2024, Example 4.2). As noted in Cox and Hinkley (1974, Example 4.16), this can be solved by reparametrization from θ to θ^q.

Rotnitzky et al. (2000) consider likelihood ratio tests of $H_0 : \theta = \theta_0$ for $\theta \in \mathbb{R}^p$, where the expected Fisher information matrix has rank $p-1$, or equivalently only $p-1$ elements of the score vector are linearly independent. They show that the rate of convergence of the distribution of the maximum likelihood estimator of θ to its limit depends strongly on the behaviour of higher-order derivatives of the log-likelihood function. In addition, the distribution of the log-likelihood ratio statistic can be either the usual χ^2_p, or an equal mixture of a χ^2_p and a χ^2_{p-1}. Bottai (2003) and Ekvall and Bottai (2022) consider confidence regions based on the score, Wald, and likelihood ratio statistics in models of this type.

1.8.4 Observed vs expected information

In regular problems the observed Fisher information $J(\widehat{\theta})$ converges in probability to the expected Fisher information $I(\theta)$, and the standardization of the score function or the maximum likelihood estimator by observed or expected information leads to the same asymptotic result. In practice, normalization by observed Fisher information usually gives more accurate approximations. Efron and Hinkley (1978) show, for scalar parameter models, that standardization of the maximum likelihood estimator by observed Fisher information approximates the conditional distribution of the maximum likelihood estimator given an approximately ancillary statistic. The emergence of observed Fisher information in the development of the p^* approximation discussed in Section 2.5.2 confirms this, for scalar and vector θ.

In the analysis of likelihood inference for stochastic processes, it is often the case that the observed Fisher information does not converge in probability to the expected Fisher information, but rather $J(\widehat{\theta})/I(\theta)$ converges in distribution to a random variable. In such models it is possible that the score or Wald statistic has an asymptotically normal distribution if standardized by the observed Fisher information, but not if standardized by the expected Fisher information.

Example 1.38 (Poisson birth process) Davison (2003, Example 4.6) considers a sequence of Poisson random variables, $Y_0, Y_1, \ldots, Y_n$, with Y_0 distributed as Poisson(θ) and Y_t, given Y_{t-1}, distributed as Poisson(θY_{t-1}). This is a curved exponential family, with log-likelihood function

$$\ell(\theta; y_0, \ldots, y_n) = \sum_{i=0}^{n} y_i \log\theta - \theta\left(\sum_{i=1}^{n-1} y_i + 1\right), \tag{1.102}$$

and sufficient statistics

$$m = \sum_{i=0}^{n} y_i, \quad s = \sum_{i=0}^{n-1} y_i.$$

The maximum likelihood estimate is $\widehat{\theta} = m/s$, the observed information is $J(\widehat{\theta}) = s^2/m$, and the expected information function is $I(\theta) = \theta^{-1}(1 + \theta + \cdots + \theta^n)]$. If $\theta \leq 1$, then $I(\theta)$ is asymptotically bounded and thus $\widehat{\theta}$ is neither consistent nor asymptotically normal; see Davison (2003, Example 4.44) and Exercise 1.26. A feature of this model is that $J(\widehat{\theta})/I(\theta)$ does not converge to 1, but rather to a random variable, as $S/\mathrm{E}_\theta(S)$ converges in distribution to an exponential random variable with rate 1. As a result $I^{1/2}(\theta)(\widehat{\theta} - \theta)$ does not have an asymptotic normal distribution, although $J^{1/2}(\widehat{\theta})(\widehat{\theta} - \theta)$ does.

□

1.9 Bibliographic notes

The construction of statistical models is both an art and a science, and depends critically on the scientific problem at hand. General overviews are provided in Cox (1990), Cox and Donnelly (2011, Chapter 6), Cox (2006, Chapter 5), and Lehmann (1990). A more formal approach is described in McCullagh (2002) and McCullagh (2022, Chapter 11). Davison (2003) gives a masterful and wide-ranging introduction to models useful for many applications; see especially Chapters 4, 5, and 6. Cox and Hinkley (1974, Chapter 1), and Barndorff-Nielsen and Cox (1994, Chapter 2) provide a range of examples that illustrate many subtleties in statistical modelling. The Poisson process examples in Section 1.1 are adapted from Davison (2003, Section 6.5), where several other examples of likelihood functions for stochastic processes are presented. For a more rigorous development of likelihood for stochastic processes, see, for example, Gill and Johansen (1990).

Barndorff-Nielsen and Cox (1994, Section 1.3) give an introduction to exponential families, with more detailed treatment in Sections 2.9 and 2.10. Davison (2003, Section 5.2) defines exponential families by exponential tilting of a base density. More technical discussions can be found in Brown (1986) and Schervish (1995, Section 2.2). Transformation models are treated in generality in Barndorff-Nielsen and Cox (1994, Section 2.8). Exact inference in the regression-scale model is the focus of Fraser (1979).

General introductions to the definition of the likelihood function and its use in inference are given in Fisher (1956), Edwards (1992) and Azzalini (1996). Direct use of the likelihood function for inference is developed further in Royall (1997); see also Burnham and Anderson (2002). The book by Pawitan (2026) is the closest in aims to this one, as it discusses many extensions of likelihood methods for complex data.

An introduction to the main principles of inference, including likelihood, sufficiency, ancillarity, and the properties of maximum likelihood estimators and likelihood ratio tests, is given in most treatments of theoretical statistics. The approach in this chapter draws heavily on Cox and Hinkley (1974), Barndorff-Nielsen and Cox (1994), and Davison (2003). The likelihood principle, which states that observations with proportional likelihood functions should lead to the same inference, has been invoked to argue for a Bayesian approach to inference; see Cox and Hinkley (1974, Section 2.3) and Cox (2006, Section 4.2) for a criticism and discussion.

Significance functions $p(\theta)$, as described in Section 1.3, were the focus of Fraser (1990), who in particular emphasized plotting the significance function as an adjunct to a plot of the log-likelihood function. The significance function is essentially the same as the confidence distribution function of Cox (1958) and Efron (1993). Xie and Singh (2013) and Schweder and Hjort (2016) take the confidence distribution and its associated density as the main object for inference, and consider pivotal quantities other than the standard likelihood-based versions in Section 1.3.

For a vector version of the asymptotic normality of the Bayesian posterior, see Walker (1969). An Edgeworth expansion for the posterior density is derived in Johnson (1978) and discussed further in Bickel and Ghosh (1990). Computational methods for Bayesian inference continue to be developed and refined. For an introduction see Davison (2003, Section 11.3) and for more recent overviews Robert (2014) and Plummer (2023).

Model selection is covered extensively in Claeskens and Hjort (2008); see also Sections 4.3, 4.4, and the bibliographic notes for Chapter 4. Tsao (2024) develops a model-selection criterion based on likelihood ratio tests, for inference about the regression parameters β in linear and generalized linear models. In Section 4.4, the Takeuchi information criterion is derived, which generalizes AIC when the assumed model is misspecified. The Takeuchi information criterion is also used with composite likelihoods; see Section 4.7.

The discussion of predictive likelihood draws heavily on Pawitan (2026, Chapter 16). Several versions of predictive likelihood, developed to improve on the profile predictive likelihood, are reviewed in Bjørnstadt (1990). For an overview of extended likelihood and h-likelihood see Lee and Nelder (2009), and Jin and Lee (2021). Lawless and Fredette (2005) develop an approach to calibrated prediction intervals based on confidence distribution functions, which are closely related to the significance functions described in Section 1.3; see also Shen et al. (2018). A review of predictive inference from a Bayesian perspective is provided in Vehtari and Ojanen (2012), including the link between prior predictive distributions and model selection.

The generalized extreme-value distribution is described in Davison (2003, Section 6.3), and properties of likelihood-based inference for its non-regular parameters are developed in Smith (1985) and Zhang and Shaby (2022). Davison et al. (2012) and Davison and Huser (2015) provide reviews of models and inference for extreme values. Huser et al. (2019) devise a stochastic version of the EM-algorithm, discussed in Section 3.5.2, in the context of model (1.84).

Careful presentations of the asymptotic properties of the maximum likelihood estimator are provided in many textbooks; we have relied heavily on Knight (2000, Chapter 4) and Lehmann and Casella (1998, Chapter 6). Scholz (2006) gives a clear and succinct overview. A more advanced account, applying also to estimators defined using estimating equations, discussed in Chapter 5, is provided in van der Vaart (1998, Chapter 5). Proving consistency of the maximum likelihood estimator under more general settings than the i.i.d. case outlined in Section 1.7.1 can be difficult, and results continue to be established, including, for example, Balabdaoui et al. (2009), McGoff et al. (2015), and Mukherjee et al. (2025). The Hauck–Donner effect in logistic regression was identified in Hauck and Donner (1977, 1980) and is reviewed in Yee (2022). DasGupta (2008) provides a detailed study of asymptotic theory for parametric and nonparametric models.

Kosmidis (2014) reviews explicit and implicit methods for bias reduction of maximum likelihood estimators. The jackknife approach to estimation is originally due to Quenouille (1956). Efron and Gong (1983) give an introduction to bootstrap and jackknife with a discussion on the use of resampling methods for bias correction. Reference monographs for the bootstrap include Efron (1982), Efron and Tibshirani (1994) and Davison and Hinkley (1997). Bias correction based on explicit asymptotic expansions was proposed in Cox and Snell (1968) and has been applied to various classes of models, including nonlinear models with normal errors (Cook et al., 1986), generalized linear models (Cordeiro and McCullagh, 1991) and generalized linear mixed models (Breslow and Lin, 1995).

The implicit bias-reduction method of Firth (1993a) has been extended to several models, including exponential family nonlinear models (Kosmidis and Firth, 2009) and cumulative link models for ordinal data (Kosmidis, 2014). Bias reduction of M-estimators is discussed in Kosmidis and Lunardon (2024). Heinze and Schemper (2002) noted that Firth's bias correction solves the problem of complete separation in logistic regression, where the maximum likelihood estimator diverges (Albert and Anderson, 1984). Kenne Pagui et al. (2017) present a modification of the score equation that reduces the median bias of the maximum likelihood estimator; in particular, their modified estimator, $\widetilde{\theta}$, has the same limiting distribution as the maximum likelihood estimator, and satisfies $\mathrm{pr}_\theta(\widetilde{\theta} \leq \theta) = 1/2 + O(n^{-3/2})$. Kosmidis et al. (2020) develop an integrated approach for mean and median bias reduction in generalized linear models.

Another implicit approach for bias-correction is indirect inference (Gouriéroux et al., 1993); see Section 4.6. Kuk (1995) uses indirect inference to obtain asymptotically unbiased estimates in generalized linear mixed models.

Overviews of likelihood-based inference in models that do not satisfy the classical regularity conditions are provided, for example, in Smith (1987), Barndorff-Nielsen and Cox (1994, Section 3.8), Davison (2003, Section 4.6), and Brazzale and Mameli (2024). Rotnitzky et al. (2000) analyse a special type of non-regular model, in which the first s terms of the Taylor-series expansion of the log-likelihood function or the score function vanish for all values of the parameter.

Likelihood inference for variance components and more general mixed effects models has a large literature; some key references include Crainiceanu and Ruppert (2004), Stram and Lee (1994), and further references in Brazzale and Mameli (2024). Battey and McCullagh (2024) show in variance component models and related scale-parameter models that inference based on the Wald statistic can be misleading in finite samples, and argue in favour of the likelihood ratio statistic.

It can be difficult in some settings to assess whether or not the parametrization of a given model is identifiable. Log-linear models for the analysis of multi-way contingency tables can be formulated as generalized linear models; see for example Davison (2003, Section 10.5). Some components of the model parameters may not be identifiable if one or more cell counts are zero. Far et al. (2021) develop a method to detect this, which they call the "parameter redundancy" approach. Their method is compared to those described in Fienberg and Rinaldo (2012) and Wang et al. (2019).

There is a large literature on tests and estimates for change-points, especially change-points in the expected value: the topic is important in many applied settings, especially involving data collected over time. Brazzale and Mameli (2024) provide several references, as do Jewell et al. (2022). The latter paper considers testing the null hypothesis of no change-point, using an estimate of the change-point, and links this to inference after model selection. When interest focusses on estimating the parameters of the response distribution, the change-point parameter is a nuisance parameter that vanishes under the null hypothesis, using the terminology of Davies (1977, 1987, 2002).

For a discussion of likelihood inference in selected stochastic processes, see Barndorff-Nielsen and Cox (1994, Section 9.3). The Poisson birth process of Example 1.38 is discussed in Barndorff-Nielsen and Cox (1994, Example 9.1) and Jensen (1987); the latter also considers higher-order approximations. Feigin and Reiser (1979) relate this to the notion of approximate ancillarity of observed information developed in Efron and Hinkley (1978). Kwan et al. (2024) study likelihood inference for the Hawkes process, which is often used to model the temporal clustering of financial time series.

1.10 Further results and exercises

1.1. In Example 1.7, verify that the likelihood function for constant rate is given by (1.12). Give an expression for the likelihood function when $\rho(t) = \exp(\theta_0 + \theta_1 t)$.

1.2. A random vector $Y = (Y_1, \ldots, Y_d)^\top$ follows a multinomial distribution with sample size m and parameter $\pi = (\pi_1, \ldots, \pi_d)^\top$ if its density is

$$f(y; \pi) = \frac{m!}{y_1! \ldots y_d!} y_1^{\pi_1} \ldots y_d^{\pi_d}, \quad \Sigma_j y_j = m, \, \Sigma_j \pi_j = 1, \, 0 \le y_j \le m.$$

(a) Show that this is an exponential family model and give an expression for the canonical parameter in terms of π.

(b) If there are non-linear constraints on π, for example $\pi = \pi(\theta)$, where $\theta \in \mathbb{R}^q$, $q < d$, show that this is a curved exponential family model.

(c) Show that equating observed proportions y_j/m to the expected probabilities π_j determines the maximum likelihood estimator of π in (a) and of θ in (b).

(d) In the curved model of (b), show that the maximum likelihood estimator of θ satisfies $y_j = m\pi_j(\widehat{\theta})$, $j = 1, \ldots, d$. *(Davison, 2003, Example 5.12)*

1.3. The von Mises–Fisher distribution on the unit sphere in $\mathbb{R}^d$ has density function

$$f(y; \mu, \kappa) = H_d(\kappa) \exp(\kappa\mu^{\mathsf{T}} y), \quad \Sigma_i y_i^2 = \Sigma_i \mu_i^2 = 1,\ \kappa \geq 0. \tag{1.103}$$

The normalizing constant $H_d(\kappa) = \kappa^{d/2-1}/\{(2\pi)^{d/2} I_{d/2-1}(\kappa)\}$, where $I_d(\kappa)$ is the modified Bessel function of the first kind of order d.

(a) Show that this exponential family model can be constructed as an exponential tilt of the uniform distribution on the sphere.

(b) Show that the density on the circle is

$$f(y; \mu, \kappa) = \{2\pi I_0(\kappa)\}^{-1} \exp\{\kappa \cos(y - \mu)\}, \quad 0 \leq y < 2\pi,\ 0 \leq \mu < 2\pi,\ \kappa > 0 \tag{1.104}$$

(c) Show that for fixed κ, the density on the circle is a transformation model under the group of rotations on the unit circle. *(Davison, 2003, Example 5.11)*

1.4. Suppose $\{Y_1, \ldots, Y_n\}$ are independent observations from the full exponential family model with density (1.16). Show that the marginal density of $S = \Sigma_i Y_i$ is again of exponential family form with cumulant generating function $K_S(\alpha) = nK_{Y_1}(\alpha) = n\{c(\alpha + \varphi) - c(\varphi)\}$.

1.5. A family of distributions on $\mathbb{R}$ is said to be a scale family if the density function $f(y; \theta)$ takes the form $\theta^{-1} f_0(y/\theta)$, for $0 < \theta < \infty$. The density $f_0(y)$ is the standard form of the density and θ is the scale parameter.

Show that the normal distribution with known expected value and unknown variance, and the gamma distribution with known shape parameter, are scale families.

1.6. A special case of the gamma scale family is the simple exponential distribution for a positive random variable Y with density

$$f(y; \theta) = \theta^{-1} \exp(-y/\theta), \qquad y > 0,\ \theta > 0. \tag{1.105}$$

Verify that the distribution of $Z = \log(Y)$ is a location family, and give an expression for its density. What is the location parameter?

1.7. Suppose $Y_1, \ldots, Y_n$ are independently distributed inverse Gaussian random variables, with density

$$f(y; \psi, \lambda) = \exp\left\{-\frac{1}{2}(\lambda y^{-1} + \psi y) + \sqrt{(\lambda\psi)} + \frac{1}{2}\log\lambda\right\} h(y).$$

(a) Show that $h(y) = y^{-3/2}(2\pi)^{-1/2}$.

(b) Show that $\overline{Y} = n^{-1}\Sigma_i Y_i$ has the inverse Gaussian distribution with parameters $(n\lambda, n\psi)$.

(c) Show that $\Sigma_i(Y_i^{-1} - \overline{Y}^{-1})$ is independent of $\overline{Y}$.

(Barndorff-Nielsen and Cox, 1994, Example 6.2)

1.8. In Example 1.11, give an expression for the log-likelihood function for θ from a sample of independent, identically distributed pairs from the special bivariate normal distribution. Show that $(S_1, S_2) = (\Sigma_i(y_{1i}^2 + y_{2i}^2)/2, \Sigma_i y_{1i}y_{2i})$ is a minimal sufficient statistic for θ. Show that while $\Sigma_i x_i^2$ and $\Sigma_i y_i^2$ are individually ancillary for θ, that S_1 is only approximately ancillary for θ, as $\mathrm{E}(S_1) = n$ is free of θ but $\mathrm{var}(S_1) = n(1+\theta^2)$.

(Barndorff-Nielsen, 1980; Efron and Hinkley, 1978)

1.9. Suppose in the generalized linear mixed model of Example 1.13, that Y_i has a normal distribution with expected value $x_i^{\mathsf{T}}\beta$ and variance σ^2. Give a closed form expression for the likelihood function (1.24). *Davison (2003, Section 9.4.2)*

1.10. Suppose $Y_1, \ldots, Y_n$ are independently distributed random variables from the location-scale model with density $\theta_2^{-1} f_0\{\theta_2^{-1}(y_i - \theta_1)\}$, and let $\overline{y} = n^{-1}\Sigma_i Y_i$, $s^2 = \Sigma_i(Y_i - \overline{y})^2$, $a_i = s^{-1}(Y_i - \overline{y}), i = 1, \ldots, n$.

(a) Show that Jacobian of the transformation from Y to $(\overline{y}, s, a)$ is proportional to s^{n-2} by first making the transformation one-to-one:

$$t_1 = \overline{y}, \quad t_2 = s, \quad t_i = a_i \qquad i = 3, \ldots, n,$$

with inverse transformation $Y_i = t_1 + t_2 t_i$, $i = 3, \ldots, n$. A geometric derivation of this result can be obtained using standardized coordinates $z_i = (y_i - \theta_1)/\theta_2$. The Jacobian of the transformation from the vector z to $\{\overline{z}, s(z), d(z)\}$ is the differential element dz in these new coordinates, which are locally orthogonal. As $d(z)$ is orthogonal to the 1-vector, and lies on a unit sphere in the $n-1$-dimensional subspace of $\mathbb{R}^n$ that is orthogonal to the 1-vector, its volume element is the surface volume on the sphere of radius $s(z)$, which is $s(z)^{n-2}du$, where du is surface volume on the unit sphere in $\mathbb{R}^{n-1}$. An explicit expression for the surface area of the unit sphere in $\mathbb{R}^d$ is $(2\pi)^{d/2}/\Gamma(d/2)$.

(b) In (a), show that the argument is essentially unchanged if t is replaced by $(\widehat{\theta}_1, \widehat{\theta}_2, a)$, where $a_i = \widehat{\theta}_2^{-1}(y_i - \widehat{\theta}_1), i = 1, \ldots, n$ and $(\widehat{\theta}_1, \widehat{\theta}_2)$ are the maximum likelihood estimates of θ_1 and θ_2. Although the coordinates $(\widehat{\theta}_1, \widehat{\theta}_2, a)$ are not locally orthogonal, the volume element of the associated parallelogram is still the product of the base and height.

(Fraser, 1979, Section 2.1)

1.11. The first two Bartlett identities are given in (1.36) and (1.39). Further such identities are readily derived, assuming the density function is differentiable with respect to the parameter to the appropriate order. Continue the argument leading to (1.39) to show that

$$I_{001}(\theta) + 3I_{11}(\theta) + I_{30}(\theta) = 0,$$
$$I_{0001}(\theta) + 4I_{101}(\theta) + 3I_{02}(\theta) + 6I_{21}(\theta) + I_{40}(\theta) = 0,$$

where

$$I_{abcd}(\theta) = \mathrm{E}\left\{\left(\frac{\partial \ell(\theta)}{\partial \theta}\right)^a \left(\frac{\partial^2 \ell(\theta)}{\partial \theta^2}\right)^b \left(\frac{\partial^3 \ell(\theta)}{\partial \theta^3}\right)^c \left(\frac{\partial^4 \ell(\theta)}{\partial \theta^4}\right)^d\right\}.$$

McCullagh (1987) uses these identities to simplify asymptotic expansions of the maximum likelihood estimator and the likelihood ratio statistic. For the multivariate versions he develops tensor notation for third- and fourth-derivative arrays.

1.12. Given sequences of constants $\{a_n\}$ and $\{b_n\}$ we say that $a_n = o(b_n)$ if $a_n/b_n \to 0$, as $n \to \infty$, and $a_n = O(b_n)$ if $a_n/b_n \to A$ as $n \to \infty$, where $|A| < \infty$ does not depend on n.

Given sequences of random variables $\{X_n\}$ we say that $X_n = o_p(a_n)$ if $X_n/a_n \to 0$ in probability, and $X_n = O_p(a_n)$ if X_n/a_n is bounded in probability. More explicitly for the latter, given $\epsilon > 0$ there exist constants k_ϵ and $n_0 = n_0(\epsilon)$ such that if $n > n_0$, $\text{pr}(|X_n| < a_n k_\epsilon) > 1 - \epsilon$.

Verify the following rules for O, o, O_p, and o_p from their definitions:

(a) $O(n^{-a})O(n^{-b}) = O(n^{-a-b})$,
(b) $O_p(n^{-a})O_p(n^{-b}) = O_p(n^{-a-b})$,
(c) $O_p(n^{-a})O(n^{-b}) = O_p(n^{-a-b})$,
(d) $O(n^{-a})o(n^{-b}) = o(n^{-a-b})$,
(e) $O_p(n^{-a})o(n^{-b}) = o_p(n^{-a-b})$,
(f) $o_p(n^{-a})o_p(n^{-b}) = o_p(n^{-a-b})$.

(Barndorff-Nielsen and Cox, 1989, Section 2.3)

1.13. Suppose $\{y_1, \ldots, y_n\}$ is a sample from the exponential density $f(y;\theta) = \theta\exp(-\theta y)$.

(a) Give explicit expressions for the pivotal quantities $s(\theta)$, $q(\theta)$, and $r(\theta)$ in (1.52), (1.53), and (1.55), respectively. Verify directly that these quantities are equivalent to first order, by verifying, for example, that $r(\theta) = q(\theta)\{1 + O_p(1/\sqrt{n})\}$ and $q(\theta) = s(\theta)\{1 + O_p(1/\sqrt{n})\}$.

(b) Compare these results to those for the gamma mean parameter in Example 1.20.

1.14. In Example 1.21, expressions are given for the Wald, score and likelihood root statistics. A different significance function based on the score function, standardized by $i(\theta)$, leads to the alternate pivotal quantity $\widetilde{s}(\theta) = \sqrt{n}(\widehat{\theta} - \theta)/\{\theta(1-\theta)\}^{1/2}$. Show that the limits of an equi-tailed $(1-\alpha)$-confidence interval for θ based on this pivotal quantity solve a quadratic equation in θ. Compare the coverage of this interval to that obtained using the exact binomial probabilities, obtained by solving for θ

$$\sum_{k=y}^{n}\binom{n}{k}\theta^k(1-\theta)^{n-k} = \alpha/2, \quad \sum_{k=0}^{y}\binom{n}{k}\theta^k(1-\theta)^{n-k} = \alpha/2,$$

with lower bound 0 if $y = 0$ and upper bound 1 if $y = n$. *(Agresti, 2002, Section 1.4)*

1.15. Suppose $Y = (Y_1, \ldots, Y_n)^{\mathsf{T}}$ is a vector of i.i.d. variables from a distribution with density function $f(y;\theta), \theta \in \mathbb{R}$.

(a) If $S = s(Y)$ is an unbiased estimate of $g(\theta)$, where $g(\cdot)$ is a differentiable function, show that

$$\text{var}_\theta(S) \geq \frac{\text{Cov}\{U(\theta), S\}^2}{\text{var}\{U(\theta)\}} = \frac{\{g'(\theta)\}^2}{I(\theta)},$$

where $U(\theta)$ is the score function $u(\theta; Y)$ based on the sample, defined at (1.35). This result is known as the Cramér–Rao lower bound.

(b) Given an estimator $\widetilde{\theta}$ of θ for which

$$\sqrt{n}(\widetilde{\theta} - \theta) \xrightarrow{d} N\{0, \sigma^2(\theta)\},$$

the asymptotic relative efficiency of $\widetilde{\theta}$, relative to the maximum likelihood estimator $\widehat{\theta}$, is defined to be

$$\text{ARE}(\widetilde{\theta}) = 1/\{I(\theta)\sigma^2(\theta)\}.$$

Show that if $\sigma^2(\theta)$ is continuous in θ, the ARE is bounded above by 1, i.e., the maximum likelihood estimator is asymptotically fully efficient. *(Knight, 2000, Section 6.5)*

1.16. In Example 1.22, suppose μ is known, so without loss of generality can be taken to be 0. The exact distribution of $\widehat{\sigma}^2$ is $(\sigma^2/n)\chi^2_n$. Compare the normal approximation to the pivotal quantities in (1.59) to the exact distribution, either through asymptotic expansions or numerical work.

1.17. Suppose $X_1, \ldots, X_n$ are independently distributed as $N(\theta, 1)$, and define $Y_i = 1\{X_i > 0\}$. Show that $\mathrm{E}(Y_i) = \Phi(\theta)$, where $\Phi(\cdot)$ is the cumulative distribution function of the standard normal. Compare the relative efficiency of $\widetilde{\psi} = \overline{Y}_i$ to that of $\widehat{\psi} = \Phi(\widehat{\theta}) = \Phi(\overline{X})$.

Loss of efficiency in dichotomizing responses in a linear regression is studied in Davison (2003, Section 10.4). In Shah et al. (2025) a similar analysis is provided for what is called there 1-bit censoring, relevant for inference using quantized signals such as arise in some areas of communication and signal processing.

1.18. The Kullback–Leibler divergence from the distribution F to the distribution G is given by

$$\mathrm{KL}(F:G) = \int \log\left\{\frac{f(y)}{g(y)}\right\} f(y)\,\mathrm{d}y, \tag{1.106}$$

where f and g are density functions with respect to Lebesgue measure. (Similar versions can be defined for discrete distributions, and for more general measures.) The K-L divergence is not symmetric in its arguments. Nomenclature differs in the literature: (1.106) is called the directed information from F to G in Barndorff-Nielsen and Cox (1994, Chapter 2), where the more general definition $\mathrm{KL}(F:G) = \int \log(dF/dG)\,\mathrm{d}F$ is used, assuming F and G are mutually absolutely continuous. Other terms in common use include "the divergence of F from G", the "divergence from G to F", and the "relative entropy of F with respect to G".

(a) In the canonical exponential family model (1.16) find an expression for the KL divergence from the model with parameter φ_1 to the model with parameter φ_2.

(b) Show in general for an i.i.d. sample of observations $y_1, \ldots, y_n$ from a parametric model with density $f(y;\theta)$ and distribution function $F(y;\theta)$, that the maximum likelihood estimator minimizes the KL divergence from $F(\cdot;\theta)$ to $\widehat{F}_n(\cdot)$, where $\widehat{F}_n(\cdot)$ is the empirical distribution function putting mass $1/n$ at each observation y_i (Section 6.1).

1.19. Find the posterior distribution for μ given an independent, identically distributed sample of size n where

(a) Y_i follows a Poisson distribution with expected value μ, and μ follows a gamma distribution with shape parameter β and rate parameter α.

(b) Y_i follows a Normal distribution with expected value μ and variance σ^2, and μ follows a normal distribution with expected value μ_0 and variance σ_0^2.

1.20. Prediction limits in the classical linear regression model can be derived directly. Suppose

$$y = X\beta + \epsilon,$$

where $y \in \mathbb{R}^n$ is the vector of observed responses, X is an $n \times p$ matrix of covariates, $\beta \in \mathbb{R}^p$, and $\epsilon \sim N(0, \sigma^2 \mathcal{I}_{n\times n})$, where $\mathcal{I}_{n\times n}$ is the $n \times n$ identity matrix. For a given covariate vector $\widetilde{x} = (\widetilde{x}_1, \ldots, \widetilde{x}_p)$ the natural predicted value for $\widetilde{y}$ is

$$\widehat{\widetilde{y}} = \widetilde{x}^{\mathsf{T}}\widehat{\beta}, \tag{1.107}$$

where $\widehat{\beta} = (X^{\mathsf{T}}X)^{-1}X^{\mathsf{T}}y$ is the usual least squares estimator. This assumes the same linear model holds for the additional observation $(\widetilde{y}, \widetilde{x})$.

(a) Show that $\mathrm{E}(\widehat{\widetilde{y}}) = \widetilde{x}^{\mathsf{T}}\beta$ and $\mathrm{var}(\widehat{\widetilde{y}}) = \sigma^2\{1 + \widetilde{x}(X^{\mathsf{T}}X)^{-1}\widetilde{x}^{\mathsf{T}}\}$.

(b) Let $S^2 = (y - X\widehat{\beta})^{\mathsf{T}}(y - X\widehat{\beta})/(n-p)$ be the estimate of σ^2 based on the residual sum of squares in the observed data, and t^{α}_{n-p} be the critical α-value for Student's t-distribution with $n-p$ degrees of freedom. Show that the interval

$$\mathrm{PI}_{\alpha}(y) = \widehat{\widetilde{y}} \pm t^{\alpha}_{n-p}[S\{1 + \widetilde{x}(X^{\mathsf{T}}X)^{-1}\widetilde{x}^{\mathsf{T}}\}^{1/2}],$$

satisfies $\mathrm{pr}\{\widetilde{y} \in \mathrm{PI}_{\alpha}(y) \mid y\} = 1 - 2\alpha$, for any value of β. This probability is computed over the joint distribution of $\widetilde{y}$ and y.

1.21. This example illustrates in extreme form why exact unbiasedness of an estimator is not always useful. Suppose Y follows the geometric distribution with probability mass function

$$f(y;\theta) = (1-\theta)^y\theta, \quad 0 \le \theta \le 1,\ y = 0, 1, \ldots .$$

Show that the unique unbiased estimator of θ is $\widetilde{\theta}(0) = 1$ if $y = 0$, and $\widetilde{\theta}(y) = 0$ for $y \ge 1$. *(Cox and Hinkley, 1974, Example 8.1).*

1.22. The Bradley–Terry model (Bradley and Terry, 1952) is a classical model for paired comparisons, used in a wide variety of applications including the statistical analysis of sport competitions. Consider a tournament-shape data set consisting in n matches between p players and suppose that the underlying tournament graph is fully connected, meaning every player has been directly or indirectly compared to every other player. The Bradley–Terry model assumes that the probability that player i beats player j is

$$\mathrm{pr}(i \text{ beats } j) = \frac{\exp(\theta_i)}{\exp(\theta_i) + \exp(\theta_j)}, \quad i \ne j \in \{1, \ldots, p\},$$

where θ_i represents the skill of player i.

(a) Show that the maximum likelihood estimator of θ_i diverges to ∞ when player i wins all of their matches, and to $-\infty$ when player i loses all of them;

(b) Show that Firth's bias-reduced maximum likelihood estimator estimation of $\theta = (\theta_1, \ldots, \theta_p)^{\mathsf{T}}$ is equivalent to a regularized approach that introduces a phantom player, who wins and loses half a match against every other player in the tournament;

(c) Verify that the bias-reduced maximum likelihood estimator of θ always exists and is finite (Kosmidis and Firth, 2021).

1.23. Suppose $Y_1, \ldots, Y_n$ are independent and identically distributed uniform random variables on $(0, \theta)$:

$$f(y_i;\theta) = \frac{1}{\theta}, \quad 0 \le y_i \le \theta,\ i = 1, \ldots, n.$$

(a) Show that the maximum likelihood estimator of θ is $\widehat{\theta} = Y_{(n)}$, the largest value of the sample.

(b) Show that the limiting distribution of $n(\widehat{\theta} - \theta)$ is exponential.

See Davison (2003, Example 4.43) for a similar example with two parameters. Barndorff-Nielsen and Cox (1994, Example 3.8) discuss a general family of models with endpoint parameters.

1.24. Suppose in the mixture density (1.101) that the component densities are both von Mises–Fisher densities on the circle, with mean direction μ_1 or μ_2, respectively, and common concentration parameter κ, as in (1.104). Show that the asymptotic distribution of the log-likelihood ratio statistic for $H_0 : p = 0$ exists and can be modified to a χ^2_1 distribution. In contrast, if the component densities are $N(0, \sigma^2)$ and $N(\mu, \sigma^2)$, the log-likelihood ratio statistic for $H_0 : p = 0$ diverges to ∞ with probability 1.

(Fu et al., 2008; Chen and Chen, 2003)

1.25. This exercise is another example of a boundary value problem. Suppose $Y_1, \ldots, Y_n$ are i.i.d., with $Y_i = |X_i|$, with $X_i \sim N(\mu, 1)$. The only identifiable parameter is $\theta = |\mu|$, since all information about the sign of μ has been lost.

(a) Show that the log-likelihood function and score function based on the sample are, respectively,

$$\ell(\theta; y) = -\frac{n}{2}\theta^2 + \sum_{i=1}^{n} \log\{\cosh(\theta y_i)\},$$

$$U(\theta) = -n\theta + \sum_{i=1}^{n} y_i \tanh(\theta y_i).$$

(b) Show that the maximum likelihood estimate is consistent for $\theta \geq 0$.

(c) Show that for $\theta > 0$ the limiting distribution of $\sqrt{n}(\widehat{\theta} - \theta)$ is normal, with variance given by the inverse of the expected Fisher information.

(d) The score function at $\theta = 0$ is identically 0, and the Taylor series expansion (1.88) breaks down. By expanding $\ell(\theta; y)$ up to the fourth derivative, conclude that $\widehat{\theta} = 0$ if $\Sigma_i y_i^2 \leq n$, and otherwise

$$\widehat{\theta^2} = 3(\Sigma_i y_i^2 - n)/(\Sigma_i y_i^4).$$

(Cox and Hinkley, 1974, Example 9.2)

1.26. Consider the Poisson birth process of Example 1.38. For a sequence $Y_0, Y_1, \ldots, Y_n$, with $Y_j \mid Y_{j-1}$ following a Poisson(θY_{j-1}) distribution, and $Y_0 \sim$ Poisson(θ), show that the maximum likelihood estimator, observed and expected Fisher information are, respectively,

$$\widehat{\theta} = \sum_{i=1}^{n} Y_i/(1 + \sum_{i=1}^{n} Y_i), \quad J(\widehat{\theta}) = \frac{1}{\widehat{\theta}^2}\sum_{i=0}^{n} Y_i, \quad I(\theta) = \frac{1}{\theta^2}(\theta + \cdots + \theta^{n+1}).$$

Show that $J(\theta)$ is finite with positive probability if $\theta < 1$, and finite with probability 1 if $\theta \geq 1$. *(Davison, 2003, Section 6.5)*

Appendix 1.A Comments on models

The theory of inference based on likelihood and likelihood-like functions, as treated in this book, is essentially a theory of inference built on probability models. To be useful for applications, these models should reflect at least some aspects of the stochastic system under investigation. In rare cases they might be derived from sets of equations either known or supposed to govern the system, for example a set of stochastic differential equations in studies of climate, or models based on economic principles. More usually, for a given application a model is selected from a family of models both flexible and familiar and is intended to provide an empirical description. This does not mean this is not useful for assessing some aspects of the system under study, and in many applications key parameters in the model would correspond to meaningful concepts in the field of application. In most of this book, models are taken as given. While this is helpful for discussing particular theoretical aspects, in applications models will always need evaluation in the context of the problem under investigation.

The necessity for model parametrization to be consistent with the method of sampling is emphasized in McCullagh (2002), and McCullagh (2022, Chapter 11).

In advanced texts, models are often formulated from a measure-theoretic point of view, using a probability triple $(\mathcal{Y}, \mathcal{A}, \mathcal{P})$, where $\mathcal{Y}$ is the sample space, often identified with $\mathbb{R}^n$ or $\mathbb{R}$, $\mathcal{A}$ is the Borel σ-field, and $\mathcal{P}$ is a probability measure, or family of probability measures, absolutely continuous with respect to a dominating measure which is typically Lebesgue or counting measure. The statistical model is technically the distribution function $F(\cdot)$ or family of distribution functions $\mathcal{F}_\theta = \{F(\cdot;\theta) : \theta \in \Theta\}$ determined by the (family of) probability measures.

Appendix 1.B Sufficiency and ancillarity

Definition 1.39 A statistic $S = s(Y)$ is sufficient for θ, in the family of models $\{f(y;\theta) : \theta \in \Theta\}$, if and only if the conditional distribution of Y given S is free of θ. □

The factorization theorem states that a statistic S is sufficient for θ in this family if and only if there exist functions g and h such that

$$f(y;\theta) = g\{s(y);\theta\}h(y),$$

for each value of θ.

The usual interpretation of sufficiency, for example in Cox and Hinkley (1974, Chapter 2), is that a statistician observing Y should come to the same conclusion about θ as a statistician observing only S, because the second statistician could recover the joint density for Y from the marginal density for S by multiplying by a fixed density, that of Y given S, which does not provide any information about the unknown parameter θ, assuming that the model is correct. As discussed in Section 1.5, the density of Y given S can be used for checking the model.

A sufficient statistic partitions the sample space $\mathcal{Y}$ into subspaces. Two elements of $\mathcal{Y}$ are in the same partition if they have the same value of $s(y)$. Typically the "size" of the partition is substantially smaller than the "size" of the sample space. For example, if Y is a random variable on $\mathbb{R}^n$, $S(Y) = \Sigma_i Y_i$ takes values on $\mathbb{R}$. The biggest reduction in dimension is obtained with the coarsest partition. The statistic inducing the coarsest possible partition, if it exists, is the minimal sufficient statistic. A formal definition of a minimal sufficient statistic is that it is a function of every other sufficient statistic. This definition is difficult to work with, but can be avoided by using the following theorem:

Theorem 1.40 The likelihood statistic is minimal sufficient. □

Proof The likelihood statistic is the partition of the sample space that puts two values of y into the same partition if they have the same likelihood function (i.e., their likelihood functions are in the same equivalence class). Thus y and y' are in the same partition if

$$\begin{aligned} L(\theta; y) &= L(\theta; y'), \\ c(y)f(y;\theta) &= c(y')f(y';\theta), \\ f(y;\theta) &= k(y', y)f(y';\theta). \end{aligned}$$

If this holds, we write $s(y) = s(y')$ to indicate that $s(\cdot)$ is the function that indexes this partition.

Suppose $T = t(Y)$ is some other sufficient statistic. By the factorization theorem

$$\begin{aligned} f(y;\theta) &= g\{t(y);\theta\}h(y), \\ f(y';\theta) &= g\{t(y');\theta\}h(y'). \end{aligned}$$

If $t(y) = t(y')$, then

$$\begin{aligned} f(y;\theta) &= g\{t(y');\theta\}h(y), \\ &= f(y';\theta)\{h(y)/h(y')\}, \\ &= k(y', y)f(y';\theta), \end{aligned}$$

so $s(y) = s(y')$, which shows that $s(\cdot)$ is a function of $t(\cdot)$. □

The sketch of the proof given here is taken from Cox and Hinkley (1974, Chapter 2); there is also an excellent discussion in Davison (2003, Chapter 4). The result is also proved in Lehmann and Romano (2005, Section 1.9, 2.6), and in Lehmann and Casella (1998, Section 1.6).

The terminology "likelihood statistic" is not strictly correct, as normally functions of both y and θ are not statistics. The term "likelihood map" is used in Fraser and Naderi (2007).

Definition 1.41 A statistic $A = a(Y)$ is ancillary for θ, in the family of models $\{f(y;\theta) : \theta \in \Theta\}$, if and only if $f(a)$ is free of θ. □

A key feature of transformation models is that the model for a sample of size n permits a reduction in dimension of the sufficient statistic, obtained by conditioning on an ancillary statistic. Thus, there exist functions of the data, say $s(Y)$ and $a(Y)$, for which we can write

$$f(y;\theta) \propto f_1\{s(y) \mid a(y);\theta\}f_2\{a(y)\},$$

where the marginal density of $a(Y)$ does not depend on the parameter θ. Furthermore, the conditional distribution with density f_1 is itself a transformation family with parameter θ.

Example 1.42 (Location model) Suppose $Y_1, \ldots, Y_n$ are independent, identically distributed observations from the location family $f_0(y - \theta)$. Letting $s(y) = y_1$ and $a_i = y_i - y_1, i = 2, \ldots, n$,

$$f(y;\theta) = \prod_{i=1}^{n} f_0(y_i - \theta) = f_1(y_1|a;\theta)f_2(a),$$

where

$$f_2(a) = \int f_0(t)f_0(a_2 + t)\ldots f_0(a_n + t)\,\mathrm{d}t, \tag{1.108}$$

and

$$f_1(y_1|a;\theta) = f_2^{-1}(a)f_0(y_1 - \theta)\ldots f_0(y_1 + a_n - \theta).$$

The numerator is just a rewriting of $\prod_i f(y_i;\theta)$, and an equivalent expression is

$$f_1(y_1|a;\theta) = \frac{f_0(y_1-\theta)\dots f_0(y_n-\theta)}{\int f_0(y_1-\theta)\dots f_0(y_n-\theta)\,\mathrm{d}\theta}. \tag{1.109}$$

The density of A does not depend on θ, and the conditional density of Y_n given A is a location family density on $\mathbb{R}$.

The functions $s(Y)$ and $a(Y)$ are not uniquely determined, but they are uniquely determined up to a location transformation. We could for example let $S = \overline{y}$ and $A_i = Y_i - \overline{y}$. In this case vector A takes values on an $n-1$-dimensional subspace of $\mathbb{R}^n$, as its components must sum to 0. The marginal density for a is again given by (1.108), and

$$f(y;\theta) \propto f(\overline{y}, a;\theta) = f_1(\overline{y} \mid a;\theta) f_2(a).$$

We might choose instead to let S be $\widehat{\theta}$, the maximum likelihood estimator of θ, and then $a = (y_1 - \widehat{\theta}, \dots, y_n - \widehat{\theta})$. □

Appendix 1.C Derivative and expectation of vector and matrix

In this book vectors are represented as column vectors. We use the following format for operations on vectors or matrices. Let $\theta = (\theta_1, \dots, \theta_p)^\mathsf{T}$ be a $p \times 1$ vector where p is a positive integer. Suppose $k(\theta)$ is a differentiable function of θ and $\Psi(\theta) = (\Psi_1(\theta), \dots, \Psi_q(\theta))^\mathsf{T}$ is a $q \times 1$ vector, where $\Psi_j(\theta)$ is a differentiable function of θ for $j = 1, \dots, q$, and q is a positive integer greater than 1. Then the derivatives of $k(\theta)$ and $\Psi(\theta)$ are

$$\frac{\partial k(\theta)}{\partial \theta} = \left(\frac{\partial k(\theta)}{\partial \theta_1}, \dots, \frac{\partial k(\theta)}{\partial \theta_p}\right)^\mathsf{T},$$

and

$$\frac{\partial \Psi(\theta)}{\partial \theta^\mathsf{T}} = \begin{bmatrix} \frac{\partial \Psi_1(\theta)}{\partial \theta_1} & \frac{\partial \Psi_1(\theta)}{\partial \theta_2} & \cdots & \frac{\partial \Psi_1(\theta)}{\partial \theta_p} \\ \frac{\partial \Psi_2(\theta)}{\partial \theta_1} & \frac{\partial \Psi_2(\theta)}{\partial \theta_2} & \cdots & \frac{\partial \Psi_2(\theta)}{\partial \theta_p} \\ \vdots & \vdots & \ddots & \vdots \\ \frac{\partial \Psi_q(\theta)}{\partial \theta_1} & \frac{\partial \Psi_q(\theta)}{\partial \theta_2} & \cdots & \frac{\partial \Psi_q(\theta)}{\partial \theta_p} \end{bmatrix},$$

respectively.

If $k(\theta)$ is also a function of a random variable (or vector), and $\Psi(\theta)$ is also a vector of functions of a random variable (or vector), then the expectation of $\partial k(\theta)/\partial\theta$ is

$$\mathrm{E}\left(\frac{\partial k(\theta)}{\partial \theta}\right) = \left(\mathrm{E}\left\{\frac{\partial k(\theta)}{\partial \theta_1}\right\}, \dots, \mathrm{E}\left\{\frac{\partial k(\theta)}{\partial \theta_p}\right\}\right)^\mathsf{T},$$

and the expectation of $\partial\Psi(\theta)/\partial\theta$ is a $q \times p$ matrix with (j,k) element

$$\mathrm{E}\left\{\frac{\partial \Psi_j(\theta)}{\partial \theta_k}\right\}.$$

If $g(v) = (g_1(v), \dots, g_q(v))^\mathsf{T}$ is a $q \times 1$ vector, where $g_j(v)$ is an integrable function for $j = 1, \dots, q$, then $\int g(v)\,\mathrm{d}v$ is the $q \times 1$ vector with jth component $\int g_j(v)\,\mathrm{d}v$ for $j = 1, \dots, q$.

2

Likelihood and Nuisance Parameters

2.1 Nuisance parameters

It is often useful to consider the parameter vector $\theta \in \mathbb{R}^p$ as composed of a set of parameters of interest and a complementary set of nuisance parameters. Which parameters are thus identified depends entirely on the application, and different parameters may be of interest at different stages of the investigation. For example in a mixed linear model where

$$Y_i = x_i^{\mathsf{T}}\beta + z_i^{\mathsf{T}}b + \epsilon_i, \tag{2.1}$$

one or more components of β may be of interest in the first instance, with the covariance matrix of the random effects vector b and the variance of the error term treated as nuisance parameters. Or, this could be reversed, with the parameters of interest the components of variance, and mean vector treated as a nuisance parameter.

In general, we write $\theta = (\psi^{\mathsf{T}}, \lambda^{\mathsf{T}})^{\mathsf{T}}$, with $\psi \in \mathbb{R}^q$ the parameter of interest and $\lambda \in \mathbb{R}^{p-q}$ the nuisance parameter. It is convenient to partition the likelihood quantities of Section 1.2 to correspond to the partition of θ, writing

$$U(\theta) = \begin{bmatrix} U_\psi(\theta) \\ U_\lambda(\theta) \end{bmatrix}, \quad I(\theta) = \begin{bmatrix} I_{\psi\psi}(\theta) & I_{\psi\lambda}(\theta) \\ I_{\lambda\psi}(\theta) & I_{\lambda\lambda}(\theta) \end{bmatrix}, \quad J(\theta) = \begin{bmatrix} J_{\psi\psi}(\theta) & J_{\psi\lambda}(\theta) \\ J_{\lambda\psi}(\theta) & J_{\lambda\lambda}(\theta) \end{bmatrix}, \tag{2.2}$$

and

$$I^{-1}(\theta) = \begin{bmatrix} I^{\psi\psi}(\theta) & I^{\psi\lambda}(\theta) \\ I^{\lambda\psi}(\theta) & I^{\lambda\lambda}(\theta) \end{bmatrix}, \qquad J^{-1}(\theta) = \begin{bmatrix} J^{\psi\psi}(\theta) & J^{\psi\lambda}(\theta) \\ J^{\lambda\psi}(\theta) & J^{\lambda\lambda}(\theta) \end{bmatrix}. \tag{2.3}$$

Recall that by convention $U(\theta)$ and $I(\theta)$ are the score and information functions for a sample $\{Y_1, \ldots, Y_n\}$. The average expected Fisher information matrix in one observation, $i(\theta) = I(\theta)/n$, can be similarly partitioned.

In some models it may be more convenient to define the parameter of interest $\psi = \psi(\theta)$ without arranging, as indicated here, that it corresponds to the first q components of θ; for example, we might constrain θ to lie on a q-dimensional curve in $\mathbb{R}^p$.

For likelihood inference any convenient parametrization that does not introduce new constraints on the parameter space can be used. In applications, parameters should have some meaning in that context, in which case inference is best expressed in those parameters if possible. Reparametrizations of the form $(\psi, \lambda) \to \{g(\psi), \eta(\psi, \lambda)\}$ are called interest-respecting; often $g(\cdot)$ is the identity.

Parameters are said to be likelihood orthogonal if the likelihood function for θ can be written as the product of two likelihood functions, each depending only on ψ or λ,

$$L(\psi, \lambda; y) \propto L_1(\psi; y) L_2(\lambda; y). \tag{2.4}$$

Example 2.1 (Random sample size) If the sample size is not fixed in advance, but follows a distribution governed by parameters unrelated to the parameters of interest, then

DOI: 10.1201/9780429090936-2

the model might have the form

$$f(y, n; \psi, \lambda) = f_Y(y \mid n; \psi) f_N(n; \lambda). \tag{2.5}$$

□

If in (2.4) both $L_1(\psi; y) = L_1\{\psi; s(y)\}$ and $L_2(\lambda; y) = L_2\{\lambda; t(y)\}$ are proportional to the density functions of $s(y)$ and $t(y)$, respectively, where $s(y)$ and $t(y)$ are statistics, i.e., observable functions of y, then inference for ψ may proceed based on L_1 following the arguments outlined in Chapter 1, and no new considerations arise. Similarly in (2.5) likelihood inference for ψ is based on the conditional density of Y given the attained sample size n.

Example 2.2 (Ratio of Poisson parameters) Suppose Y_1 and Y_2 are independent observations following Poisson distributions with expected values λ and $\psi\lambda$, respectively, so the parameter of interest is the ratio of their expected values. Then

$$f(y_1, y_2; \theta) = f(y_2 \mid y_1 + y_2; \psi) f(y_1 + y_2; \eta), \tag{2.6}$$

where $\eta = \lambda(\psi + 1)$. The marginal density of $Y_1 + Y_2$ is Poisson with mean η, and the conditional density of Y_2, given $y_1 + y_2$, is binomial with sample size $y_1 + y_2$ and probability of success $\psi/(\psi + 1)$. The parameter space for (ψ, η) is $\mathbb{R}^+ \times \mathbb{R}^+$, as is the parameter space for (ψ, λ). As in Example 2.1, likelihood inference for ψ is based on the conditional density $f(y_2 \mid y_1 + y_2; \psi)$. □

Likelihood orthogonality is quite strong, and obtains only in very few models. A less restrictive concept is parameter orthogonality with respect to expected Fisher information.

Definition 2.3 The parameters ψ and λ are said to be orthogonal if

$$i_{\psi\lambda}(\theta) = \frac{1}{n} \mathrm{E}_\theta \left\{ -\frac{\partial^2 \ell(\theta; Y)}{\partial\psi \partial\lambda^{\mathsf{T}}} \right\} = 0, \tag{2.7}$$

for every value of $\theta \in \Theta$. If the weaker condition $i_{\psi\lambda}(\theta_0) = 0$ holds the parameters are said to be (locally) orthogonal at θ_0. The parameter vector θ is totally orthogonal if $i(\theta)$ is diagonal. □

From a geometric point of view, the family of distributions with densities $\{f(\cdot; \theta), \theta \in \Theta\}$ can be regarded as a manifold. Rao (1945) showed that the Riemannian metric tensor for this manifold is given by the Fisher information. In this sense parameter orthogonality of sub-vectors is essentially the same as ordinary Euclidean orthogonality, but using a different inner product.

When ψ is a scalar parameter, it is always possible to find an orthogonal nuisance parametrization $\eta = \eta(\psi, \lambda)$ by solving for η the partial differential equation

$$\sum_{r=1}^{p-1} i_{\lambda_r \lambda_s}(\psi, \lambda) \frac{\partial \lambda_r(\psi, \eta)}{\partial \psi} = -i_{\psi\lambda_s}(\psi, \lambda), \quad s = 1, \ldots, p-1. \tag{2.8}$$

This result is established in Exercise 2.1 and illustrated in Exercises 2.2 and 2.5. The differential equation (2.8) does not have a unique solution, and in particular one-to-one functions of an orthogonal parameter are themselves orthogonal. In many examples there is a natural simple choice. If ψ is a vector it is not possible in general to find an orthogonal parametrization, although it may be possible in special cases; see Exercise 2.1.

Example 2.4 (Exponential regression) Suppose Y_i follows an exponential density with expected value $\lambda\exp(-\psi x_i)$, and we have an independent sample of size n. Then $i_{\lambda\psi}(\theta) = \lambda^{-1}\Sigma_i x_i/n$, and the parameters ψ and λ are orthogonal if $\Sigma_i x_i = 0$. Similarly, if the expected value is $\lambda\exp(-\psi x_i - \beta z_i)$ then λ and β are orthogonal if $\Sigma_i z_i = 0$. We can orthogonalize the two regression parameters ψ and β by replacing x_i by its residual after regression on z_i, giving

$$\mathrm{E}(Y_i) = \lambda\exp[-\psi\{x_i - z_i(S_{xz}/S_{zz})\} - \eta z_i], \tag{2.9}$$

where $S_{xz} = \Sigma_i x_i z_i$ and $S_{zz} = \Sigma_i z_i^2$. □

Example 2.5 (Weibull distribution) A single random variable from the Weibull distribution has density

$$f(y;\psi,\lambda) = \frac{\psi}{\lambda}\left(\frac{y}{\lambda}\right)^{\psi-1}\exp\left\{-\left(\frac{y}{\lambda}\right)^{\psi}\right\}; \tag{2.10}$$

here we take the shape parameter ψ as the parameter of interest. Since $i_{\lambda\lambda}(\theta) = (\psi/\lambda)^2$ and $i_{\lambda\psi}(\theta) = -\Gamma'(2)/\lambda$, the orthogonality equation is

$$\left(\frac{\psi}{\lambda}\right)^2\frac{\partial\lambda}{\partial\psi} = \frac{\Gamma'(2)}{\lambda}, \tag{2.11}$$

where $\Gamma(\alpha) = \int_0^\infty x^{\alpha-1}\exp(-x)\,dx$ is the gamma function, and $\Gamma'(\alpha)$ its derivative. One solution is $\eta = \lambda\exp\{\Gamma'(2)/\psi\}$. The survivor function in the parametrization (ψ,η) is $\exp[-(y/\eta)^\psi\exp\{\Gamma'(2)\}]$ which is approximately 0.22 at $y = \eta$; thus η represents approximately the 80th percentile of the Weibull distribution.

Smith (1987) discusses orthogonal parametrization in the three-parameter Weibull distribution and notes that the particular choice of solution can have an impact on the resulting inference. □

2.2 Marginal and conditional likelihood

Given a sample y with density $f(y;\psi,\lambda)$, if there exist statistics $s(y)$, $t(y)$, such that

$$f(y;\psi,\lambda) \propto f_{\mathrm{m}}\{s(y);\psi\}f_{\mathrm{c}}\{t(y) \mid s(y);\psi,\lambda\}, \tag{2.12}$$

then the marginal likelihood function for ψ is

$$L_{\mathrm{m}}(\psi;y) \propto f_{\mathrm{m}}\{s(y);\psi\}.$$

If we have instead

$$f(y;\psi,\lambda) \propto f_{\mathrm{m}}\{t(y);\psi,\lambda\}f_{\mathrm{c}}\{s(y) \mid t(y);\psi\}, \tag{2.13}$$

then the conditional likelihood function for ψ is

$$L_{\mathrm{c}}(\psi;y) \propto f_{\mathrm{c}}\{s(y) \mid t(y);\psi\}.$$

The utility of these definitions derives from their role in inference. Inference based on either $L_{\mathrm{m}}(\psi)$ or $L_{\mathrm{c}}(\psi)$ can proceed along the lines outlined in Section 1.2. In particular, as long as these marginal or conditional likelihoods satisfy the regularity conditions outlined in Section 1.7 the asymptotic results and associated significance functions follow directly.

Example 2.6 (Normal-theory linear regression) In the model $Y_i = x_i^{\mathsf{T}}\beta + \epsilon_i, i = 1, \ldots, n$, if we assume that ϵ_i are independently distributed as $N(0, \psi)$, then the maximum likelihood estimate of (β, ψ) is $(\widehat{\beta}, \widehat{\psi}) = \{(X^{\mathsf{T}}X)^{-1}X^{\mathsf{T}}y, S^2/n\}$, where X is the $n \times p$ matrix with ith row x_i^{T} and $S^2 = (y - X\widehat{\beta})^{\mathsf{T}}(y - X\widehat{\beta})$. Direct calculation shows that the density of S^2 is proportional to that of a χ^2_{n-p} distribution, where p is the dimension of the vector β, so for the parameter ψ we have

$$f_{\mathrm{m}}(\widehat{\psi}; \psi) = \frac{1}{\Gamma\{(n-p)/2\}2^{(n-p)/2}} \exp\left(-\frac{n\widehat{\psi}}{2\psi}\right)\left(\frac{n\widehat{\psi}}{\psi}\right)^{(n-p)/2}\frac{1}{\widehat{\psi}}, \tag{2.14}$$

and this does not depend on the nuisance parameter β. Using this marginal density as a likelihood function for inference for ψ leads to the unbiased estimate

$$\widehat{\psi}_{\mathrm{m}} = (y - X\widehat{\beta})^{\mathsf{T}}(y - X\widehat{\beta})/(n-p).$$

The likelihood function from (2.14) is a marginal likelihood functionfor ψ, based on the sample $y = (y_1, \ldots, y_n)^{\mathsf{T}}$. In fact, it is also a conditional likelihood, because $\widehat{\psi}$ and $\widehat{\beta}$ are independently distributed in the normal model. □

One concern about inference based on marginal or conditional likelihood functions is that there might be information about the parameter of interest ψ contained in the part of the likelihood that is discarded; the second factor in (2.12) and the first factor in (2.13). It has proved rather difficult to make this precise, although there is a literature on partial information; see the bibliographic notes.

Example 2.7 (Normal distribution on the circle) Suppose (Y_1, Y_2) are independently distributed as $N(\mu_1, 1)$ and $N(\mu_2, 1)$, respectively, and that the parameter of interest is $\psi = (\mu_1^2 + \mu_2^2)^{1/2}$. With the transformation on the sample space $y_1 = r\cos z$, $y_2 = r\sin z$, and corresponding transformation $\mu_1 = \psi\cos\lambda, \mu_2 = \psi\sin\lambda$, the marginal density of $R = (Y_1^2 + Y_2^2)^{1/2}$ is

$$f(r; \psi) = r\exp\left\{-\frac{1}{2}(r^2 + \psi^2)\right\} 2\pi\, I_0(\psi r),$$

where $I_0(\psi) = (2\pi)^{-1}\int_0^{2\pi}\exp\{\psi\cos(x)\}\,dx$ is the modified Bessel function of the first kind. This defines a marginal likelihood for ψ. The conditional distribution of Z, given R, is the von Mises–Fisher distribution, the analogue of the normal distribution on the circle:

$$f(z \mid r; \psi, \lambda) = \frac{1}{2\pi I_0(\psi r)}\exp\{\psi r\cos(z - \lambda)\}, \quad 0 \le z \le 2\pi.$$

In this distribution λ is a location parameter and ψr determines the spread of the distribution around the circle. □

Example 2.8 (Comparison of two binomials) Suppose Y_1 and Y_2 are independent binomial observations with indices m_1 and m_2 and probabilities of success p_1 and p_2, respectively. Suppose the parameter of interest is the log-odds ratio $\psi = \log[p_1(1-p_2)/\{(1-p_1)p_2\}]$, and the nuisance parameter is $\lambda = \log\{p_2/(1-p_2)\}$. As a function of these parameters the joint density is

$$f(y_1, y_2; \psi, \lambda) = \binom{m_1}{y_1}\binom{m_2}{y_2}\exp(\psi y_1)\exp\{\lambda(y_1+y_2)\}\{1+\exp(\psi+\lambda)\}^{-m_1}\{1+\exp(\lambda)\}^{-m_2}, \tag{2.15}$$

from which we can see that the conditional density for Y_1, given $t = y_1 + y_2$, is

$$f_c(y_1 \mid t; \psi) = c^{-1}(\psi, t)\binom{m_1}{y_1}\binom{m_2}{t-y_1}\exp(\psi y_1), \quad \max(0, t-m_2) \le y_1 \le \min(m_1, t), \tag{2.16}$$

where

$$c(\psi, t) = \sum_{y_1=\max(0,t-m_2)}^{\min(m_1,t)} \binom{m_1}{y_1}\binom{m_2}{t-y_1}\exp(\psi y_1).$$

This density is free of λ, an example of factorization (2.13). If we are interested in testing $p_1 = p_2$, i.e., $\psi = 0$, the conditional distribution is the hypergeometric distribution, which is the basis for Fisher's exact test of independence in a 2×2 table. The marginal distribution of t depends on both ψ and λ. Zhu and Reid (1994) showed that the partial Fisher information about ψ, after adjusting for λ, is very nearly 0, except when (p_1, p_2) is close to $(1, 0)$ or $(0, 1)$. Schervish et al. (2025) show that the loss of partial information by conditioning approaches 0 as the sample size approaches ∞. □

Example 2.9 (Logistic regression) Example 2.8 is a special case of logistic regression. Suppose for $i = 1, \ldots, n$ that Y_i are independent observations from the binomial distribution, with sample size m_i and probability of success p_i, where

$$\log\left(\frac{p_i}{1-p_i}\right) = x_i^{\mathsf{T}}\beta, \tag{2.17}$$

for a p-vector of covariates x_i. The joint density for $Y = \{Y_1, \ldots, Y_n\}$ is

$$\begin{aligned} f(y; \beta) &\propto \exp\left[\sum_{i=1}^{n} x_i^{\mathsf{T}}\beta y_i - m_i \log\{1 + \exp(x_i^{\mathsf{T}}\beta)\}\right], \\ &= \exp\left[\beta_1 \sum_{i=1}^{n} x_{i1}y_i + \cdots + \beta_p \sum_{i=1}^{n} y_i x_{ip} - m_i \log\{1 + \exp(x_i^{\mathsf{T}}\beta)\}\right], \end{aligned}$$

and from general results for exponential family models (see Section 2.3 below) we can show that if the parameter of interest is, for example, β_p, that the likelihood factors as (2.13), and the conditional likelihood function is proportional to the density of $s(Y) = \Sigma_i Y_i x_{ip}$, given $t(y) = (\Sigma_i y_i x_{i1}, \ldots, \Sigma_i y_i x_{i,p-1})$.

If the y_i are binary, $m_i = 1$, and the covariates x_i are also binary, then this conditional density may for small sample sizes have very few points of support, and the asymptotic theory of Chapter 1 may lead to poor finite-sample approximations. Improved approximations are the focus of Brazzale et al. (2007, Chapter 4).

Furthermore, if y_i are binary, and if some linear combination of the covariates completely separates the responses with $y_i = 1$ from those with $y_i = 0$, then the likelihood function has a supremum as some linear combination of β approaches ∞; i.e., the maximum likelihood estimator cannot be identified. With p large relative to n this separation is increasingly likely to occur in finite samples.

Firth (1993a) proposed an additive correction to the score equation, designed to reduce the bias of the maximum likelihood estimator. For exponential families in the canonical parametrization, this correction term corresponds to multiplying the likelihood function by Jeffreys' invariant prior; see Section 1.7.3. For logistic regression, this takes the form

$$\widetilde{\ell}(\beta) = \ell(\beta) + \frac{1}{2}\log|X^{\mathsf{T}}W(\beta)X|, \tag{2.18}$$

where $W(\beta)$ is a diagonal matrix with ith diagonal element $m_i p_i(\beta)\{1 - p_i(\beta)\}$, where $p_i(\beta)$ is defined in (2.17). Maximizing $\widetilde{\ell}(\beta)$ leads to finite estimates of the regression coefficients, even in the case of separation of the responses, if X has full rank (Kosmidis and Firth, 2021). □

2.3 Exponential and transformation models

Recall from Example 1.10 that if $y = \{y_1, \dots, y_n\}$ is an i.i.d. sample from an model with $\varphi \in \mathbb{R}^p$, that

$$f(y;\varphi) = \exp\{\varphi^{\mathsf{T}} s(y) - nc(\varphi) - h(y)\}, \tag{2.19}$$

where $s(y) = \Sigma_i s(y_i)$ is a vector of length p; see (1.18). The marginal density of s is

$$f(s;\varphi) = \exp\{\varphi^{\mathsf{T}} s - nc(\varphi) - \widetilde{h}(s)\}, \tag{2.20}$$

where $\exp -\{\widetilde{h}(s)\} = \int \exp\{-h(y)\}\,dy$, the integral being over the set in $\mathbb{R}^n$ of all y that give the same value $s = s(y)$.

Now suppose that the canonical parameter φ is partitioned as (ψ, λ), where ψ is the parameter of interest and λ is a nuisance parameter. We partition the sufficient statistic accordingly,

$$f(s_1, s_2; \psi, \lambda) = \exp\{\psi^{\mathsf{T}} s_1 + \lambda^{\mathsf{T}} s_2 - nc(\psi, \lambda) - \widetilde{h}(s_1, s_2)\}. \tag{2.21}$$

The conditional distribution of S_1, given $S_2 = s_2$ is again an exponential family, and depends only on the parameter of interest ψ:

$$f(s_1 \mid s_2; \psi) = \exp\{\psi^{\mathsf{T}} s_1 - c_2(\psi) - h_2(s_1)\}, \tag{2.22}$$

where both $c_2(\psi)$ and $h_2(s_1)$ will typically depend on s_2, and the dependence on n of c_2 and h_2 has been suppressed for ease of notation.

Examples 2.4, 2.8 and 2.9 are in this class of exponential families; the parameter of interest there was defined to be a component, or more generally a linear function, of the canonical parameter φ. In the comparison of two binomial probabilities this requires the parameter of interest to be the log-odds ratio; there is no exact conditional likelihood function for the difference $p_1 - p_2$, for example.

Example 2.10 (Gamma shape parameter) Given a random sample $Y = \{Y_1, \dots, Y_n\}$ from the gamma distribution with shape parameter ψ and rate parameter λ, the joint density of Y is

$$\begin{aligned} f(y;\psi,\lambda) &= \prod_{i=1}^{n} \frac{1}{\Gamma(\psi)} y_i^{\psi-1} \lambda^{\psi} \exp(-\lambda y_i), \\ &= \exp\{(\psi - 1)\Sigma_i \log y_i - \lambda \Sigma_i y_i + n\psi \log\lambda - n\log\Gamma(\psi)\}, \end{aligned} \tag{2.23}$$

and $(\Sigma_i \log Y_i = S_1, \Sigma_i Y_i = S_2)$ is sufficient for (ψ, λ). The density of (S_1, S_2) is

$$f(s_1, s_2; \psi) = \exp\{(\psi - 1)s_1 - \lambda s_2 + n\psi\log\lambda - n\log\Gamma(\psi) - \widetilde{h}(s_1, s_2)\}, \tag{2.24}$$

where $\exp\{-\widetilde{h}(s_1, s_2)\}$ is the integral of (2.23) over the range of y, constrained by fixing $\Sigma_i \log y_i = s_1$ and $\Sigma_i y_i = s_2$. The conditional density of S_1, given $S_2 = s_2$, is

$$\begin{aligned} f(s_1 \mid s_2; \psi) &= \frac{\exp\{\psi s_1 - \lambda s_2 - n\psi \log \lambda - n \log \Gamma(\psi) - \widetilde{h}(s_1, s_2)\}}{\int \exp\{\psi s_1 - \lambda s_2 - n\psi \log \lambda - n \log \Gamma(\psi) - \widetilde{h}(s_1, s_2)\}\, ds_1}, \\ &= \frac{\exp\{\psi s_1 - \widetilde{h}(s_1, s_2)\}}{\int \exp\{\psi s_1 - \widetilde{h}(s_1, s_2)\}\, ds_1}, \\ &= \exp\{\psi s_1 - c_2(\psi) - h_2(s_1)\}, \end{aligned} \tag{2.25}$$

where $\exp\{c_2(\psi)\}$ is the integral of the denominator in (2.25) and $h_2(s_1) = \widetilde{h}(s_1, s_2)$. In this example we can derive $\exp\{c_2(\psi)\}$ explicitly, as the marginal distribution of $S_2 = \Sigma_i Y_i$ is known to follow a $\Gamma(n\psi, \lambda)$ distribution, with density function

$$f(s_2; \psi, \lambda) = \exp\{-\lambda s_2 - \log \Gamma(n\psi) + n\psi \log \lambda + (n\psi - 1) \log s_2\}, \tag{2.26}$$

from which we see that $c_2(\psi) = n \log \Gamma(\psi) - \log \Gamma(n\psi) + n\psi \log s_2$. From (2.26) we also see that the marginal density of s_2 depends on both ψ and λ, as in (2.13). □

With some exceptions, the only models for which a conditional likelihood function is available, as at (2.13), are exponential families with the parameter of interest a linear function of the canonical parameter. In these cases, the asymptotic theory of Section 1.7 applied to the conditional likelihood function gives normal approximations to conditional maximum likelihood estimators, conditional likelihood ratio tests, and conditional score functions. As in Example 2.10, the exact distribution of the sufficient statistic may not be explicitly available, but it is often possible to compute a saddlepoint approximation to its density, leading to very accurate conditional inference for this class of models; see the bibliographic notes.

Transformation models are the main class of models admitting construction of marginal likelihood functions for parameters of interest, illustrated in Examples 2.7 and 2.11.

The location, scale, and location-scale families discussed in Example 1.15 are transformation models: families of parametric models for which a transformation on the sample space has a corresponding transformation on the parameter space that leaves the density function unchanged. For example, if Y has the density $f_0(y - \theta)$, then $Z = Y + a$ has the density $f_0(z - a - \theta) = f_0(z - \theta')$ so the family of densities for Y, $\{f_0(y - \theta); \theta \in \mathbb{R}\}$, is the same as that for $Z = Y + a$. Similarly, the family of location-scale densities is unchanged under location and scale transformations: if $f(y; \theta) = \theta_2^{-1} f_0((y - \theta_1)/\theta_2)$, then $Z = cY + a$ has density $f(z; \theta') = \theta_2'^{-1} f_0((z - \theta_1')/\theta_2')$ where $\theta_1' = c\theta_1 + a$ and $\theta_2' = c\theta_2$.

In general, g denotes a transformation on the sample space $\mathcal{Y}$, and g^* the induced transformation on the parameter space. Then the formalization of the above invariance is

$$\text{pr}(gY \in A; \theta) = \text{pr}(Y \in A; g^*\theta).$$

If $g^*\Theta = \Theta$, the family of densities indexed by $\theta \in \Theta$ is invariant under the transformation g on $\mathcal{Y}$. Let $\mathcal{C}$ be a class of transformations on $\mathcal{Y}$ satisfying this condition, and $\mathcal{G}$ the smallest class containing $\mathcal{C}$ that is a group. Then $\mathcal{G}^* = \{g^* \text{ induced by } g \in \mathcal{G}\}$ is a group on Θ. Transformation models are reviewed in detail in Barndorff-Nielsen and Cox (1994, Section 2.8).

Example 2.11 (Regression-scale model) Suppose we have a regression model for a vector $Y \in \mathbb{R}^n$:

$$Y = X\beta + \sigma\epsilon, \tag{2.27}$$

where $\epsilon = (\epsilon_1, \ldots, \epsilon_n)$ and ϵ_i are assumed to be independent, identically distributed with density $f_0(\cdot)$. Then $gY = Xb + cY$, with $b \in \mathbb{R}^p$ and $c > 0$ follows the model

$$gY = X(b + c\beta) + c\sigma\epsilon,$$

which is a member of the same family, as long as the parameter space is $\mathbb{R}^p \times \mathbb{R}^+$. The associated transformation on $\Theta = \mathbb{R}^p \times \mathbb{R}^+$ is

$$g^*(\beta, \sigma) = (c\beta + b, c\sigma).$$

Let $\widetilde{\beta}$ and $\widetilde{\sigma}$ be equivariant estimators of β and σ,

$$\widetilde{\beta}(cY + Xb) = c\widetilde{\beta}(Y) + b,$$
$$\widetilde{\sigma}(cY + Xb) = c\widetilde{\sigma}(Y),$$

where X is the $n \times p$ matrix with rows x_i^{T}, $c > 0$ is an arbitrary re-scaling, and $b = (c_1, \ldots, c_p)$ is an arbitrary vector. The maximum likelihood estimators, for example, are equivariant, as are $(\widehat{\beta}, \widetilde{\sigma}^2)$, where $\widehat{\beta}$ is the least squares estimator, and $\widetilde{\sigma}^2$ is the usual unbiased estimator of the variance.

The likelihood function for (β, σ) can be expressed in terms of $\widetilde{\beta}, \widetilde{\sigma}$, and $a = (Y - X\widetilde{\beta})/\widetilde{\sigma}$ as

$$L(\beta, \sigma; \widetilde{\beta}, \widetilde{\sigma}, a) \propto \widetilde{\sigma}^{n-p-1} \prod_{i=1}^{n} f_0 \left\{ \frac{\widetilde{\sigma}}{\sigma} \left(a_i + \frac{x_i^{\mathsf{T}}(\widetilde{\beta} - \beta)}{\widetilde{\sigma}} \right) \right\}. \tag{2.28}$$

From this expression it can be established that $A = a(Y)$ is ancillary for (β, σ) (Appendix 1.B), and the conditional density of $(\widetilde{\beta}, \widetilde{\sigma})$, given $A = a$ is obtained by renormalizing the likelihood function. The location-model version of this result is (1.109), and the location-scale version is derived in Exercise 1.10.

For inference about component parameters, a generalization of Exercise 2.7 shows that $T_j = (\widetilde{\beta}_j - \beta_j)/\widetilde{\sigma}$ is a pivotal quantity, i.e., its distribution is known. This distribution thus defines a marginal likelihood function for inference about β_j, following (2.12). In the special case that $f_0(\epsilon)$ is standard normal, T_j follows a Student t-distribution with $n - p$ degrees of freedom. □

2.4 Profile likelihood

If an exact marginal or conditional likelihood is not available, some other approach to eliminating nuisance parameters is needed. One of the simplest is to maximize over the nuisance parameter. The resulting function of the parameter of interest is called the profile likelihood function; in econometrics it is often called the concentrated likelihood function. It is not a likelihood function in the sense of the definition in Section 1.1, as it is not proportional to the density of an observable random variable, except in very special settings. We write $\widehat{\lambda}_\psi$ for the constrained maximum likelihood estimator of the nuisance parameter, and $\ell_{\mathrm{p}}(\psi)$ for the profile log-likelihood function:

$$\ell_{\mathrm{p}}(\psi) = \ell(\psi, \widehat{\lambda}_\psi) = \ell(\psi, \widehat{\lambda}_\psi; y), \tag{2.29}$$

with the last equation simply to serve as a reminder that the log-likelihood function depends on the parameter and the data. The first and second derivatives of $\ell_{\mathrm{p}}(\psi)$ are denoted by $\ell_{\mathrm{p}}'(\psi)$ and $\ell_{\mathrm{p}}''(\psi)$; when ψ is a vector these will be a column vector and a matrix, respectively.

The profile score function $\ell_{\rm p}'(\psi)$ does not have expected value 0, but it is easy to check using the chain rule that solving the profile score equation $\ell_{\rm p}'(\psi) = 0$ gives the same solution, $\widehat{\psi}$, as does solving the pair of equations

$$\frac{\partial}{\partial \psi}\ell(\psi,\lambda)\bigg|_{(\widehat{\psi},\widehat{\lambda})} = 0, \quad \frac{\partial}{\partial \lambda}\ell(\psi,\lambda)\bigg|_{(\widehat{\psi},\widehat{\lambda})} = 0, \tag{2.30}$$

in regular models where $\widehat{\lambda}_\psi$ is the solution of

$$\frac{\partial}{\partial \lambda}\ell(\psi,\lambda)\bigg|_{\widehat{\lambda}_\psi} = 0. \tag{2.31}$$

Differentiating (2.31) with respect to ψ gives

$$\frac{\partial \widehat{\lambda}_\psi}{\partial \psi} = -J_{\lambda\lambda}^{-1}(\psi,\widehat{\lambda}_\psi)J_{\lambda\psi}(\psi,\widehat{\lambda}_\psi), \tag{2.32}$$

and applying the chain rule to (2.29) and recalling (2.3),

$$\begin{aligned} J_{\rm p}(\psi) \equiv -\ell_{\rm p}''(\psi) &= J_{\psi\psi}(\psi,\widehat{\lambda}_\psi) - J_{\psi\lambda}(\psi,\widehat{\lambda}_\psi)J_{\lambda\lambda}^{-1}(\psi,\widehat{\lambda}_\psi)J_{\lambda\psi}(\psi,\widehat{\lambda}_\psi), \\ &= \{J^{\psi\psi}(\psi,\widehat{\lambda}_\psi)\}^{-1}. \end{aligned} \tag{2.33}$$

In Section 2.5.1 we outline the derivation of limiting results analogous to those in Chapter 1.2 for the profile log-likelihood function. In particular,

$$w_u(\psi_0) = \ell_{\rm p}'(\psi_0)^{\sf T} J_{\rm p}^{-1}(\widehat{\psi})\ell_{\rm p}'(\psi_0) \xrightarrow{d} \chi_q^2, \tag{2.34}$$

$$w_{\rm e}(\psi_0) = (\widehat{\psi} - \psi_0)^{\sf T} J_{\rm p}(\widehat{\psi})(\widehat{\psi} - \psi_0) \xrightarrow{d} \chi_q^2, \tag{2.35}$$

$$w(\psi_0) = 2\{\ell_{\rm p}(\widehat{\psi}) - \ell_{\rm p}(\psi_0)\} \xrightarrow{d} \chi_q^2, \tag{2.36}$$

all under sampling from the true model $f(y;\psi_0,\lambda_0)$. In (2.34) the asymptotic result still holds if instead we standardize by $I^{\psi\psi}(\widehat{\psi},\widehat{\lambda})$, and similarly for (2.39) if we standardize by its inverse. We could also evaluate the information quantities at $(\psi_0,\widehat{\lambda}_{\psi_0})$ without changing the limiting distribution. Higher-order expansions suggest that the standardizations in (2.34) and (2.35) may be more accurate in finite samples. For some models the expected information may be simpler to compute, and with the score function in particular, if we are testing just a single value ψ_0, it may be more convenient to evaluate the observed information at $(\psi_0,\widehat{\lambda}_{\psi_0})$.

The expression for $w(\psi_0)$ in (2.36) is a version of the (generalized) likelihood ratio statistic, which in many texts is defined in the context of testing a composite null hypothesis $H_0 : \theta \in \Theta_0$ against a composite alternative $H_1 : \theta \in \Theta_1$, with $\Theta = \Theta_0 \cup \Theta_1$, as

$$\Lambda_n = \frac{\sup_{\theta\in\Theta_0} L(\theta; y_1,\ldots,y_n)}{\sup_{\theta\in\Theta} L(\theta; y_1,\ldots,y_n)}. \tag{2.37}$$

The result that $-2\log\Lambda_n \xrightarrow{d} \chi_q^2$, where q is the dimension of Θ_0, is often referred to as Wilks' theorem. In (2.36), $\Theta_0 = (\psi_0,\lambda)$ is the composite null hypothesis being tested.

In the special case that $q = 1$, we have three analogues to the scalar parameter pivotal quantities (1.52), (1.53) and (1.54):

$$s_{\rm p}(\psi) = \ell_{\rm p}'(\psi)J_{\rm p}^{-1/2}(\widehat{\psi}) \mathrel{\dot\sim} N(0,1), \tag{2.38}$$

$$q_{\rm p}(\psi) = (\widehat{\psi} - \psi)J_{\rm p}^{1/2}(\widehat{\psi}) \mathrel{\dot\sim} N(0,1), \tag{2.39}$$

$$r_{\rm p}(\psi) = \text{sign}(\widehat{\psi} - \psi)[2\{\ell_{\rm p}(\widehat{\psi}) - \ell_{\rm p}(\psi)\}]^{1/2} \mathrel{\dot\sim} {\rm N}(0,1). \tag{2.40}$$

As in Section 1.3, $\Phi(s_{\rm p})$, $\Phi(q_{\rm p})$, and $\Phi(r_{\rm p})$ each provide an approximate significance function for ψ. In Section 2.5.2 we describe higher-order approximations to significance functions.

Example 2.12 (Exponential regression, cont.) In an exponential regression model with two covariates the log-likelihood function from a sample of size n is

$$\ell(\psi, \lambda) = -n \log \lambda - \frac{1}{\lambda} \sum_{i=1}^{n} y_i \exp(\psi x_i + \beta z_i).$$

The equations determining the constrained maximum likelihood estimates, for fixed ψ, are

$$\widehat{\lambda}_\psi = \frac{1}{n} \sum_{i=1}^{n} y_i \exp(\psi x_i + \widehat{\beta}_\psi z_i),$$

$$0 = \sum_{i=1}^{n} y_i z_i \exp(\psi x_i + \widehat{\beta}_\psi z_i),$$

giving the profile log-likelihood function

$$\ell_{\rm p}(\psi) = -n \log \left\{ \sum_{i=1}^{n} y_i \exp(\psi x_i + \widehat{\beta}_\psi z_i) \right\},$$

ignoring additive constants. Construction of significance functions for ψ based on (2.38), (2.39), or (2.40), can be computed directly, although not explicitly. Using the right-hand side of (2.33) may be simpler than differentiating the profile log-likelihood function twice. □

Example 2.13 (Exponential family, cont'd) For the exponential family in (2.21) with partitioned parameter $\varphi = (\psi, \lambda)^{\mathsf{T}}$, the score equation is

$$\begin{bmatrix} s_1 \\ s_2 \end{bmatrix} = \begin{bmatrix} c_\psi(\widehat{\psi}, \widehat{\lambda}) \\ c_\lambda(\widehat{\psi}, \widehat{\lambda}) \end{bmatrix}, \tag{2.41}$$

using the shorthand $c_\psi = \partial c(\psi, \lambda)/\partial \psi$ and so on. As the equation defining the constrained maximum likelihood estimator $\widehat{\lambda}_\psi$ is

$$nc(...) = s_2\, c_\lambda(\psi, \widehat{\lambda}_\psi) = s_2, \tag{2.42}$$

we conclude that $c_\lambda(\widehat{\psi}, \widehat{\lambda}) = c_\lambda(\psi, \widehat{\lambda}_\psi)$, so if we reparametrize the model to (ψ, η), where $\eta = c_\lambda(\psi, \lambda)$, then

$$\widehat{\eta} = \widehat{\eta}_\psi. \tag{2.43}$$

In Exercise 2.2 we show that the parameter η is orthogonal to ψ, with respect to (observed or expected) Fisher information, although (2.43) is a stronger condition than orthogonality. □

Example 2.14 (Weibull distribution, cont.) The log-likelihood function for a sample of size n from the Weibull density (2.10) is

$$\ell(\psi, \lambda) = n \log(\psi) - n\psi \log(\lambda) + \psi \sum_{i=1}^{n} \log y_i - \sum_{i=1}^{n} (y_i/\lambda)^\psi,$$

from which we have

$$\widehat{\lambda}_\psi = \left(\frac{1}{n} \sum_{i=1}^{n} y_i^\psi \right)^{1/\psi},$$

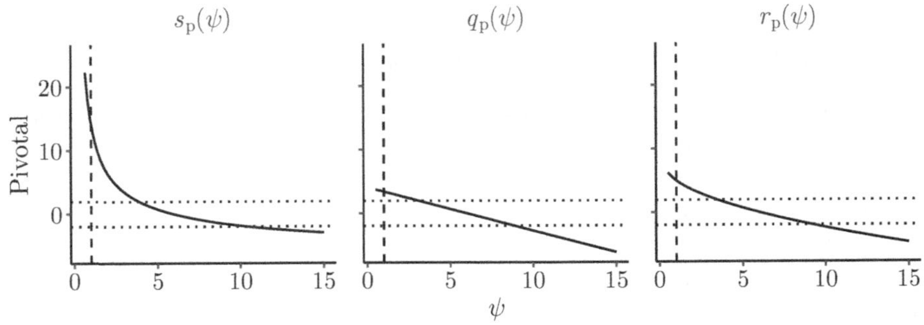

FIGURE 2.1
Approximate pivotal quantities for a Weibull model fit to the data $y = (225, 171, 198, 189, 189, 135, 162, 135, 117, 162)$, from Davison (2003, Ex. 1.2). The horizontal dotted lines identify 95% confidence intervals for ψ. The vertical dashed line corresponds to the exponential distribution, $\psi = 1$.

and

$$\ell_p(\psi) = n \log \psi - n \log \left(\sum_{i=1}^{n} y_i^{\psi} \right) + \psi \sum_{i=1}^{n} \log y_i.$$

As an illustration, Figure 2.1 shows the three pivotal quantities for the shape parameter of the Weibull, using the stress-strength data of Example 1.2 in Davison (2003). The dashed lines at ± 1.96 determine approximate 95% confidence intervals for ψ based on each pivotal. The value $\psi = 1$, which would correspond to a simple exponential distribution, is not supported by this data, using any of the three pivotal quantities. □

Example 2.15 (Comparison of two binomials, cont.) Following Example 2.8, the score equation for λ is

$$t = m_1 \frac{\exp(\psi + \widehat{\lambda}_\psi)}{1 + \exp(\psi + \widehat{\lambda}_\psi)} + m_2 \frac{\exp(\widehat{\lambda}_\psi)}{1 + \exp(\widehat{\lambda}_\psi)}, \tag{2.44}$$

and from (2.15) we have

$$\ell_p(\psi) = \psi y_1 + \widehat{\lambda}_\psi t - m_1 \log\{1 + \exp(\psi + \widehat{\lambda}_\psi)\} - m_2 \log\{1 + \exp(\widehat{\lambda}_\psi)\}. \tag{2.45}$$

In this case it will be easier to calculate the full information matrix and the first expression in (2.33) than to obtain $J_p(\psi)$ by differentiation of (2.45). As described in Example 2.8, likelihood inference based on the conditional distribution of y_1, given t has the advantage of not requiring computation of $\widehat{\lambda}_\psi$, although depending on the values of m_1 and m_2, calculation of the normalizing constant may be cumbersome. □

Example 2.16 (Normal variance in regression, cont.) As in Example 2.6 we assume the variance $\psi = \sigma^2$ is the parameter of interest. The profile likelihood function is

$$L_p(\psi) = \psi^{-n/2} \exp\left(-\frac{n\widehat{\psi}}{\psi} \right), \tag{2.46}$$

where $\widehat{\psi} = (y - X\widehat{\beta})^{\mathsf{T}}(y - X\widehat{\beta})/n$ is the maximum likelihood estimate of ψ. This is the kernel of a χ^2_n density: there is no adjustment of degrees of freedom for the estimation of β.

This is a general feature of profile likelihood, pointing to a need for a more refined version to adequately accommodate nuisance parameters. As shown in Example 2.6, there is a marginal likelihood function for ψ based on the χ^2_{n-p} density, and maximizing this gives the usual unbiased estimate $\widehat{\psi}_{\mathrm{m}} = (y - X\widehat{\beta})^{\mathsf{T}}(y - X\widehat{\beta})/(n-p)$.

If we allow the dimension of the regression parameter β to increase with n, then the maximum likelihood estimator of ψ is not even consistent. This does not contradict the asymptotic results, as the regularity conditions in Section 1.7 do not apply for parameter space that depends on the sample size. For example, let $(Y_{11}, Y_{12}), \ldots, (Y_{n1}, Y_{n2})$ be n pairs of independent observations from the normal distribution, where each pair has expected value $\lambda_i, i = 1, \cdots, n$ and variance ψ. The constrained maximum likelihood estimate of the nuisance parameter is $\widehat{\lambda}_{i,\psi} = \widehat{\lambda}_i = (y_{i1} + y_{i2})/2$, leading to the profile log-likelihood function,

$$\ell_{\mathrm{p}}(\psi) = -n \log \psi - \frac{1}{4\psi} \sum_{i=1}^{n} (y_{i1} - y_{i2})^2, \tag{2.47}$$

and the maximum likelihood estimator $\widehat{\psi} = (4n)^{-1}\Sigma_i(Y_{i1} - Y_{i2})^2$. As $n \to \infty$, $\widehat{\psi} \xrightarrow{p} \psi/2$. □

2.5 Asymptotic theory

2.5.1 Limiting distributions

In this section we verify the χ^2 approximation to the distribution of the score, Wald, and likelihood ratio tests given in Section 2.4. For $\theta = (\psi, \lambda)$, with ψ the parameter of interest and λ the complementing nuisance parameter, we partition the related likelihood quantities as at (2.2) and (2.3). Then, from (1.50),

$$\sqrt{n} \begin{bmatrix} \widehat{\psi} - \psi_0 \\ \widehat{\lambda} - \lambda_0 \end{bmatrix} \xrightarrow{d} N\left(\begin{bmatrix} 0 \\ 0 \end{bmatrix}, \begin{bmatrix} i^{\psi\psi}(\theta_0) & i^{\psi\lambda}(\theta_0) \\ i^{\lambda\psi}(\theta_0) & i^{\lambda\lambda}(\theta_0) \end{bmatrix} \right), \tag{2.48}$$

and thus

$$\sqrt{n}(\widehat{\psi} - \psi_0) \xrightarrow{d} N\{0, i^{\psi\psi}(\theta_0)\},$$

where as in Section 1.2 $i(\theta)$ is the expected Fisher information in a single observation. The arguments in Section 1.7 require consistency of the maximum likelihood estimator $\widehat{\theta}$: we assume here that this has already been established.

For vector parameters of interest the quadratic form version may be more useful:

$$(\widehat{\psi} - \psi_0)^{\mathsf{T}} \{I^{\psi\psi}(\psi_0, \lambda_0)\}^{-1} (\widehat{\psi} - \psi_0) \xrightarrow{d} \chi^2_q. \tag{2.49}$$

The χ^2_q approximation to the distribution of $w_e(\psi_0)$ in (2.35) follows from (2.49) on applying (2.33), the weak law of large numbers for $J(\theta)$, and the assumed consistency of $\widehat{\theta}$.

To establish (2.34),

$$\begin{aligned} w_u(\psi_0) &= \ell_{\mathrm{p}}'(\psi_0)^{\mathsf{T}} J_{\mathrm{p}}^{-1}(\widehat{\psi}) \ell_{\mathrm{p}}'(\psi_0) \\ &= U_\psi(\psi_0, \widehat{\lambda}_{\psi_0})^{\mathsf{T}} J^{\psi\psi}(\widehat{\psi}, \widehat{\lambda}) U_\psi(\psi_0, \widehat{\lambda}_{\psi_0}), \\ &= U_\psi(\psi_0, \widehat{\lambda}_{\psi_0})^{\mathsf{T}} I^{\psi\psi}(\psi_0, \lambda_0) U_\psi(\psi_0, \widehat{\lambda}_{\psi_0}) \{1 + o_p(1)\}. \end{aligned} \tag{2.50}$$

Furthermore,

$$U_\psi(\psi_0, \widehat{\lambda}_{\psi_0}) = U_\psi(\psi_0, \lambda_0) - J_{\psi\lambda}(\psi_0, \lambda_0)(\widehat{\lambda}_{\psi_0} - \lambda_0) + O_p(1), \tag{2.51}$$

where the first two terms are both $O_p(\sqrt{n})$. In the model with ψ fixed, we have again from (1.50) that

$$\widehat{\lambda}_{\psi_0} - \lambda_0 = I_{\lambda\lambda}^{-1}(\psi_0, \lambda_0)U_\lambda(\psi_0, \lambda_0)\{1 + o_p(1)\}; \tag{2.52}$$

using this in (2.50), with $J_{\psi\lambda}(\psi_0, \lambda_0) = I_{\psi\lambda}(\psi_0, \lambda_0)\{1 + o_p(1)\}$ gives

$$\begin{aligned} w_u(\psi_0) = \{&U_\psi(\psi_0, \lambda_0) - I_{\psi\lambda}(\psi_0, \lambda_0)I_{\lambda\lambda}^{-1}(\psi_0, \lambda_0)U_\lambda(\psi_0, \lambda_0)\}^\mathsf{T} I^{\psi\psi}(\psi_0, \lambda_0) \\ &\{U_\psi(\psi_0, \lambda_0) - I_{\psi\lambda}(\psi_0, \lambda_0)I_{\lambda\lambda}^{-1}(\psi_0, \lambda_0)U_\lambda(\psi_0, \lambda_0)\}\{1 + o_p(1)\}. \end{aligned}$$

While the notation is somewhat cumbersome, this expression is simply a quadratic form in $U_\psi - aU_\lambda$, the residual after regression of U_ψ on U_λ, and the variance of this residual is $[I^{\psi\psi}]^{-1}$. The limiting distribution of U is normal, from which we have the limiting χ^2_q distribution for $w_u(\psi_0)$. The residual $U_\psi - aU_\lambda$ is asymptotically uncorrelated with U_λ. For more general estimating functions this property is sometimes called Neyman orthogonality.

To establish (2.36),

$$\begin{aligned} w(\psi_0) &= 2\{\ell(\widehat{\psi}, \widehat{\lambda}) - \ell(\psi_0, \lambda_0)\} - 2\{\ell(\psi_0, \widehat{\lambda}_{\psi_0}) - \ell(\psi_0, \lambda_0)\} \\ &= U(\psi_0, \lambda_0)^\mathsf{T} I^{-1}(\psi_0, \lambda_0)U(\psi_0, \lambda_0)\{1 + o_p(1)\} \\ &\quad - U_\lambda(\psi_0, \lambda_0)^\mathsf{T} I_{\lambda\lambda}^{-1}(\psi_0, \lambda_0)U_\lambda(\psi_0, \lambda_0)\{1 + o_p(1)\}, \end{aligned} \tag{2.53}$$

using the multivariate version of the asymptotic equivalence established in (1.93). Using the formulae for the inverse of a partitioned matrix and collecting terms, this simplifies to

$$w(\psi_0) = w_u(\psi_0)\{1 + o_p(1)\}. \tag{2.54}$$

This argument repeatedly relies on the consistency of the full and constrained maximum likelihood estimators, i.e., $(\widehat{\psi}, \widehat{\lambda}) \xrightarrow{p} (\psi_0, \lambda_0)$ and $\widehat{\lambda}_{\psi_0} \xrightarrow{p} \lambda_0$, and the law of large numbers for the information matrices:

$$J(\widehat{\psi}, \widehat{\lambda})I^{-1}(\psi_0, \lambda_0) \xrightarrow{p} \mathcal{I}_{p\times p}, \quad J(\psi_0, \widehat{\lambda}_{\psi_0})I^{-1}(\psi_0, \lambda_0) \xrightarrow{p} \mathcal{I}_{p\times p}, \tag{2.55}$$

where $\mathcal{I}_{p\times p}$ is the $p \times p$ identity matrix.

A more general version of the residual score function $U_\psi - aU_\lambda$ can be developed based on unbiased estimating functions, which are the subject of Chapter 5. Chernozhukov et al. (2018) described Neyman orthogonalization of estimating functions, in the context of inference for a low-dimensional parameter of interest in the presence of high-dimensional nuisance parameters. A similar approach is developed in a likelihood context in Ning and Liu (2017).

2.5.2 Higher-order approximations

Although the three statistics w_u, w_e, and w in (2.34)–(2.36) have the same limiting distribution, they will be different in finite samples, except in the special case of inference for a normal mean. The score statistic w_u and the likelihood ratio statistic w are invariant under interest-respecting reparametrizations; the maximum likelihood estimator is asymptotically efficient, but confidence regions based on $w_e(\psi)$ do not adapt to skewness in the log-likelihood function so will not be, for example, regions of highest likelihood.

There is a large literature developed mainly in the 1980s and 1990s on using Edgeworth, saddlepoint, and Laplace expansions to obtain higher-order approximations for inference, leading to more accurate calculation of significance functions and confidence regions. Here we present some of the main results.

The normal approximation to the distribution of the maximum likelihood estimator can be improved by use of the p^*-approximation, originally presented in Barndorff-Nielsen (1983). This approximation is

$$p^*(\widehat{\theta} \mid a; \theta) = c(\theta, a)|J(\widehat{\theta})|^{1/2} \exp\{\ell(\theta; \widehat{\theta}, a) - \ell(\widehat{\theta}; \widehat{\theta}, a)\}. \tag{2.56}$$

This is an approximation to the conditional distribution of $\widehat{\theta}$, given a statistic a, and it is implicitly assumed in (2.56) that the sample $y = (y_1, \ldots, y_n)$ has been transformed to the pair $(\widehat{\theta}, a)$, where $\widehat{\theta} = \widehat{\theta}(y)$ is the maximum likelihood estimator, and a is an auxiliary statistic that makes the transformation one-to-one. If instead of y we first reduce to a sufficient statistic s, say, then we assume that the transformation from s to $(\widehat{\theta}, a)$ is one-to-one. The function $c(\theta, a)$ is a normalizing constant, and as usual $J(\widehat{\theta})$ is the observed Fisher information in the sample.

It has been established in fairly wide generality that (2.56) approximates the exact conditional density of the maximum likelihood estimator with relative error $O(n^{-3/2})$. It can be verified that the approximation is exact in some families, including the normal, gamma, and inverse Gaussian. In order that it be useful for inference about θ, the marginal distribution of a should not contain any, or at least very much, information about θ. This will be the case if a is ancillary, i.e., $f(a; \theta) = f(a)$ is free of θ. However for the p^*-approximation to be accurate to $O(n^{-3/2})$ it is enough for the approximation that the density of a is free of θ to order $O(n^{-1})$. It is not always obvious how to find such an approximately ancillary statistic, but in exponential family models it is not needed, as the maximum likelihood estimator is sufficient, and in transformation family models the ancillary is typically the maximal invariant for the relevant group of transformations. To use (2.56) for inference it is more useful to be able to approximate the tail integral of (2.56), which provides a more accurate significance function than those in Sections 1.3 and 2.4 based on the normal approximation.

If θ is a scalar parameter, and if the likelihood root $r(\theta; y)$ defined in (1.55) is monotone in $\widehat{\theta}(y)$, then the distribution of the pivotal quantity

$$r^*(\theta) = r(\theta) + \frac{1}{r(\theta)} \log\left\{\frac{Q(\theta)}{r(\theta)}\right\}, \tag{2.57}$$

can be shown to be approximated by the standard normal distribution, with relative error $O(n^{-3/2})$:

$$\mathrm{pr}(r^*(\theta) \leq r^0) = \Phi(r^0)\{1 + O(n^{-3/2})\},$$

with

$$r(\theta) = \mathrm{sign}(\widehat{\theta} - \theta)[2\{\ell(\widehat{\theta}) - \ell(\theta)\}]^{1/2},$$
$$Q(\theta) = J^{-1/2}(\widehat{\theta})\{\ell_{;\widehat{\theta}}(\widehat{\theta}; \widehat{\theta}, a) - \ell_{;\widehat{\theta}}(\theta; \widehat{\theta}, a)\}. \tag{2.58}$$

In (2.57) $r(\theta)$ is the log-likelihood root defined in (1.55), $Q(\theta)$ is a Wald-type statistic, and $\ell_{;\widehat{\theta}}(\theta; \widehat{\theta}, a) = \partial\ell(\theta; \widehat{\theta}, a)/\partial\widehat{\theta}$ is a sample-space derivative of the log-likelihood function. The dependence of r and Q on the data has been suppressed for ease of notation.

If $\theta \in \mathbb{R}^p$, and $\theta = (\psi, \lambda)$ with $\psi \in \mathbb{R}$, the relevant pivotal quantity is of the same form:

$$r^*(\psi) = r_{\mathrm{p}}(\psi) + \frac{1}{r_{\mathrm{p}}(\psi)} \log\left\{\frac{Q(\psi)}{r_{\mathrm{p}}(\psi)}\right\}, \tag{2.59}$$

where $r_{\rm p}(\psi)$ is the profile log-likelihood root defined in (2.40), and

$$Q(\psi) = \frac{|\ell_{;\widehat{\theta}}(\widehat{\theta}) - \ell_{;\widehat{\theta}}(\widehat{\theta}_\psi) \quad \ell_{\lambda;\widehat{\theta}}(\widehat{\theta}_\psi)|}{|\ell_{\theta;\widehat{\theta}}(\widehat{\theta})|}\left\{\frac{|J(\widehat{\theta})|}{|J_{\lambda\lambda}(\widehat{\theta}_\psi)|}\right\}^{1/2}. \tag{2.60}$$

In (2.60), the matrix in the numerator of the first factor is $p \times p$, with the first column determined from $\ell_{;\widehat{\theta}}(\theta)$; $\ell_{\lambda;\widehat{\theta}}$ is a $p \times (p-1)$ matrix. As with the p^*-approximation, the dependence of the log-likelihood function on the data is expressed through $(\widehat{\theta}, a)$.

In exponential family models where ψ is a component of the canonical parameter the expression for Q simplifies to

$$Q(\psi) = (\widehat{\psi} - \psi) J_{\rm p}^{1/2}(\widehat{\psi}) \frac{|J_{\lambda\lambda}(\widehat{\theta})|^{1/2}}{|J_{\lambda\lambda}(\widehat{\theta}_\psi)|^{1/2}} = q_{\rm p}(\psi)\rho(\widehat{\psi}, \psi), \tag{2.61}$$

where $q_{\rm p}(\psi)$ is the Wald statistic defined at (2.39), and $\rho^2(\widehat{\psi}, \psi)$ is the ratio of information determinants. In the regression-scale model of Example 2.11, with $\theta = (\beta, \log\sigma)$, it can be shown that

$$Q(\psi) = \ell'_{\rm p}(\psi) J_{\rm p}^{-1/2}(\widehat{\psi})\rho^{-1}(\widehat{\psi}, \psi) = s_{\rm p}(\psi)\rho^{-1}(\widehat{\psi}, \psi), \tag{2.62}$$

where $s_{\rm p}(\psi)$ is the score statistic defined at (2.38), the parameter of interest ψ is any component of θ and the nuisance parameter λ is the remaining components.

Outside of these special cases the derivation of $Q(\psi)$ can be difficult. A version that approximates the sample-space derivatives and avoids specification of the approximate ancillary statistic was developed in Skovgaard (1996); this version is exact in exponential families. Fraser (1990) developed a different approach to the calculation of the supplementary statistic Q based on approximate pivotal quantities.

If the parameter of interest is a vector, higher-order approximations to the distribution of the log-likelihood ratio $w(\psi)$ defined in (2.36) have been developed. The most direct is Bartlett correction, using the result that

$$\mathrm{E}_\theta\{w(\psi)\} = q\{1 + B(\theta)/n + O(n^{-2})\}, \tag{2.63}$$

and dividing $w(\psi)$ by an estimate of $B(\theta)$. The result, originally derived in Lawley (1956) by direct calculation, is that the χ^2_q approximation to the distribution of

$$w_B(\psi) = w(\psi)/\{1 + B(\psi, \widehat{\lambda}_\psi)\},$$

is accurate to $O(n^{-2})$, whereas the χ^2_q approximation to the distribution of $w(\psi)$ is accurate to just $O(n^{-1})$. The so-called Bartlett factor, $\{1 + B(\theta)/n\}$, is related to the normalizing constant in the p^* approximation (Barndorff-Nielsen and Cox, 1994, Section 6.5). A related approximation that is generally more accurate in finite samples, and easier to compute than $B(\theta)$ was developed in Skovgaard (2001) as a generalization of the r^* approximation.

2.6 Adjusted and modified profile likelihood

The model considered at the end of Example 2.16 is one of a family of problems that have come to be called Neyman–Scott problems, after Neyman and Scott (1948), in which independent vector observations $y_i = (y_{ij}, j = 1, \ldots, m_i)$ are modelled with a density

$f(y_i; \psi, \lambda_i)$; in (2.47) $m_i = 2$ for all i. Because we have more than one observation of the ith unit, there is information available to estimate the nuisance parameter λ_i, so it seems reasonable to expect that we can estimate the common parameter of interest reasonably well as n increases. Example 2.16 shows that this is not always the case.

Example 2.17 (Weighted normal mean) A closely related normal-theory example, also investigated by Neyman and Scott, has common expected value within unequally-sized groups: $\{Y_{i1}, \ldots, Y_{im_i}\}$ are independently distributed as $N(\psi, \lambda_i)$, $i = 1, \ldots, n$. The maximum likelihood estimate of ψ is the solution of the score equation

$$\sum_{i=1}^{n} \frac{m_i^2(\bar{y}_{i\bullet} - \psi)}{\sum_{j=1}^{m_i}(y_{ij} - \psi)^2} = 0. \tag{2.64}$$

While it can be shown that this solution is consistent for ψ, the solution of the equation

$$\sum_{i=1}^{n} \frac{m_i(m_i - 2)(\bar{y}_{i\bullet} - \psi)}{\sum_{j=1}^{m_i}(y_{ij} - \psi)^2} = 0 \tag{2.65}$$

leads to an estimator of ψ with smaller asymptotic variance. Thus the maximum likelihood estimator is not fully efficient. □

Example 2.18 (Product of exponential means) Suppose y_{i1} and y_{i2}, $i = 1, \cdots, n$, are independent observations from an exponential density with means $\psi\lambda_i$ and ψ/λ_i, respectively. The profile log-likelihood function is

$$l_{\mathrm{p}}(\psi) = -2n \log(\psi) - \frac{2}{\psi} \sum_{i=1}^{n} (y_{i1} y_{i2})^{1/2}, \tag{2.66}$$

with maximum likelihood estimate $\widehat{\psi} = n^{-1}\Sigma_i (y_{i1}y_{i2})^{1/2}$, which converges in probability to $(\pi/4)\psi$, as $n \to \infty$. □

Neyman–Scott models have by construction a parameter dimension that increases with the sample size. More general versions of such problems have come to be called "small n, large p" problems, "p" here denoting the dimension of the parameter, and n the number of observations. The structure of Neyman–Scott models makes them more easily studied than the general problem; see the bibliographic notes and Example 2.21.

In Section 2.2 it was noted that inference based on the asymptotic distribution of quantities derived from the profile log-likelihood can have poor behaviour in finite samples, especially if the number of nuisance parameters is large relative to the sample size. In Neyman–Scott problems, where the number of nuisance parameters increases with the sample size, profile log-likelihood inference can have poor limiting behaviour as well. When available, inference based on conditional or marginal likelihood will usually be preferred. It is natural then to ask if there is a way to achieve this approximately, in more general families of models.

One of the simplest approaches in this direction is an adjusted profile log-likelihood function proposed in Cox and Reid (1987). This is defined as

$$\ell_{\mathrm{a}}(\psi) = \ell_{\mathrm{p}}(\psi) - \frac{1}{2} \log |J_{\lambda\lambda}(\psi, \widehat{\lambda}_\psi)| = \ell(\psi, \widehat{\lambda}_\psi; y) - \frac{1}{2} \log |J_{\lambda\lambda}(\psi, \widehat{\lambda}_\psi)|, \tag{2.67}$$

where $J_{\lambda\lambda}(\theta)$ is the sub-block of the observed Fisher information matrix corresponding to the nuisance parameter; see (1.40) and (2.3). The adjustment term can be loosely described as an adjustment for errors in the estimation of the nuisance parameter. When y is a vector of

n independent observations, the profile log-likelihood function is $O_p(n)$ and the adjustment term is $O_p(1)$.

One difficulty introduced by this adjustment is a lack of parametrization invariance. To mitigate this lack of invariance, we require λ and ψ to be orthogonal with respect to expected Fisher information, as defined in (2.7). When ψ is a scalar this can always be achieved by a transformation of the original non-orthogonal parametrization. The adjusted log-likelihood function is still not invariant, since one-to-one functions of an orthogonal parameter are still orthogonal, but the lack of invariance is reduced to $O(n^{-1})$.

Example 2.19 (Linear regression, cont.) Suppose $Y_i = x_i^{\mathsf{T}}\beta + \sigma\epsilon_i$, where x_i is a $p \times 1$ vector and $\epsilon_1, \ldots, \epsilon_n$ are independent standard normal random variables. We saw in Examples 2.6 and 2.16 that the maximum likelihood estimate of σ^2 is biased, but that an unbiased estimate can be obtained from the marginal likelihood for σ^2 based on the least-squares residuals. It is readily verified that β is orthogonal to σ^2, and further calculation shows that

$$\begin{aligned} \ell_{\mathrm{a}}(\sigma^2) &= -\frac{n}{2}\log\sigma^2 - \frac{(y - X\widehat{\beta})^{\mathsf{T}}(y - X\widehat{\beta})}{2\sigma^2} - \frac{1}{2}\log\left|\frac{1}{\sigma^2}X^{\mathsf{T}}X\right|, \\ &= -\frac{n}{2}\log\sigma^2 - \frac{n\widehat{\sigma}^2}{2\sigma^2} + \frac{p}{2}\log\sigma^2 - \frac{1}{2}\log|X^{\mathsf{T}}X|, \\ &= -\frac{n-p}{2}\log\sigma^2 - \frac{n\widehat{\sigma}^2}{2\sigma^2}, \end{aligned} \tag{2.68}$$

ignoring constants in the last equality. This is the kernel of the χ^2_{n-p} distribution, so we recover the exact marginal likelihood for σ^2 (see Example 2.6), and in particular the estimate $\widehat{\sigma}^2_{\mathrm{a}}$ obtained by maximizing (2.68) is the usual unbiased estimate of σ^2. □

Example 2.20 (Linear exponential family) For an independent sample of size n from the linear exponential family (2.21) the likelihood function is

$$\ell(\psi, \lambda; y) = \psi\, s_1(y) + \lambda^{\mathsf{T}} s_2(y) - nc(\psi, \lambda), \tag{2.69}$$

where we assume here that the parameter of interest ψ is scalar. An orthogonal re-parametrization is given by (ψ, η), where $\eta = c_\lambda(\psi, \lambda)$, see (2.43), so the adjusted profile log-likelihood function is

$$\begin{aligned} \ell_{\mathrm{a}}(\psi) &= \ell_{\mathrm{p}}(\psi) - (1/2)\log|J_{\eta\eta}(\psi, \widehat{\eta}_\psi)|, \\ &= \ell_{\mathrm{p}}(\psi) + (1/2)\log|J_{\lambda\lambda}(\psi, \widehat{\lambda}_\psi)|, \end{aligned} \tag{2.70}$$

where the last equality follows (up to an additive constant) as $\partial\eta/\partial\lambda = c_{\lambda\lambda}(\psi, \lambda)$. □

Example 2.21 (Product of exponential means, cont.) The profile log-likelihood function based on n pairs of independent observations (y_{i1}, y_{i2}) from exponential distributions with expected values ψ/λ_i and $\psi\lambda_i$ is given in Example 2.18 at (2.66), and $|J_{\lambda\lambda}(\psi, \widehat{\lambda}_\psi)| \propto \psi^{-n}$, giving the adjusted profile log-likelihood function

$$\ell_{\mathrm{a}}(\psi) = -\frac{3n}{2}\log(\psi) - \frac{2}{\psi}\sum_{i=1}^{n}(y_{i1}y_{i2})^{1/2}, \tag{2.71}$$

from which we have the estimator $\widehat{\psi}_{\mathrm{a}} = (4/3)\widehat{\psi} = (4/3)n^{-1}\Sigma_i(Y_{i1}Y_{i2})^{1/2}$. This estimator, like $\widehat{\psi}$, is not consistent for ψ; it converges to $(\pi/3)\psi$, but this limit is at least closer to ψ than the limit of the maximum likelihood estimator, $\pi/4$. The information adjustment has gone some way to accommodating the large number of nuisance parameters. □

In general, the limiting distributions of quantities based on $\ell_{\mathrm{a}}(\psi)$ are the same as those based on $\ell_{\mathrm{p}}(\psi)$, as outlined in Section 2.4. From this point of view there seems no advantage in using the adjusted profile likelihood function for inference, yet the examples above suggest that the finite sample properties of inferential quantities based on $\ell_{\mathrm{a}}(\psi)$ may be expected to be better. While no general results are available to make this precise, this does seem to be the case in a number of examples.

The adjusted profile log-likelihood function is a simplification of an approximate conditional or marginal log-likelihood function developed using the theory of higher-order asymptotics. This theory leads, in fairly wide generality, to the modified profile log-likelihood function

$$\ell_{\mathrm{M}}(\psi) = \ell_{\mathrm{p}}(\psi) - \frac{1}{2}\log|J_{\lambda\lambda}(\psi, \widehat{\lambda}_{\psi})| + M(\psi), \tag{2.72}$$

where the additional term $M(\psi)$, defined below, is $O_p(n^{-1})$, in the usual setting where we have a sample of independent observations. The function $M(\psi)$ ensures that the modified profile likelihood function is invariant under interest-respecting re-parametrizations; thus $\ell_{\mathrm{M}}(\psi)$ does not need to be calculated in the orthogonal parametrization. The last term in (2.72) is

$$M(\psi) = \log|J_{\lambda\lambda}(\psi, \widehat{\lambda}_{\psi})| - \log|\ell_{\lambda;\widehat{\lambda}}(\psi, \widehat{\lambda}_{\psi}; y)|, \tag{2.73}$$

where the notation in the second term of $M(\psi)$ is shorthand for

$$\left.\frac{\partial^2 \ell(\psi, \lambda; \widehat{\psi}, \widehat{\lambda}, a)}{\partial\lambda\partial\widehat{\lambda}}\right|_{\lambda=\widehat{\lambda}_{\psi}}. \tag{2.74}$$

This expression assumes that there is a one-to-one transformation from y to $(\widehat{\psi}, \widehat{\lambda}, a)$, with a either exactly or approximately ancillary. The derivative in (2.74) is a mixed derivative with respect to the parameter and with respect to the data. Sample-space derivatives of the log-likelihood function arise naturally in the theory of higher order asymptotics, but this particular mixed derivative can be difficult to compute, because it requires the log-likelihood function to be written as a function of $(\widehat{\psi}, \widehat{\lambda}, a)$ and this is rarely available explicitly.

In the special case of a linear exponential family, the transformation from the minimal sufficient statistic (s_1, s_2) to $(\widehat{\psi}, \widehat{\lambda})$ is explicitly available, see (1.41), and $\ell_{\mathrm{M}}(\psi)$ simplifies to $\ell_{\mathrm{a}}(\psi)$ given in Example 2.20.

2.7 Partial likelihood

Cox (1975) introduced the partial likelihood function as a tool to be used when conditional or marginal likelihood functions are not available, but there are sufficiently many nuisance parameters that inference based on the profile likelihood may be unsatisfactory. The general setting assumes we have two sequences of observations; $\{X_1, S_1, X_2, S_2, \ldots, X_i, S_i, \ldots, X_n, S_n\}$. The joint distribution of the observations can be factored as

$$f(x, s; \theta) = \prod_{i=1}^{n} f(x_i \mid x_{(i-1)}, s_{(i-1)}; \theta) \prod_{i=1}^{n} f(s_i \mid x_{(i)}, s_{(i-1)}; \theta), \tag{2.75}$$

where $x_{(i)} = (x_1, \ldots, x_i)$, $s_{(i)} = (s_1, \ldots, s_i)$, and for $i = 1$ the factors are $f(x_1; \theta)$ and $f(s_1 \mid x_1; \theta)$. As above, we assume $\theta = (\psi, \lambda)$, where ψ is the parameter of interest and λ is a nuisance parameter. The partial likelihood function of Cox (1975) is the second factor in (2.75),

$$L_{\text{part}}(\psi, \lambda) = \prod_{i=1}^{n} f(s_i \mid x_{(i)}, s_{(i-1)}; \psi, \lambda), \tag{2.76}$$

and the goal is to find such a sequence for the problem at hand for which L_{part} depends only on ψ, and the omitted part has little information about ψ. Cox (1975, Section 2) writes:

> *(i) no omitted factor ... should contain important information about the parameters of interest ...*
> *(ii) ... so far as possible nuisance parameters, should not occur in the partial likelihood.*

Example 2.22 (Discrete time stochastic process) Cox (1975) illustrates partial likelihood for a discrete time stochastic process with states $\{0, 1, ...\}$, where interest focusses on the special state 0. Assume that if state 0 is occupied at time n, the probability that it is occupied at time $n+1$ is $p_n(\theta)$, independently of the history of the process. Assume further that when the system exits state 0, the probability that determines the path through other states until the return to 0 is potentially complex or unknown. Then the variable X_i can be identified with the path taken between the $(i-1)$st and ith return to state 0, and S_i with the number of periods until the next exit from state 0. The partial likelihood based on S is

$$L_{\text{part}}(\theta) = \prod_{i=1}^{n} \left[\{1 - p_{e_i+r_i}(\theta)\} \prod_{k=1}^{r_i-1} p_{e_i+k}(\theta) \right], \tag{2.77}$$

where e_i is the entry time of the ith visit to state 0, and $e_i + r_i$ is the time of exit from state 0. □

Example 2.23 (Proportional hazards regression) A prominent application of partial likelihood is to the proportional hazards regression model, discussed in detail in Section 3.6.3. In Example 1.7 we considered the likelihood function for a non-homogeneous Poisson process, with rate function $\lambda(t)$. Consider now a vector of observations $(y_1, \ldots, y_n)^{\mathsf{T}}$ that represent times to an event, for example survival times, observed on n independent units. We assume each unit has an associated vector of explanatory variables x_i, and further that each y_i may either represent a genuine failure time t_i, or a censoring time c_i, and that the type of event is recorded with an indicator variable δ_i. The observations are

$$\{(y_1, \delta_1), \ldots, (y_i, \delta_i)\}, \tag{2.78}$$

where $y_i = \min(t_i, c_i)$ and $\delta_i = 1(y_i = t_i)$. We assume that the failure times T_i are independently distributed with cumulative distribution functions $F(t_i)$, and the censoring times C_i are uninformative about F, in the sense that when $y_i = c_i$, the only information about the random variable T_i is that $T_i > c_i$. The censoring times can be either fixed constants, or observations of a random variable C_i with distribution $G(\cdot)$, say. In the latter case we assume that $G(\cdot)$ has no parameters in common with $F(\cdot)$, and that T_i and C_i are independently distributed. This is often called the random censorship model; see also Section 3.6.1. The assumption of uninformative censoring cannot be tested from the set of observations (2.78).

The hazard function associated with F is, as in the Poisson process, the instantaneous failure rate

$$\lambda(t)\,\mathrm{d}t = \mathrm{pr}\{T \in [t, t + \mathrm{d}t) \mid T \geq t\} = \frac{f(t)\,\mathrm{d}t}{1 - F(t)}; \tag{2.79}$$

see also (3.61). The proportional hazards regression model assumes the hazard function takes the form

$$\lambda(t; x, \beta) = \lambda_0(t) \exp(x^{\mathsf{T}}\beta), \tag{2.80}$$

where $\lambda_0(\cdot)$ is a baseline hazard function. In principle any positive function $r(x; \beta)$, say, could replace $\exp(x^{\mathsf{T}}\beta)$, but this is the most commonly used version. Without any assumptions on the form of the baseline hazard function this is a semi-parametric model.

The likelihood function for the unknown parameters β and $\lambda_0(\cdot)$, from the observations (2.78) is

$$L\{\beta, \lambda_0(\cdot)\} \propto \prod_{i=1}^{n} \{\lambda_0(y_i) \exp(x_i^{\mathsf{T}}\beta)\}^{\delta_i} \exp\{-\exp(x_i^{\mathsf{T}}\beta)\Lambda_0(y_i)\}, \tag{2.81}$$

where $\Lambda_0(y) = \int_0^y \lambda_0(t)\,dt$ is the cumulative baseline hazard function.

The partial likelihood function for β is

$$L_{\text{part}}(\beta) = \prod_{i=1}^{n} \left\{ \frac{\exp(x_i^{\mathsf{T}}\beta)}{\sum_{k\in\mathcal{R}_i} \exp(x_k^{\mathsf{T}}\beta)} \right\}^{\delta_i}, \tag{2.82}$$

where $\mathcal{R}_i$ is the risk set of individuals still alive and uncensored at time y_i^-.

The partial likelihood function is the product of the conditional probabilities that individual i fails at time y_i, given the set of units available to fail just before y_i and the information that a unit failed at time y_i. In the general formulation of (2.75), S_i records the event that it was unit i that failed, and X_i the events occurring in the previous time interval, and the information that a unit failed at y_i. In Cox (1972) this was called a conditional likelihood, but the conditioning events change with the index i, so the partial likelihood function is not proportional to the density of an observable random variable. An exception is the special case that there is no censoring, in which case the partial likelihood function is proportional to the marginal distribution of the ranks of the failure times (Kalbfleisch and Prentice, 1972).

There are many applications of partial likelihood to more complex models for the analysis of failure time data and multi-state processes; see the bibliographic notes.

In the model (2.80) an estimate of the nuisance function $\lambda_0(\cdot)$ is needed to estimate the probability of failure for specific values of x. As shown in Section 3.6.3 and Section 6.6, a nonparametric estimator of the cumulative baseline hazard function $\Lambda_0(\cdot)$ can be derived by assuming that the hazard function is constant between observed failure times, and maximizing the full likelihood function (2.81) over these constants. This leads to the constrained maximum likelihood estimate

$$\widehat{\Lambda}_0(y_i; \beta) = \frac{\delta_i}{\sum_{\{j: y_j \le y_i\}} \exp(x_j^{\mathsf{T}}\beta)}, \qquad i = 1, \ldots, n. \tag{2.83}$$

It is shown in Section 3.6.3 that the partial likelihood function for β is identical to the profile likelihood function (3.66), so the partial likelihood estimator of β is in fact the full maximum likelihood estimator. This is an unusual property of the proportional hazards model. It is also an example of the empirical likelihood argument discussed in Section 6.6. The partial likelihood function is also an example of a type of composite likelihood function treated in Chapter 4; see Example 4.10. □

2.8 Marginal posterior distributions

Bayesian inference for a parameter of interest is based on the marginal posterior

$$\pi(\psi \mid y) = \int \pi(\psi, \lambda \mid y)\,\mathrm{d}\lambda = \frac{\int L(\psi, \lambda; y)\pi(\psi, \lambda)\,\mathrm{d}\lambda}{\int L(\psi, \lambda; y)\pi(\psi, \lambda)\,\mathrm{d}\psi\,\mathrm{d}\lambda}. \tag{2.84}$$

In Section 1.4 we showed that the posterior distribution for θ is approximately normal, and thus the marginal posterior distribution for any parameter of interest is also approximately normal, with covariance matrix given by the relevant portion of the inverse Fisher information; see (1.66).

Example 2.24 (Multinomial-Dirichlet) Suppose Y follows a Multinomial(n, p) distribution, where $p = (p_1, \ldots, p_k)$ satisfy $0 \leq p_j \leq 1$ and $\Sigma_j p_j = 1$. The conjugate prior for this distribution is the Dirichlet, with probability function

$$\pi(p; \alpha) \propto \prod_{j=1}^{k} p_j^{\alpha_j - 1},$$

for some choice of hyperparameters α with $0 \leq \alpha \leq 1$ and $\Sigma_j \alpha_j = 1$. The posterior distribution of p given $Y = y$ is again Dirichlet:

$$\pi(p \mid y) \propto \prod_{j=1}^{k} p_j^{y_j + \alpha_j - 1}, \quad \Sigma_j y_j = n, \Sigma_j p_j = 1.$$

The marginal posterior distribution of y_1, say, is readily verified to follow a Beta distribution with parameters $(y_1 + \alpha_1, n - y_1 + \Sigma_{j=1}^{k} \alpha_j)$. □

Example 2.25 (Many normal means) This example is due to Stein (1959) and discussed in Cox and Hinkley (1974, Example 2.39). It illustrates the potential impact of the prior in a high-dimensional setting. Suppose for each $i = 1, \ldots, k$, Y_i follows a normal distribution with expected value μ_i and variance 1, independently over i. With the flat prior $\pi(\mu_i) \propto 1$, the posterior for μ_i is a normal distribution with posterior expected value y_i and variance 1. If the parameter of interest is $\psi = \Sigma_i \mu_i^2$, the marginal posterior density of ψ is a non-central χ_k^2 distribution, with non-centrality parameter $\Sigma_{i=1}^{k} y_i^2$. A plausible estimate of ψ is the posterior expected value, $\widetilde{\psi}_B = k + \Sigma_i y_i^2$. Under the model for the observations, $\mathrm{E}(\widetilde{\psi}_B) = 2k + \psi$ and $\mathrm{var}(\widetilde{\psi}_B) = 2k + 4\psi$. Thus a seemingly non-informative prior for μ has a large impact on the posterior for ψ. This is a general feature of Bayesian marginal inference, related to the marginalization paradox of Dawid et al. (1973). As $k \to \infty$, the Bayes posterior expectation will not be a consistent estimator of ψ under the model. This does not contradict the asymptotic theory of Section 1.4, as the regularity conditions there do not apply if the dimension of the parameter space increases with the sample size. Similar calculations show that credible intervals based on the posterior distribution will not have frequentist coverage at the stated level.

In a variation of this example, in which we assume the variance of each Y_i is $1/n$, the estimator $\widetilde{\psi}_B = k/n + \Sigma_i y_i^2$ has expected value under the model $2k/n + \psi$, so is consistent for ψ if $k/n \to 0$ as $n \to \infty$. □

Evaluating the integrals in the marginal posterior analytically is usually possible only for special choices of model/prior combinations, typically exponential family models and so-called conjugate priors. Evaluating the integrals computationally is also not feasible if the dimension of the parameter is more than three or four. As a result, a number of approximate techniques have been developed. One of the most well-known is Markov chain Monte Carlo (MCMC) simulation, which provides samples from the posterior. The method of approximate Bayesian computation (ABC) extends this to models for which the likelihood function is not available in closed form.

An analytical approximation to the marginal posterior can be obtained using the Laplace approximation. In both the numerator and the denominator of (2.84), the likelihood function is expanded in a Taylor series about its maximum: in the numerator treated as a function of λ with ψ fixed, and in the denominator treated as a function of (ψ, λ). The resulting expression is

$$\pi(\psi \mid y) \doteq \left(\frac{1}{\sqrt{2\pi}}\right)^q \exp\{\ell_{\mathrm{p}}(\psi) - \ell_{\mathrm{p}}(\widehat{\psi})\} \frac{|J_{\lambda\lambda}(\psi, \widehat{\lambda}_\psi)|^{-1/2}}{|J(\widehat{\theta})|^{-1/2}} \frac{\pi(\psi, \widehat{\lambda}_\psi)}{\pi(\widehat{\theta})}, \tag{2.85}$$

and can be shown to approximate the exact marginal posterior with relative error $O(n^{-3/2})$ after renormalization.

The validity of this approximation requires that the log-likelihood function have a unique maximum determined by the score equation, for both the constrained and the full maximum likelihood estimators, so is not valid for multi-modal likelihood functions. Perhaps for this reason it has been less widely used in applications than MCMC methods. Laplace approximation for Bayesian inference in hierarchical spatial models has been developed under the name INLA, for integrated, nested Laplace approximation; see the bibliographic notes.

2.9 High-dimensional parameters

Models with complex probability structure often have very large numbers of parameters, and the log-likelihood function may have several local maxima. This arises for example in neural network models with several hidden layers. One strategy in these settings is to approximate the log-likelihood function by a convex function. Another is to run the optimization algorithm from many different starting points. Depending on the details of the problem, it may be that a local maximum would be adequate for some purposes, for example, providing reasonable predictions. Loh and Wainwright (2015) show that at least in some settings the statistical error is larger than the gap between local optima, in which case any local optimum could be a suitable point estimator.

There is a rapidly developing literature on likelihood-based inference in the setting often called "large p, small n", or "$p > n$". For studying asymptotic properties, this is formulated as "p increasing with n", and the rate of increase is then of interest. While terminology is still evolving, one classification is the following, where $p = p_n$ is the dimension of the parameter space and n is the number of observations, or more generally the order at which the Fisher information accumulates:

- classical: p is fixed as $n \to \infty$ so that $p/n \to 0$.
- semi-classical: $p_n = O(n^\alpha), 0 < \alpha < 1$.
- large dimension: $p_n/n \to C$ as $n \to \infty$.
- high dimension: $p_n/n \to \infty$.

The large dimension case is also called fixed-aspect asymptotics in the context of p indexing columns and n indexing rows of a matrix of data. Sur and Candès (2019) study logistic regression with $p/n \to \kappa < 1$, and describe this as moderate dimension. The high-dimension case above is also sometimes divided into polynomial and exponential growth: $p = O(n^\alpha)$ with $\alpha > 1$, and $p = O(e^n)$, respectively. Sometimes the latter models are called ultra-high dimensional.

The classical case leads to the asymptotic theory summarized in Sections 1.2 and 2.4.

The semi-classical case is illustrated by the Neyman–Scott problems discussed in Section 2.6. Using slightly different notation to distinguish the sample size more carefully, suppose our model is for doubly indexed responses $y_{ij}, i = 1, \ldots, q, j = 1, \ldots, m$, with densities $f(y_{ij}; \psi, \lambda_i)$. We assume the dimension of ψ is fixed, but that $q \to \infty$, so the dimension of the parameter increases with the sample size $n = mq$, whether m is fixed or m increases with n. The examples in Section 2.6 consider m fixed, and although the usual asymptotic theory of likelihood inference is not valid, there is a conditional or marginal likelihood function for which it is. Sartori (2003) considers likelihood inference with both m and q increasing, and shows that the classical inferential results are obtained for inference based on the modified profile log-likelihood function if $1/m = o(q^{-1})$, or $q = o(n^{1/2})$, whereas inference based on the profile likelihood function is asymptotically correct only if $1/m = o(q^{-1/3})$, or $q = o(n^{3/4})$.

Fan et al. (2019) study asymptotic normality of the maximum likelihood estimator in generalized linear models, and show that for fixed X, the usual asymptotic normality holds for $p = o(n^{1/2})$, as in Portnoy (1988), under some conditions on the design matrix. They also show, for logistic regression, and assuming a particular distribution for the covariate matrix X, that the usual theory breaks down at $p = O(n^{2/3})$.

For large, high, and ultra-high dimensional parameters, the asymptotic theory is more tentative, and still being developed. We illustrate this with some examples.

Example 2.26 (High-dimensional linear regression) We consider the linear regression model $Y = X\beta + \epsilon$, where $Y, \epsilon \in \mathbb{R}^n$, and X is an $n \times p$ matrix, each column corresponding to a potential explanatory variable. If $p > n$ the model is over-determined, the least-squares estimator is not unique, and the model is overfitted, in the sense that the minimized value of the least squares function is zero; see Section 7.4. In Chapter 7 various methods of regularizing the estimator are discussed. Two widely used estimators are the ridge regression estimate

$$\widehat{\beta}_{\text{ridge}} = (X^{\mathsf{T}}X + \lambda I)^{-1}X^{\mathsf{T}}y,$$

which minimizes $(y - X\beta)^{\mathsf{T}}(y - X\beta) + \lambda\Sigma_j\beta_j^2$, and the lasso estimate, which minimizes $(y - X\beta)^{\mathsf{T}}(y - X\beta) + \lambda\Sigma_j|\beta_j|$, in each case with λ to be chosen; see Section 7.6.

The distribution of $\widehat{\beta}_{\text{ridge}}$ for fixed λ follows directly, and if we assume the errors are normally distributed, has an exact normal distribution; in general, $\widehat{\beta}_{\text{ridge}}$ will be asymptotically normally distributed, as it is a linear function of y. The lasso estimator is not a linear function of y and its exact or approximate distribution is not readily available. The lasso estimate has the property that many elements of $\widehat{\beta}_{\text{lasso}}$ are identically 0, so in that sense estimation of the regression parameter also functions as a method of model selection. In studying the inferential properties of the lasso estimator, the questions that arise include whether or not it identifies the true underlying model, what is the distribution of the estimated coefficients, and what is the distribution of the predictor $X\widehat{\beta}_{\text{lasso}}$. Model selection and prediction is discussed in Sections 7.2 and 7.4. Asymptotic theory for the lasso estimator and estimators constructed with other regularization constraints on β is discussed in Section 7.4. □

Example 2.27 (Logistic regression, cont'd) As in Example 2.9, suppose the log-likelihood function is

$$\ell(\beta) = \sum_{i=1}^{n}[y_i x_i^{\mathsf{T}}\beta - m_i \log\{1 + \exp(x_i^{\mathsf{T}}\beta)\}].$$

For binary data with $m_i = 1$, the maximum likelihood estimate of β will be infinite if some linear combination of the covariates separates the 1 and 0 responses completely. Maximizing the penalized likelihood of Firth (1993a), $\widetilde{\ell}(\beta)$ in (2.18), leads to finite estimates of the regression coefficients, even in the case of separation, as long as the covariate matrix X is of full rank (Kosmidis and Firth, 2021).

When $p > n$, X cannot be of full rank, and indeed a different asymptotic theory applies. Sur and Candès (2019) and Sur et al. (2019) study the case that $p_n/n \to \kappa$, with $0 < \kappa < 1$, and derive the limiting distribution of the maximum likelihood estimator and the likelihood ratio statistic. These results are generalized in Zhao et al. (2022), and we present that version here. Assuming that the covariate vectors x_i are independent and identically distributed as $N(0, \Sigma)$ for some positive definite matrix Σ, and that the signal strength, $\text{var}(x_i^\top \beta) \to \gamma^2 < \infty$, there is a phase transition depending on κ and γ that determines whether or not the components of the maximum likelihood estimate of the regression parameter β are all finite (Candès and Sur, 2020). In the region where the maximum likelihood estimator exists, the limiting distribution of the estimate of a component β_j is

$$\frac{\sqrt{n}(\widehat{\beta}_j - \alpha_* \beta_j)}{\sigma_* \tau_j} \xrightarrow{d} N(0,1),$$

where $\tau_j^2 = \text{var}(x_{ij} \mid x_{i(-j)})$ is the conditional variance of the jth column of the design matrix given the remaining columns, and α_* and σ_* depend on κ and γ.

The log-likelihood ratio statistic for testing $H : \beta_j = \beta_{0j}$ has limiting distribution

$$2\{\ell_{\text{p}}(\widehat{\beta}) - \ell_{\text{p}}(\widetilde{\beta}_{0(j)})\} \xrightarrow{d} \frac{\kappa \sigma_*^2}{\lambda_*} \chi_1^2,$$

where here $\widetilde{\beta}_{0(j)}$ is the constrained maximum likelihood estimator of β with the jth component fixed at β_{0j}, and λ_* depends on κ and γ. The constants α_*, σ_* and λ_* arise as ingredients in the optimization algorithm, described, for example, in Zhao et al. (2022, Equation (3.4)).

This is a different asymptotic result than in the classical regime; the maximum likelihood estimator of β_j is asymptotically biased, and its asymptotic variance is not the appropriate element of the inverse Fisher information matrix. The constants α_*, σ_* and λ_* need to be estimated from the data. Kosmidis and Zietkiewicz (2026) study via simulations the bias-adjusted maximum likelihood estimator of Firth (1993a) defined in (2.18) in the large-dimension regime, and relate this to the work of Sur and Candès (2019) and Zhao et al. (2022). □

Outside of these two examples, the theory of likelihood inference for high-dimensional models is still evolving. In any given application with fixed $p < n$, it would not be clear whether p could be regarded as fixed, or as increasing with n, and whether inference should be based on classical maximum likelihood theory or this new theory; this was part of the motivation for the simulation studies in Kosmidis and Zietkiewicz (2026). Tang and Reid (2025), with a similar motivation, studied the higher-order approximation based on r^* discussed in Section 2.5.2, for $p = O(n^\alpha), 0 < \alpha < 1/2$. One observation that emerged from that paper is that the biggest improvement from first-order to higher-order approximation is due to the more refined treatment of nuisance parameters. This is consistent with the finding of Sartori (2003) that modified profile likelihood has better asymptotic properties than profile likelihood, in the semi-classical regime.

There is a close connection between high-dimensional parameter spaces and nonparametric methods; the latter are discussed in Chapter 6. The Bernstein–Von Mises theorem for the posterior distribution with nonparametric models is discussed in Spokoiny and Panov (2025).

2.10 Bibliographic notes

Parameter orthogonality was defined in Jeffreys (1961) and explored further in Huzurbazar (1950). The role of parameter orthogonality in inference is the subject of Cox and Reid (1987) and the discussion in Section 2.1 is taken from that paper. Inference based on marginal and conditional likelihood functions was introduced in Bartlett (1936a,b, 1937). Definitions of sufficiency and ancillarity in the presence of nuisance parameters are discussed in Barndorff-Nielsen and Cox (1994, Chapter 2) and Reid (1995).

Pace and Salvan (1997, Chapter 7) give a thorough discussion of transformation families and the related group theory, as well as inference based on the marginal likelihood function.

The regression-scale analysis in Example 2.11 is developed in generality in Fraser (1979). The group structure of the model plays an essential role, both in conditioning on the ancillary statistic, and in the use of the pivotal quantities for marginal inference about the parameters of interest. Marginal likelihood inference in Weibull and extreme value regression models is described in Lawless (2003, Chapter 6). Approximations to the integral defining a marginal likelihood are described and illustrated in Brazzale et al. (2007, Chapters 5,8). These approximations are also implemented in the `R` package `hoa` (Brazzale, 2005).

As described in most treatments, including the one here, marginal or conditional likelihood functions emerge from detailed analysis of particular models. Battey et al. (2024) in contrast consider deriving the likelihood directly by imposing the constraints implied by (2.12) and (2.13).

An interesting construction of a conditional likelihood using stopping rules is applied to extreme value analysis in Barlow et al. (2020). Analysis of data is sometimes prompted by extreme events; in this case large flooding events, and the authors introduce stochastic stopping criteria to try to accommodate the bias that would otherwise result.

McCullagh and Tresoldi (2021) illustrate the use of the profile likelihood function for choosing among alternative quantile transformations of the original observations.

The derivation of the limiting distributions of Section 2.5.1 follows Davison (2003) and Barndorff-Nielsen and Cox (1994). Lehmann and Romano (2005, Section 14) gives the version of Wilks' theorem described at (2.37). Barndorff-Nielsen (1987) showed that the p^*-approximation discussed in Section 2.5.2 approximates the appropriate marginal or conditional distribution, when these are available. An overview of the p^*-approximation is given in Reid and Battey (2026) and its relationship to the saddlepoint approximation is discussed in Reid (1988). The derivations of the higher-order asymptotic results discussed in Section 2.5.2 are available in several books, including Barndorff-Nielsen and Cox (1994), Pace and Salvan (1997), Severini (2000), and Brazzale et al. (2007). Pierce and Bellio (2017) review higher-order likelihood inference based on the version of the r^* approximation developed by Skovgaard (1996), provide an implementation in the `R` package `likelihoodAsy` (Bellio and Pierce, 2020). Conditional inference in exponential family models using the saddlepoint approximation is implemented in Brazzale (2005) in the package `hoa`, and several examples are given in Brazzale et al. (2007, Chapter 4).

The derivation of the partial likelihood for the proportional hazards model as a profile likelihood appears in Johansen (1983) and is further discussed in Section 3.6. Kalbfleisch and Schnaubel (2023) describe partial likelihood functions for modelling survival data with competing risks and for the analysis of recurrent events. Aalen et al. (2008), Cook and Lawless (2018) and Cook and Lawless (2022) describe partial likelihood functions for event history and other multistate processes. Partial likelihood for change-point regression models is described in Takeishi (2023). A 2024 special issue of the *Journal of the Royal Statistical Society, Series A* (Keogh, 2024) has a series of papers on the proportional hazards regression model, very often called the Cox model.

The modified profile likelihood function of Section 2.6 was derived in Barndorff-Nielsen (1983), along with the p^*-approximation to the distribution of the maximum likelihood estimator (2.56). A closely related adjusted likelihood function was developed in Fraser (2003), and compared there to modified profile likelihood.

A Bayesian interpretation of the adjusted profile likelihood is given in Sweeting (1987). The Laplace approximation to the marginal posterior distribution is derived in Tierney and Kadane (1986) and Kass et al. (1987); see also Davison (2003, Chapter 11).

Integrated likelihoods of the form $L(\psi) = \int L(\psi, \lambda)\pi(\lambda \mid \psi)\,\mathrm{d}\lambda$ are reviewed in Berger et al. (1999). One interpretation of $\pi(\lambda \mid \psi)$ is as a conditional prior density for the nuisance parameter, but the paper also considers other weight functions, including the uniform, $\pi(\lambda \mid \psi) = 1$ and various default priors, such as the reference prior reviewed in Berger and Bernardo (1992).

Some discussions of Laplace approximation, especially in hierarchical models, refer to Taylor series expansion of the normalizing constant only, so the resulting approximations are normal densities. A good introduction is Rue et al. (2009). Laplace approximation in high-dimensional models, in the sense of (2.85), is discussed in Tang and Reid (2025).

Limiting theory for two-index asymptotics discussed in Section 2.9 was developed in Portnoy (1988), motivated by similar analysis of M-estimators in regression (Huber, 1967). Sartori (2003) established the rates for inference based on the modified profile likelihood function and the profile likelihood function.

Likelihood inference for stochastic processes requires careful attention to the properties of the process: in particular it is necessary for the usual asymptotic theory that the nature of the dependence allows sufficient accumulation of information. Sant et al. (2022) build on general results of Ibragimov and Has'minski (1981) to show that for a particular diffusion process the maximum likelihood estimator is consistent and asymptotically normal. At the time of writing asymptotic theory for likelihood inference in stochastic processes is an active area of research. Lejay and Mazzonetto (2024), for example, show that the rate of convergence of the maximum likelihood estimator in skew Brownian motion is $O(n^{-1/4})$, and the limiting distribution is a normal mixture.

A review of inference after model selection in high-dimensional regression is given in Kuchibhotla et al. (2022).

2.11 Further results and exercises

2.1. Suppose $Y \in \mathbb{R}$ has a density $f(y;\theta)$, and $\theta = (\psi, \lambda_1, \dots, \lambda_{p-1})^{\mathsf{T}}$, with $\psi \in \mathbb{R}$. We wish to find an interest-respecting reparametrization $\phi = (\psi, \eta_1, \dots, \eta_{p-1})$ so that η is orthogonal to ψ with respect to expected Fisher information $i(\theta)$. By showing that $i(\phi) = (\partial\theta/\partial\phi^{\mathsf{T}})i(\theta)(\partial\theta^{\mathsf{T}}/\partial\phi)$, deduce that the condition for orthogonality is

$$\sum_{s=1}^{p-1} \frac{\partial\lambda_s}{\partial\eta_t}\left\{ i_{\psi\lambda_s}(\theta) + \sum_{r=1}^{p-1} i_{\lambda_s\lambda_r}(\theta)\frac{\partial\lambda_r}{\partial\psi} \right\} = 0, \quad t = 1, \dots, p-1, \tag{2.86}$$

and since the transformation is required to be one-to-one, this is equivalent to (2.8).

If $\psi \in \mathbb{R}^2$, say, then there will be two equations of this form, one for ψ_1 and one for ψ_2, and there is no guarantee that they are compatible. However Shen et al. (2025) present some examples where global orthogonality is indeed achievable. *(Cox and Reid, 1987)*

2.2. Suppose $f(y;\psi,\lambda)$ is a linear exponential family as at (2.21). Show that $\eta(\psi,\lambda) = E_{\psi,\lambda}(S_2) = \partial c(\psi,\lambda)/\partial\lambda$ is orthogonal to ψ with respect to Fisher information, thus verifying (2.70). Show further that $\widehat{\eta}_\psi = \widehat{\eta}$, although $\widehat{\lambda}_\psi \neq \widehat{\lambda}$. Hence or otherwise show that in the gamma distribution, the expected value parameter is orthogonal to the shape parameter.
(Staicu and Reid, 2008)

2.3. Show that the constrained maximum likelihood estimator, $\widehat{\lambda}_\psi$, is related to the full maximum likelihood estimator by

$$\widehat{\lambda}_\psi = \widehat{\lambda} + O_p(1/\sqrt{n}),$$

but that if ψ and λ are orthogonal,

$$\widehat{\lambda}_\psi = \widehat{\lambda} + O_p(1/n).$$

Show that this result does not hold if ψ and λ are only locally orthogonal at θ_0.
(Cox and Reid, 1987)

2.4. A random variable Z follows the skew-normal distribution with skewness parameter α if it has density

$$f_{\text{SN}}(z;\alpha) = 2\phi(z)\Phi(\alpha z), \quad -\infty < z, \alpha < \infty, \tag{2.87}$$

where $\phi(\cdot)$ and $\Phi(\cdot)$ are the standard normal density and distribution function, respectively. The standard skew-normal density can be extended to a three-parameter version with a location and scale transformation $Y = \omega Z + \xi$, with density $f_{\text{SN}}(y;\xi,\omega,\alpha)$. For a sample $y = (y_1,\ldots,y_n)$ from the density in (2.87), find an expression for the log-likelihood function $\ell(\xi,\omega,\alpha;y)$, and show that the expected Fisher information matrix is singular at $\alpha = 0$. Show however that the Fisher information is non-singular in the parametrization (μ,σ,γ_1), where

$$Y = \mu + \sigma\left[\frac{Z - \text{E}(Z)}{\{1-\text{E}(Z)^2\}^{1/2}}\right],$$

and γ_1 is the standardized skewness of Z. The skew-normal family has been extended in many directions, including the multivariate skew-normal and a class of skew-symmetric distributions. Skewed versions of Gaussian processes have also been developed; see the bibliographic notes.
(Azzalini, 1985, 2005)

2.5. Consider the AR(p) model

$$y_i = \phi_1 y_{i-1} + \phi_2 y_{i-2} + \cdots + \phi_p y_{i-p} + \epsilon_i, \quad i = 1,\ldots,n, \tag{2.88}$$

where $\epsilon_i \sim N(0,\sigma^2)$ and y_0 is from the stationary distribution. Find the transformation of $(\phi_1,\ldots,\phi_{p-1})$ that is locally orthogonal to ϕ_p. *(Tong, 1988; Shen et al., 2025)*

2.6. Suppose in Example 2.10 that the parameter of interest is λ. Find the conditional density of t_2, given t_1. If the parameter of interest is $\mu = \text{E}(y_i)$ an exact conditional density depending only on μ is not available.

2.7. Suppose $y = (y_1,\ldots,y_n)$ is a sample from the location-scale family

$$y_i = \mu + \sigma\epsilon_i \tag{2.89}$$

where ϵ_i has density $f_0(\cdot)$. In Exercise 1.10 it was established that the joint distribution of y can be factored as

$$f(y;\mu,\sigma) = f_1(\widehat{\mu},\widehat{\sigma} \mid a;\mu,\sigma) f_2(a), \tag{2.90}$$

where $(\widehat{\mu}, \widehat{\sigma})$ are the maximum likelihood estimators of μ and σ, $a_i = (y_i - \widehat{\mu})/\widehat{\sigma}$, and $a = (a_1, \ldots, a_n)$ is ancillary. In this model, all the information about $\theta = (\mu, \sigma)$ is contained in the conditional density f_1, although the marginal density f_2 can be used to assess the fit of the model.

(a) Show that

$$f_1(\widehat{\mu}, \widehat{\sigma} \mid a; \mu, \sigma) \propto \sigma^{-n}\widehat{\sigma}^{n-2} \prod_{i=1}^{n} f_0\{(a_i\widehat{\sigma} + \widehat{\mu} - \mu)/\sigma\}. \tag{2.91}$$

(b) Re-express this density in terms of $t_1 = (\widehat{\mu} - \mu)/\widehat{\sigma}$ and $t_2 = \widehat{\sigma}/\sigma$. Show that the Jacobian of the transformation from $(\widehat{\mu}, \widehat{\sigma})$ to (t_1, t_2) is $\widehat{\sigma}\sigma$, and the joint density of (t_1, t_2) is

$$f(t_1, t_2 \mid a) = c(a)t_2^{n-1} \prod_{i=1}^{n} f_0(a_i t_2 + t_1 t_2), \tag{2.92}$$

where $t_1 = t_1(\mu)$ and $t_2 = t_2(\sigma)$.

(c) Deduce from this expression that the marginal density of t_1 is free of σ, and the marginal density of t_2 is free of μ. *(Fraser, 1979, Section 2.1)*

2.8. Suppose we have a one-way random effects model

$$y_{ij} = \mu + b_i + \epsilon_{ij}, \quad j = 1, \ldots, m; i = 1, \ldots, n,$$

where we assume b_i and ϵ_{ij} are independent, with expected values 0 and variances σ_b^2 and σ^2, respectively. This is a special case of a linear mixed effects model; see Example 3.3. Let $y = (y_{11}, \ldots, y_{mn})^{\mathsf{T}}$ be the complete vector of responses.

(a) Show that the joint density $f(y; \mu, \sigma_b^2, \sigma^2)$ has the form

$$f(y; \mu, \sigma_b^2, \sigma^2) \propto L_1(\mu, \sigma_b^2, \sigma^2; \bar{y}_{\bullet\bullet})L_2(\sigma_b^2, \sigma^2; \mathrm{MS_b}, \mathrm{MS_w}),$$

where $\bar{y}_{\bullet\bullet}$ is the grand mean of the observations, and

$$\mathrm{MS_b} = m\sum_{i=1}^{n}(\bar{y}_{i\bullet} - \bar{y}_{\bullet\bullet})^2/(n-1), \quad \mathrm{MS_w} = \sum_{i=1}^{n}\sum_{j=1}^{m}(y_{ij} - \bar{y}_{i\bullet})^2/\{n(m-1)\},$$

are the mean-squared deviations between and within groups, respectively. Thus inference about the components of variance can be based on the marginal likelihood function $L_2(\sigma_b^2, \sigma^2)$. This is also known as the REML likelihood in mixed linear models; see Section 3.2.5. *(Patterson and Thompson, 1971; Harville, 1977)*

(b) Show that $(n-1)\mathrm{MS_b} \sim (\sigma^2 + m\sigma_b^2)\chi_{n-1}^2$ and $n(m-1)\mathrm{MS_w} \sim \sigma^2\chi_{n(m-1)}^2$.

(c) For the hypothesis $H_0 : \sigma_b^2 = 0$ show that the Wald and likelihood ratio test statistics (2.35) and (2.36) constructed from the marginal log-likelihood function $\log L_2(\sigma_b^2, \sigma^2)$ have the forms

$$w_{\mathrm{e}}^2 = \left(\frac{F-1}{F}\right)^2 \frac{n(n-1)(m-1)}{2(mn-1)},$$

$$w_{\mathrm{p}} = (nm-1)\log\left\{1 + \frac{(n-1)(F-1)}{nm-1}\right\} - (n-1)\log F,$$

where $F = \mathrm{MS_b}/\mathrm{MS_w}$ is the F-statistic for testing H_0. As noted in Example 1.35, with probability 1/2 the two statistics will both be zero. Battey and McCullagh (2024) study the difference between the nominal χ^2 approximations to the distributions of w_e and $w_{\rm p}$, in the case where both are positive. They show how the geometry of the Wald statistic (and the score statistic) can lead to severe underestimation of precision in finite samples and recommend use of the likelihood ratio statistic.

2.9. Complete the asymptotic argument sketched in Section 2.5.1, using the partitioned form of the score and information matrix.

(a) Show that $U_\psi(\theta) - I_{\psi\lambda}(\theta)I_{\lambda\lambda}^{-1}(\theta)U_\lambda(\theta)$ is uncorrelated with $U_\lambda(\theta)$.

(b) Verify (2.54).

2.10. A generalization of the generalized likelihood ratio statistic (2.37), called the split likelihood ratio statistic, was proposed in Wasserman et al. (2020). For a sample of observations $y = \{y_1, \ldots, y_n\}$ from a model with density $f(y;\theta)$, create two sub-samples of size $n/2$, $y_{(1)}$ and $y_{(2)}$, say. Use the second sub-sample to estimate θ by $\widehat{\theta}_2(y_{(2)})$; this need not be the maximum likelihood estimate. Use the first sub-sample to construct a likelihood ratio of the form

$$T_n(\theta; y) = \frac{L(\widehat{\theta}_2; y_{(1)})}{L(\theta; y_{(1)})}. \tag{2.93}$$

The authors used T_n to suggest a universal confidence set for θ of the form

$$C_U(y) = \{\theta \in \Theta : T_n(\theta; y) \le 1/\alpha\}.$$

(a) Show that

$$\mathrm{E}_{\theta_0}\left\{\frac{L(\theta; Y_{(1)})}{L(\theta_0; Y_{(1)})}\right\} \le 1,$$

and hence

$$\mathrm{E}_{\theta_0}\{T_n(\theta; y) \mid y_{(2)}\} \le 1.$$

Statistics bounded by unity in expectation are called *e*-values in the recent literature; for a review see Vovk and Wang (2023). Maximum likelihood *e*-values are defined in Blier-Wong and Wang (2026).

(b) Show that an application of Markov's inequality gives

$$\mathrm{pr}_\theta\{\theta \in C_U(Y)\} \ge 1 - \alpha,$$

for any fixed n.

(c) Show that the test of $H_0 : \theta \in \Theta_0 \subset \Theta$ which rejects H_0 when $U_n > 1/\alpha$ has size $\le \alpha$, where

$$U_n = \frac{L(\widehat{\theta}_2; y_{(1)})}{L(\widehat{\theta}_1; y_{(1)})},$$

$\widehat{\theta}_2$ is any estimate of θ constructed from $y_{(2)}$, and $\widehat{\theta}_1 = \arg\sup_\theta L(\theta; y_{(1)})$ is the maximum likelihood estimate based on $y_{(1)}$.

Wasserman et al. (2020) note that T_n can be defined for any model for which there is a likelihood function, and thus can be applied to various nonregular problems. While the size of the test and coverage of the confidence set are guaranteed, they might be overly conservative. This is discussed for example in Tse and Davison (2022). Strieder and Drton (2022) consider optimal splitting proportions and study the asymptotic distribution of T_n. Shi and Drton (2025) establish asymptotic results for Gaussian mixture models, and show that it is possible in this setting to ensure that the split likelihood ratio test is not conservative.

2.11. In Example 2.17 where $\{y_{i1}, \ldots, y_{im_i}\}$ are independently distributed as $N(\psi, \lambda_i)$, $i = 1, \ldots, n$, show that (2.64) defines the maximum likelihood estimator, and that (2.65) is the solution of the score equation based on the adjusted profile log-likelihood function $\ell_{\mathrm{a}}(\psi)$ (2.67).
(Cox and Reid, 1987)

2.12. In Example 2.18, (y_{i1}, y_{i2}) are independently distributed as exponential, with expected values $\psi\lambda_i$ and ψ/λ_i, respectively. Assume we have a sample of n such pairs, independently distributed. Show that the maximum likelihood estimator $\widehat{\psi}$ converges in probability to $(\pi/4)\psi$, and the adjusted maximum likelihood estimator derived from (2.71) converges in probability to $(\pi/3)\psi$.
(Cox and Reid, 1992)

2.13. There are examples where modified profile log-likelihood function coincides with the adjusted profile log-likelihood function, even if the nuisance parameters are not orthogonal to the parameter of interest.

(a) Show that in a location-scale model $\sigma^{-1}f\{\sigma^{-1}(y-\mu)\}$, with $f(\cdot)$ known but not necessary symmetric, that $j_{\mu\mu}(\mu, \sigma) = \partial^2\ell(\mu, \sigma; \widehat{\mu}, \widehat{\sigma}, a)/\partial\mu\partial\widehat{\mu}$, and thus $\ell_{\mathrm{M}}(\psi) = \ell_{\mathrm{a}}(\psi)$, even though μ and σ may not be orthogonal. This is an example of a composite transformation model, and this result holds for such models.

(b) This equivalence can also hold in a model without a group structure. Suppose that $Y \sim N_k(A_\psi\lambda, \mathcal{I}_{k\times k})$, and $Z \mid Y \sim f(z; \psi, y)$, where A_ψ is a matrix with determinant 1, and $f(z; \psi, y)$ is an arbitrary density. Show that $|\partial\widehat{\lambda}/\partial\widehat{\lambda}_\psi| = 1$. Show also that the parameters $\eta = A_\psi\lambda$ and ψ are orthogonal with respect to observed Fisher information.
(Barndorff-Nielsen and McCullagh, 1993)

2.14. Suppose $\{(y_{i1}, y_{i2}), i = 1, \ldots, n\}$ are independent binary random variables with probabilities of success p_{i1}, p_{i2}, respectively. We assume the log-odds ratio for each table is the same: $\psi = \log\{p_{i1}(1-p_{i2})/(1-p_{i1})p_{i2}\}$, and $\lambda_i = \log\{p_{i2}/(1-p_{i2})\}$.

(a) Give an expression for the adjusted profile likelihood $\ell_{\mathrm{a}}(\psi)$; note that the modified profile likelihood is the same function.

(b) The conditional likelihood function is derived in Example 2.8: compare inference based on the same pivotal quantities derived from $\ell_{\mathrm{c}}(\psi)$ to that based on those derived from $\ell_{\mathrm{a}}(\psi)$.
(Sartori, 2003)

2.15. Student's t distribution with ν degrees of freedom approaches the normal distribution as $\nu \to \infty$. With the reparametrization $\psi = 1/\nu$ the density is

$$f(y; \mu, \sigma, \psi) = \frac{\Gamma\{(\psi^{-1}+1)/2\}\psi^{1/2}}{(\sigma^2\pi)^{1/2}\Gamma\{1/(2\psi)\}}\{1 + \psi(y-\mu)/\sigma^2\}^{-(\psi^{-1}+1)/2}, \quad \psi, \sigma > 0; y, \mu \in \mathbb{R},$$

and as $\psi \to 0$ this converges to the density for $N(\mu, \sigma^2)$.

(a) Write $z = (y-\mu)/\sigma$, expand $\ell(\psi; z) = \log f(z; \psi)$ in a Taylor series around $\psi = 0$, and deduce that the expected Fisher information matrix for a single observation is

$$i(\mu, \sigma, 0) = \begin{bmatrix} \sigma^{-2} & 0 & 0 \\ 0 & \sigma^{-4}/2 & \sigma^{-2} \\ 0 & \sigma^{-2} & 7/2 \end{bmatrix},$$

showing that the score vector at $\psi = 0$ is asymptotically normal with expected value 0 and covariance matrix as above.

(b) For a sample $Y = \{Y_1, \ldots, Y_n\}$, show that the score vector $u(\mu, \sigma, \psi; Y)$ at $\psi = 0$ is asymptotically normal with expected value 0 and covariance matrix $I(\theta) = ni(\theta)$.

(c) Show then that

$$\frac{\partial \log f(y; \widehat{\mu}_0, \widehat{\sigma}_0, 0)}{\partial \psi} \stackrel{.}{\sim} N(0, 3/2),$$

and conclude that the likelihood ratio statistic

$$W_{\mathrm{p}}(0) = 2\{\log f(y; \widehat{\mu}, \widehat{\sigma}^2, \widehat{\psi}) - \log f(y; \widehat{\mu}_0, \widehat{\sigma}_0^2, 0)\},$$

is asymptotically distributed as a 50% mixture of a point mass at 0 and a χ_1^2 random variable. *(Davison, 2003, Example 4.39)*

2.16. With reference to Example 2.25, suppose Y_i are independently distributed as $N(\mu_i, 1/n), i = 1, \ldots, k$.

(a) Show that under the improper prior $\pi(\mu) \propto 1$, the marginal posterior for $\psi = \Sigma_{i=1}^k \mu_i^2$ is proportional to a noncentral χ_k^2 distribution, with noncentrality parameter $n\Sigma_{i=1}^k y_i^2$.

(b) Show that the maximum likelihood estimator of ψ has a distribution proportional to a noncentral χ_k^2 distribution with noncentrality parameter $n\psi$.

(c) Conclude from these results that the coverage under the normal model of a posterior $(1-\alpha)$ credibility bound for ψ diverges from $1-\alpha$ as $k \longrightarrow \infty$ with n fixed, but converges to $1-\alpha$ as $n \longrightarrow \infty$ with k fixed. *(Reid and Sun, 2010)*

3

Likelihood for Complex Models

3.1 Introduction

The likelihood-based methods of Chapters 1 and 2 can be extended to handle data with more complex features. In this chapter, we examine likelihood formulations for a variety of models that have been developed to account for such data, including longitudinal data (Section 3.2), time series (Section 3.3), data with spatial variation (Section 3.4), missing data (Section 3.5), and survival data (Section 3.6).

The models discussed in this chapter are not exhaustive but serve to illustrate how likelihood-based methods can be adapted to diverse types of complex data. In subsequent chapters, we discuss computational or theoretical challenges that arise when applying likelihood inference for complex modelling scenarios. The emphasis is on frequentist approaches, although there is a large literature on Bayesian inference for many of the models described here.

As noted after Example 1.12, likelihood formulations for regression models are constructed from the conditional distribution of the responses given the covariates. When the covariates are fixed at their observed values, we omit them from the notation for simplicity, except in settings where the covariates are treated as random or when we want to stress the dependence on covariates.

3.2 Longitudinal data

Longitudinal data are repeated observations on each subject over a period of time, typically at multiple discrete time points. A key characteristic of longitudinal data is that the observations within each subject or unit exhibit some degree of dependence. In this section, we explore modelling strategies and likelihood-based approaches for analyzing such data. The methods can also be applied to other types of data that have a group structure, including clustered or grouped data, or clustered longitudinal data, where observations within a cluster or group are expected to be correlated. In longitudinal and clustered data, observations on different subjects or different clusters are typically assumed to be independent.

For longitudinal data on n subjects, $Y_i = (Y_{i1}, \ldots, Y_{im_i})^\top$ denotes the vector of repeated response measurements for subject i measured at m_i time points, and X_i denotes the

DOI: 10.1201/9780429090936-3

associated $m_i \times p$ matrix of covariates for subject i, $i = 1, \ldots, n$,

$$X_i = \begin{bmatrix} x_{i11} & \ldots & x_{i1p} \\ \vdots & & \vdots \\ x_{ij1} & \ldots & x_{ijp} \\ \vdots & & \vdots \\ x_{im_i1} & \ldots & x_{im_ip} \end{bmatrix} = \begin{bmatrix} x_{i1}^{\mathsf{T}} \\ \vdots \\ x_{im_i}^{\mathsf{T}} \end{bmatrix}.$$

3.2.1 Regression models

One useful approach is to model the conditional expected value of the response by a linear or nonlinear regression model, and to couple this with some assumption on the distribution of the response. To illustrate, consider the linear regression model,

$$Y_{ij} = \mu_{ij} + \epsilon_{ij} = x_{ij}^{\mathsf{T}}\beta + \epsilon_{ij}, \quad j = 1, \ldots, m_i, i = 1, \ldots, n, \tag{3.1}$$

where β is a $p \times 1$ vector of unknown parameters. For simplicity, no intercept is explicitly included in the parameter vector β; however, if needed, the intercept can be included by adding an $m_i \times 1$ unit vector as the first column of the matrix X_i. The total number of observations is $M = \sum_{i=1}^{n} m_i$.

We can consider several distributional assumptions for $\epsilon_i = (\epsilon_{i1}, \ldots, \epsilon_{im_i})^{\mathsf{T}}$ to model dependence within subjects, as illustrated in the following examples.

Example 3.1 (Independence) The simplest assumption is $\epsilon_i \sim N(0, \sigma^2 \mathcal{I}_{m_i \times m_i})$. The likelihood function for (β, σ^2) is readily derived, and the maximum likelihood estimators of β and σ^2 are

$$\widehat{\beta} = \left(\sum_{i=1}^{n} X_i^{\mathsf{T}} X_i \right)^{-1} \sum_{i=1}^{n} X_i^{\mathsf{T}} y_i, \quad \widehat{\sigma}^2 = \frac{1}{M} \sum_{i=1}^{n} (y_i - X_i\widehat{\beta})^{\mathsf{T}} (y_i - X_i\widehat{\beta});$$

see also Example 2.6. □

Example 3.2 (Correlation) One way to model within-subject correlation is to assume $\epsilon_i \sim N\{0, \sigma^2 R_i(\alpha)\}$, where α parametrizes the form of the correlation. Some commonly used specifications for $R_i(\alpha)$ are:

(a) exchangeable: $R_i(\alpha) = \alpha 1_{m_i} 1_{m_i}^{\mathsf{T}} + (1 - \alpha)\mathcal{I}_{m_i \times m_i}$, where 1_{m_i} is the vector of m_i ones, so $\mathrm{cor}(\epsilon_{ij}, \epsilon_{ik}) = \alpha$ for any $j \neq k$, where $-1/(m_i - 1) < \alpha < 1$, $i = 1, \ldots, n$;

(b) autoregressive of order one, AR(1): the (j, k) element of $R_i(\alpha)$ is $\mathrm{cor}(\epsilon_{ij}, \epsilon_{ik}) = \alpha^{|j-k|}$, for $|\alpha| < 1$. This is a symmetric Toeplitz matrix;

(c) moving average of order one, MA(1): the (j, k) element of $R_i(\alpha)$, $\mathrm{cor}(\epsilon_{ij}, \epsilon_{ik})$, is equal to one if $|j - k| = 0$, to α if $|j - k| = 1$ and to zero if $|j - k| > 1$, where $|\alpha| < 1/2$;

(d) unstructured: an arbitrary symmetric correlation matrix $R_i(\alpha)$ for which $\mathrm{cor}(\epsilon_{ij}, \epsilon_{ik}) = \mathrm{cor}(\epsilon_{ik}, \epsilon_{ij}) = \alpha_{jk}$ for any $j \neq k$, where $0 < \alpha_{jk} < 1$, and α includes all α_{jk} with $j \neq k$.

The log-likelihood function for $\theta = (\beta^{\mathsf{T}}, \sigma^2, \alpha^{\mathsf{T}})^{\mathsf{T}}$ is

$$\ell(\theta) = -\frac{M}{2} \log \sigma^2 - \frac{1}{2} \sum_{i=1}^{n} \log |R_i(\alpha)| - \frac{1}{2\sigma^2} \sum_{i=1}^{n} (y_i - X_i\beta)^{\mathsf{T}} R_i^{-1}(\alpha)(y_i - X_i\beta). \tag{3.2}$$

For fixed α, the constrained maximum likelihood estimates of β and σ^2 are

$$\widehat{\beta}_\alpha = \left\{\sum_{i=1}^{n} X_i^{\mathsf{T}} R_i^{-1}(\alpha) X_i\right\}^{-1} \sum_{i=1}^{n} X_i^{\mathsf{T}} R_i^{-1}(\alpha) y_i;$$

$$\widehat{\sigma}^2_\alpha = \frac{1}{M}\sum_{i=1}^{n}(y_i - X_i\widehat{\beta}_\alpha)^{\mathsf{T}} R_i^{-1}(\alpha)(y_i - X_i\widehat{\beta}_\alpha).$$

The maximum likelihood estimate of α is obtained by maximizing the profile log-likelihood function

$$\ell_{\mathrm{p}}(\alpha) = -\frac{M}{2}\log \widehat{\sigma}^2(\alpha) - \frac{1}{2}\sum_{i=1}^{n} \log |R_i(\alpha)|.$$

Explicit expressions for the inverse and determinant are available for the exchangeable correlation matrix in (a) above:

$$R_i^{-1}(\alpha) = \frac{1}{(1-\alpha)}\left\{\mathcal{I}_{m_i\times m_i} - \frac{\alpha}{1+(m_i-1)\alpha} 1_{m_i} 1_{m_i}^{\mathsf{T}}\right\},$$
$$|R_i(\alpha)| = (1-\alpha)^{m_i-1}\left\{1+(m_i-1)\alpha\right\}. \tag{3.3}$$

These expressions simplify computation of the maximum likelihood estimates. The exchangeable correlation model requires $-1/(m_{(1)}-1) < \alpha < 1$, where $m_{(1)} = \min\{m_i : i = 1, \ldots, n\}$.

Expressions for the inverse and determinant of the AR(1) correlation matrix in (b) above are also available:

$$R_i^{-1}(\alpha) = \frac{1}{1-\alpha^2}\begin{bmatrix} 1 & -\alpha & 0 & \cdots & \cdots & 0 \\ -\alpha & 1+\alpha^2 & -\alpha & \cdots & \cdots & 0 \\ 0 & -\alpha & 1+\alpha^2 & \cdots & \cdots & 0 \\ \vdots & \vdots & \ddots & \ddots & \ddots & 0 \\ 0 & \cdots & \cdots & -\alpha & 1+\alpha^2 & -\alpha \\ 0 & \cdots & \cdots & 0 & -\alpha & 1 \end{bmatrix},$$
$$|R_i(\alpha)| = (1-\alpha^2)^{m_i-1}. \tag{3.4}$$

□

The validity of inference based on the classical theory of Chapter 2 in the model (3.1) relies on the appropriate specification of the correlation matrix, and the correct distributional assumption for ϵ_i. Chapter 4 discusses inference when aspects of the model, such as the correlation structure, are misspecified. For example, if the assumption of independence in Example 3.1 is only tentative, while $\widehat{\beta}$ is unbiased for β, its variance must account for within-subject dependence; see Example 4.8.

For robust inference against violations of the distributional assumption for ϵ_i, such as the normality assumption considered here, marginal models that focus on partial features of the response distribution, such as the expected value and variance, can be used. Quasi-likelihood functions and estimating equations are then particularly useful; these are discussed in Chapter 5.

The interpretation of μ_{ij} as the conditional expected value of the response Y_{ij}, given the covariates, needs some care when the covariate values are also changing with time, or more generally with the index j. Here we assume $\mathrm{E}(Y_{ij} \mid X_i) = \mathrm{E}(Y_{ij} \mid x_{ij})$, i.e., at time point j the expected value of the response cannot depend on values of the covariates at other time-points. Without this assumption, the score function may not have expected value zero; see Example 5.5 and the bibliographic notes for Chapter 5.

3.2.2 Random effects models

An alternative approach for modelling longitudinal data introduces random effects into the linear predictor to account for unobserved heterogeneity, as in Example 1.13. These effects induce correlation between repeated measurements within the same unit and can be combined with the correlation models discussed in Example 3.2.

Let b_i be an unobserved random effect or a vector of random effects associated with the ith response, capturing subject-specific variability beyond what is explained by the covariates. We interpret μ_{ij} as the conditional expected value of Y_{ij}, that is $\mu_{ij} = \mathrm{E}(Y_{ij} \mid x_{ij}, b_i)$. In a generalized linear mixed model, the responses Y_{ij} are assumed to be conditionally independent given random effects b_i and the covariates, for $j = 1, \ldots, m_i$. This implies that dependence among repeated measurements is fully captured through the random effects and covariates. In longitudinal data analysis, random effects b_i are often assumed to be independent across subjects. We extend the generalized linear model (1.20) by incorporating a linear predictor with random effects:

$$g(\mu_{ij}) = x_{ij}^{\mathsf{T}}\beta + z_{ij}^{\mathsf{T}} b_i, \tag{3.5}$$

where g is the link function, and z_{ij} either reflects the study design or is a known covariate vector, typically a subset of the covariate vector x_{ij}. A common distributional assumption for the random effects is $b_i \sim N\{0, D(\alpha)\}$, where $D(\alpha)$ is a covariance matrix depending on a vector of variance component parameters α.

The components of β are referred to as fixed effects or fixed covariate effects. The kth component, β_k, describes the change of the transformed expected response $g(\mu_{ij})$ as covariate x_{ijk} increases by one unit, with other covariates held fixed. The model may also include a dispersion parameter ϕ, as in standard generalized linear models.

3.2.3 Conditional likelihood inference

In some random effects models, it is possible to identify a sufficient statistic S_i for the random effect b_i, effectively treating the b_i as unknown parameters. Consequently, inference on the regression coefficients can be based on the conditional likelihood function computed from the conditional distribution of the responses given the observed value of the sufficient statistic $s = (s_1, \ldots, s_n)^{\mathsf{T}}$ and the covariates, as discussed in Section 2.2.

Example 3.3 (Linear mixed effects models) An important special case where a sufficient statistic for the random effect is available is the linear model with random intercept:

$$Y_{ij} = \mu_{ij} + \epsilon_{ij} = \beta_0 + x_{ij}^{\mathsf{T}}\beta + b_i + \epsilon_{ij},$$

where ϵ_{ij} are independent normal errors with zero expected value and variance σ^2, and independent of X_i, and the intercept β_0 is explicitly included. The statistic $S_i = Y_{i\bullet} = \Sigma_j Y_{ij}$ is sufficient for $\gamma_i = \beta_0 + b_i$, indicating that the fixed and random intercept are not separately estimable. In the more general linear mixed model,

$$\mu_{ij} = \beta_0 + x_{ij}^{\mathsf{T}}\beta + z_{ij}^{\mathsf{T}} b_i,$$

the sufficient statistic for b_i is $S_i = \Sigma_j z_{ij} Y_{ij}$. If z_{ij} includes a constant term, its coefficient is confounded with β_0 in the conditional likelihood function. □

Example 3.4 (Logistic regression with random intercept) Let Y_{ij} be a binary response taking values 0 or 1, with probabilities $1 - \mu_{ij}$ and μ_{ij}, respectively, and suppose

$$\text{logit}\, \mu_{ij} = \beta_0 + x_{ij}^{\mathsf{T}}\beta + b_i.$$

Let $\gamma_i = \beta_0 + b_i$, for $i = 1, \ldots, n$. The conditional distribution of the responses is

$$f(y;\beta,\gamma) \propto \prod_{i=1}^{n} \exp\left[\gamma_i \sum_{j=1}^{m_i} y_{ij} + \sum_{j=1}^{m_i} y_{ij} x_{ij}^{\mathsf{T}}\beta - \sum_{j=1}^{m_i} \log\{1 + \exp(x_{ij}^{\mathsf{T}}\beta + \gamma_i)\}\right],$$

and $S_i = Y_{i\bullet}$ is sufficient for γ_i. The conditional likelihood function given $S = (S_1, \ldots, S_n)^{\mathsf{T}}$ is

$$L_{\mathrm{c}}(\beta) = \prod_{i=1}^{n} \frac{\exp\left(\sum_{j=1}^{m_i} y_{ij} x_{ij}^{\mathsf{T}}\beta\right)}{\sum_{\{y_i' \in \mathcal{A}_i\}} \exp\left(\sum_{j=1}^{m_i} y_{ij}' x_{ij}^{\mathsf{T}}\beta\right)},$$

where $\mathcal{A}_i$ is the set of vectors $y_i' \in \{0,1\}^{m_i}$ for which $y_{i\bullet}' = y_{i\bullet}$. This model, without a random intercept, was discussed in Example 2.9.

□

Example 3.5 (Log-linear model with random intercept) Consider count variables Y_{ij} that follow a Poisson distribution with the conditional mean μ_{ij}, modeled by a log-linear model with a random intercept,

$$\log \mu_{ij} = \beta_0 + x_{ij}^{\mathsf{T}}\beta + b_i.$$

Again $S_i = Y_{i\bullet} = \Sigma_j Y_{ij}$ is sufficient for $\beta_0 + b_i$, and the conditional likelihood function for β is

$$L_{\mathrm{c}}(\beta) = \prod_{i=1}^{n}\left[\binom{y_{i\bullet}}{y_{i1}, \ldots, y_{im_i}} \prod_{j=1}^{m_i}\left\{\frac{\exp(x_{ij}^{\mathsf{T}}\beta)}{\sum_{k=1}^{m_i}\exp(x_{ik}^{\mathsf{T}}\beta)}\right\}^{y_{ij}}\right],$$

which is a multinomial model with $y_{i\bullet}$ trials and class probabilities

$$p_{ij} = \frac{\exp(x_{ij}^{\mathsf{T}}\beta)}{\sum_{k=1}^{m_i}\exp(x_{ik}^{\mathsf{T}}\beta)}, \quad j = 1, \ldots, m_i,\ i = 1, \ldots, n.$$

This result is a specific instance of the so-called "Poisson trick", which allows multinomial models to be estimated via a Poisson likelihood (McCullagh and Nelder, 1989, Section 6.4).

□

3.2.4 Integrating over random effects

Inference based on the conditional likelihood function in random effects models has several limitations. Sufficient statistics for the random effects may not be available, and even when they are, the intercept often cannot be identified. This creates problems in interpretation, especially for categorical covariates, where the intercept typically represents a reference level. Furthermore, the conditional likelihood approach does not allow estimation of the parameters of the random effects distribution. In some cases, these parameters are of interest, for example, for predicting future responses.

For these reasons, inference in generalized linear random effects models is often based on the likelihood function obtained by integrating out the random effects from the joint distribution of Y_i and b_i. Suppose Y_{ij} follows a generalized linear model, with conditional density given b_i as in (1.20) and linear predictor (3.5). The likelihood function for $\theta = (\beta^{\mathsf{T}}, \phi, \alpha^{\mathsf{T}})^{\mathsf{T}}$ is

$$L(\theta) = \prod_{i=1}^{n} \int_{\mathbb{R}^d} f(b_i;\alpha) \prod_{j=1}^{m_i} f(y_{ij} \mid b_i, x_i, z_i; \beta, \phi)\, \mathrm{d}b_i, \tag{3.6}$$

where d is the dimension of the random effects. Here, conditional independence among the components in Y_i is assumed given random effects b_i and subject-level covariates $X_i = x_i, Z_i = z_i$. For time-varying covariates, an additional assumption is often made, such as

$$f(y_{ij} \mid b_i, x_i, z_i; \beta, \phi) = f(y_{ij} \mid b_i, x_{ij}, z_{ij}; \beta, \phi).$$

This reflects the focus of regression models on covariates at the same time point as Y_{ij}, rather than on past or future values. When regression models incorporate the dependence of the mean response Y_{ij} on its historical covariates up to and including time point j, a similar assumption is made by replacing x_{ij} and z_{ij} with their respective histories, $\{x_{i1}, \ldots, x_{i,j-1}, x_{ij}\}$ and $\{z_{i1}, \ldots, z_{i,j-1}, z_{ij}\}$. Without such an assumption, the likelihood function in (3.6) cannot be constructed by directly multiplying $f(y_{ij} \mid b_i, x_{ij}, z_{ij}; \beta, \phi)$ over all time points $j = 1, \ldots, m_i$. This assumption is often implicit or overlooked and aligns with the interpretation of μ_{ij}, noted at the end of Section 3.2.1.

When the marginal distribution of the random effects is conjugate to the conditional distribution of the responses given the random effects and covariates, (3.6) has a closed form. An important special case is the linear mixed effects model with normal random effects.

Example 3.6 (Linear mixed effects model, cont'd) Consider a more general form of the linear mixed effects model of Example 3.3:

$$\begin{aligned} Y_i &= X_i\beta + Z_i b_i + \epsilon_i, \\ b_i &\sim N\{0, D(\alpha)\}, \quad \epsilon_i \sim N(0, \Sigma_i), \quad i = 1, \ldots, n, \end{aligned} \tag{3.7}$$

where X_i is the $m_i \times (p+1)$ fixed-effects design matrix, and Z_i is the $m_i \times d$ random-effects design matrix. The within-subject error terms are usually assumed to be independent, i.e., $\Sigma_i = \sigma^2 \mathcal{I}_{m_i \times m_i}$. The marginal distribution of Y_i is multivariate normal with expected value $X_i\beta$ and covariance matrix

$$V_i(\psi) = Z_i D(\alpha) Z_i^{\mathsf{T}} + \sigma^2 \mathcal{I}_{m_i \times m_i}, \tag{3.8}$$

with $\psi = (\sigma^2, \alpha^{\mathsf{T}})^{\mathsf{T}}$. The corresponding likelihood function for $\theta = (\beta^{\mathsf{T}}, \psi^{\mathsf{T}})^{\mathsf{T}}$ is

$$L(\theta) \propto \prod_{i=1}^{n} |V_i(\psi)|^{-1/2} \exp\left\{-\frac{1}{2}(y_i - X_i\beta)^{\mathsf{T}} V_i^{-1}(\psi)(y_i - X_i\beta)\right\}.$$

Maximizing the likelihood function with respect to β for fixed ψ gives

$$\widehat{\beta}_\psi = \left\{\sum_{i=1}^{n} X_i^{\mathsf{T}} V_i^{-1}(\psi) X_i\right\}^{-1} \sum_{i=1}^{n} X_i^{\mathsf{T}} V_i^{-1}(\psi) y_i. \tag{3.9}$$

The profile likelihood $L_{\mathrm{p}}(\psi) = L(\widehat{\beta}_\psi, \psi)$ needs to be maximized numerically to compute the maximum likelihood estimate $\widehat{\psi} = (\widehat{\sigma}^2, \widehat{\alpha}^{\mathsf{T}})^{\mathsf{T}}$. □

Beyond the linear mixed model, assuming normally distributed random effects does not allow for analytical calculation of the integrals in (3.6). One alternative is to use the Laplace approximation to the integral, introduced in Section 1.4.

Example 3.7 (Laplace approximation) The first-order Laplace approximation to the integral in (3.6) is the basic method used in most software for fitting generalized mixed effects models. The ith integral in (3.6) is written in more concise notation:

$$L_i(\theta) = \int_{\mathbb{R}^d} \exp\{-h_i(b_i; \theta)\}\, \mathrm{d}b_i, \quad i = 1, \ldots, n, \tag{3.10}$$

where

$$h_i(b_i;\theta) = -\log f(y_i, b_i;\theta) = -\log\left\{f(b_i;\alpha)\prod_{j=1}^{m_i} f(y_{ij} \mid b_i, x_{ij};\beta,\phi)\right\}.$$

The Laplace approximation to the d-dimensional integral is based on a Taylor expansion of $h_i(b_i;\theta)$ for fixed θ. Let $\widehat{b}_i$ be the value that minimizes $h_i(b_i;\theta)$ and let $h_i''(\widehat{b}_i;\theta)$ be the Hessian of $h_i(b_i;\theta)$ evaluated at $\widehat{b}_i$. The dependence of $\widehat{b}_i$ on θ is suppressed for ease of notation. Then

$$\begin{aligned} L_i(\theta) &\doteq \int_{\mathbb{R}^d} \exp\left\{-h_i(\widehat{b}_i;\theta) - \frac{1}{2}(b_i - \widehat{b}_i)^{\mathsf{T}} h_i''(\widehat{b}_i;\theta)(b_i - \widehat{b}_i)\right\} \mathrm{d}b_i \\ &= (2\pi)^{d/2}|h_i''(\widehat{b}_i;\theta)|^{-1/2}\exp\{-h_i(\widehat{b}_i;\theta)\}, \end{aligned}$$

and the corresponding Laplace approximation to the integrated log-likelihood function is, ignoring additive constants,

$$\widetilde{\ell}(\theta) \doteq -\frac{1}{2}\sum_{i=1}^{n}\log|h_i''(\widehat{b}_i;\theta)| - \sum_{i=1}^{n} h_i(\widehat{b}_i;\theta).$$

This is equal to the exact marginal likelihood function in the linear mixed model with normally distributed random effects. If $h_i(\cdot)$ is a smooth function of b_i with a unique minimizer, and the dimension d of the random effect is fixed, then the approximation error is $|\widetilde{\ell}(\theta) - \ell(\theta)| = O(\Sigma_i m_i^{-1})$. The approximation may be poor if the m_i are not very large. In principle, more accurate results can be obtained using a higher-order Laplace approximation, derived from a Taylor expansion with $k > 2$ terms. The corresponding Laplace approximation to the log-likelihood function will have an approximation error of $O(\Sigma_i m_i^{-k})$. □

Example 3.8 (Crossed random effects) Crossed random effects can be used to model data classified according to two or more factors. For example, in recommendation systems, observations may be classified in terms of customers and items sold. Let Y_{ij} be a response variable with indices corresponding to a row effect $i = 1,\ldots,r$ and a column effect $j = 1,\ldots,c$. The linear predictor for a generalized linear model with crossed random effects is

$$g(\mu_{ij}) = x_{ij}^{\mathsf{T}}\beta + a_i + b_j,$$

where $\mu_{ij} = \mathrm{E}(Y_{ij} \mid x_{ij}, a_i, b_j)$. For example, in recommendation systems the random effects a_i induce correlation between reviews of two items from the same customer, and the random effects b_j induce correlation between reviews of the same item by two customers. We assume Y_{ij} are independent conditional on a_i, b_j, and the covariates. Row and column random effects are usually assumed to be mutually independent and normally distributed, $a = (a_1,\ldots,a_r)^{\mathsf{T}} \sim N(0,\sigma_a^2\mathcal{I}_{r\times r})$ and $b = (b_1,\ldots,b_c)^{\mathsf{T}} \sim N(0,\sigma_b^2\mathcal{I}_{c\times c})$.

The likelihood function for $\theta = (\beta^{\mathsf{T}},\phi,\sigma_a^2,\sigma_b^2)^{\mathsf{T}}$ based on the marginal distribution of the responses is a cumbersome integral of dimension $d = r + c$,

$$L(\theta) = \int_{\mathbb{R}^d}\left\{\prod_{i=1}^{r}\prod_{j=1}^{c} f(y_{ij} \mid a_i, b_j;\beta,\phi)^{\delta_{ij}}\right\}\prod_{i=1}^{r} f(a_i;\sigma_a^2)\prod_{j=1}^{c} f(b_j;\sigma_b^2)\,\mathrm{d}a\,\mathrm{d}b,$$

where $\delta_{ij} = 1$ if the cross-classified pair (i,j) is observed and 0 otherwise. The total number of observations is $n = \Sigma_{ij}\delta_{ij}$. Approximating this high-dimensional integral is substantially more complicated than approximating the n low-dimensional integrals in the previous example. For example, the validity of the Laplace approximation to the marginal likelihood function requires $d = o(n^{1/3})$, but in typical applications of cross-classified data, $n \ll rc$, which makes $\max\{r,c\} \geq n^{1/2}$ and thus $d = (r+c) \geq n^{1/2}$.

□

It is sometimes of interest to consider estimation of the individual random effects, although the term prediction is often preferred to distinguish the problem from estimation of parameters. In the linear mixed model of Example 3.6, the best linear unbiased predictor minimizes the mean-squared error

$$\mathrm{E}[\{\widetilde{b}(y) - b\}^{\mathsf{T}}\{\widetilde{b}(y) - b\}]; \tag{3.11}$$

see Exercise 3.5. This will be a function of the parameters in the model (3.7), which will need to be estimated. Mean-squared error for prediction is considered in more detail in Section 7.6.

Generalized linear mixed models in which the distribution of the random effects is not restricted to be normal are called hierarchical generalized linear models in Lee and Nelder (1996). The joint likelihood function for the unknown parameters and the random effects is an extended likelihood function, as defined in Section 1.5, also called an h-likelihood function. Unlike a likelihood function for a parameter, the h-likelihood function changes with monotone transformations of the random effects, and the resulting inference can change as well. Jin and Lee (2021) discussed this in detail and described canonical choices for modelling the random effects in different generalized linear models.

3.2.5 Restricted maximum likelihood

Consider again the linear mixed effects model (3.7). In Example 3.6, maximum likelihood estimates of β and the covariance parameters $\psi = (\sigma^2, \alpha^{\mathsf{T}})^{\mathsf{T}}$ were computed from the marginal distribution of the observations Y_i. The restricted maximum likelihood approach constructs a likelihood function for ψ based on the marginal distribution of a linear transformation of the full $M \times 1$ response vector Y ($M = \Sigma_i m_i$) created by stacking the column vectors $Y_1, \dots, Y_n$.

Suppose there exists a full-rank matrix A of dimension $M \times \{M - (p+1)\}$ with columns orthogonal to the columns of the matrix X formed by stacking the $m_i \times (p+1)$ matrices $X_1, \dots, X_n$. Then the random vector $T = A^{\mathsf{T}}Y$ follows a normal distribution with expected value zero and covariance matrix $A^{\mathsf{T}}V(\psi)A$, where $V(\psi)$ is a block diagonal matrix, with ith diagonal $V_i(\psi)$. The marginal distribution of T is free of β, and thus provides a marginal likelihood as in Section 2.2. The marginal log-likelihood function is

$$\ell_{\mathrm{m}}(\psi) = \ell_{\mathrm{p}}(\psi) - \frac{1}{2}\log\left|\sum_{i=1}^{n} X_i^{\mathsf{T}} V_i^{-1}(\psi) X_i\right|, \tag{3.12}$$

where $\ell_{\mathrm{p}}(\psi) = \ell(\widehat{\beta}_\psi, \psi)$ is the profile log-likelihood function for ψ, and $\widehat{\beta}_\psi$ is given in (3.9).

The marginal log-likelihood function does not depend on the choice of A as long as A is full-rank and its columns are orthogonal to those of X. The maximizer $\widehat{\psi}_{\mathrm{REML}}$ of (3.12) is called the restricted maximum likelihood estimator. Once $\widehat{\psi}_{\mathrm{REML}}$ is obtained, the REML estimator of β is the generalized least squares estimator

$$\widehat{\beta}_{\mathrm{REML}} = \left\{\sum_{i=1}^{n} X_i^{\mathsf{T}} V_i^{-1}(\widehat{\psi}_{\mathrm{REML}}) X_i\right\}^{-1} \sum_{i=1}^{n} X_i^{\mathsf{T}} V_i^{-1}(\widehat{\psi}_{\mathrm{REML}}) Y_i.$$

The marginal log-likelihood function (3.12) is also a penalized profile log-likelihood function with penalty term

$$-\frac{1}{2}\log\left|\sum_{i=1}^{n} X_i^{\mathsf{T}} V_i^{-1}(\psi) X_i\right| = \frac{1}{2}\log|\mathrm{cov}(\widehat{\beta}_\psi)|.$$

The determinant $|\text{cov}(\widehat{\beta}_\psi)|$ is the generalized variance of $\widehat{\beta}_\psi$ and is an overall summary of the estimation uncertainty in $\widehat{\beta}_\psi$. In this sense the penalty term in the marginal likelihood function adjusts inference on ψ for the effective degrees of freedom, as in Example 2.6. Penalized likelihood is the topic of Chapter 7.

Example 3.9 (Bias of maximum likelihood estimator) As a simple comparison of restricted maximum likelihood estimation to maximum likelihood estimation, suppose in the linear effects model (3.7) that there are no random effects, and that the cluster sizes are all identical, $m_i = m$ for $i = 1, \ldots, n$. Then $\psi = \sigma^2$, and the maximum likelihood estimators of β and σ^2 are

$$\widehat{\beta} = \left(\sum_{i=1}^n X_i^\mathsf{T} X_i\right)^{-1} \sum_{i=1}^n X_i^\mathsf{T} Y_i, \quad \widehat{\sigma}^2 = \frac{1}{nm} \sum_{i=1}^n (Y_i - X_i\widehat{\beta})^\mathsf{T} (Y_i - X_i\widehat{\beta}).$$

The maximum likelihood estimator $\widehat{\sigma}^2$ is biased, as

$$E(\widehat{\sigma}^2) = \left\{\frac{mn - (p+1)}{mn}\right\} \sigma^2. \tag{3.13}$$

In contrast, the restricted maximum likelihood estimator of σ^2 replaces mn by the residual degrees of freedom $\{mn - (p+1)\}$ in the denominator of $\widehat{\sigma}^2$, leading to an unbiased estimator:

$$\widehat{\sigma}^2_{\text{REML}} = \frac{1}{mn - (p+1)} \sum_{i=1}^n \sum_{j=1}^m (Y_{ij} - x_{ij}^\mathsf{T} \widehat{\beta})^2.$$

In this simple example, the restricted maximum likelihood estimate of β coincides with the maximum likelihood estimate. □

Example 3.10 (Meta-analysis) Random effect models are frequently used in meta-analysis for research synthesis. Let Y_i be the estimator of a target quantity μ in study $i = 1, \ldots, n$. The random-effects meta-analysis model is

$$Y_i = \mu_i + \epsilon_i, \quad \epsilon_i \sim N(0, \sigma_i^2),$$
$$\mu_i = \mu + \eta_i, \quad \eta_i \sim N(0, \tau^2),$$

where μ_i, the quantity actually measured in study i, is allowed to differ from the target quantity μ. The two error terms ϵ_i and η_i are usually assumed to be independent. In many applications, the sample size in each study is large enough to assume that the within-study variability σ_i^2 is practically equal to its estimate. This meta-analysis model corresponds to the linear mixed model $Y_i = \beta + b_i + \epsilon_i$ where $\beta = \mu$ and $b_i = \eta_i$.

The model parameters are β and the between-study variance τ^2. For fixed τ^2, the maximum likelihood estimate of β is the weighted average

$$\widehat{\beta}_{\tau^2} = \frac{\sum_{i=1}^n w_i(\tau^2) y_i}{\sum_{i=1}^n w_i(\tau^2)}, \quad w_i(\tau^2) = \frac{1}{\sigma_i^2 + \tau^2}. \tag{3.14}$$

Estimation of τ^2 is difficult in the case where the number of studies n is small. A traditional estimate of τ^2 based on the method of moments is

$$\widehat{\tau}^2 = \max\left\{0, \frac{Q(y) - (n-1)}{\sum_{i=1}^n w_i(0) - \sum_{i=1}^n w_i^2(0) / \sum_{i=1}^n w_i(0)}\right\}, \quad w_i(0) = \frac{1}{\sigma_i^2},$$

where $Q(y) = \Sigma_i w_i(0)(y_i - \widehat{\mu}_{\text{FE}})^2$, and $\widehat{\mu}_{\text{FE}} = \Sigma_i w_i(0) y_i / \Sigma_i w_i(0)$ is the fixed-effects estimate obtained assuming $\tau^2 = 0$. This estimate is easy to compute, but is known to give inaccurate confidence intervals and p-values for β when used in (3.14). The marginal log-likelihood function for τ^2 based on (3.12) is

$$\ell_{\mathrm{m}}(\tau^2) = -\frac{1}{2}\sum_{i=1}^{n} w_i(\tau^2)(y_i - \widehat{\beta}_{\tau^2})^2 + \frac{1}{2}\sum_{i=1}^{n} \log w_i(\tau^2) - \frac{1}{2}\log\left\{\sum_{i=1}^{n} w_i(\tau^2)\right\}.$$

The restricted maximum likelihood estimator of τ^2 obtained by maximizing this function is preferred to the method-of-moments estimator or the maximum likelihood estimator, which can be badly biased if n is small. □

3.3 Time series

Time series have a natural ordering that motivates most of the methods developed for their analysis. Time series can be observed at regularly- or irregularly-spaced time points. Statistical analysis of irregularly-spaced time series is often performed with the geostatistical methods discussed in Section 3.4.1. In the remainder of this section only time series sampled at regularly-spaced discrete times are considered.

Most models and inferential methods for time series rely on replication, usually coming from an assumption of stationarity. Consider a time series $\{Y_t : t \in \mathbb{Z}\}$ with $\mathrm{E}(Y_t^2) < \infty$ for all $t \in \mathbb{Z}$. Strict stationarity, i.e., complete invariance of the distribution to time shifts, is too restrictive an assumption and is difficult to assess from a single time series. For this reason, it is usual to make an assumption of weak or second-order stationarity. The mean function of the time series is denoted by $\mu(t) = \mathrm{E}(Y_t)$ and the autocovariance function by $\gamma(s,t) = \mathrm{cov}(Y_s, Y_t)$. A time series $\{Y_t\}$ is second-order stationary if $\mu(t) = \mu$ for all $t \in \mathbb{Z}$ and the autocovariance function depends only on the time lag between any two pairs of observations, $\gamma(s,t) = \gamma(|s-t|)$, for all s and $t \in \mathbb{Z}$. Stationary processes are often described in terms of their autocorrelation function $\rho(h) = \gamma(h)/\gamma(0)$.

3.3.1 Autoregressive and moving average models

Linear time series models provide a general framework for the analysis of stationary processes, as Wold's decomposition theorem ensures that any second-order stationary process is either linear or can be rendered linear through the removal of a deterministic component (Brockwell and Davis, 2002, p.51).

A useful class of linear time series processes is the class of autoregressive and moving average models, in which

$$Y_t = \phi_1 Y_{t-1} + \cdots + \phi_p Y_{t-p} + \epsilon_t + \omega_1 \epsilon_{t-1} + \cdots + \omega_q \epsilon_{t-q}, \tag{3.15}$$

where p and q are non-negative integers, $\phi = (\phi_1, \ldots, \phi_p)^{\mathsf{T}}$ and $\omega = (\omega_1, \ldots, \omega_q)^{\mathsf{T}}$ are unknown parameters, and ϵ_t are assumed to be independent variables with $\mathrm{E}(\epsilon_t) = 0$ and $\mathrm{var}(\epsilon_t) = \sigma^2$. Model (3.15) is commonly abbreviated to ARMA(p,q). In the special case that $q = 0$, it is an autoregressive model of order p, denoted by AR(p). Conversely when $p = 0$, it is a moving-average model of order q, denoted by MA(q).

Basic properties of the ARMA(p,q) processes are derived from the autoregressive and moving average polynomials

$$\phi(z) = 1 - \phi_1 z - \cdots - \phi_p z^p, \quad \phi_p \neq 0, \tag{3.16}$$

$$\omega(z) = 1 + \omega_1 z + \cdots + \omega_q z^q, \quad \omega_q \neq 0. \tag{3.17}$$

The ARMA(p, q) process has a unique stationary solution if $\phi(z) \neq 0$ for all $|z| = 1$. A stationary process is causal if Y_t can be expressed as a function of the errors $\{\epsilon_s, s \leq t\}$. ARMA processes are causal if the roots of $\phi(z) = 0$ lie outside the unit circle $|z| > 1$.

Two conditions are needed to ensure that the ARMA model is identifiable. The first condition is that the ARMA process is invertible: $\omega(z) \neq 0$ for all $|z| \leq 1$. This condition is symmetric to that of causality, here applied to the moving average polynomial. The second condition is that the polynomials $\phi(z)$ and $\omega(z)$ must not have common factors, which can be discarded without modifying the model.

Example 3.11 (Parameter redundancy of MA(1) model) Consider the MA(1) process with normal errors,

$$Y_t = \epsilon_t + \omega_1 \epsilon_{t-1}, \quad \epsilon_t \sim N(0, \sigma^2).$$

The autocovariance function is

$$\gamma(h) = \begin{cases} (1 + \omega_1^2)\sigma^2, & h = 0, \\ \omega_1 \sigma^2, & |h| = 1, \\ 0, & |h| > 1. \end{cases} \tag{3.18}$$

Since the process is Gaussian, two models with the same autocovariance function are indistinguishable. For example, we cannot distinguish between $(\omega_1 = 1/5, \sigma^2 = 25)$ and $(\omega_1 = 5, \sigma^2 = 1)$. To resolve this ambiguity and make the model identifiable, the invertible model corresponding to $\omega_1 = 1/5$ is chosen. □

Example 3.12 (Parameter redundancy of ARMA(1, 1) model) Consider the ARMA process with $p = q = 1$,

$$Y_t = \phi_1 Y_{t-1} + \epsilon_t + \omega_1 \epsilon_{t-1}.$$

This is a stationary, causal, and invertible process if $|\phi_1| < 1$ and $|\omega_1| < 1$. These two conditions alone are not sufficient to ensure that the model can be used to analyse a time series. If the series is a pure white noise, $Y_t = \epsilon_t$, then $\phi_1 Y_{t-1} = \phi_1 \epsilon_{t-1}$ and $Y_t - \phi_1 Y_{t-1} = \epsilon_t - \phi_1 \epsilon_{t-1}$, which is an ARMA$(1, 1)$ process with $\phi_1 = -\omega_1$,

$$Y_t = \phi_1 Y_{t-1} + \epsilon_t - \phi_1 \epsilon_{t-1}.$$

If an ARMA$(1, 1)$ model is fitted to white noise, the estimate $\widehat{\phi}_1$ may be very different from zero, but there is no serial correlation. This over-parametrization of the ARMA$(1, 1)$ model can be deduced from the equality of the autoregressive and moving average polynomials, $\phi(z) = (1 - \phi_1 z)$ and $\omega(z) = (1 - \phi_1 z)$. Discarding their common factor leads to the conclusion that the model is actually white noise. This argument extends to the ARMA(p, p) models: if $\phi_j = -\omega_j$, for all $j = 1, \ldots, p$, then the ARMA(p, p) process is white noise (Shumway and Stoffer, 2016, Example 3.7). □

Consider a vector of observations $y = (y_1, \ldots, y_n)^\top$ from a stationary, causal, and invertible ARMA(p, q) process whose polynomials $\phi(z)$ and $\omega(z)$ have no common factors. The full parameter is $\theta = (\psi^\top, \sigma^2)^\top$, where $\psi = (\phi^\top, \omega^\top)^\top$ is the parameter of interest and σ^2 is the nuisance parameter. The likelihood function for θ is readily computed using the fact that a causal ARMA process has the infinite MA representation

$$Y_t = \sum_{j=0}^{\infty} \tau_j \epsilon_{t-j}, \tag{3.19}$$

whose coefficients are determined from the autoregressive and moving average polynomials (3.16)–(3.17) by solving

$$\tau(z) = \sum_{j=0}^{\infty} \tau_j z^j = \frac{\omega(z)}{\phi(z)}, \quad |z| \leq 1,$$

see, for example, Shumway and Stoffer (2016, p. 95). If the errors are normally distributed, the log-likelihood function for $\theta = (\psi^{\mathsf{T}}, \sigma^2)^{\mathsf{T}}$ is

$$\ell(\theta) = -\frac{1}{2} \log |\Gamma(\psi, \sigma^2)| - \frac{1}{2} y^{\mathsf{T}} \Gamma^{-1}(\psi, \sigma^2) y, \tag{3.20}$$

where $\Gamma(\psi, \sigma^2)$ is the autocovariance matrix of the ARMA process with (i, j) entry $\gamma(i, j) = \gamma(|i - j|)$. From the infinite moving average representation (3.19), $\gamma(h) = \sigma^2 \sum_{j=0}^{\infty} \tau_j \tau_{j+h}$, where the dependence of τ_j on ψ is implicit. Write the autocovariance matrix $\Gamma(\psi, \sigma^2) = \sigma^2 R(\psi)$, where $R(\psi)$ is the autocorrelation matrix. The constrained maximum likelihood estimate of σ^2 is $\widehat{\sigma}^2_\psi = n^{-1} y^{\mathsf{T}} R^{-1}(\psi) y$, and the profile log-likelihood function for ψ is

$$\ell_{\mathrm{p}}(\psi) = -\frac{n}{2} \log\{y^{\mathsf{T}} R^{-1}(\psi) y\} - \frac{1}{2} \log |R(\psi)|. \tag{3.21}$$

This expression for the profile likelihood is not particularly useful because computing the inverse and the determinant of $R(\psi)$ requires $O(n^3)$ operations with a standard Cholesky decomposition; see the discussion of this in the spatial case in Example 3.22. There are recursive algorithms, such as the innovations algorithm (Brockwell and Davis, 1987, Section 8.3) and the Kalman filter (Example 3.17), that enable efficient computation of the likelihood function for an ARMA(p, q) model with a linear cost of $O(n)$.

In general, $\ell_{\mathrm{p}}(\psi)$ needs to be maximized numerically to find the maximum likelihood estimate $\widehat{\psi}$ of ψ. The asymptotic distribution of $\widehat{\psi}$ is

$$\sqrt{n}(\widehat{\psi} - \psi) \xrightarrow{d} N_{p+q}\{0, \sigma^2 i^{-1}(\psi)\}, \quad i(\psi) = \begin{bmatrix} \Gamma_{\phi\phi} & \Gamma_{\phi\omega} \\ \Gamma_{\omega\phi} & \Gamma_{\omega\omega} \end{bmatrix}. \tag{3.22}$$

The $p \times p$ matrix $\Gamma_{\phi\phi}$ is the autocovariance matrix for the AR(p) process $Y_t = \sum_{i=1}^{p} \phi_i Y_{t-i} + \epsilon_t$. The $q \times q$ matrix $\Gamma_{\omega\omega}$ is the autocovariance matrix for an auxiliary AR(q) process $X_t = -\sum_{j=1}^{q} \omega_j X_{t-j} + \epsilon_t$. The $p \times q$ matrix $\Gamma_{\phi\omega}(\psi)$ is the cross-covariance between the two autoregressive processes Y_t and X_t. See Exercise 3.6.

Example 3.13 (Overfitting the AR(1) model) The asymptotic distribution (3.22) can be used to illustrate the effect of overfitting. Suppose the series is generated from an AR(1) model but the bigger AR(2) model was fitted. The asymptotic variance of $\widehat{\phi}_1$ from the incorrect AR(2) model is $n^{-1}(1 - \phi_2^2)$; see (3.76) in Exercise 3.6. This asymptotic variance is n^{-1} as $\phi_2 = 0$. The asymptotic variance of $\widehat{\phi}_1$ from the correct AR(1) model is $n^{-1}(1 - \phi_1^2)$, so the effect of overfitting is to increase the variance.

Suppose the fitted model is the ARMA(1, 1), and the true model is AR(1). The asymptotic variance of $\widehat{\phi}_1$ from formula (3.77) in Exercise 3.6 is

$$\mathrm{avar}(\widehat{\phi}_1) = \frac{(1 + \phi_1 \omega_1)^2}{(\phi_1 + \omega_1)^2} \frac{(1 - \phi_1^2)}{n} = \frac{1 - \phi_1^2}{n \phi_1^2},$$

because $\omega_1 = 0$; this is not only uniformly larger than the asymptotic variance of $\widehat{\phi}_1$ from the correct AR(1) model but also explodes as ϕ_1 approaches zero. □

Many models for non-stationary time series are constructed as the composition of a non-stationary term and a zero-mean stationary term. The simplest case, but also the most widely used in applications, is the linear model with stationary errors

$$Y_t = \mu_t + \epsilon_t. \tag{3.23}$$

When μ_t is a deterministic or stochastic trend, the process is typically brought back to stationarity by differencing. The d-order difference operator is denoted by ∇^d. For $d = 1$, $\nabla Y_t = Y_t - Y_{t-1}$. For $d = 2$, $\nabla^2 Y_t = \nabla(\nabla Y_t) = \nabla(Y_t - Y_{t-1}) = Y_t - 2Y_{t-1} + Y_{t-2}$. If the trend is linear,

$$\mu_t = \beta_0 + \beta_1 t,$$

then the first-order difference is stationary,

$$\nabla Y_t = Y_t - Y_{t-1} = \beta_1 + \nabla \epsilon_t.$$

If μ_t is the d-th order polynomial $\mu_t = \sum_{j=0}^{d} \beta_j t^j$, then $\nabla^d Y_t$ is stationary.

Differencing is used also to remove stochastic forms of non-stationarity. If the mean is a random walk, $\mu_t = \mu_{t-1} + \eta_t$, where η_t is stationary, then ∇Y_t is stationary. Similarly, in the stochastic trend model

$$\begin{aligned} \mu_t &= \mu_{t-1} + \nu_t \\ \nu_t &= \nu_{t-1} + \eta_t, \end{aligned}$$

the twice-differenced series is stationary.

Autoregressive integrated moving average models (ARIMA) of order p, d and q are ARMA(p, q) models fitted to series differenced d times. Likelihood inference for these models is typically conducted as for standard ARMA(p, q) models, assuming that the choice of the differencing order d that makes the model stationary is fixed. A more general case is when the non-stationarity component of the model is a function of some exogenous covariates.

Example 3.14 (Linear regression with ARMA errors) Exogenous covariates can be included in (3.23) through a linear model for the response,

$$Y_t = \beta_0 + \beta_1 x_{t1} + \cdots + \beta_k x_{tk} + \epsilon_t,$$

where the errors ϵ_t follow a stationary, causal and invertible ARMA(p, q) process. Let X denote the $n \times (k+1)$ matrix of the covariates for the n observations. The profile log-likelihood function $\ell_{\mathrm{p}}(\psi)$ for the linear regression with ARMA errors has the same form as (3.21) with the vector responses y replaced by residuals $y - X\widehat{\beta}_\psi$ where $\widehat{\beta}_\psi$ are the generalized least-squares estimates

$$\widehat{\beta}_\psi = \{X^\top R^{-1}(\psi) X\}^{-1} X^\top R^{-1}(\psi) y,$$

where $R(\psi)$ is the autocorrelation matrix of the ARMA errors. Scalable computation of the profile log-likelihood function can be obtained with the innovations algorithm or the Kalman Filter, applied to the residuals. □

3.3.2 State space models

A very flexible class of linear time series models is that of state space models, which consist of two components: the state process that describes the evolution over time of the underlying signal and the observation process that relates the observed series to the signal. The signal is assumed to follow a vector autoregressive process according to the state equation

$$X_t = \Phi X_{t-1} + \eta_t, \quad \eta_t \sim N_p(0, \Gamma), \tag{3.24}$$

where X_t is a random vector of length p, Φ is a $p \times p$ matrix and Γ is a $p \times p$ covariance matrix. Observations are related to states according to the observation equation,

$$Y_t = A_t X_t + \epsilon_t, \quad \epsilon_t \sim N_m(0, \Sigma), \tag{3.25}$$

where A_t is a $m \times p$ matrix and Σ is a $m \times m$ covariance matrix. Many time series models can be written in state space form. For example, any causal and invertible ARMA process can be represented as a stationary state space model.

Example 3.15 (ARMA(1, 1) model) Consider the causal and invertible ARMA(1, 1) model

$$Y_t = \phi_1 Y_{t-1} + \omega_1 \epsilon_{t-1} + \epsilon_t, \quad |\phi_1| < 1, \ |\omega_1| < 1.$$

A state space form of the model is

$$\begin{aligned} X_t &= \phi_1 X_{t-1} + \eta_t, \\ Y_t &= X_t + \epsilon_t, \end{aligned}$$

under the constraint $\eta_t = (\phi_1 + \omega_1)\epsilon_{t-1}$. □

The main objective of state space modelling is usually to reconstruct the state X_t of the underlying signal. An efficient approach to this is provided by the Kalman filter. Assume that η_t and ϵ_t are independent white noise processes, and that the initial state $X_0 \sim N_p(\mu_{0|0}, \Omega_{0|0})$. For $s \leq t$, define the conditional expectation and variance of X_t as $\mu_{t|s} = \mathrm{E}(X_t \mid \mathcal{F}_s)$ and $\Omega_{t|s} = \mathrm{var}(X_t \mid \mathcal{F}_s)$, respectively, where $\mathcal{F}_s$ is the history of the series to time s.

The Kalman filter iterates between a forecast and a filter step as shown in Figure 3.1. The forecast step consists in computing the optimal predictor of X_t given $\mathcal{F}_{t-1}$. Under the assumption that the state and observations are normally distributed, this optimal predictor is $\mu_{t|t-1}$ with variance $\Omega_{t|t-1}$. If the state and observations are not normally distributed, $\mu_{t|t-1}$ is the best linear unbiased predictor of X_t. The forecast equations follow directly from the state equation (3.24),

$$\begin{aligned} \mu_{t|t-1} &= \Phi \mu_{t-1|t-1}, \\ \Omega_{t|t-1} &= \Phi \Omega_{t-1|t-1} \Phi^{\mathsf{T}} + \Gamma. \end{aligned}$$

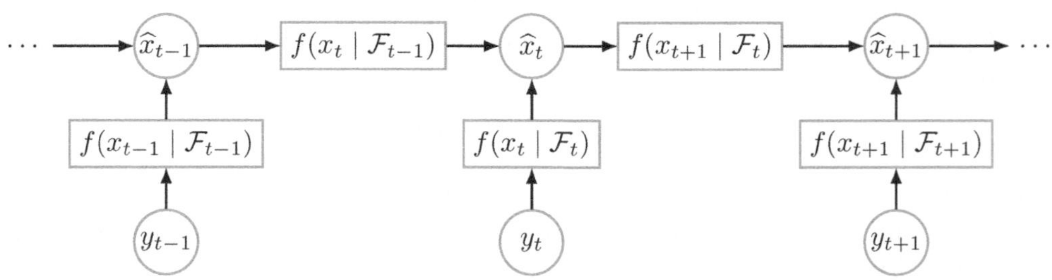

FIGURE 3.1
The Kalman filter algorithm. The forecast step is represented along the horizontal path in which the unknown states X_t are predicted using information available up to time $t-1$. In the filter step, represented along the vertical paths, the predicted states $\mu_{t|t-1}$ are updated with the current observations y_t to obtain the filtered estimates $\widehat{x}_t$.

In the filter step, the predictor of X_t is updated with observation y_t, and we denote the filtered state as $\widehat{x}_t = \mu_{t|t}$ with variance $\Omega_{t|t}$. The filter equations are

$$\mu_{t|t} = \mu_{t|t-1} + K_t(y_t - A_t\mu_{t|t-1}),$$
$$\Omega_{t|t} = (\mathcal{I}_{p\times p} - K_t A_t)\Omega_{t|t-1},$$

where $K_t = \Omega_{t|t-1}A_t^\mathsf{T}(A_t\Omega_{t|t-1}A_t^\mathsf{T} + \Sigma)^{-1}$ is called the Kalman gain coefficient.

Example 3.16 (AR(1) model with additive error) Consider the one-dimensional state space model

$$X_t = \phi X_{t-1} + \eta_t, \quad \eta_t \sim N(0, \gamma^2),$$
$$Y_t = X_t + \epsilon_t, \quad \epsilon_t \sim N(0, \sigma^2).$$

When $\phi = 1$, the state equation corresponds to a random walk and the state space model is called the local level model. Assume the initial state $X_0 \sim N(\mu_{0|0}, \omega_{0|0}^2)$. The forecast density at $t = 1$ is the marginal density

$$f(x_1) = \int_{-\infty}^{\infty} f(x_1 \mid x_0) f(x_0)\, \mathrm{d}x_0,$$

which is the density of a normal variable with expected value and variance

$$\mu_{1|0} = \phi\mu_{0|0}, \quad \omega_{1|0}^2 = \phi^2\omega_{0|0}^2 + \gamma^2.$$

In the next step of the Kalman filter observation y_1 enters. The updated prediction of X_1 given y_1 is derived from the filter density

$$f(x_1 \mid y_1) \propto f(y_1 \mid x_1) f(x_1),$$

where the observation density $f(y_1 \mid x_1)$ is normal with expected value x_1 and variance σ^2; $f(x_1)$ was computed at the previous step. The filter density is that of a normal random variable with expected value and variance

$$\mu_{1|1} = \mu_{1|0} + k_1(y_1 - \mu_{1|0}), \quad \omega_{1|1}^2 = (1 - k_1)\omega_{1|0}^2,$$

where $k_1 = \omega_{1|0}^2/(\omega_{1|0}^2 + \sigma^2)$ is the Kalman gain coefficient. At $t = 2$, the filter density is

$$f(x_2 \mid y_1, y_2) \propto f(y_2 \mid x_2) f(x_2 \mid y_1),$$

and the forecast density $f(x_2 \mid y_1)$ is

$$f(x_2 \mid y_1) = \int_{-\infty}^{\infty} f(x_1, x_2 \mid y_1)\, \mathrm{d}x_1$$
$$\propto \int_{-\infty}^{\infty} f(x_2 \mid x_1) f(x_1 \mid y_1)\, \mathrm{d}x_1.$$

The resulting filter density is normal with expected value and variance, respectively,

$$\mu_{2|2} = \mu_{2|1} + k_2(y_2 - \mu_{2|1}), \quad \omega_{2|2}^2 = (1 - k_2)\omega_{2|1}^2,$$

with $k_2 = \omega_{2|1}^2/(\omega_{2|1}^2 + \sigma^2)$. At time t, the filter density is proportional to the product of the observation and the forecast densities

$$f(x_t \mid y_1, \ldots, y_t) \propto f(y_t \mid x_t) f(x_t \mid y_1, \ldots, y_{t-1}),$$

and the forecast density is the integral

$$\begin{aligned} f(x_t \mid y_1, \ldots, y_{t-1}) &= \int_{-\infty}^{\infty} f(x_t, x_{t-1} \mid y_1, \ldots, y_{t-1}) \, \mathrm{d}x_{t-1} \\ &= \int_{-\infty}^{\infty} f(x_t \mid x_{t-1}) f(x_{t-1} \mid y_1, \ldots, y_{t-1}) \, \mathrm{d}x_{t-1}, \end{aligned}$$

which depends on the filter density computed at the previous step of the Kalman algorithm. □

Example 3.17 (Likelihood for state-space models) Let θ denote the parameter vector of the state space model. The likelihood function for θ based on a sample $\{y_1, \ldots, y_n\}$ is

$$L(\theta) = f(y_1; \theta) \prod_{t=2}^{n} f(y_t \mid \mathcal{F}_{t-1}; \theta).$$

For sufficiently long series the contribution from the first observation can be ignored and θ is estimated by maximizing the likelihood function with $f(y_1; \theta)$ excluded. If the series is not sufficiently long, then the marginal density of Y_1 is

$$f(y_1; \theta) = \int_{-\infty}^{\infty} f(y_1 \mid x_1; \theta) f(x_1; \theta) \, \mathrm{d}x_1,$$

where $f(y_1 \mid x_1; \theta)$ is the density of the observation equation and $f(x_1; \theta)$ is a normal density with mean $\mu_{1|0}$ and variance $\Omega_{1|0}$ that can be computed with the Kalman filter. From properties of the normal distribution, $f(y_1; \theta)$ is the density of a normal distribution with mean and variance

$$\mathrm{E}(Y_1) = A_1 \mu_{1|0}, \quad \mathrm{var}(Y_1) = A_1 \Omega_{1|0} A_1^{\mathsf{T}} + \Sigma.$$

The density of Y_t given $\mathcal{F}_{t-1}$ is also computed from the Kalman filter as

$$\begin{aligned} f(y_t \mid y_{t-1}, \ldots, y_1; \theta) &= \int_{-\infty}^{\infty} f(y_t, x_t \mid y_{t-1}, \ldots, y_1; \theta) \, \mathrm{d}x_t \\ &= \int_{-\infty}^{\infty} f(y_t \mid x_t; \theta) f(x_t \mid y_{t-1}, \ldots, y_1; \theta) \, \mathrm{d}x_t. \end{aligned}$$

Using again the properties of the normal distribution, it follows that Y_t given $\mathcal{F}_{t-1}$ is normally distributed, with expected value and variance

$$\mathrm{E}(Y_t \mid \mathcal{F}_{t-1}) = A_t \mu_{t|t-1}, \quad \mathrm{var}(Y_t \mid \mathcal{F}_{t-1}) = A_t \Omega_{t|t-1} A_t^{\mathsf{T}} + \Sigma.$$

Writing $r_t(\theta) = y_t - \mathrm{E}(Y_t \mid \mathcal{F}_{t-1})$ and $V_t(\theta) = \mathrm{var}(Y_t \mid \mathcal{F}_{t-1})$, with the convention that $\mathcal{F}_0$ is the empty set, the log-likelihood function is

$$\ell(\theta) = -\frac{1}{2} \sum_{t=1}^{n} \left\{ \log |V_t(\theta)| + r_t^{\mathsf{T}}(\theta) V_t^{-1}(\theta) r_t(\theta) \right\}. \tag{3.26}$$

□

The Kalman filter enables computation of the likelihood function in $O(n)$ operations. Since any causal and invertible ARMA process can be represented as a stationary state space model, the Kalman filter can be used for scalable computation of the likelihood function of an ARMA(p, q) model.

3.3.3 Nonlinear time series models

Statistical inference for nonlinear time series raises difficulties of varying degrees, and sometimes the likelihood function is intractable. Even in the simplest cases where the calculation of the likelihood function is straightforward, obtaining the maximum likelihood estimates of the model parameters may not be straightforward.

Example 3.18 (Nonlinear moving average model) Consider the nonlinear moving-average model

$$Y_t = \epsilon_t + \alpha\epsilon_{t-1} + \beta\epsilon_t\epsilon_{t-1}, \tag{3.27}$$

where the ϵ_t are independent and identically distributed random variables with expectation 0 and variance σ^2. Model (3.27) may be of interest to describe stationary time series with sudden jumps, because consecutive large values of ϵ_t may generate very large values of Y_t (Robinson, 1977).

For a sample of observations $\{y_1, \ldots, y_n\}$ and parameter $\theta = (\alpha, \beta, \sigma^2)^\mathsf{T}$, assume $\epsilon_0 = 0$, and invert the model equation (3.27) to define the recursive scaled residuals

$$e_1 = y_1, \quad e_t(\alpha, \beta) = \frac{y_t - \alpha e_{t-1}(\alpha, \beta)}{1 + \beta e_{t-1}(\alpha, \beta)}, \quad t \geq 2. \tag{3.28}$$

The Jacobian determinant for the transformation from $y = (y_1, \ldots, y_n)^\mathsf{T}$ to $e = (e_1, \ldots, e_n)^\mathsf{T}$ is

$$\left|\frac{\partial(e_1, \ldots, e_n)}{\partial(y_1, \ldots, y_n)}\right| = \prod_{t=2}^{n} \frac{1}{1 + \beta e_{t-1}(\alpha, \beta)}.$$

This gives the log-likelihood function

$$\ell(\theta) = \sum_{t=1}^{n} \log f\{e_t(\alpha, \beta); \sigma^2\} - \sum_{t=2}^{n} \log\{1 + \beta e_{t-1}(\alpha, \beta)\}.$$

Solving the likelihood equations for this model is not straightforward due to the involvement of recursive expressions. To illustrate this, suppose ϵ_t is Gaussian white noise with variance σ^2. The score functions are

$$\begin{aligned}
\frac{\partial \ell(\theta)}{\partial \alpha} &= -\sum_{t=2}^{n-1} \frac{\partial e_t}{\partial \alpha}\left(\frac{e_t}{\sigma^2} + \frac{\beta}{1 + \beta e_t}\right) - \frac{e_n}{\sigma^2}\frac{\partial e_n}{\partial \alpha}, \\
\frac{\partial \ell(\theta)}{\partial \beta} &= -\sum_{t=2}^{n-1} \frac{\partial e_t}{\partial \beta}\left(\frac{e_t}{\sigma^2} + \frac{\beta}{1 + \beta e_t}\right) - \frac{e_n}{\sigma^2}\frac{\partial e_n}{\partial \beta} - \sum_{t=2}^{n} \frac{e_{t-1}}{1 + \beta e_{t-1}}, \\
\frac{\partial \ell(\theta)}{\partial \sigma^2} &= \frac{1}{2\sigma^2}\sum_{t=1}^{n}\left(\frac{e_t^2}{\sigma^2} - 1\right),
\end{aligned} \tag{3.29}$$

where the dependence of e_t on (α, β) has been suppressed for notational convenience. The maximum likelihood estimator of σ^2 is $\widehat{\sigma}^2 = n^{-1}\Sigma_t \widehat{e}_t^2$ from (3.29), where $\widehat{e}_t$ is determined by the recursive identities (3.28), with α and β replaced by their maximum likelihood estimators. The second derivatives are cumbersome but straightforward, and the observed or expected Fisher information matrix provides estimates of the standard error of the maximum likelihood estimators as usual. Both matrices are block diagonal, as σ^2 is orthogonal to (α, β).

The development here illustrates that although it may be simple to calculate the likelihood function using recursive identities, computing the maximum likelihood estimate can be difficult, even under normality. In Chapter 5, estimation of θ is approached from a different perspective, by relinquishing some distributional assumptions. By retaining certain moment properties of the data, sensible estimating functions for θ can be formulated; see Example 5.4. □

Several approaches have been developed to extend generalized linear models (1.20) for time series data. Cox (1981) distinguished between observation-driven models and parameter-driven models. In observation-driven models, the state of the series at a given time is described as a deterministic function of lagged observations, as well as contemporaneous and lagged covariates. In parameter-driven models, the past dependence is expressed through an unobserved latent stochastic process.

Example 3.19 (Observation-driven models) In a generalized linear autoregressive moving average model, Y_t follows a generalized linear model (Example 1.12) with linear predictor

$$g(\mu_{t|t-1}) = x_t^{\mathsf{T}}\beta + Z_t,$$

where $g(\cdot)$ is a suitable link function, and Z_t is a function of past observations and covariates. In generalized linear autoregressive moving average models, Z_t is assumed to follow the ARMA(p, q) recursion

$$Z_t = \sum_{i=1}^{p} \phi_i(Z_{t-i} + e_{t-i}) + \sum_{j=1}^{q} \omega_j e_{t-j}, \tag{3.30}$$

where $e_t = (Y_t - \mu_{t|t-1})/\nu_{t|t-1}$ are called predictive scaled residuals. The scaling sequence $\nu_{t|t-1}$ depends on past information. Common choices for $\nu_{t|t-1}$ are σ_t, σ_t^2, or 1, where $\sigma_t^2 = \text{var}(Y_t \mid \mathcal{F}_{t-1})$. With these choices the predictive scaled residuals correspond to Pearson residuals, score-type residuals, and identity residuals, respectively (Dunsmuir and Scott, 2015).

Observation-driven models for time series of counts are developed in Davis et al. (2003). Suppose that Y_t given the (infinite) past is distributed as Poisson with expected value $\mu_{t|t-1} = \exp(x_t^{\mathsf{T}}\beta + Z_t)$, and the scaling sequence for Z_t is $\nu_{t|t-1}^2 = \mu_{t|t-1}$, with initial conditions $e_s = 0$ and $Y_s = 0$, for $s \leq 0$. The predictive residuals form a martingale difference sequence, and e_t are uncorrelated random variables with expected value 0 and variance 1. Approximating the marginal distribution of Z_t as a zero-mean Gaussian process, we obtain

$$\begin{aligned}
\text{E}(Y_t) &\approx \exp(x_t^{\mathsf{T}}\beta)\text{E}\{\exp(Z_t)\} \\
&= \exp\left\{x_t^{\mathsf{T}}\beta + \frac{1}{2}\text{var}(Z_t)\right\} \\
&= \exp\left\{x_t^{\mathsf{T}}\beta + \frac{1}{2}\sum_{i=1}^{\infty} \tau_i^2(\phi, \omega)\right\},
\end{aligned}$$

where $\tau_i(\phi, \omega)$ are the coefficients of the infinite moving average representation of the ARMA(p, q) model (3.30). For example, in the simplest case that Z_t follows an AR(1) model, $\tau_i = \phi^i$ and $\text{E}(Y_t) = \exp\{x_t^{\mathsf{T}}\beta + 1/(1 - \phi^2)\}$. The expression for $\text{E}(Y_t)$ shows that serial correlation inflates the expected number of counts compared with the Poisson regression model for independent observations.

Inference is typically based on the likelihood function for the data, conditioned on the initial values,

$$\ell(\theta) = \sum_{t=2}^{n} \log f(y_t \mid \mathcal{F}_{t-1}; \theta) = \sum_{t=2}^{n} [y_t\eta_t(\theta) - \exp\{\eta_t(\theta)\}], \tag{3.31}$$

where $\theta = (\beta, \phi, \omega)^\top$ and $\eta_t(\theta) = x_t^\top \beta + Z_t(\theta)$ with the dependence of Z_t on θ made explicit. Computation of $\ell(\theta)$ and the score function requires recursive expressions for $Z_t(\theta)$ and its derivatives using (3.30). Davis et al. (2003) showed that martingale central limit theorem ensures that the maximizer $\widehat{\theta}$ of $\ell(\theta)$ is consistent and $n^{1/2}(\widehat{\theta} - \theta_0) \xrightarrow{d} N\{0, i^{-1}(\theta_0)\}$, where the per-observation Fisher information $i(\theta)$ is

$$i(\theta) = \lim_{n\to\infty} \frac{1}{n} \sum_{t=2}^{n} \exp\{\eta_t(\theta)\} \frac{\partial \eta_t(\theta)}{\partial \theta} \frac{\partial \eta_t(\theta)}{\partial \theta^\top}.$$

□

Example 3.20 (Parameter-driven models) Likelihood inference is considerably more complicated in parameter-driven models where observations Y_t are assumed to be independent realizations of a generalized linear model conditional on a serially correlated latent process $\{b_t\}$. The linear predictor for Y_t given b_t is

$$g(\mu_{t|t-1}) = x_t^\top \beta + b_t, \tag{3.32}$$

for a suitable link function $g(\cdot)$, and the random intercept b_t is assumed to follow a zero-mean stationary and invertible ARMA(p, q) process,

$$b_t = \sum_{i=1}^{p} \phi_i b_{t-i} + \sum_{j=1}^{q} \omega_j \epsilon_{t-j} + \epsilon_t, \quad \epsilon_t \sim N(0, \sigma^2). \tag{3.33}$$

This is an example of generalized linear mixed model (Example 1.13) with serially correlated random effects.

If Y_t given b_t is distributed as Poisson, then the expected number of counts at time t is

$$\begin{aligned} \mathrm{E}(Y_t) &= \mathrm{E}\{\mathrm{E}(Y_t \mid b_t)\} \\ &= \exp(x_t^\top \beta)\mathrm{E}\{\exp(b_t)\} \\ &= \exp\left\{ x_t^\top \beta + \frac{\sigma^2}{2} \sum_{i=0}^{\infty} \tau_i^2(\phi, \omega) \right\}, \end{aligned}$$

where $\tau_i(\phi, \omega)$ are the coefficients of the infinite moving average representation of the ARMA$(p,\ q)$ model for the latent process (3.33). As in observation-driven models, the expected counts are inflated by the presence of serial correlation.

An important theoretical advantage of parameter-driven models is that ergodicity and mixing properties are easily derived. A disadvantage is that the likelihood function involves a high-dimensional integral. Let θ be the parameter vector that includes both the parameters in the conditional distribution of Y_t given b_t and the parameters of the assumed ARMA model for b_t. The likelihood function is

$$L(\theta) = \int_{\mathbb{R}^n} \prod_{t=1}^{n} f(y_t \mid b_t; \theta) f(b_t, \mid b_{t-1} \ldots, b_1; \theta)\, \mathrm{d}b, \tag{3.34}$$

where $b = (b_1, \ldots, b_n)^\top$. A direct application of Laplace approximation to the integral is not valid as the dimension of the integral is n; see Shun and McCullagh (1995) and Tang and Reid (2025). Several Monte Carlo methods have been proposed to efficiently approximate the likelihood function. □

Example 3.21 (Financial time series) Observation and parameter-driven models are used for modelling volatility in financial time series. Generalized autoregressive conditional heteroscedasticity models (Bollerslev, 1986) of orders p and q are observation-driven models of the form

$$Y_t = \sigma_t \epsilon_t, \quad \epsilon_t \sim N(0,1),$$
$$\sigma_t^2 = \phi_0 + \sum_{i=1}^{p} \phi_i Y_{t-i}^2 + \sum_{j=1}^{q} \omega_j \sigma_{t-j}^2,$$

where $\phi_0 > 0$ and the coefficients ϕ_i and ω_j are constrained to ensure $\sigma_t^2 > 0$ and stationarity.

The parallel parameter-driven approach to financial time series is given by stochastic volatility models (Taylor, 1994),

$$Y_t = \exp(b_t/2)\epsilon_t, \quad \epsilon_t \sim N(0,1), \tag{3.35}$$

where b_t is a latent stationary and invertible ARMA(p, q) process. As discussed in Examples 3.19 and 3.20, computation of the likelihood function does not pose difficulties in observation-driven models, whereas high-dimensional integrals need to be approximated for parameter-driven stochastic volatility models.

Common extensions allow the innovations ϵ_t to follow a standardized distribution with heavy tails or asymmetry, such as a standardized Student's t or a skew-t distribution. □

3.4 Spatial data

A spatial process is a collection of random variables $\{Y(x) : x \in \mathcal{R}\}$ indexed by spatial coordinates $x \in \mathcal{R} \subset \mathbb{R}^m$, where m is a positive integer. When $m > 1$, a spatial process is also called a random field. It is common to classify spatial processes based on the set $\mathcal{R}$ that identifies the type of spatial coordinates of interest for scientific study. In geostatistical processes (Section 3.4.1), $\mathcal{R}$ is a known set of volume greater than zero, on which the locations vary continuously. In lattice processes (Section 3.4.2), $\mathcal{R}$ is also a known set but locations vary discretely over it and the number of locations is countable, so its volume is zero. Finally in point processes (Section 3.4.3), $\mathcal{R}$ is a random set of points in $\mathbb{R}^m$.

3.4.1 Geostatistics

Unless replicates of the data are available, some form of stationarity is required for statistical inference. A spatial process is said to be strictly stationary when its distribution is invariant to shifts, that is the joint distribution of $\{Y(x_1), \ldots, Y(x_d)\}$ is the same as the joint distribution of $\{Y(x_1 + h), \ldots, Y(x_d + h)\}$ for $d \in \mathbb{N}$ and $h \in \mathbb{R}^m$. A weaker version, commonly used for geostatistical processes, is second-order or weak stationarity, which requires that the mean of the process is constant,

$$\mathrm{E}\{Y(x)\} = \mu, \quad x \in \mathcal{R},$$

and the covariance between any two observations depends only on the differences between their coordinates,

$$\mathrm{cov}\{Y(x), Y(x+h)\} = \gamma(h), \quad x \in \mathcal{R},\ h \in \mathbb{R}^m,$$

where $\gamma(\cdot)$ is a positive-definite function. Weakly stationary processes are isotropic when their covariance depends only on the Euclidean distance between the observations,

$$\text{cov}\{Y(x), Y(x+h)\} = \gamma(\|h\|_2),$$

where $\|\cdot\|_2$ is Euclidean distance or L2 norm. Weakly stationary processes that are not isotropic are said to be anisotropic; their correlation varies with direction. A process with a constant mean function is intrinsically stationary if

$$\text{var}\{Y(x+h) - Y(x)\} = 2\delta(h),$$

for a positive-definite function $\delta(\cdot)$ that depends only on h, called the semi-variogram. Weakly stationary implies intrinsically stationary but the converse is not true (Exercise 3.9). In the geostatistical literature it is more common to specify spatial processes in terms of their semi-variogram than their correlation function, because knowledge of the variogram is sufficient for optimal linear prediction at unknown locations (ordinary kriging), without the need to estimate the process mean.

An important class of random fields is that of Gaussian random fields, whose finite-dimensional distributions are all multivariate normal. Since Gaussian random fields are completely specified by their first two moments, strict and weak stationarity are equivalent. The covariance function of a stationary Gaussian random field has the form

$$\gamma(h) = \text{cov}\{Y(x), Y(x+h)\} = \sigma^2\rho(h),$$

for some positive-definite correlation function $\rho : \mathbb{R}^m \to [-1, 1]$. If $\rho(\cdot)$ depends on h only through its Euclidean norm $\|h\|_2$, then the Gaussian random field is isotropic on $\mathbb{R}^m$. The Matérn class of isotropic correlation functions is often used, where

$$\rho(h; \alpha) = \frac{2^{1-\nu}}{\Gamma(\nu)} \left(\frac{\|h\|_2}{\alpha}\right)^{\nu} K_\nu\left(\frac{\|h\|_2}{\alpha}\right), \quad \alpha > 0,$$

where $\Gamma(\cdot)$ is the Gamma function, $K_\nu(\cdot)$ is the modified Bessel function of the second kind of order $\nu > 0$ and α is the so-called range parameter. The parameter ν controls the smoothness of the random field, or more precisely the order of differentiability of a random field, whereas α is often called the range and controls how quickly the correlation decays with distance.

Special cases of the Matérn class are exponential correlation $\rho(h; \alpha) = \exp(-\|h\|_2/\alpha)$, with $\nu = 0.5$, Whittle correlation, with $\nu = 1$, and Gaussian correlation $\rho(h; \alpha) = \exp(-\|h\|_2^2/\alpha^2)$, with $\nu \to \infty$. Exponential and Gaussian correlations are also special cases of the powered exponential class $\rho(h; \alpha) = \exp(-\|h\|_2^\nu/\alpha^\nu)$, $\nu \in (0, 2]$. Other correlation functions are the spherical correlation

$$\rho(h; \alpha) = 1 - \frac{3}{2}\left(\frac{\|h\|_2}{\alpha}\right) + \frac{1}{2}\left(\frac{\|h\|_2}{\alpha}\right)^3, \quad 0 \leq \|h\|_2 < \alpha,$$

and the wave correlation

$$\rho(h; \alpha) = \frac{\sin(\|h\|_2/\alpha)}{(\|h\|_2/\alpha)}.$$

Figure 3.2 illustrates these correlation functions with the range parameter α chosen so that the correlation is less than 0.05 for distances of at least 0.75, for all examples except the wave correlation. This distance is called the practical range in the spatial statistics literature. For the wave correlation α was chosen so that the correlation is 0.05 at a distance of 0.75.

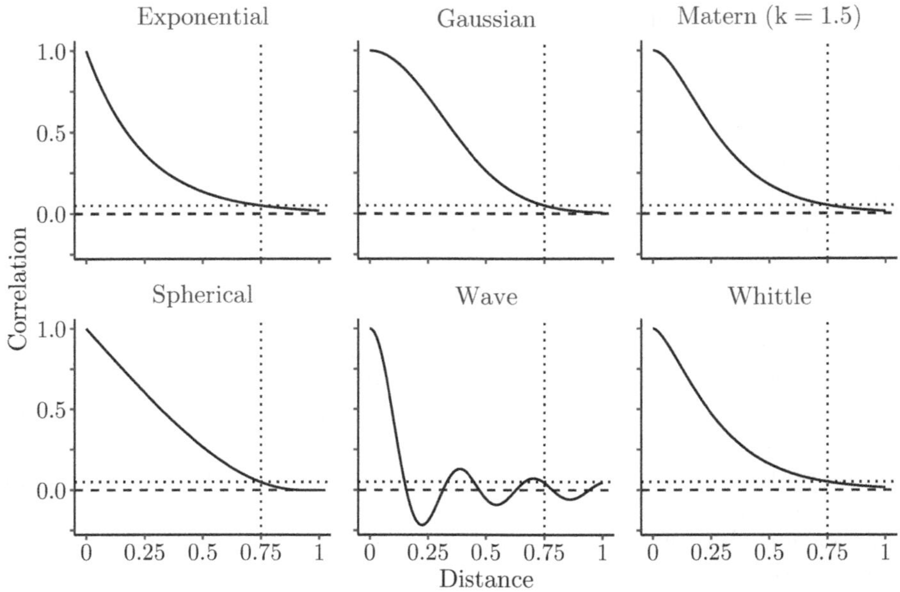

FIGURE 3.2
Some spatial isotropic correlation functions. The range parameter α is chosen so the practical range is 0.75 for all correlation functions except the wave correlation, where α is chosen so that the correlation is 0.05 at distance 0.75.

In geostatistical applications, it is common to observe a discontinuity at distance 0, called the nugget effect. Statistically, the nugget effect can be described as extra variability due to repeated measurements at or near a given location;

$$\operatorname{var}\{Y(x)\} = \sigma^2 + \delta, \qquad \lim_{\|h\|_2 \to 0^+} \operatorname{cov}\{Y(x), Y(x+h)\} = \sigma^2,$$

where $\delta \geq 0$ is the nugget parameter.

Example 3.22 (Stationary Gaussian random field) Statistical inference for geostatistical models may be difficult even for a stationary Gaussian random field observed at d locations $\{x_1, \ldots, x_d\}$ within region $\mathcal{R}$. Let $y = (y_1, \ldots, y_d)^\top$ be the vector of the d observations with $y_i = y(x_i)$, $i = 1, \ldots, d$. The likelihood function for $\theta = (\mu, \sigma^2, \delta, \alpha^\top)^\top$ is

$$\ell(\theta) = -\frac{1}{2}\log|\delta\mathcal{I}_{d\times d} + \sigma^2 R(\alpha)| - \frac{1}{2}(y - \mu 1_d)^\top\{\delta\mathcal{I}_{d\times d} + \sigma^2 R(\alpha)\}^{-1}(y - \mu 1_d),$$

where $\mathcal{I}_{d\times d}$ is the $d \times d$ identity matrix, 1_d is the column vector of d ones and $R(\alpha)$ is the correlation matrix of the spatial stationary process with (i, j) element $\rho(\|x_i - x_j\|_2; \alpha)$.

Consider the reparametrization $\theta = (\mu, \sigma^2, \tau, \alpha^\top)^\top$, where $\tau = \delta/\sigma^2$. The log-likelihood function in this parametrization is

$$\ell(\theta) = -\frac{d}{2}\log\sigma^2 - \frac{1}{2}\log|\tau\mathcal{I}_{d\times d} + R(\alpha)| - \frac{1}{2\sigma^2}(y - \mu 1_d)^\top\{\tau\mathcal{I}_{d\times d} + R(\alpha)\}^{-1}(y - \mu 1_d).$$

Write $\Gamma(\psi) = \tau\mathcal{I}_{d\times d} + R(\alpha)$, with $\psi = (\tau, \alpha^\top)^\top$. Maximizing $\ell(\theta)$ with respect to μ and σ^2 gives

$$\widehat{\mu}_\psi = \{1_d^\top\Gamma^{-1}(\psi)1_d\}^{-1}1_d^\top\Gamma^{-1}(\psi)y, \quad \widehat{\sigma}^2_\psi = \frac{1}{d}(y - \widehat{\mu}_\psi 1_d)^\top\Gamma^{-1}(\psi)(y - \widehat{\mu}_\psi 1_d). \tag{3.36}$$

The profile log-likelihood function for ψ is

$$\ell_{\mathrm{P}}(\psi) = -\frac{d}{2} \log \widehat{\sigma}^2_{\psi} - \frac{1}{2} \log |\Gamma(\psi)|, \tag{3.37}$$

which needs to be maximized numerically. For typical spatial correlation functions, such as Matérn, powered and wave, the computational cost of computing the inverse and determinant of Γ is $O(d^3)$, which is prohibitive in many applications. Even when computation of the inverse and the determinant are possible, maximization of the profile likelihood (3.37) is difficult because the function may be multimodal, see, for example, Mardia and Watkins (1989). Example 4.26 discusses a simpler computational approach to estimating θ using composite likelihood. □

Stationary Gaussian random fields often appear as components of non-stationary geostatistical models, as briefly described in the next three examples.

Example 3.23 (Geostatistical linear regression) Consider the linear regression model

$$Y(x) = z^{\mathsf{T}}(x)\beta + \epsilon(x), \quad x \in \mathcal{R},$$

where $z(x)$ is a p-dimensional vector of covariates observed at location x and $\epsilon(x)$ is a stationary Gaussian random field with zero mean, nugget parameter δ, and stationary correlation function depending on some parameter α. As in Example 3.22, define $\tau = \delta/\sigma^2$.

For a sample of observations at d locations $\{x_1, \ldots, x_d\}$, the constrained maximum likelihood estimators of β and σ^2 are

$$\widehat{\beta}_{\psi} = \{Z^{\mathsf{T}}\Gamma^{-1}(\psi)Z\}^{-1}Z^{\mathsf{T}}\Gamma^{-1}(\psi)y,$$
$$\widehat{\sigma}^2_{\psi} = \frac{1}{d}\{y - Z\widehat{\beta}_{\psi}\}^{\mathsf{T}}\Gamma^{-1}(\psi)\{y - Z\widehat{\beta}_{\psi}\},$$

where $\psi = (\tau, \alpha^{\mathsf{T}})^{\mathsf{T}}$, Z is the $d \times p$ matrix of covariates with ith row $z^{\mathsf{T}}(x_i)$, and $\Gamma(\psi) = \tau\mathcal{I}_{d\times d} + R(\alpha)$. The profile likelihood function for ψ has the same form as (3.37) and poses the same computational difficulties. □

Example 3.24 (Generalized linear spatial models) One approach to modelling non-normally distributed spatial data combines generalized linear models (1.20) with geostatistics (Diggle et al., 1998). Generalized linear spatial models assume that the response variables $Y(x)$ are independent conditionally on an unobserved spatial process with conditional expectation

$$\mathrm{E}\{Y(x) \mid b(x)\} = g^{-1}\{z^{\mathsf{T}}(x)\beta + b(x)\}, \quad x \in \mathcal{R},$$

where $g(\cdot)$ is a link function, $z(x)$ is covariate vector observed at x, $\{b(x) : x \in \mathcal{R}\}$ is a zero-mean stationary Gaussian random field with covariance matrix $\sigma^2 R(\alpha)$ that may include a nugget effect δ. Generalized linear spatial models are generalized linear mixed models (Example (1.13)) with a spatial random intercept.

Given a vector of d observations at locations $\{x_1, \ldots, x_d\} \in \mathcal{R}$, the likelihood function for $\theta = (\beta^{\mathsf{T}}, \phi, \delta, \sigma^2, \alpha^{\mathsf{T}})^{\mathsf{T}}$ is

$$L(\theta) = \int_{\mathbb{R}^d} f(b; \delta, \sigma^2, \alpha) \prod_{i=1}^{d} f(y_i \mid b_i; \beta, \phi) \, \mathrm{d}b,$$

where $y_i = y(x_i)$, $b_i = b(x_i)$ and $b = (b_1, \ldots, b_d)^{\mathsf{T}}$ is multivariate normal random vector with zero mean and variance $\delta\mathcal{I}_{d\times d} + \sigma^2 R(\alpha)$. Inference based on this likelihood function is complicated because in addition to the $O(d^3)$ cost of computing $f(b; \delta, \sigma^2, \alpha)$ it is necessary to approximate with sufficient accuracy an integral whose dimension is equal to the dimension of the data. The standard Laplace approximation of $L(\theta)$ is not asymptotically valid for this likelihood function (Shun and McCullagh, 1995; Tang and Reid, 2025). The solutions most often adopted in the literature use Monte Carlo simulation, as in the next example. □

Example 3.25 (Maps of disease prevalence) An important application of generalized spatial linear models is the survey-based mapping of disease prevalence in a region. A simple disease-prevalence survey samples m_i individuals in each of d communities within a study region $\mathcal{R}$.

Let Y_i be the number of individuals testing positive at community i located at x_i. The prevalence of the disease at x is $\text{pr}\{Y(x) = 1\}$. The scientific interest is in estimating $\text{pr}\{Y(x) = 1\}$ at a possibly unsampled location x within $\mathcal{R}$. Diggle and Giorgi (2016) suggested drawing maps of disease prevalence using the logistic geostatistical model

$$\text{pr}\{Y(x) = 1 \mid t(x)\} = \frac{\exp\{t(x)\}}{1 + \exp\{t(x)\}}, \quad t(x) = z^{\mathsf{T}}(x)\beta + b(x) + \epsilon(x), \tag{3.38}$$

where $z(x)$ is a vector of covariates at location x, $\epsilon(x)$ are uncorrelated zero-mean normal variables with variance δ and $\{b(x) : x \in \mathcal{R}\}$ is a zero-mean stationary Gaussian random field with covariance function $\text{cov}\{b(x), b(x+h)\} = \sigma^2\rho(h;\alpha)$. Here the nugget parameter δ captures possible extra-binomial variation that may be due to associations between individuals within the same community, whereas $b(x)$ ensures that the covariate-adjusted prevalence varies smoothly over the study region $\mathcal{R}$.

Giorgi and Diggle (2017) discussed the following Monte Carlo estimation of the likelihood function for this logistic model. The likelihood function for $\theta = (\beta^{\mathsf{T}}, \delta, \sigma^2, \alpha^{\mathsf{T}})^{\mathsf{T}}$ is

$$L(\theta) = \int_{\mathbb{R}^d} \frac{f(t;\theta)f(y \mid t)}{f(t;\widetilde{\theta})f(y \mid t)} f(t, y; \widetilde{\theta})\, \mathrm{d}t \propto \int_{\mathbb{R}^d} \frac{f(t;\theta)}{f(t;\widetilde{\theta})} f(t \mid y; \widetilde{\theta})\, \mathrm{d}t = \mathrm{E}_{T|y}\left\{\frac{f(t;\theta)}{f(t;\widetilde{\theta})}\right\},$$

where $\widetilde{\theta}$ is a predefined value of θ and $f(y \mid t) = \prod_{i=1}^d f(y_i \mid t_i)$ does not depend on θ; see (3.38). A Markov chain Monte Carlo algorithm is then used to obtain k samples $t^{(1)}, \ldots, t^{(k)}$ from $f(t \mid y; \widetilde{\theta})$, giving an approximation to the likelihood function,

$$\widetilde{L}(\theta) = \frac{1}{k}\sum_{j=1}^{k} \frac{f(t^{(j)};\theta)}{f(t^{(j)};\widetilde{\theta})}. \tag{3.39}$$

The optimal choice for $\widetilde{\theta}$ is the maximum likelihood estimator $\widehat{\theta}$, which would guarantee that $\max_\theta \widetilde{L}(\theta) \to 1$, as $k \to \infty$. Since the maximum likelihood estimator of θ is not known a priori, the quality of the approximation of $L(\theta)$ by $\widetilde{L}(\theta)$ is measured by the distance of $\max_\theta \widetilde{L}(\theta)$ from 1. This algorithm is implemented via an iterative procedure where the maximizer of the approximate likelihood at one step is used as the updated guess $\widetilde{\theta}$ for the next step, until a convergence criterion is met. □

3.4.2 Lattice data

A lattice is a countable set of known spatial locations $\mathcal{R} = \{x_1, x_2, \ldots\} \subset \mathbb{R}^m$ such as a grid of nodes, pixels or areal data. In lattice processes, spatial dependence is expressed in terms of neighbourhood associations instead of the correlation or variogram functions used in geostatistics. The set of neighbours of x_i is $\mathcal{N}(x_i) \subset \mathcal{R} \setminus \{x_i\}$. This set contains positions spatially close to x_i.

Neighbours are identified by a spatial weights matrix $W = [w_{ij}]$, where w_{ij} is non-zero only if $x_j \in \mathcal{N}(x_i)$. It is assumed that locations are not included in their own neighbourhood, so the diagonal elements of W are zero. Nondiagonal elements w_{ij} of W may be inversely proportional to the Euclidean distance $\|x_i - x_j\|_2$ or reflect some other measure of spatial proximity. A particular case of a spatial dependence matrix is the adjacency matrix, which is a symmetric matrix whose entries are 1 if a neighbourhood relation exists and 0 otherwise.

The most commonly used spatial process on a lattice is a Markov random field; this assumes the conditional distribution of $Y(x)$ depends on the observations at other locations only through its neighbours,

$$f\{Y(x) \mid Y(-x)\} = f\{Y(x) \mid Y(u) : u \in \mathcal{N}(x)\}, \quad x \in \mathcal{R},$$

where $Y(-x) = \{Y(u) : u \in \mathcal{R} \setminus \{x\}\}$.

Specification of the conditional distributions of Markov random fields does not guarantee the existence of a joint distribution over the set of locations $\mathcal{R}$. A clique c is any set of locations such that if x_i and x_j belong to c, then x_i is a neighbour of x_j and, consequently, x_j is a neighbour of x_i. The set of all possible cliques defined on $\mathcal{R}$ is $\mathcal{C}$. The Hammersley–Clifford theorem (Besag, 1974) states that if the distribution of a Markov random field is positive with respect to all possible configurations of $\mathcal{R}$, then its joint distribution has the Gibbs form

$$f(y; \theta) = \frac{1}{Z(\theta)} \exp\left\{-\frac{V(y;\theta)}{T}\right\}, \quad V(y;\theta) = \sum_{c \in \mathcal{C}} V_c(y;\theta), \tag{3.40}$$

where $V(y;\theta)$ is often called the energy function, T the system temperature and $V_c(y;\theta)$ the clique potentials. Computation of the normalization $Z(\theta)$, or partition function, is a formidable task as it requires integration, or summation, over all possible realizations of the spatial process,

$$Z(\theta) = \int_{\mathcal{Y}} \exp\left\{-\frac{1}{T}\sum_{c \in \mathcal{C}} V_c(y;\theta)\right\} \mathrm{d}y.$$

Example 3.26 (Conditional autoregressive spatial models) An important class of Markov random processes is the conditional autoregressive process introduced by Besag (1974). These Gaussian random fields have conditional means and variances

$$\mathrm{E}\{Y(x_i) \mid Y(u) : u \in \mathcal{N}(x_i)\} = \rho \sum_{j=1}^{d} w_{ij} Y(x_j),$$
$$\mathrm{var}\{Y(x_i) \mid Y(u) : u \in \mathcal{N}(x_i)\} = \sigma_i^2,$$

where ρ is the spatial autoregressive coefficient, w_{ij} are spatial weights and σ_i^2 are the conditional scale parameters. Let $W = [w_{ij}]$ be the spatial weights matrix with zero diagonal entries, and let $D = \mathrm{diag}(\sigma_1^2, \ldots, \sigma_d^2)$. If the precision matrix $D^{-1}(I - \rho W)$ is symmetric and positive-definite, then the joint distribution of $Y = (Y(x_1), \ldots, Y(x_d))^\mathsf{T}$ is multivariate normal with mean zero and covariance matrix

$$\Sigma = (I - \rho W)^{-1} D.$$

□

Example 3.27 (Autologistic model) A spatial binary process is said to be autologistic when the joint distribution of $Y = (Y(x_1), \ldots, Y(x_d))^\mathsf{T}$ has the Gibbs form with energy function

$$V(y; \alpha, \rho) = -\alpha \sum_{i=1}^{d} y(x_i) - \rho \sum_{j>k}^{d} y(x_j) y(x_k) w_{jk},$$

where ρ is the spatial interaction parameter, w_{jk} are spatial weights and the system temperature is $T = 1$. The joint distribution of Y corresponds to a logistic regression with covariates given by the observations themselves and the pairwise interaction between neighbours. This Gibbs distribution is called the Boltzmann distribution in statistical physics, and the autologistic model is called the Ising model; see also Example 1.32. It was originally proposed to describe the behaviour of magnets.

The normalization constant $Z(\alpha, \beta)$ of the autologistic model is a sum over 2^d terms. The computational difficulties associated with the normalization constant motivated Besag (1974) to base statistical inference on a pseudo-likelihood function, constructed from the product of the conditional distributions of $Y(x_i)$ given the neighbourhood observations. This is equivalent to fitting to the observations $y(x_i)$ the logistic regression model with probability of success

$$\pi(x_i) = \frac{\exp\{\alpha + \rho \sum_j y(x_j) w_{ij}\}}{1 + \exp\{\alpha + \rho \sum_j y(x_j) w_{ij}\}},$$

where $\pi(x_i) = \text{pr}[Y(x_i) = 1 \mid \{Y(u) : u \in \mathcal{N}(x_i)\}]$. Maximizing the likelihood function gives consistent, although inefficient, estimates of α and ρ. The inefficiency is due to the fact that the Besag pseudo-likelihood is built under the working assumption that the conditional distributions of $Y(x_i)$ given the neighbours are independent. The stronger the degree of dependence in the random field, the lower will be the efficiency of the estimates of α and ρ. Besag's pseudo-likelihood is an example of composite likelihood, which is the topic of Section 4.7. □

3.4.3 Spatial point data

A spatial point process is a stochastic process that describes the locations of events in a region of interest $\mathcal{R} \subset \mathbb{R}^m$. The position of the events and their number are both assumed to be random. Consider the counting process $\{N(A) : A \in \mathcal{R}\}$, where $N(A)$ denotes the number of random points that fall within $A \subset \mathcal{R}$. The basic model for $N(A)$ is the Poisson point process under which $N(A)$ has a Poisson distribution with expected value $\mu(A)$,

$$\text{pr}\{N(A) = n\} = \frac{1}{n!}\{\mu(A)\}^n \exp\{-\mu(A)\}, \quad A \subset \mathcal{R},$$

where the expected number of points in A is

$$\mu(A) = \int_A \lambda(x)\,\mathrm{d}x, \quad \lambda(x) > 0,\ x \in \mathcal{R}.$$

The function $\lambda(x)$ is the first-order intensity function, defined by

$$\lambda(x) = \lim_{|\delta x| \to 0^+} \frac{\mathrm{E}\{N(\delta x)\}}{|\delta x|},$$

where δx is a small set centred at $x \in \mathcal{R}$ with volume $|\delta x|$. In Poisson point processes, the number of points in two non-overlapping regions are independent,

$$\text{pr}\{N(A_1) = n_1, N(A_2) = n_2\} = \text{pr}\{N(A_1) = n_1\}\text{pr}\{N(A_2) = n_2\}, \quad \text{if } A_1 \cap A_2 = \emptyset.$$

Poisson point processes are said to be homogeneous when $\lambda(x) = \lambda$ for all $x \in \mathcal{R}$. In homogeneous Poisson point processes, events are equally likely to occur anywhere within the study area, regardless of the locations of other events. For this reason, the homogeneous Poisson point process describes complete spatial randomness and is used as the null model in tests for the absence of nonrandom patterns in the distribution of events. Nonhomogeneous Poisson point process are often described in terms of spatial covariates $z(x)$ via the log-linear model $\lambda(x) = \exp\{z^{\mathsf{T}}(x)\beta\}$. Poisson processes for event times are discussed in Examples 1.7 and 1.8.

Example 3.28 (Cox processes) Cox processes, introduced in Cox (1955), are doubly stochastic processes, consisting of an inhomogeneous Poisson process with a stochastic intensity function. Consider a point pattern consisting of d events recorded at locations $\{y_1, \ldots, y_d\}$ within a region $\mathcal{R}$. Let $\{b(u) : u \in \mathcal{R}\}$ be a stationary Gaussian random field with zero mean and covariance function $\sigma^2 R(\alpha)$. The log-Gaussian Cox process assumes that, conditionally on $b(u)$, the first-order intensity function is

$$\lambda\{u \mid b(u)\} = \exp\{x^{\mathsf{T}}(u)\beta + b(u)\}, \quad u \in \mathcal{R},$$

where $x(u)$ is the vector of covariates at location $u \in \mathcal{R}$. The marginal first-order intensity of the log-Gaussian Cox process is $\lambda(u) = \exp\{x^{\mathsf{T}}(u)\beta + \sigma^2/2\}$.

As with spatial generalized linear models, (Example 3.24), the likelihood function for the log-Gaussian Cox process involves high-dimensional integration:

$$L(\beta, \sigma^2, \alpha) = \int f(y \mid b; \beta) f(b; \sigma^2, \alpha)\, \mathrm{d}b,$$

where the integration is taken over all possible realizations of the random field $\{b(u) : u \in \mathcal{R}\}$. Unlike standard mixed models where the random effect is a finite vector, here b is a stochastic process. The term $\mathrm{d}b$ thus represents integration over the infinite-dimensional space of functions defined on the region $\mathcal{R}$. The conditional density of the observed point pattern y given b is

$$f(y \mid b; \beta) = \prod_{i=1}^{d} \lambda(y_i \mid b_i) \exp\left[- \int_{\mathcal{R}} \lambda\{u \mid b(u)\}\, \mathrm{d}u \right].$$

where $b_i = b(y_i)$. The intractability of this likelihood function has motivated various Monte Carlo and approximation methods, including composite likelihood, which is discussed in Chapter 4. □

3.5 Missing data

The discussions in Sections 3.2–3.4 highlight the application of likelihood methods for the analysis of non-independent data. In this section we consider data that may consist of independent but incomplete observations. When dealing with such data, likelihood methods also prove to be effective tools.

3.5.1 Analysis of missing data with joint modelling

For a response variable Y_i with a covariate vector X_i, in a scenario where the response variable for subject i may be missing, let R_i be the missing data indicator, taking value 1 if Y_i is observed, and 0 otherwise.

Regression analysis with incomplete responses usually involves modelling the conditional distribution of Y_i and R_i given $X_i = x_i$. Two commonly used strategies are based on the factorizations

$$f(y_i, r_i) = f(r_i \mid y_i)f(y_i), \tag{3.41}$$

$$f(y_i, r_i) = f(y_i \mid r_i)f(r_i), \tag{3.42}$$

where $f(\cdot)$ and $f(\cdot \mid \cdot)$ denote probability density or mass functions indicated by the corresponding random variables, and are assumed to be correctly specified; conditioning on the covariates is suppressed in the notation, and model parameters are omitted for notational simplicity. Models derived from (3.41) are often called selection models and those from (3.42) are referred to as pattern-mixture models. These factorizations offer dual perspectives on the relationship between Y_i and R_i. The selection model first conditions on Y_i, modelling the missing data mechanism via $f(r_i \mid y_i)$, whereas the pattern-mixture model conditions on R_i, describing the response distribution via $f(y_i \mid r_i)$.

Alternatively, instead of conditioning on either Y_i or R_i, a third approach treats them on equal footing and introduces a latent variable, U_i to induce dependence. Given $U_i = u_i$ and the covariates, Y_i and R_i are assumed to be conditionally independent, so that

$$f(y_i, r_i \mid u_i) = f(y_i \mid u_i)f(r_i \mid u_i). \tag{3.43}$$

Example 3.29 (Pattern-mixture model) Consider the modelling strategy (3.42), where for subject i the response variable Y_i may be missing, while the associated covariate vector x_i is assumed to be always observed. Suppose that the distribution of Y_i differs for the observed and unobserved measurements,

$$\begin{aligned} Y_i \mid R_i = 1 &\sim \text{Poisson}(\mu_1), \quad \text{with } \log \mu_1 = \beta_0 + x_i^{\mathsf{T}}\beta_1, \\ Y_i \mid R_i = 0 &\sim \text{Poisson}(\mu_0), \quad \text{with } \log \mu_0 = \beta_0 + x_i^{\mathsf{T}}\beta_2. \end{aligned}$$

Assume further that the distribution of the missing data indicator is $R_i \sim \text{Bernoulli}(\pi)$, $0 < \pi < 1$, and let $\theta = (\beta_0, \beta_1^{\mathsf{T}}, \beta_2^{\mathsf{T}}, \pi)^{\mathsf{T}}$. For subject i with $R_i = 1$, the likelihood contribution is

$$\begin{aligned} L_i^{\text{obs}}(\theta) &= \text{pr}(Y_i = y_i, R_i = 1) \\ &= \text{pr}(Y_i = y_i \mid R_i = 1)\text{pr}(R_i = 1) \\ &= \frac{\mu_1^{y_i}\exp(-\mu_1)}{y_i!}\pi. \end{aligned}$$

For subject j with $R_j = 0$, the likelihood contribution is

$$L_j^{\text{mis}}(\theta) = \sum_{y=0}^{\infty} \text{pr}(Y_j = y, R_j = 0) = \text{pr}(R_j = 0) = 1 - \pi.$$

The likelihood function based on the sample is

$$L(\theta) = \prod_{i=1}^{n} \{L_i^{\text{obs}}(\theta)\}^{r_i}\{L_i^{\text{mis}}(\theta)\}^{1-r_i} = \prod_{i=1}^{n} \left\{\frac{\mu_1^{y_i}\exp(-\mu_1)}{y_i!}\pi\right\}^{r_i}(1-\pi)^{1-r_i},$$

with the log-likelihood function, up to an additive constant,

$$\ell(\theta) = \sum_{i=1}^{n} \left\{y_i r_i(\beta_0 + x_i^{\mathsf{T}}\beta_1) - r_i\exp(\beta_0 + x_i^{\mathsf{T}}\beta_1) + r_i\log\pi + (1-r_i)\log(1-\pi)\right\}.$$

Solving the score equations leads to $\widehat{\pi} = n^{-1}\Sigma_i r_i$ and the system of equations

$$\sum_{i=1}^{n} r_i\{y_i - \exp(\widehat{\beta}_0 + x_i^{\mathsf{T}}\widehat{\beta}_1)\} = 0,$$
$$\sum_{i=1}^{n} r_i x_i\{y_i - \exp(\widehat{\beta}_0 + x_i^{\mathsf{T}}\widehat{\beta}_1)\} = 0.$$

Although β_2 is part of the model above, it is not identifiable as there is no information in the data on the conditional distribution $\mathrm{pr}(Y_i = y \mid R_i = 0)$. This example illustrates that in the presence of missing observations, some parameters may not be identifiable. □

Example 3.30 (Shared random effects) Here we consider the modelling strategy (3.43) for a logistic regression problem with n observations. Suppose that conditional on random effects $U_i = u_i$,

$$Y_i \mid u_i \sim \text{Bernoulli}(p_i), \quad \text{with} \;\; \text{logit}\; p_i = u_i + x_i^{\mathsf{T}}\beta,$$
$$R_i \mid u_i \sim \text{Bernoulli}(\pi_i), \quad \text{with} \;\; \text{logit}\; \pi_i = u_i + x_i^{\mathsf{T}}\alpha,$$

and that $U_i \sim \text{Bernoulli}(p)$, $0 < p < 1$. For subject i with $R_i = 1$, the contribution to the likelihood function for $\theta = (\beta^{\mathsf{T}}, \alpha^{\mathsf{T}}, p)^{\mathsf{T}}$ is

$$\begin{aligned}
L_i^{\text{obs}}(\theta) &= \mathrm{pr}(Y_i = y_i, R_i = 1) \\
&= \sum_{u_i=0}^{1} \mathrm{pr}(Y_i = y_i \mid u_i)\mathrm{pr}(R_i = 1 \mid u_i)\mathrm{pr}(U_i = u_i) \\
&= \sum_{u_i=0}^{1} p_i^{y_i}(1-p_i)^{1-y_i}\pi_i p^{u_i}(1-p)^{1-u_i} \\
&= \frac{(1-p)\exp(x_i^{\mathsf{T}}\beta y_i + x_i^{\mathsf{T}}\alpha)}{\{1+\exp(x_i^{\mathsf{T}}\beta)\}\{1+\exp(x_i^{\mathsf{T}}\alpha)\}} + \frac{p\exp\{(1+x_i^{\mathsf{T}}\beta)y_i + (1+x_i^{\mathsf{T}}\alpha)\}}{\{1+\exp(1+x_i^{\mathsf{T}}\beta)\}\{1+\exp(1+x_i^{\mathsf{T}}\alpha\}}.
\end{aligned}$$

For subject j with $R_j = 0$, the likelihood contribution is

$$\begin{aligned}
L_j^{\text{mis}}(\theta) &= \sum_{y_j=0}^{1} \mathrm{pr}(Y_j = y_j, R_j = 0) \\
&= \sum_{y_j=0}^{1}\sum_{u_j=0}^{1} \mathrm{pr}(Y_j = y_j \mid u_j)\mathrm{pr}(R_j = 0 \mid u_j)\mathrm{pr}(U_j = u_j) \\
&= \sum_{y_j=0}^{1}\sum_{u_j=0}^{1} p_j^{y_j}(1-p_j)^{1-y_j}(1-\pi_j)p^{u_j}(1-p)^{1-u_j} \\
&= 1 - \frac{(1-p)\exp(x_j^{\mathsf{T}}\alpha)}{1+\exp(x_j^{\mathsf{T}}\alpha)} - \frac{p\exp(1+x_j^{\mathsf{T}}\alpha)}{1+\exp(1+x_j^{\mathsf{T}}\alpha)}.
\end{aligned}$$

The log-likelihood function based on the sample is

$$\ell(\theta) = \sum_{i=1}^{n}\left\{r_i \log L_i^{\text{obs}}(\theta) + (1-r_i)\log L_i^{\text{mis}}(\theta)\right\}.$$

Estimation of θ can proceed by solving the score equations, as in Example 3.29. However, the two examples differ in several respects. In Example 3.29, the likelihood estimator of the nuisance parameter π has a closed form, determined by the proportion of observed responses. The regression parameter β_2 is not identifiable, while the estimable model parameters β_0 and β_1 are estimated solely from the complete cases, i.e., subjects without missing values. In this example, estimation of the model parameter β and the nuisance parameters α and p are intertwined. The contributions from subjects with and without missing data cannot be separated in the estimation procedure, and all parameters are identifiable. □

Example 3.31 (Selection model) In contrast to the univariate response considered in the previous two examples, we now examine the case of longitudinal responses subject to missingness under the scheme (3.41). Define Y_i as the response vector $(Y_{i1}, \ldots, Y_{im_i})^{\mathsf{T}}$, where Y_{ij} is the response variable for subject i at time point $j = 1, \ldots, m_i$, and $m_i \geq 2$ is the number of the measurements for subject i. Similarly, $R_i = (R_{i1}, \ldots, R_{im_i})^{\mathsf{T}}$, where R_{ij} is the missing data indicator for response component Y_{ij}. For the models $f(y_i)$ and $f(r_i \mid y_i)$ in (3.41), let β be the parameter of interest associated with the first model, and let α be the parameter for the latter model, treated as a nuisance parameters. These parameters are assumed to be distinct, or functionally independent.

Partition y_i, a realization of Y_i, into a sub-vector y_i^{obs} of observed responses and a sub-vector y_i^{mis} of missing components. The likelihood contribution from the ith subject is

$$f(y_i^{\text{obs}}, r_i; \beta, \alpha) = \int f(y_i^{\text{obs}}, y_i^{\text{mis}}; \beta) f(r_i \mid y_i^{\text{obs}}, y_i^{\text{mis}}; \alpha) \, dy_i^{\text{mis}}, \tag{3.44}$$

where the integral is replaced by summation for the discrete case.

In general, inference for β requires modelling both the response distribution and the missing data distribution. However, under some assumptions, modelling the missing data process can be bypassed, allowing inference for β to be performed using only the model for the response distribution. For example, if the second factor in the integrand of (3.44) satisfies

$$f(r_i \mid y_i^{\text{obs}}, y_i^{\text{mis}}; \alpha) = f(r_i \mid y_i^{\text{obs}}; \alpha), \tag{3.45}$$

then the likelihood function factors as $f(y_i^{\text{obs}}; \beta) f(r_i \mid y_i^{\text{obs}}; \alpha)$, with each function involving either β or α alone. As a result, the contribution to the log-likelihood function from (3.44) is

$$\log f(y_i^{\text{obs}}; \beta) + \log f(r_i \mid y_i^{\text{obs}}; \alpha),$$

and inference for β is thereby based on the observed log-likelihood function $\Sigma_i \log f(y_i^{\text{obs}}; \beta)$, which is obtained from the response model alone,

$$f(y_i^{\text{obs}}; \beta) = \int f(y_i^{\text{obs}}, y_i^{\text{mis}}; \beta) \, dy_i^{\text{mis}},$$

and modelling the missing data process is not needed.

The property (3.45) is called missing at random, or MAR. An important special case arises when $f(r_i \mid y_i^{\text{obs}}, y_i^{\text{mis}}; \alpha) = f(r_i; \alpha)$, which is called missing completely at random, or MCAR. If $m_i = 1$, meaning no repeated measurements are available for the ith subject, Y_i is a scalar rather than a vector, and the missing data mechanism cannot be described using (3.45). In that case, the term missing at random is used if

$$f(r_i \mid y_i; \alpha) = f(r_i; \alpha).$$

Kenward and Molenberghs (1998) reviewed the application of likelihood inference methods to data that are missing at random. An important point raised there is that the observed information should be used to estimate the variance of the maximum likelihood estimator. □

3.5.2 The EM algorithm

The discussion in Example 3.31 evaluates the likelihood function for the observed data by integrating out the missing responses. This integration can sometimes be simplified by using the expectation-maximization algorithm. Begin by decomposing the model for the joint distribution of Y_i and R_i as

$$f(y_i^{\mathrm{obs}}, y_i^{\mathrm{mis}}, r_i; \theta) = f(y_i^{\mathrm{mis}} \mid y_i^{\mathrm{obs}}, r_i; \theta) f(y_i^{\mathrm{obs}}, r_i; \theta),$$

where $\theta = (\alpha^{\mathsf{T}}, \beta^{\mathsf{T}})^{\mathsf{T}}$, and the dependence on the covariates is suppressed. Taking the logarithm and then applying it to random variables $\{Y_i^{\mathrm{obs}}, Y_i^{\mathrm{mis}}, R_i\}$ yields

$$\log f(Y_i^{\mathrm{obs}}, R_i; \theta) = \log f(Y_i^{\mathrm{obs}}, Y_i^{\mathrm{mis}}, R_i; \theta) - \log f(Y_i^{\mathrm{mis}} \mid Y_i^{\mathrm{obs}}, R_i; \theta).$$

For a given θ^*, taking conditional expectations with respect to $f(y_i^{\mathrm{mis}} \mid y_i^{\mathrm{obs}}, r_i; \theta^*)$ gives

$$\begin{aligned}
\log f(y_i^{\mathrm{obs}}, r_i; \theta) &= \int \log f(y_i^{\mathrm{obs}}, r_i; \theta) f(y_i^{\mathrm{mis}} \mid y_i^{\mathrm{obs}} r_i; \theta^*)\, \mathrm{d}y_i^{\mathrm{mis}} \\
&= \int \log f(y_i^{\mathrm{obs}}, y_i^{\mathrm{mis}} r_i; \theta) f(y_i^{\mathrm{mis}} \mid y_i^{\mathrm{obs}}, r_i; \theta^*)\, \mathrm{d}y_i^{\mathrm{mis}} \\
&\quad - \int \log f(y_i^{\mathrm{mis}} \mid y_i^{\mathrm{obs}}, r_i; \theta) f(y_i^{\mathrm{mis}} \mid y_i^{\mathrm{obs}}, r_i; \theta^*)\, \mathrm{d}y_i^{\mathrm{mis}}.
\end{aligned}$$

Applying this to the sample $\{(y_i^{\mathrm{obs}}, r_i) : i = 1, \ldots, n\}$ we obtain the log-likelihood function based on the observed data

$$\ell_{\mathrm{obs}}(\theta) = \sum_{i=1}^{n} \log f(y_i^{\mathrm{obs}} r_i; \theta) = Q(\theta; \theta^*) - H(\theta; \theta^*), \tag{3.46}$$

where

$$Q(\theta; \theta^*) = \sum_{i=1}^{n} \mathrm{E}_{\theta^*} \left\{ \log f(Y_i^{\mathrm{obs}}, Y_i^{\mathrm{mis}}, R_i; \theta) \right\}, \tag{3.47}$$

$$H(\theta; \theta^*) = \sum_{i=1}^{n} \mathrm{E}_{\theta^*} \left\{ \log f(Y_i^{\mathrm{mis}} \mid Y_i^{\mathrm{obs}}, R_i; \theta) \right\},$$

with the expectations evaluated with respect to $f(y_i^{\mathrm{mis}} \mid y_i^{\mathrm{obs}}, r_i; \theta^*)$.

The formulation (3.46) represents $\ell_{\mathrm{obs}}(\theta)$ as the difference of two functions that depend not only on θ but also on the value θ^*; for iterative computations, θ^* is replaced by $\theta^{(k)}$. The increment in the observed log-likelihood function is

$$\ell_{\mathrm{obs}}(\theta^{(k+1)}) - \ell_{\mathrm{obs}}(\theta^{(k)}) = \{Q(\theta^{(k+1)}; \theta^{(k)}) - Q(\theta^{(k)}; \theta^{(k)})\} - \{H(\theta^{(k+1)}; \theta^{(k)}) - H(\theta^{(k)}; \theta^{(k)})\}.$$

From Jensen's inequality, $H(\theta; \theta^*) \le H(\theta^*; \theta^*)$ for any $\theta \in \Theta$, so

$$-\{H(\theta^{(k+1)}; \theta^{(k)}) - H(\theta^{(k)}; \theta^{(k)})\} \ge 0.$$

Thus, if $\theta^{(k+1)}$ is chosen such that $Q(\theta^{(k+1)}; \theta^{(k)}) - Q(\theta^{(k)}; \theta^{(k)}) \ge 0$, the log-likelihood function increases at each iteration

$$\ell_{\mathrm{obs}}(\theta^{(k+1)}) \ge \ell_{\mathrm{obs}}(\theta^{(k)}).$$

This is the expectation–maximization, or EM, algorithm for maximizing the observed log-likelihood function. It iterates until convergence between two operations; the E-step, computing $Q(\theta; \theta^{(k)})$ using (3.47), and the M-step, maximizing $Q(\theta; \theta^{(k)})$ with respect to θ to obtain the next iterate $\theta^{(k+1)}$.

The EM algorithm is described here in the context of Example 3.31, where the maximization of the observed log-likelihood function (3.46) is reformulated as an iterative procedure for maximizing the function $Q(\theta; \theta^{(k)})$ in (3.47). The complete data likelihood function is

$$L_{\text{comp}}(\theta) = \prod_{i=1}^{n} f(y_i^{\text{obs}}, y_i^{\text{mis}}, r_i; \theta); \tag{3.48}$$

and $Q(\theta; \theta^{(k)})$ represents the expectation of its logarithm over the conditional distribution of the missing data, given the observed data.

Example 3.32 (Multinomial data) Following Dempster et al. (1977), we illustrate the EM algorithm for a random sample $Y = (Y_1, Y_2, Y_3, Y_4)^\mathsf{T}$ from the multinomial distribution of size $n = \Sigma_{i=1}^4 Y_i$ and probability vector $p = ((1-\theta)/2, \theta/4, \theta/4, 1/2)^\mathsf{T}$, with $0 < \theta < 1$. The observed data $Y^{\text{obs}} = (Y_1, Y_2, Y_3 + Y_4)$ follows the multinomial distribution of size n and probability vector $q = ((1-\theta)/2, \theta/4, \theta/4 + 1/2)^\mathsf{T}$. The likelihood function for the complete data $y = (y_1, y_2, y_3, y_4)^\mathsf{T}$ is

$$L_{\text{comp}}(\theta) = \frac{n!}{y_1! y_2! y_3! y_4!} \left(\frac{1-\theta}{2}\right)^{y_1} \left(\frac{\theta}{4}\right)^{y_2+y_3} \left(\frac{1}{2}\right)^{y_4},$$

with the log-likelihood function

$$\ell_{\text{comp}}(\theta) = y_1 \log(1-\theta) + (y_2 + y_3) \log \theta,$$

up to an additive constant.

The expectation of $\ell_{\text{comp}}(\theta)$ with respect to the conditional distribution of the missing data given the observed data, evaluated at $\theta = \theta^{(k)}$, is

$$\begin{aligned} Q(\theta; \theta^{(k)}) &= \mathrm{E}_{\theta^{(k)}}\{\ell_{\text{comp}}(\theta) \mid Y^{\text{obs}} = y^{\text{obs}}\} \\ &= y_1 \log(1-\theta) + y_2 \log \theta + \mathrm{E}_{\theta^{(k)}}(Y_3 \mid Y^{\text{obs}} = y^{\text{obs}}) \log \theta. \end{aligned}$$

Maximizing $Q(\theta; \theta^{(k)})$ with respect to θ gives the updated estimate of θ:

$$\theta^{(k+1)} = \frac{y_2 + \mathrm{E}_{\theta^{(k)}}(Y_3 \mid Y^{\text{obs}} = y^{\text{obs}})}{y_1 + y_2 + \mathrm{E}_{\theta^{(k)}}(Y_3 \mid Y^{\text{obs}} = y^{\text{obs}})}.$$

The conditional distribution of Y_3 given $Y^{\text{obs}} = y^{\text{obs}}$ is

$$\begin{aligned} \mathrm{pr}(Y_3 = y_3 \mid Y^{\text{obs}} = y^{\text{obs}}) &= \mathrm{pr}(Y_3 = y_3 \mid Y_1 = y_1, Y_2 = y_2, Y_3 + Y_4 = m) \\ &= \binom{m}{y_3} (1-r)^{y_3} r^{m-y_3}, \end{aligned}$$

where $r = (1/2)/(\theta/4 + 1/2)$. From this $\mathrm{E}_{\theta^{(k)}}(Y_3 \mid Y^{\text{obs}} = y^{\text{obs}}) = m(1-r)$. If the observed counts are $y^{\text{obs}} = (38, 34, 125)^\mathsf{T}$,

$$\theta^{(k+1)} = \frac{34 + \mathrm{E}_{\theta^{(k)}}(Y_3 \mid Y^{\text{obs}} = y^{\text{obs}})}{72 + \mathrm{E}_{\theta^{(k)}}(Y_3 \mid Y^{\text{obs}} = y^{\text{obs}})}. \tag{3.49}$$

At convergence, $\widehat{\theta}$ solves the quadratic equation $197\widehat{\theta}^2 - 15\widehat{\theta} - 68 = 0$, giving the solution $\widehat{\theta} = 0.627$.

□

Example 3.33 (Mixture models) An important application of the EM algorithm is the estimation of finite mixture models. Consider a sample of independent and identically distributed random variables $\{Y_1, \ldots, Y_n\}$ that follow a mixture distribution with density function

$$f(y_i; \omega) = \sum_{j=1}^{d} \omega_j f_j(y_i), \quad i = 1, \ldots, n,$$

where for $j = 1, \ldots, d$, the component density $f_j(y_i)$ is assumed known, while the weight vector $\omega = (\omega_1, \ldots, \omega_d)^\mathsf{T}$ is unknown, with $\omega_j \geq 0$ and $\Sigma_j \omega_j = 1$.

Directly maximizing the log-likelihood function for the weight vector ω is difficult due to the additive structure inside the logarithm,

$$\ell(\omega) = \sum_{i=1}^{n} \log \left\{ \sum_{j=1}^{d} \omega_j f_j(y_i) \right\}.$$

The EM algorithm provides a convenient and elegant alternative for obtaining the maximum likelihood estimator of ω. Specifically, we introduce latent indicator variables R_{ij} to denote whether observation Y_i originates from component j,

$$R_{ij} = \begin{cases} 1, & Y_i \text{ comes from } f_j, \\ 0, & \text{otherwise.} \end{cases}$$

and assume that $\text{pr}(R_{ij} = 1) = \omega_j$. The complete-data log-likelihood function is

$$\ell_{\text{comp}}(\omega) = \sum_{i=1}^{n} \sum_{j=1}^{d} r_{ij} \log \omega_j + \sum_{i=1}^{n} \sum_{j=1}^{d} r_{ij} \log f_j(y_i).$$

The second term does not involve ω and is thus irrelevant for its estimation. At iteration $(k+1)$ of the E-step, the unobserved indicators R_{ij} are replaced by their conditional expectations given the current parameter estimate $\omega^{(k)}$,

$$\text{pr}(R_{ij} = 1 \mid y_i; \omega^{(k)}) = \frac{\omega_j^{(k)} f_j(y_i)}{\sum_{l=1}^{d} \omega_l^{(k)} f_l(y_i)} = z_{ij}^{(k)}, \text{ say.}$$

In the M-step, the weights are updated using the closed-form expression

$$\omega_j^{(k+1)} = \frac{1}{n} \sum_{i=1}^{n} z_{ij}^{(k)}.$$

At convergence, $\widehat{\omega}$ is the nonparametric maximum likelihood estimator of the mixing distribution. Nonparametric maximum likelihood estimation is discussed in more general terms in Chapter 6.

If the component densities $f_j(y_i)$ are also unknown, they can be modeled parametrically as $f_j(y_i; \theta)$ using an additional parameter θ. The EM algorithm then alternates between updating ω and θ; see Example 6.3. □

The EM algorithm applies more generally to settings with incomplete data or latent variables, which, in the discussion above, correspond to the missing data. Let $\{\{Y_i, Z_i\} : i = 1, \ldots, n\}$ be a random sample, with Y_i and Z_i denoting observed and unobserved latent variables, respectively. If the joint distribution follows a parametric model $f(y, z; \theta)$, then the observed data likelihood function is the integrated likelihood function

$$L_{\text{obs}}(\theta) = \prod_{i=1}^{n} f(y_i; \theta) = \prod_{i=1}^{n} \int f(y_i, z_i; \theta)\, \mathrm{d}z_i.$$

When directly maximizing $L_{\text{obs}}(\theta)$ is difficult due to the integral, the EM algorithm instead maximizes the expected completed data log-likelihood function,

$$Q(\theta; \theta^{(k)}) = \sum_{i=1}^{n} \mathrm{E}_{\theta^{(k)}} \left\{ \log f(Y_i, Z_i; \theta) \mid Y_i = y_i \right\},$$

where the expectation is taken with respect to the conditional distribution of the missing data Z_i given the observed data Y_i, using the parameter estimate $\theta^{(k)}$ at the kth iteration. At the M-step for iteration $(k+1)$, the parameter θ is updated by maximizing $Q(\theta; \theta^{(k)})$. While the EM maximization step can often be simpler, the convergence rate of the EM algorithm is typically slower than that of direct likelihood maximization.

Example 3.34 (Monte Carlo EM algorithm) The EM algorithm can also be used to compute the maximum likelihood estimator in generalized linear mixed models discussed in Section 3.2. Consider the random effects model with the linear predictor given in (3.5). Instead of directly maximizing the likelihood function (3.6), the maximum likelihood estimator $\widehat{\theta}$ can be computed using the EM algorithm, treating the random effects as missing observations.

The complete data log-likelihood function for subject i is based on the joint distribution of $Y_i = (Y_{i1}, \ldots, Y_{im_i})^{\mathsf{T}}$ and the random effect b_i,

$$\ell_i(\theta; y_i, b_i) = \log f(b_i; \alpha) + \sum_{j=1}^{m_i} \log f(y_{ij} \mid b_i; \beta, \phi),$$

with x_i and z_i suppressed in the notation. At iteration $(k+1)$, given the current estimate $\widehat{\theta}^{(k)}$, the expectation step evaluates

$$\begin{aligned} Q(\theta; \widehat{\theta}^{(k)}) &= \sum_{i=1}^{n} \int \log f(b_i; \alpha) f(b_i \mid y_i; \widehat{\theta}^{(k)})\, \mathrm{d}b_i \\ &\quad + \sum_{i=1}^{n} \sum_{j=1}^{m_i} \int \log f(y_{ij} \mid b_i; \beta, \phi) f(b_i \mid y_i; \widehat{\theta}^{(k)})\, \mathrm{d}b_i. \end{aligned}$$

Computational and numerical efficiency considerations suggest approximating the integrals above using a sampling method. For example, the Monte Carlo estimate of $Q(\theta; \widehat{\theta}^{(k)})$ is given by the average

$$\widehat{Q}(\theta; \widehat{\theta}^{(k)}) = \frac{1}{s} \sum_{r=1}^{s} \sum_{i=1}^{n} \log f(b_i^{(r)}; \alpha) + \frac{1}{s} \sum_{r=1}^{s} \sum_{i=1}^{n} \sum_{j=1}^{m_i} \log f(y_{ij} \mid b_i^{(r)}; \beta, \phi),$$

where $b_i^{(1)}, \ldots, b_i^{(s)}$ are drawn from $f(b_i \mid y_i; \widehat{\theta}^{(k)})$, for each $i = 1, \ldots, n$. The function $\widehat{Q}(\theta; \widehat{\theta}^{(k)})$ is maximized over θ to obtain $\widehat{\theta}^{(k+1)}$. The maximization step conveniently separates estimation of α from estimation of β and ϕ. Inference for θ should account for the effect of the Monte Carlo error, as $\widehat{Q}(\theta; \widehat{\theta}^{(k)})$ represents a misspecified likelihood function, a topic discussed in Chapter 4. Other sampling methods, such as importance sampling or Markov chain Monte Carlo sampling could also be used to approximate $Q(\theta; \widehat{\theta}^{(k)})$. □

Estimation of the variance of the maximum likelihood estimator requires the observed information matrix. One advantage of the EM algorithm is that it avoids direct computation of the observed likelihood function, but this also makes it challenging to compute the Hessian of the log-likelihood. However, Louis (1982) noted that the observed information matrix admits the decomposition

$$J_{\text{obs}}(\widehat{\theta}) = K_{\text{comp}}(\widehat{\theta}) - I_{\text{mis}|\text{obs}}(\widehat{\theta}), \tag{3.50}$$

where $J_{\text{obs}}(\theta)$ is the observed information matrix, $K_{\text{comp}}(\theta) = \text{E}\{J_{\text{comp}}(\theta) \mid y^{\text{obs}}, r; \theta\}$, and $I_{\text{mis}|\text{obs}}(\theta) = \text{var}\{U_{\text{comp}}(\theta) \mid y^{\text{obs}}, r; \theta\}$, with $U_{\text{comp}}(\theta)$ denoting the score function associated with the complete data likelihood (3.48). Matrix $I_{\text{mis}|\text{obs}}(\theta)$ is often referred to as the 'missing information'.

Louis' identity (3.50) is established as follows. Differentiating the observed score $U_{\text{obs}}(\theta) = \text{E}\{U_{\text{comp}}(\theta) \mid y^{\text{obs}}, r; \theta\}$ with respect to θ^{T} by applying the rule for differentiation under the integral sign, we obtain

$$\begin{aligned} -J_{\text{obs}}(\theta) = & \int \frac{\partial U_{\text{comp}}(\theta)}{\partial \theta^{\mathsf{T}}} f(y^{\text{mis}} \mid y^{\text{obs}}, r; \theta)\, \mathrm{d}y^{\text{mis}} \\ & + \int U_{\text{comp}}(\theta) U^{\mathsf{T}}_{\text{mis}|\text{obs}}(\theta) f(y^{\text{mis}} \mid y^{\text{obs}}, r; \theta)\, \mathrm{d}y^{\text{mis}}. \end{aligned}$$

By substituting the identity $U_{\text{mis}|\text{obs}}(\theta) = U_{\text{comp}}(\theta) - U_{\text{obs}}(\theta)$, the second integral can be written as

$$\text{E}\{U_{\text{comp}}(\theta) U^{\mathsf{T}}_{\text{comp}}(\theta) \mid y^{\text{obs}}, r; \theta\} - U_{\text{obs}}(\theta) U^{\mathsf{T}}_{\text{obs}}(\theta),$$

which corresponds to $\text{var}\{U_{\text{comp}}(\theta) \mid y^{\text{obs}}, r; \theta\}$. Replacing θ with $\widehat{\theta}$ yields the Louis identity (3.50). Both $K_{\text{comp}}(\widehat{\theta})$ and $I_{\text{mis}|\text{obs}}(\widehat{\theta})$ can be directly obtained from the final iteration of the EM algorithm used to obtain $\widehat{\theta}$.

3.6 Survival data

Section 3.5 presents the use of likelihood-based methods for analyzing incomplete data, where observations are either fully available or entirely missing. With survival data, observations are often censored, and these censored observations provide some information about the unobserved survival times. For example, with right censoring, we know the survival time is greater than the observed value; with interval censoring, we know the survival time falls between two known values.

3.6.1 Right-censored data

Let T be a non-negative random variable denoting the survival time of an individual, with density function $f(t)$. The associated survivor function

$$S(t) = \text{pr}(T > t) = \int_t^{\infty} f(v)\, \mathrm{d}v, \quad t \geq 0, \tag{3.51}$$

is the probability that a subject's survival time exceeds a given time point t.

Survival times are often subject to right censoring, either because the study ends before some subjects experience the event of interest or because subjects drop out of the study. To account for right censoring, an additional random variable, C, is introduced to represent the censoring time. For each subject in the study, the observed time is recorded as $Y = \min(T, C)$, along with the censoring indicator $\delta = 1(T \leq C)$.

Consider a study involving n individuals observed from $t = 0$ until the study concludes. For each individual $i = 1, \ldots, n$, let T_i denote the survival time, C_i the censoring time, and $\delta_i = 1(T_i \leq C_i)$ the censoring indicator variable. The observed time for subject i is then given by y_i, a realization of $\min(T_i, C_i)$.

Under the assumption that the censoring time is independent of the survival time, the joint distribution of the sample $\{(y_i, \delta_i) : i = 1, \ldots, n\}$ yields the likelihood function

$$L(\theta) = \prod_{i=1}^{n} \{f(y_i; \theta)\}^{\delta_i} \{S(y_i; \theta)\}^{1-\delta_i}, \tag{3.52}$$

where, in principle, the parameter θ could be infinite-dimensional, leading to a nonparametric likelihood; see Chapter 6. It is assumed here that any parameters in the distribution of the censoring time are not related to θ. While this is often considered a reasonable assumption, it cannot be verified using the observed data.

Example 3.35 (Log-logistic distribution) Suppose the survival time T follows a log-logistic distribution with the probability density function

$$f(t; \beta_1, \beta_2) = \frac{(\beta_2/\beta_1)(t/\beta_1)^{\beta_2 - 1}}{\{1 + (t/\beta_1)^{\beta_2}\}^2}, \quad t > 0,$$

where β_1 and β_2 are positive parameters. The survivor function is

$$S(t) = \{1 + (t/\beta_1)^{\beta_2}\}^{-1},$$

and the likelihood function (3.52) based on a sample $\{(y_i, \delta_i) : i = 1, \ldots, n\}$ is

$$L(\beta_1, \beta_2) = \prod_{i=1}^{n} \left[\frac{(\beta_2/\beta_1)(y_i/\beta_1)^{\beta_2 - 1}}{\{1 + (y_i/\beta_1)^{\beta_2}\}^2}\right]^{\delta_i} \{1 + (y_i/\beta_1)^{\beta_2}\}^{-(1-\delta_i)}.$$

The methods discussed in Chapters 1 and 2 can be used directly for inference about the model parameters β_1 and β_2. □

Example 3.36 (Shifted exponential distribution) Suppose that the probability density function of the survival time T is

$$f(t; \mu, \sigma) = \frac{1}{\sigma} \exp\left(-\frac{t - \mu}{\sigma}\right), \quad t \geq \mu,$$

where $\theta = (\mu, \sigma)^{\mathsf{T}}$ is the vector of parameters. The log-likelihood function for θ, based on a right-censored random sample $\{(y_i, \delta_i) : i = 1, \ldots, n\}$, is given by

$$\ell(\theta) = -n_0 \log \sigma - \frac{1}{\sigma} \sum_{i=1}^{n} (y_i - \mu),$$

with $n_0 = \sum_{i=1}^{n} \delta_i$, assumed to be positive without loss of generality.

Let $y_{(1)} = \min\{y_i : i = 1, \ldots, n\}$. Since $y_{(1)} \geq \mu$, it is easily seen that $\ell(\theta)$ is maximized with respect to μ at $\widehat{\mu} = y_{(1)}$. Fixing μ at $\widehat{\mu}$ and maximizing over σ gives

$$\widehat{\sigma} = \sum_{i=1}^{n} (y_i - y_{(1)})/n_0.$$

The usual normal-theory asymptotic results do not apply, as μ is an endpoint parameter; see Exercise 1.23. □

Example 3.37 (Transformation models) Transformation models, described in Example 1.15, can be used to model survival data by applying a log-transformation to the survival time. Let $W = \log T$. A location-scale model for W has density

$$f(w; \mu, \sigma) = \frac{1}{\sigma} f_0 \left(\frac{w - \mu}{\sigma} \right), \quad -\infty < w < \infty, \tag{3.53}$$

where $f_0(w)$ is assumed known, $-\infty < \mu < \infty$ is a location parameter, and $\sigma > 0$ is a scale parameter. In Figure 3.3, we illustrate the density function (3.53) and the survivor function for two values of μ, where $f_0(w)$ is the density of the standard extreme value distribution

$$f_0(w) = \exp(-w) \exp\{-\exp(-w)\}, \quad -\infty < w < \infty;$$

and then the corresponding survivor function is

$$S_0(w) = 1 - \exp\{-\exp(-w)\}, \quad -\infty < w < \infty.$$

For a sample $\{(y_i, \delta_i) : i = 1, \ldots, n\}$, from (3.52) and (3.53), the likelihood function is

$$L(\theta) = \prod_{i=1}^{n} \left\{ \frac{1}{\sigma} f_0 \left(\frac{w_i - \mu}{\sigma} \right) \right\}^{\delta_i} \left\{ S_0 \left(\frac{w_i - \mu}{\sigma} \right) \right\}^{1-\delta_i},$$

where $\theta = (\mu, \sigma)^{\mathsf{T}}$, $w_i = \log y_i$, and the term not depending on θ is omitted. Let $\widetilde{w}_i = (w_i - \mu)/\sigma$, and $n_0 = \Sigma_{i=1}^{n} \delta_i$. The log-likelihood function is

$$\ell(\theta) = -n_0 \log \sigma + \sum_{i=1}^{n} \{\delta_i \log f_0(\widetilde{w}_i) + (1 - \delta_i) \log S_0(\widetilde{w}_i)\},$$

yielding the score functions

$$\frac{\partial \ell(\theta)}{\partial \mu} = -\frac{1}{\sigma} \sum_{i=1}^{n} \left\{ \delta_i \frac{\partial \log f_0(\widetilde{w}_i)}{\partial \widetilde{w}_i} + (1 - \delta_i) \frac{\partial \log S_0(\widetilde{w}_i)}{\partial \widetilde{w}_i} \right\};$$

$$\frac{\partial \ell(\theta)}{\partial \sigma} = -\frac{n_0}{\sigma} - \frac{1}{\sigma} \sum_{i=1}^{n} \left\{ \delta_i \widetilde{w}_i \frac{\partial \log f_0(\widetilde{w}_i)}{\partial \widetilde{w}_i} + (1 - \delta_i) \widetilde{w}_i \frac{\partial \log S_0(\widetilde{w}_i)}{\partial \widetilde{w}_i} \right\}.$$

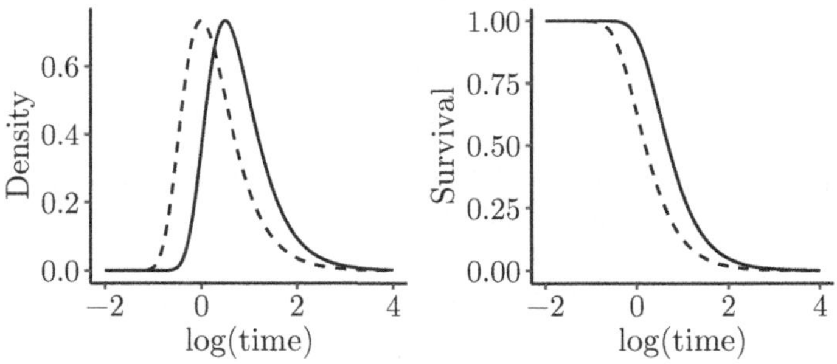

FIGURE 3.3
The left panel shows the density functions of the transformation model (3.53) for $\mu = 0$ (dashed line) or $\mu = 0.5$ (solid line) with $\sigma = 0.5$ in both cases. The right panel shows the associated survivor functions.

The maximum likelihood estimator $\widehat{\theta} = (\widehat{\mu}, \widehat{\sigma})^{\mathsf{T}}$ is obtained by solving the score equations

$$\frac{\partial \ell(\theta)}{\partial \mu} = 0, \quad \frac{\partial \ell(\theta)}{\partial \sigma} = 0. \tag{3.54}$$

The limiting distribution of $\widehat{\theta}$ in (1.50) can be applied for inference about the parameters. Alternatively, the result (1.51) for the likelihood ratio test discussed in Section 1.2 can be used to test the hypotheses $H_0 : \theta = \theta_0$ versus $H_1 : \theta \neq \theta_0$, where, for example, $\theta_0 = (0, 1)^{\mathsf{T}}$. Specifically,

$$\begin{aligned} 2\{\ell(\widehat{\theta}) - \ell(\theta_0)\} = &-2n_0 \log \widehat{\sigma} + 2\sum_{i=1}^{n} \delta_i \left\{ \log f_0\left(\frac{w_i - \widehat{\mu}}{\widehat{\sigma}}\right) - \log f_0(w_i) \right\} \\ &+ 2\sum_{i=1}^{n} (1 - \delta_i) \left\{ \log S_0\left(\frac{w_i - \widehat{\mu}}{\widehat{\sigma}}\right) - \log S_0(w_i) \right\}, \end{aligned} \tag{3.55}$$

and its limiting distribution is χ^2_2. When $f_0(t)$ is the density of the standard extreme value distribution, the likelihood ratio test statistic (3.55) simplifies to

$$\begin{aligned} 2\{\ell(\widehat{\theta}) - \ell(\theta_0)\} = &-2n_0 \log \widehat{\sigma} + 2\sum_{i=1}^{n} \delta_i \left\{ w_i - \frac{w_i - \widehat{\mu}}{\widehat{\sigma}} - \exp(-w_i) + \exp\left(-\frac{w_i - \widehat{\mu}}{\widehat{\sigma}}\right) \right\} \\ &+ 2\sum_{i=1}^{n} (1 - \delta_i) \log \left[1 - \exp\left\{ -\exp\left(\frac{w_i - \widehat{\mu}}{\widehat{\sigma}}\right) \right\} \right] \\ &- 2\sum_{i=1}^{n} (1 - \delta_i) \log \left[1 - \exp\left\{ -\exp(w_i) \right\} \right]. \end{aligned}$$

□

Example 3.38 (Cure rate model) Consider a scenario where the population is comprised of two sub-populations in proportions p and $1 - p$, where $p \in (0, 1)$. The probability density functions for the survival times of individuals in these sub-populations are denoted by $f_1(t)$ and $f_2(t)$, respectively. Then the survival time for a randomly selected individual has density function

$$f(t) = pf_1(t) + (1 - p)f_2(t),$$

and survivor function

$$S(t) = pS_1(t) + (1 - p)S_2(t), \tag{3.56}$$

where $S_j(t)$ is the survivor function for the jth sub-population with $j = 1, 2$.

If there is a sub-population of individuals who will never experience the event of interest, so that $S_2(t) = 1$ for all $t \geq 0$, then (3.56) becomes

$$S(t) = pS_1(t) + (1 - p).$$

This model is useful for scenarios involving a sub-population with extremely long lifetimes. If the density function $f_1(t)$ is a Weibull density,

$$f_1(t; \beta) = \frac{\beta_2}{\beta_1^{\beta_2}} t^{\beta_2 - 1} \exp\{-(t/\beta_1)^{\beta_2}\},$$

then the corresponding survivor function is $S_1(t; \beta) = \exp\left\{-(t/\beta_1)^{\beta_2}\right\}$, and the log-likelihood function for $\theta = (p, \beta_1, \beta_2)^{\mathsf{T}}$ is

$$\ell(\theta) = \sum_{i=1}^{n} [\delta_i \log\{pf_1(y_i; \beta)\} + (1 - \delta_i) \log\{pS_1(y_i; \beta) + (1 - p)\}].$$

□

3.6.2 Interval censored data

Interval censoring is another form of missing data that may be encountered in survival analysis. For each subject $i = 1, \dots, n$, the survival time T_i is not observed; we know only that T_i falls within the time interval $(a_i, b_i]$, where a_i and b_i are known values.

Consider a sample of n individuals subject to independent interval censoring, i.e., we observe only that $t_i \in (a_i, b_i]$, for $i = 1, \dots, n$. Assuming the survival times T_i are independent and identically distributed with survivor function $S(t)$, the joint probability is

$$\prod_{i=1}^{n} \mathrm{pr}(a_i < T_i \leq b_i) = \prod_{i=1}^{n} \{S(a_i) - S(b_i)\}. \tag{3.57}$$

If we assume a parametric model with parameter θ for the survival times, then the log-likelihood function is

$$\ell(\theta) = \sum_{i=1}^{n} \log\{S(a_i;\theta) - S(b_i;\theta)\},$$

with score function

$$\frac{\partial \ell(\theta)}{\partial \theta} = \sum_{i=1}^{n} \left(\frac{1}{\Delta S_i}\right) \left\{\frac{\partial(\Delta S_i)}{\partial \theta}\right\},$$

where $\Delta S_i = S(a_i;\theta) - S(b_i;\theta)$. The maximum likelihood estimator $\widehat{\theta}$ of θ can then be obtained by solving the score equation, and inference for θ can proceed using standard likelihood-based methods.

On the other hand, as the likelihood function depends only on the survivor function evaluated at a sequence of points $\{a_i, b_i : i = 1, \dots, n\}$, it is natural to consider nonparametric inference. Let $\{c_1, \dots, c_{m-1}\}$ denote the distinct values of $\{a_i, b_i : i = 1, \dots, n\}$, arranged in increasing order $0 = c_0 < c_1 < \cdots < c_{m-1} < c_m = \infty$.

Define $p_j = S(c_{j-1}) - S(c_j)$ and $z_{ij} = 1\{(c_{j-1}, c_j] \subset (a_i, b_i]\}$ for $j = 1, \dots, m$ and $i = 1, \dots, n$. Let $p = (p_1, \dots, p_m)^{\mathsf{T}}$. The likelihood function (3.57) is then

$$L(p) = \prod_{i=1}^{n} \sum_{j=1}^{m} z_{ij} p_j. \tag{3.58}$$

The maximum of (3.58) can be obtained using the EM algorithm, discussed in Section 3.5.2, by framing it as an incomplete data problem related to a multinomial distribution. For $i = 1, \dots, n$, let $W_i = (W_{i1}, \dots, W_{im})^{\mathsf{T}}$ denote a multinomial random vector with a single trial and probability vector $p = (p_1, \dots, p_m)^{\mathsf{T}}$. The log-likelihood function for this multinomial is

$$\ell_{\mathrm{W}}(p) = \sum_{i=1}^{n} \sum_{j=1}^{m} W_{ij} \log p_j, \tag{3.59}$$

where W_{ij} is either 0 or 1 for $j = 1, \dots, m$, and $\Sigma_j W_{ij} = 1$.

The likelihood function (3.58), based on $Z = \{z_{ij} : i = 1, \dots, n, j = 1, \dots, m\}$, is that for the incomplete data; the complete data are $W = \{W_{ij} : i = 1, \dots, n, j = 1, \dots, m\}$, with log-likelihood function (3.59). In the expectation step at iteration k, we calculate

$$\mathrm{E}_{p^{(k)}}\{\ell_{\mathrm{W}}(p) \mid Z\} = \sum_{i=1}^{n} \sum_{j=1}^{m} \mathrm{E}_{p^{(k)}}(W_{ij} \mid Z) \log p_j, \tag{3.60}$$

where

$$E_{p^{(k)}}(W_{ij} \mid Z) = \frac{z_{ij}p_j^{(k)}}{\sum_{l=1}^{m} z_{il}p_l^{(k)}}.$$

In the maximization step, we maximize $E_{p^{(k)}}\{\ell_W(p) \mid Z\}$ over p, with solution

$$p_j^{(k+1)} = \frac{1}{n}\sum_{i=1}^{n} \frac{z_{ij}p_j^{(k)}}{\sum_{l=1}^{m} z_{il}p_l^{(k)}}, \quad j = 1, \ldots, m.$$

The expectation and maximization steps are iterated until convergence. A variance estimate of the resulting estimator of p can be obtained using the formula in Louis (1982), as noted at the end of Section 3.5.2.

3.6.3 Proportional hazards regression

The preceding subsections (except Example 3.38) describe the use of likelihood-based methods to analyze survival data from a homogeneous population, where a single distribution for the survival process is assumed. In contrast, for nonhomogeneous populations, where individual survival depends on covariates, regression models are commonly used. One approach is to transform the survival time (e.g., by taking the logarithm) to remove its positivity constraint, and then use linear regression models to describe the dependence of the transformed survival time on the covariates.

Alternatively, the focus can shift to describing how the hazard function is associated with the covariates. Conditional on the covariate vector $X_i = x_i$ for subject i, the hazard function, also referred to as the instantaneous failure rate, is defined as

$$\lambda(t; x_i) = \lim_{\Delta t \to 0^+} \frac{\text{pr}(t \leq T < t + \Delta t \mid T \geq t, X_i = x_i)}{\Delta t}, \quad t \geq 0, \tag{3.61}$$

and the conditional survivor function $S(t; x_i) = \text{pr}(T_i > t \mid X_i = x_i)$ is

$$S(t; x_i) = \exp\left\{-\int_0^t \lambda(v; x_i)\, dv\right\}. \tag{3.62}$$

As discussed in Section 2.7, the proportional hazards model links the hazard function (3.61) to the covariates through the model

$$\lambda(t; x_i) = \lambda_0(t) \exp(x_i^{\mathsf{T}}\beta), \tag{3.63}$$

where β is a vector of regression parameters, and $\lambda_0(t)$ is a baseline hazard function. Unlike earlier regression models (e.g., those in Section 3.2.3), there is no intercept term in β, since any such term is not identifiable separately from $\lambda_0(t)$.

Using (3.62) and a corresponding expression for the probability density function similar to (3.51), the conditional survivor and probability density functions are given by

$$S(t; x_i) = \exp\left\{-\Lambda_0(t)\exp(x_i^{\mathsf{T}}\beta)\right\}, \quad f(t; x_i) = \lambda_0(t)\exp(x_i^{\mathsf{T}}\beta)\exp\left\{-\Lambda_0(t)\exp(x_i^{\mathsf{T}}\beta)\right\},$$

where $\Lambda_0(t) = \int_0^t \lambda_0(v)\, dv$ is the baseline cumulative hazard function.

For a right-censored sample $\{\{y_i, \delta_i\} : i = 1, \ldots, n\}$ as described in Section 3.6.1, and assuming independent censoring, the log-likelihood function for the sample is, from (3.52),

$$\ell\{\beta, \lambda_0(\cdot)\} = \sum_{i=1}^{n} \left[\delta_i\{\log \lambda_0(y_i) + x_i^{\mathsf{T}}\beta\} - \Lambda_0(y_i)\exp(x_i^{\mathsf{T}}\beta)\right]. \tag{3.64}$$

For inference about β, the baseline hazard function $\lambda_0(t)$ can be modelled either parametrically or nonparametrically. Here we consider the nonparametric approach, and leave $\lambda_0(t)$ unspecified but restrict the baseline cumulative function $\Lambda_0(t)$ to have jumps only at the observed values $y_1, \ldots, y_n$; the sizes of these jumps, $\lambda_1, \ldots, \lambda_n$, are treated as unknown parameters, along with the parameter β. For ease of exposition, we assume that there are no tied survival times in the sample. For the case with tied survival times, see Exercise 3.19 and Davison (2003, Section 5.4).

As a function of (β, λ), the log-likelihood function is

$$\ell(\beta, \lambda) = \sum_{i=1}^{n} \Big\{ \delta_i (x_i^\mathsf{T}\beta + \log \lambda_i) - \lambda_i \sum_{j: y_j \geq y_i} \exp(x_j^\mathsf{T}\beta) \Big\},$$

and

$$\frac{\partial \ell(\beta, \lambda)}{\partial \lambda_i} = \frac{\delta_i}{\lambda_i} - \sum_{j: y_j \geq y_i} \exp(x_j^\mathsf{T}\beta), \quad i = 1, \ldots, n. \tag{3.65}$$

Setting (3.65) to zero gives the constrained maximum likelihood estimate

$$\widehat{\lambda}_{i,\beta} = \frac{\delta_i}{\sum_{j: y_j \geq y_i} \exp(x_j^\mathsf{T}\beta)}, \qquad i = 1, \ldots, n,$$

and substituting this into (3.64) gives the profile likelihood function for β:

$$L_\mathrm{p}(\beta) \propto \prod_{i=1}^{n} \left[\frac{\exp(x_i^\mathsf{T}\beta)}{\sum_{j: y_j \geq y_i} \exp(x_j^\mathsf{T}\beta)} \right]^{\delta_i}, \tag{3.66}$$

which is the same as the partial likelihood function $L_\mathrm{part}(\beta)$ (2.82) discussed in Section 2.7. The set $\{j : y_j \geq y_i\}$ is the risk set at time y_i, i.e., the set of individuals who have not failed or been censored before time y_i. A key advantage of the partial likelihood is that the baseline hazard function, $\lambda_0(t)$, can remain unspecified.

The estimator of β obtained by maximizing (3.66) is consistent and asymptotically normal, with asymptotic variance estimated by the inverse negative Hessian of $\log L_\mathrm{p}$ in the usual way; see for example Kalbfleisch and Prentice (2002, Section 4.2) and Lawless (2003, Section 7.1). This is somewhat surprising, as the asymptotic theory for inference based on the profile likelihood discussed in Chapter 2 does not apply for models where the number of parameters increases with the sample size; the result is related to that for empirical likelihood, discussed in Section 6.6.

The usual nonparametric estimator of the baseline cumulative hazard function $\Lambda_0(\cdot)$ is the Nelson–Aalen estimator, reviewed in Borgan (2014). This is closely related to the Kaplan–Meier estimator discussed in Section 6.1.

The proportional hazards model was introduced in the groundbreaking and highly cited paper by Cox (1972). That paper derived the partial likelihood function through an ingenious argument motivated by the theory of Poisson processes. Cox argued that at each ordered failure time y_i, an appropriate contribution to the likelihood function was the conditional probability that subject i was observed to fail, conditional on the information that there was a failure at that time, and that with an arbitrary hazard function there could be no information about the regression parameter available from the process between failure times. This is related to the ancillary conditioning described in Section 2.2.

The derivation of the partial likelihood as a profile likelihood appears in Johansen (1983), and was formally verified in Murphy and van der Vaart (2000). Aalen (1978) set out the basic martingale theory for the study of failure time processes. Andersen and Gill (1988) used Aalen's formulation to show that the limiting distribution of the partial maximum likelihood estimator $\widehat{\beta}_{\rm part}$ is normal, and to justify the use of the usual pivotal quantities. In fact, $\widehat{\beta}_{\rm part}$ is also asymptotically fully efficient in the semi-parametric model (2.76), and has also been shown to have relatively high efficiency against several parametric models (Efron, 1977; Oakes, 1977). This is surprising in light of the failure of profile likelihood in Neyman–Scott problems, as we are maximizing over n_0 nuisance parameters in the hazard function, where n_0 is the number of the observed failure times. The same phenomenon emerges in empirical likelihood; see Chapter 6.

3.6.4 Bivariate survival models

Suppose the sample includes n clusters, each with two survival times (e.g., survival times for n couples). For $i = 1, \ldots, n$ and $j = 1, 2$, let T_{ij} denote the jth survival time in cluster i and let C_{ij} and X_{ij} be the corresponding censoring time and associated covariate vector, respectively. Define $\delta_{ij} = 1(T_{ij} \leq C_{ij})$. Let y_{ij} denote a realization of $\min(T_{ij}, C_{ij})$ and $R_{ij}(t) = 1(y_{ij} \geq t)$ indicate whether the jth subject in cluster i remains at risk at time t.

Write $y_i = (y_{i1}, y_{i2})^{\mathsf{T}}$, $\delta_i = (\delta_{i1}, \delta_{i2})^{\mathsf{T}}$, $T_i = (T_{i1}, T_{i2})^{\mathsf{T}}$, $C_i = (C_{i1}, C_{i2})^{\mathsf{T}}$, and $X_i = (X_{i1}^{\mathsf{T}}, X_{i2}^{\mathsf{T}})^{\mathsf{T}}$. Within cluster i, the survival times T_{i1} and T_{i2} may be correlated, while the sets $\{T_i, C_i, X_i\}$ are assumed to be independent across clusters $i = 1, \ldots, n$. Conditional on the covariates X_i, survival times T_i and censoring times C_i are assumed to be independent for each i.

Let $S(t_1, t_2; \theta) = \mathrm{pr}(T_{i1} > t_1, T_{i2} > t_2 \mid X_i = x_i)$ represent the joint survivor function of T_{i1} and T_{i2}, parametrized by $\theta = (\beta^{\mathsf{T}}, \phi^{\mathsf{T}})^{\mathsf{T}}$, where the dependence on x_i is suppressed in $S(t_1, t_2; \theta)$ for simplicity, β includes all the parameters associated with the marginal survivor functions $S_j(t; \beta) = \mathrm{pr}(T_{ij} > t; \beta)$, $j = 1, 2$, and ϕ comprises the association parameters. The likelihood function contribution from cluster i is

$$L_i(\theta) = \{f(y_{i1}, y_{i2}; \theta)\}^{\delta_{i1}\delta_{i2}} \left\{-\frac{\partial S(y_{i1}, y_{i2}; \theta)}{\partial y_{i1}}\right\}^{\delta_{i1}(1-\delta_{i2})} \left\{-\frac{\partial S(y_{i1}, y_{i2}; \theta)}{\partial y_{i2}}\right\}^{(1-\delta_{i1})\delta_{i2}} \times \{S(y_{i1}, y_{i2}; \theta)\}^{(1-\delta_{i1})(1-\delta_{i2})}, \tag{3.67}$$

where $f(y_{i1}, y_{i2}; \theta) = \partial^2 S(y_{i1}, y_{i2}; \theta)/\partial y_{i1}\partial y_{i2}$.

Maximizing (3.67) may in some applications be computationally cumbersome. To address this, a two-stage estimation algorithm can be employed to simplify computation, treating the model parameters differently. In the first stage, the dependence between T_{i1} and T_{i2} is ignored, and a marginal model is used to construct a pseudo-likelihood for estimating β. In the second stage, ϕ is estimated using the joint model with β replaced by its estimate obtained from the first stage. The first-stage pseudo-likelihood contribution from cluster i is

$$L_i^*(\beta) = L_{i1}^*(\beta) L_{i2}^*(\beta),$$

where

$$L_{ij}^*(\beta) = \left\{\frac{-\partial S_j(y_{ij}; \beta)}{\partial y_{ij}}\right\}^{\delta_{ij}} \{S_j(y_{ij}; \beta)\}^{1-\delta_{ij}}$$

is the marginal likelihood function for β based on the distribution of T_{ij}, $j = 1, 2$; this construction treats T_{i1} and T_{i2} as independent.

The first stage estimate $\widehat{\beta}$ is obtained by solving for β

$$\sum_{i=1}^{n} \frac{\partial \log L_i^*(\beta)}{\partial \beta} = 0. \tag{3.68}$$

In the second stage, an estimate $\widehat{\phi}$ is obtained by solving

$$\sum_{i=1}^{n} \frac{\partial \log L_i(\beta, \phi)}{\partial \phi} = 0, \tag{3.69}$$

with β replaced by $\widehat{\beta}$. The validity of this procedure can be justified from the perspective of composite likelihood discussed in Section 4.7, or alternatively, from the estimating function viewpoint. By considering the derivatives of the log pseudo-likelihood function as unbiased estimating functions for the marginal model parameters, Theorem 5.3 in Section 5.4 implies the consistency of $\widehat{\beta}$. Estimating ϕ in the second stage uses the profile method discussed in Section 5.9.

Variance estimates for the estimator $\widehat{\theta} = (\widehat{\beta}^{\mathsf{T}}, \widehat{\phi}^{\mathsf{T}})^{\mathsf{T}}$ can be obtained by constructing an unbiased estimating function for parameters β and ϕ simultaneously using the functions in (3.68) and (3.69),

$$\Psi_i(\theta) = \left(\frac{\partial \log L_i^*(\beta)}{\partial \beta^{\mathsf{T}}}, \frac{\partial \log L_i(\beta, \phi)}{\partial \phi^{\mathsf{T}}} \right)^{\mathsf{T}},$$

Applying Theorem 5.3 to $\Psi_i(\theta)$ yields the asymptotic covariance matrix of $\sqrt{n}(\widehat{\theta} - \theta)$,

$$\left[\mathrm{E}\left\{ \frac{\partial \Psi_i(\theta)}{\partial \theta} \right\} \right]^{-1} \mathrm{E}\{\Psi_i(\theta)\Psi_i^{\mathsf{T}}(\theta)\} \left[\mathrm{E}\left\{ \frac{\partial \Psi_i(\theta)}{\partial \theta^{\mathsf{T}}} \right\} \right]^{-1}.$$

Example 3.39 (Copula models) Dependence among the failure times T_{ij} can be modelled using copula functions. Let $C(\cdot, \cdot; \phi)$ be a copula function indexed by a scalar or vector parameter ϕ. The joint survivor function of T_{i1} and T_{i2} can be expressed as

$$S(t_1, t_2) = C\{S_1(t_1), S_2(t_2); \phi\}, \tag{3.70}$$

where $S_j(\cdot)$ is the marginal survivor function for T_{ij}, $j = 1, 2$. Consequently, the joint probability density function of (T_{i1}, T_{i2}) is

$$f(t_1, t_2) = c\{S_1(t_1), S_2(t_2); \phi\} f_1(t_1) f_2(t_2)$$

where $c(\cdot; \phi)$ is the density function corresponding to the copula $C(\cdot, \cdot; \phi)$, and $f_j(\cdot)$ is the marginal density function of T_{ij}, $j = 1, 2$.

The family of Archimedean copula functions is given by

$$C(u_1, u_2; \phi) = \psi^{[-1]}\{\psi(u_1; \phi) + \psi(u_2; \phi)\},$$

where $\psi : [0, 1] \to [0, \infty]$ is a continuous, strictly decreasing, and convex function indexed by parameter ϕ, satisfying $\psi(1) = 0$. The pseudo-inverse $\psi^{[-1]}$ is defined by

$$\psi^{[-1]}(u) = \begin{cases} \psi^{-1}(u), & 0 \le u \le \psi(0), \\ 0, & \psi(0) \le u \le \infty. \end{cases}$$

When $\psi(0) = \infty$, $\psi^{[-1]} = \psi^{-1}$. Here, ψ^{-1} denotes the inverse function of ψ (Nelsen, 2007, Sections 4.1 and 4.3).

Three commonly used Archimedean copula families are:

(a) the Clayton copula: $\psi(u; \phi) = (u^{-\phi} - 1)/\phi, \quad \phi \in [-1, 0) \cup (0, \infty)$;

(b) the Frank copula: $\psi(u; \phi) = -\log[\{\exp(-\phi u) - 1\}/\{\exp(-\phi) - 1\}], \quad \phi \neq 0$;

(c) the Gumbel family: $\psi(u; \phi) = (-\log u)^{\phi}, \quad \phi \ge 1$.

Models for univariate survival times can be used for $S_j(\cdot)$ for $j = 1, 2$. The estimation of all model parameters can then be based on either the full likelihood function (3.67) or the pseudo-likelihood method using (3.68) and (3.69). □

3.6.5 Joint analysis of survival and longitudinal data

For each subject $i = 1, \ldots, n$, let T_i and C_i be the survival time and censoring time, respectively, and y_i a realization of $\min(T_i, C_i)$, as in Section 3.6.1. Associated with each subject's survival, there is a longitudinal covariate process $\{X_i(t) : t > 0\}$ and a scalar time-invariant covariate Z_i.

The longitudinal process $\{X_i(t) : t > 0\}$ is not usually fully observed, but only measured at specific time points; it may also be measured with error. In most studies the measurements of $X_i(t)$ terminate at the censoring or event time. Let $0 \leq t_{i1} < \cdots < t_{im_i} \leq y_i$ be the observation times, and $X_i^*(t_{ij})$ the observed measurement of $X_i(t_{ij})$, where $j = 1, \ldots, m_i$, and m_i is the number of longitudinal measurements for individual i.

We assume that the observed and the true longitudinal measurements are linked by the model

$$X_i^*(t_{ij}) = X_i(t_{ij}) + e_{ij}, \quad t_{i1} < \cdots < t_{im_i} \leq y_i, \tag{3.71}$$

where e_{ij} are assumed to be i.i.d. $N(0, \sigma_e^2)$, and to be independent of $\{T_i, C_i, Z_i, X_i(\cdot)\}$. We further assume that the true covariate follows a random effects model

$$X_i(t) = u_{0i} + u_{1i}t + \gamma Z_i. \tag{3.72}$$

The random effects u_{0i} and u_{1i} are assumed independent of each other, e_{ij}, and $\{T_i, C_i, Z_i\}$, and distributed as $N(0, \sigma_0^2)$ and $N(0, \sigma_1^2)$, respectively. We also assume that the censoring and longitudinal measurement processes are noninformative and independent of the future covariates and random effects.

For the survival process, we consider the Cox regression model

$$\lambda(t; X_i(t), Z_i) = \lambda_0(t) \exp\{\beta X_i(t) + \alpha Z_i\}. \tag{3.73}$$

Figure 3.4 gives a visualization of the relationship among the variables and their associated parameters. For example, if Z_i is a treatment indicator, and $X_i(t)$ is a longitudinal marker for patient i, then β indicates how the longitudinal marker may be associated with the survival time, α captures the direct treatment effect on a patient's survival time, and γ in (3.72) captures the effect of treatment on the longitudinal marker. A value of $\beta = 0$ suggests no association between the longitudinal marker and the survival time; negative β implies a decreasing hazard rate, which in turn suggests an increasing survival time with an increasing value of the longitudinal marker.

The treatment also exerts an indirect effect on the survival time through the longitudinal marker, expressed as $\gamma\beta$, using (3.72). Therefore, the total treatment effect on a patient's survival is given by $\alpha + \gamma\beta$.

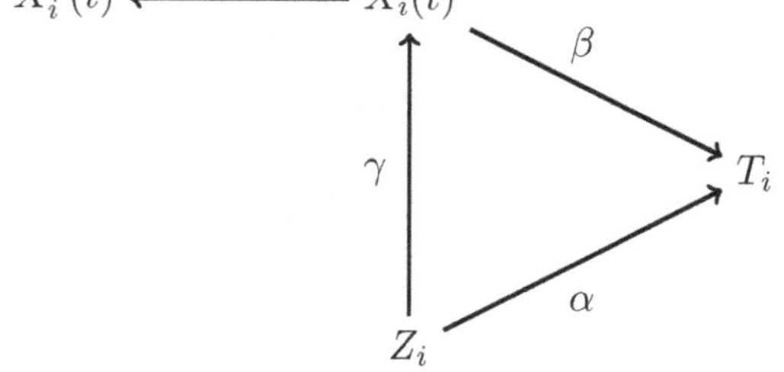

FIGURE 3.4
A display of the relationship among the variables with one mis-measured longitudinal covariate $X(t)$, and one time-invariant covariate Z, as described by the model in (3.71) – (3.73).

Now we discuss a method of estimating the associated parameters using the observed data, where variances σ_0^2 and σ_1^2 in model (3.72) are assumed known for simplicity. Suppose $\lambda_0(t)$ is modeled parametrically with a Weibull distribution

$$\lambda_0(t;\beta_1,\beta_2) = \beta_1\beta_2(\beta_1 t)^{\beta_2-1}. \tag{3.74}$$

Let $\theta = (\beta_1, \beta_2, \beta, \alpha, \vartheta^\mathsf{T})^\mathsf{T}$, where $\vartheta = (\sigma_e^2, \gamma)^\mathsf{T}$. From (3.72), (3.73), and (3.74), the likelihood contribution of the ith observation (y_i, δ_i) is

$$\begin{aligned} L_{\mathrm{S}i}(\theta) &= [\lambda_0(y_i)\exp\{\beta X_i(y_i) + \alpha Z_i\}]^{\delta_i} \\ &\quad \times \exp\left[-\int_0^{y_i} \lambda_0(v)\exp\{\beta X_i(v) + \alpha Z_i\}\,\mathrm{d}v\right] \\ &= [\lambda_0(y_i)\exp\{\beta(u_{0i} + u_{1i}y_i) + (\beta\gamma+\alpha)Z_i\}]^{\delta_i} \\ &\quad \times \exp\left[-\beta_2\beta_1^{\beta_2}\exp\{\beta u_{0i} + (\beta\gamma+\alpha)Z_i\}\Gamma(y_i;u_{1i},\beta_2,\beta)\right], \end{aligned}$$

with $\Gamma(y_i;u_{1i},\beta_2,\beta) = \int_0^{y_i} v^{\beta_2-1}\exp(\beta u_{1i}v)\,\mathrm{d}v$. Let $L_{\mathrm{L}i}(\vartheta)$ be the likelihood function for parameters governing the longitudinal components,

$$L_{\mathrm{L}i}(\vartheta) = \frac{1}{(2\pi\sigma_e^2)^{m_i/2}}\exp\left[-\frac{1}{2\sigma_e^2}\sum_{j=1}^{m_i}\{X_i^*(t_{ij}) - u_{0i} - u_{1i}t_{ij} - \gamma Z_i\}^2\right].$$

Then the joint likelihood function based on the observed data is

$$L(\theta,\vartheta) = \prod_{i=1}^n L_i(\theta,\vartheta) = \prod_{i=1}^n \iint L_{\mathrm{S}i}(\theta)L_{\mathrm{L}i}(\vartheta)f(u_{0i})f(u_{1i})\,\mathrm{d}u_{0i}\,\mathrm{d}u_{1i}. \tag{3.75}$$

This expression is analytically intractable, even for simple baseline hazards such as when $\beta_2 = 1$, but may be approximated using, for example, a Monte Carlo method.

3.7 Bibliographic notes

Reference books on statistical methods for longitudinal data include Diggle et al. (2013), Verbeke and Molenberghs (2000), Molenberghs and Verbeke (2005), and Fitzmaurice et al. (2011). Monographs on generalized linear mixed models include McCulloch and Searle (2004) and Jiang and Nguyen (2021). Inference for generalized linear mixed models based on the h-likelihood and its approximations is reviewed in Lee and Nelder (2009) and Jin and Lee (2021).

Conditional likelihood inference in generalized linear mixed models is discussed in Sartori and Severini (2004). Their approach extends previous work on conditional likelihood inference for generalized linear models in Davison (1988). A popular technique for approximate inference in generalized linear mixed models is the penalized quasi-likelihood method proposed by Breslow and Clayton (1993) and Wolfinger and O'Connell (1993), building on earlier work by Schall (1991). The Laplace approximation and adaptive Gauss–Hermite quadrature for mixed effects models are discussed in Pinheiro and Bates (1995) and implemented in the popular `lme4` package (Bates et al., 2015) for the R language (R Core Team, 2025). The Laplace approximation when the number of parameters increases with the sample size is discussed in Shun and McCullagh (1995) and Tang and Reid (2025). Approximate maximum likelihood estimation in generalized linear mixed models can also be obtained using a variety of Monte Carlo methods, including importance sampling, the Monte Carlo EM algorithm, and Markov chain Monte Carlo methods; see McCulloch (1997). A method for computing profile likelihood functions for parameters of multi-level models that avoids high-dimensional integration is discussed in Lele (2023); it builds on a method of data-doubling proposed by Efron (1993).

Bayesian analysis of generalized linear mixed models is commonly performed using Markov chain Monte Carlo methods. Seminal examples include the approximate Gibbs sampler discussed by Zeger and Karim (1991) and the Metropolis–Hastings algorithms presented by Clayton (1996) and Gamerman (1997). Bayesian inference for mixed-effects models has been greatly facilitated by the availability of flexible, general-purpose software such as WinBUGS (Lunn et al., 2000), JAGS (Plummer, 2003), and Stan (Carpenter et al., 2017).

The computational burden of Markov chain Monte Carlo methods has motivated the development of alternative non-simulation approaches to Bayesian inference, such as the integrated nested Laplace approximation (Rue et al., 2009) and variational inference (Jordan et al., 1999). In the context of generalized linear mixed models, Fong et al. (2010) discussed the integrated nested Laplace approximation, while Ormerod and Wand (2012) presented variational Gaussian inference.

Jiang (2013) addressed the consistency of maximum likelihood estimators in generalized linear mixed models, with an emphasis on crossed random effects. Asymptotic normality of the estimators is established in Jiang (2025). There is considerable interest in computationally efficient fitting of crossed random effects models, as they arise in large-scale applications such as recommender systems. Approximate likelihood methods for scalable inference in crossed random effects models are proposed in Xu et al. (2024) and Bellio et al. (2025), while a Bayesian approach is developed in Papaspiliopoulos et al. (2023).

The effect of misspecification of the random effects distribution is discussed in various articles, including Verbeke and Lesaffre (1997), Heagerty and Kurland (2001), McCulloch and Neuhaus (2011), and Hui et al. (2021).

Joint mean-covariance models in longitudinal data are discussed in Pourahmadi (1999), Pan and Mackenzie (2003), Daniels and Zhao (2003), and Zhang et al. (2015), among others.

Restricted maximum likelihood estimation of variance components was suggested by Patterson and Thompson (1971) and further developed in a variety of papers, in particular Harville (1974, 1977) and Corbeil and Searle (1976). See also Verbeke and Molenberghs (2000, Section 5.3). Maestrini et al. (2026) discussed extensions of restricted maximum likelihood estimation for generalized linear mixed models and their connections to inference based on h-likelihood.

The random-effects model for meta-analysis was first proposed by DerSimonian and Laird (1986) and has since become a standard approach for describing heterogeneity among studies in meta-analysis. Higgins et al. (2009) provided several insights into the limitations of standard methods used in random-effects meta-analysis. One of the major difficulties arises when only a limited number of studies are combined or when some studies have small sample sizes. In these cases, first-order likelihood methods can be highly inaccurate. Guolo (2012) used the approximation of Skovgaard (1996), mentioned in Section 2.5.2, to obtain second-order accurate inference in random-effects meta-analysis. Kosmidis et al. (2017) showed that methods reducing the asymptotic bias of the maximum likelihood estimator of the variance component can also substantially improve inference about the mean effect size. Guolo and Varin (2017) discussed and compared various likelihood and non-likelihood methods for meta-analysis when the number of studies is limited.

Autoregressive moving average processes were introduced independently by Yule and Slutsky (Yule, 1926, 1927; Slutsky, 1927). The theoretical foundations of autoregressive and moving average models come from the decomposition theorem introduced in Wold's

PhD thesis (Wold, 1938). Autoregressive and moving average models are closely associated with the influential monograph of Box and Jenkins (1970); their integrated methodology for fitting such models to time series data is commonly referred to as the Box–Jenkins approach.

There are many texts on theory and methods for linear and nonlinear time series. Popular introductory books on time series include Chatfield (1996) and Brockwell and Davis (2002). More advanced treatments can be found in Fuller (1976), Brockwell and Davis (1987), and Hamilton (1994). At an intermediate level, Shumway and Stoffer (2016) covered frequency and time-domain time series methods. Durbin and Koopman (2012) focused on state space models. Nonlinear times series are discussed in Tong (1990), from which Example 3.18 is taken.

Davis et al. (2021) provided an overview of several approaches to time series of counts, including the observation- and parameter-driven models mentioned in Examples 3.19 and 3.20, dynamic Bayesian models (West et al., 1985), integer autoregressive models (McKenzie, 1986), and hidden Markov models (Zucchini and MacDonald, 2009).

There is a vast literature on financial time series; see, for example, Tsay (2010) and references therein. The autoregressive conditional heteroscedastic model was introduced in Engle (1982) and extended in various directions, most notably in Bollerslev (1986). See the review by Bollerslev et al. (1994) and the monograph by Gouriéroux (2012). Simulation approaches for approximate likelihood inference in stochastic volatility models are discussed in Kim et al. (1998) and Pitt and Shephard (1999).

Reference texts on spatial statistics include Ripley (1981) and Cressie (1993). The term geostatistics is associated with the pioneering work of Matheron (1963) and the development of kriging methods (Cressie, 1990). A rigorous mathematical treatment of the theory of kriging is given in Stein (1999). Diggle et al. (1998) developed a general approach to spatial prediction which combines geostatistics and generalized linear mixed models; see also Diggle and Ribeiro (2007). The monograph of Cressie and Wikle (2011) treats spatio-temporal models; see also Porcu et al. (2021) and Chen et al. (2021) for reviews of spatio-temporal covariance structures.

The Monte Carlo maximum likelihood algorithm mentioned in Example 3.25 is an example of a more general class of simulation approaches to approximate likelihood inference developed in Geyer and Thompson (1992) and Geyer (1994).

The fundamental equivalence between Markov random fields and Gibbs distributions is established by the Hammersley–Clifford theorem, a seminal result that first appeared in an unpublished note in 1971; see Clifford (1990). The importance of Markov random fields in spatial statistics and image analysis owes much to the influential contributions of Besag (1974), Geman and Geman (1984), and Besag (1986). Gaussian Markov random fields are used in Rue and Held (2005) for Bayesian inference. Lindgren et al. (2011) explained how Gaussian geostatistical models can be approximated by Gaussian Markov random fields, thus reducing the computational burden of the standard simulation approaches for geostatistics.

Cox and Lewis (1966) and Cox and Isham (1980) are seminal accounts of point processes. The theory of statistical inference for point processes is the focus of Karr (1991). Monographs on spatial point processes include Møller and Waagepetersen (2003), Diggle (2013), and Baddeley et al. (2015); see also Gaetan and Guyon (2010, Chapter 3). The log-Gaussian Cox process for spatial data is reviewed in Diggle et al. (2013) and Møller et al. (1998).

The application of spatial statistics to model the geographical distribution of disease, a field often known as spatial epidemiology, is the topic of several monographs, including Elliott et al. (2001), Waller and Gotway (2004), and Lawson (2018).

A fundamental contribution to the formal statistical analysis of missing data is Rubin (1976); a comprehensive reference for this approach to missing data analysis is Little and Rubin (1987). A useful review of the assumptions regarding missing data mechanisms, which are fundamental to analyzing incomplete data, is provided by Little (2021).

An important form of informative missing data is when studies are removed in meta-analysis, as this removal can arise in publication bias. Selection models to handle publication bias are discussed in Copas and Li (1997) and Copas (1999). Rothstein et al. (2006) gave an account of methodology developed for publication bias in meta-analysis.

The expectation-maximization algorithm was initially developed by Dempster et al. (1977) for likelihood inference with missing data. It has been remarkably widely used for a variety of incomplete data problems, and many modifications and extensions have been developed, including Meng and Van Dyk (1998) and Booth and Hobert (1999). The convergence properties of the EM algorithm are studied in Wu (1983). A book-length treatment is McLachlan and Krishnan (2008). The EM algorithm is a special case of the more general class of majorize-minimize algorithms summarized in Hunter and Lange (2004).

Kalbfleisch and Prentice (2002) and Lawless (2003) emphasize likelihood methods for survival data analysis, with many examples. Semi-parametric regression models for multivariate interval-censored data are discussed in Zeng et al. (2017); the introductory section gives a helpful overview of the literature. They model the fixed and random covariate effects parametrically, but estimate the failure time distribution nonparametrically. Kalbfleisch and Prentice (2002, Section 3.9) discussed the use of the EM algorithm to maximize (3.58). See also Lawless (2003, Section 3.5). The Kaplan–Meier nonparametric estimator of the survival distribution is discussed in Example 6.1. High dimensional regression for survival data is reviewed in Salerno and Li (2023), with emphasis on penalization of the partial likelihood function and model selection. Penalized likelihood is discussed in Chapter 7.

Overviews of joint modelling of longitudinal and survival data include Tsiatis and Davidian (2004) and Papageorgiou et al. (2019); see also Rizopoulos (2012).

In many applications it is possible to describe how to generate data from an assumed model, but not possible to write the likelihood function in closed form. The field of Approximate Bayesian Computation (ABC) addresses Bayesian inference in this setting. There is a large and growing literature on this topic. For an introduction to ABC methods, see Marin et al. (2012). Bernton et al. (2019) presented an approach to ABC based on approximating the Wasserstein distance between the sample and synthetic observations. Approximate Bayesian computation may be interpreted as a form of indirect inference; see Section 4.6.

3.8 Further results and exercises

3.1. Verify the expressions (3.3) for the inverse and the determinant of the exchangeable correlation matrix.

3.2. Perform a computer experiment to compare the computational cost of inverting the AR(1) correlation matrix numerically or using (3.4). Consider AR(1) correlation matrices of size $m = 50, 100, 500, 1000, 5000$.

3.3. Consider the equicorrelation model $Y_i = \mu + \epsilon_i$, where $\epsilon = (\epsilon_1, \ldots, \epsilon_n)^\mathsf{T}$ follows a multivariate normal distribution with zero expected value and covariance matrix $\sigma^2 R(\alpha)$, with $R(\alpha) = \alpha 1_n 1_n^\mathsf{T} + (1-\alpha)\mathcal{I}_{n\times n}$. Let $\widehat{\alpha}$ be the maximum likelihood estimator of α.

(a) Show that $\mathrm{E}(\widehat{\alpha}) = (n-1)n^{-1}\alpha$, that is, the maximum likelihood estimator $\widehat{\alpha}$ is biased for α.

(b) Suppose that μ is known. Find the maximum likelihood estimator of α, and show that it is an unbiased estimator of α.

(c) Consider the constraint matrix $C = \mathcal{I}_{n\times n} - n^{-1}1_n1_n^{\mathsf{T}}$. Show that $\mathrm{E}(CY) = 0$ and $CY \sim N\{0, \sigma^2(1-\alpha)C\}$.

(d) Find the maximum likelihood estimator of α by using the result in (c).

(e) Compare the estimators of α in (a) and (d).

(f) Can you find another $n \times n$ matrix D such that $\mathrm{E}(DY) = 0$?

(Schabenberger and Gotway, 2005, Section 4.5)

3.4. The Gauss–Hermite quadrature rule is used in Statistics to approximate integrals with a normal weight function as

$$\int_{-\infty}^{\infty} \phi(y;\mu,\sigma^2) f(y)\,\mathrm{d}y \approx \sum_{i=1}^{m} \frac{w_i}{\sqrt{\pi}} f(\mu + \sqrt{2}\sigma t_i),$$

where $\phi(\cdot;\mu,\sigma^2)$ is the density of the $N(\mu,\sigma^2)$ distribution. The weights $\{w_1,\ldots,w_m\}$ and nodes $\{t_1,\ldots,t_m\}$ in the quadrature approximation are derived from Hermite polynomials in such a way that the approximation is exact if $f(y)$ is a polynomial of order $2m-1$ or less.

(a) Show how Gauss–Hermite quadrature can be used to approximate the likelihood integrals (3.10).

(b) Explain why Gauss–Hermite quadrature approximation in random effects models is unlikely to be accurate enough for inference in generalized linear mixed models.

(c) Adaptive Gauss–Hermite quadrature (Liu and Pierce, 1994) approximates the likelihood integrals (3.10) by centring and scaling the node points, using the Laplace approximation discussed in Example 3.7. Write down the expression of the adaptive Gauss–Hermite quadrature and explain why this adaptive quadrature improves upon the standard Gauss–Hermite quadrature approximation. *(Liu and Pierce, 1994).*

3.5. In the linear mixed-effects model of Example 3.6, show that $\mathrm{E}(b \mid y)$ minimizes the mean-squared error (3.11). Show further that

$$\begin{aligned}\mathrm{E}(b \mid y) &= \{Z^{\mathsf{T}}Z\sigma^{-2} + D^{-1}(\alpha)\}^{-1}Z^{\mathsf{T}}\sigma^{-2}(y - X\beta),\\ \mathrm{var}(b \mid y) &= \{Z^{\mathsf{T}}Z\sigma^{-2} + D^{-1}(\alpha)\}^{-1}.\end{aligned}$$

(Davison, 2003, Section 9.4)

3.6. Using (3.22), verify the following results for the maximum likelihood estimator in stationary causal and invertible ARMA(p, q) models with $p + q \leq 2$:

(a) For the AR(1) model, $Y_t = \phi_1 Y_{t-1} + \epsilon_t$,

$$\widehat{\phi}_1 \mathrel{\dot\sim} N\{\phi_1, (1-\phi_1^2)/n\}.$$

(b) For the MA(1) model, $Y_t = \epsilon_t + \omega_1\epsilon_{t-1}$,

$$\widehat{\omega}_1 \mathrel{\dot\sim} N\{\omega_1, (1-\omega_1^2)/n\}.$$

(c) For the AR(2) model, $Y_t = \phi_1 Y_{t-1} + \phi_2 Y_{t-2} + \epsilon_t$,

$$\begin{bmatrix}\widehat{\phi}_1\\ \widehat{\phi}_2\end{bmatrix} \mathrel{\dot\sim} N_2\left(\begin{bmatrix}\phi_1\\ \phi_2\end{bmatrix}, n^{-1}\begin{bmatrix}1-\phi_2^2 & -\phi_1(1+\phi_2)\\ -\phi_1(1+\phi_2) & 1-\phi_2^2\end{bmatrix}\right). \tag{3.76}$$

(d) For the MA(2) model, $Y_t = \epsilon_t + \omega_1\epsilon_{t-1} + \omega_2\epsilon_{t-2}$,

$$\begin{bmatrix}\widehat{\omega}_1\\ \widehat{\omega}_2\end{bmatrix} \mathrel{\dot\sim} N_2\left(\begin{bmatrix}\omega_1\\ \omega_2\end{bmatrix}, n^{-1}\begin{bmatrix}1-\omega_2^2 & \omega_1(1+\omega_2)\\ \omega_1(1+\omega_2) & 1-\omega_2^2\end{bmatrix}\right).$$

(e) For the ARMA(1,1) model, $Y_t = \phi_1 Y_{t-1} + \epsilon_t + \omega_1\epsilon_{t-1}$,

$$\begin{bmatrix}\widehat{\phi}_1\\ \widehat{\omega}_1\end{bmatrix} \mathrel{\dot\sim} N_2\left(\begin{bmatrix}\phi_1\\ \omega_1\end{bmatrix}, \frac{(1+\phi_1\omega_1)}{n(\phi_1+\omega_1)^2}\begin{bmatrix}(1-\phi_1^2)(1+\phi_1\omega_1) & -(1-\phi_1^2)(1-\omega_1^2)\\ -(1-\phi_1^2)(1-\omega_1^2) & (1-\omega_1^2)(1+\phi_1\omega_1)\end{bmatrix}\right). \quad (3.77)$$

(Shumway and Stoffer, 2016, Example 3.32)

3.7. Consider an invertible MA(1) model $Y_t = \epsilon_t + \omega\epsilon_{t-1}$.

(a) Code the log-likelihood function (3.20) using the autocovariance function (3.18).

(b) Write the MA(1) model in state space form and code the log-likelihood function (3.26) using the Kalman Filter.

(c) Carry out a simulation study to compare the computational costs of these two implementations.

3.8. Consider an invertible MA(1) model $Y_t = \epsilon_t + \omega\epsilon_{t-1}$.

(a) Show that the method-of-moments estimator of ω is

$$\widehat{\omega}_{\mathrm{mm}} = \frac{1-\sqrt{1-4\widehat{\rho}(1)^2}}{2\widehat{\rho}(1)}, \quad \text{for } |\widehat{\rho}(1)| < 1/2,$$

where $\widehat{\rho}(1)$ is the sample autocorrelation at lag 1.

(b) Using the central limit theorem for the autocovariance function (Shumway and Stoffer, 2016, Theorem A.6) and the delta method, show that, as $n \to \infty$,

$$\sqrt{n}(\widehat{\omega}_{\mathrm{mm}} - \omega) \xrightarrow{d} N\left\{0, \frac{1+\omega^2+4\omega^4+\omega^6+\omega^8}{(1-\omega^2)^2}\right\}.$$

(c) Carry out a simulation study to assess the asymptotic variance obtained at (b).

(d) Compare the asymptotic precision of $\widehat{\omega}_{\mathrm{mm}}$ with that of the maximum likelihood estimator $\widehat{\omega}_{\mathrm{mle}}$ (see Exercise 3.6) by plotting the asymptotic relative efficiency, $\mathrm{ARE}(\widehat{\omega}_{\mathrm{mm}}) = \mathrm{avar}(\widehat{\omega}_{\mathrm{mle}})/\mathrm{avar}(\widehat{\omega}_{\mathrm{mm}})$ as a function of ω.

3.9. Consider a spatial process $Y(x)$ as defined on page 94.

(a) Prove that if $Y(x)$ is weakly stationary, then it is also intrinsically stationary.

(b) Show that the converse is not true. Provide a counterexample of a process that is intrinsically stationary but fails to be weakly stationary

3.10. Let $Y = (Y(x_1), \ldots, Y(x_n))^\mathsf{T}$ be a vector of count random variables associated with n spatial locations, $x_1, \ldots, x_n$. Suppose that Y follows an auto-Poisson model (Besag, 1974) defined by the conditional distributions

$$Y_i \mid \{y_j : j \neq i\} \sim \mathrm{Poisson}(\lambda_i),\ \log(\lambda_i) = \alpha_i + \rho\sum_{j=1}^{n} w_{ij}y_j 1\{x_j \in \mathcal{N}(x_i)\}, \quad i = 1, \ldots, n,$$

where w_{ij} are fixed spatial weights satisfying $w_{ij} = w_{ji}$ and $w_{ii} = 0$, and ρ is an unknown parameter.

(a) Using the Hammersley-Clifford theorem, derive the joint distribution of Y.

(b) Explain why the joint distribution obtained in (a) is valid only if $\rho \leq 0$.

(c) Discuss the type of spatial interaction that the auto-Poisson model can represent given the restriction $\rho < 0$. How does the value of a neighbour y_j affect the expected value of Y_i?

3.11. Suppose that for $i = 1, \ldots, n$, the response variable Y_i is subject to missingness while the associated covariate X_i is always observed. Let R_i be the indicator variable $R_i = 1(Y_i \text{ is observed})$. Suppose that the distribution of Y_i is different for the observed and unobserved measurements,

$$Y_i \mid (R_i = 1, X_i) \sim \text{Bernoulli}(p_1), \text{with } \text{logit}(p_1) = \beta_0 + \beta_1 X_i;$$
$$Y_i \mid (R_i = 0, X_i) \sim \text{Bernoulli}(p_0), \text{with } \text{logit}(p_0) = \beta_0 + \beta_2 X_i;$$

where β_0, β_1 and β_2 are unknown parameters. Suppose further that R_i is independent of X and satisfies

$$R_i \sim \text{Bernoulli}(\pi), \quad 0 < \pi < 1.$$

Discuss likelihood inference for $\theta = (\beta_0, \beta_1, \beta_2, \pi)^{\mathsf{T}}$, including which components of θ are identifiable from the observed data.

3.12. Suppose that for $i = 1, \ldots, n$, the bivariate response vector $Y_i = (Y_{i1}, Y_{i2})^{\mathsf{T}}$ is subject to missingness. Let R_i be the categorical variable indicating the missingness status with

$$R_i = \begin{cases} 1, & Y_{i1}, Y_{i2} \text{ both observed}, \\ 2, & Y_{i1} \text{ observed}, Y_{i2} \text{ missing}, \\ 3, & Y_{i1} \text{ missing}, Y_{i2} \text{ observed}. \end{cases}$$

Suppose that the distribution of Y_i varies among different missingness patterns,

$$Y_i \mid R_i = r \sim N(\mu_r 1_2, \sigma_r^2 \mathcal{I}_{2\times 2}), \quad r = 1, 2, 3,$$

where $-\infty < \mu_r < \infty$ and $\sigma_r > 0$. Suppose that $\text{pr}(R_i = r) = \pi_r$, where $0 \leq \pi_r \leq 1$ and $\pi_1 + \pi_2 + \pi_3 = 1$. Let $\beta = (\mu_1, \mu_2, \sigma_1^2, \sigma_2^2)^{\mathsf{T}}$, $\pi = (\pi_1, \pi_2, \pi_3)^{\mathsf{T}}$, and $\theta = (\beta^{\mathsf{T}}, \pi^{\mathsf{T}})^{\mathsf{T}}$.

(a) Discuss inference based on the maximum likelihood estimator of θ.

(b) Suppose the missingness probabilities are determined by design, so that π is known. Discuss inference procedures based on the maximum likelihood estimator of β.

(c) Compare the estimators of β obtained from (a) and (b).

3.13. Suppose that for $i = 1, \ldots, n$, the second component of the response vector $Y_i = (Y_{i1}, Y_{i2})^{\mathsf{T}}$ is subject to missingness, while the first component Y_{i1} is always observed. Suppose also that for $i = 1, \ldots, n$, an associated scalar covariate variable X_i is always observed. Let R_i be the indicator variable $1(Y_{i2} \text{ is observed})$.

Suppose that the distribution of Y_i varies across missingness patterns,

$$Y_i \mid R_i = r \sim N(\mu_r 1_2, \sigma_r^2 \mathcal{I}_{2\times 2}), \quad \mu_r \in \mathbb{R}, \sigma_r > 0; \ r = 0, 1.$$

Suppose that the distribution of the missing data indicator is

$$R_i \mid X_i \sim \text{Bernoulli}(\pi), \quad \pi = \Phi(\alpha_0 + \alpha_1 X_i),$$

where $\Phi(\cdot)$ is the cumulative distribution function of the standard normal distribution. Let $\theta = (\mu_0, \mu_1, \sigma_0^2, \sigma_1^2, \alpha_0, \alpha_1)^{\mathsf{T}}$.

(a) Derive the observed data likelihood function for θ.

(b) Calculate the score function for estimating θ.

(c) Find the asymptotic distribution of the maximum likelihood estimator of θ.

3.14. This problem modifies Example 3.32. Suppose that the distribution of the random sample $Y = (Y_1, Y_2, Y_3, Y_4)^\mathsf{T}$ is still a multinomial distribution but with different probabilities: $Y \sim \text{Mult}(n, p^*)$, where $n = \Sigma_{i=1}^4 Y_i$, and $p^* = (1/2 - \theta/4, 1/2 - \theta/2, \theta/2, \theta/4)^\mathsf{T}$ with $0 < \theta < 1$.

(a) Suppose that the observed data are the vector $Y^{(o)} = (Y_1, Y_2, Y_3 + Y_4)$, and $Y^{(o)}$ follows the multinomial distribution $\text{Mult}(n, q^*)$, where $q^* = (1/2 - \theta/4, 1/2 - \theta/2, 3\theta/4)^\mathsf{T}$, with $0 < \theta < 1$. Show that

$$\text{E}(Y_3 \mid Y^{(o)}; \theta) = 2(Y_3 + Y_4)/3, \quad \text{E}(Y_4 \mid Y^{(o)}; \theta) = (Y_3 + Y_4)/3,$$

which are free of the parameter θ.

(b) If the observed value of $Y^{(o)}$ in (a) is $y^{(o)} = (38, 34, 125)^\mathsf{T}$, find the estimate of θ obtained from the EM algorithm.

(c) Use the likelihood function for the observed data in (a) to find the maximum likelihood estimator of θ. Compare this estimator with that in (b).

(d) Suppose the observed data in (a) are changed to $Y^{(o)} = (Y_1 + Y_2, Y_3 + Y_4)$ and $Y^{(o)}$ follows the multinomial distribution $\text{Mult}(n, q^{**})$ where $q^{**} = (1 - 3\theta/4, 3\theta/4)^\mathsf{T}$ with $0 < \theta < 1$. Can you find $\text{E}(Y_j \mid Y^{(o)}; \theta)$ for $j = 1, 2, 3, 4$? Can you find an estimator of θ using the EM algorithm? Can you use the likelihood function based on the observed data to find the maximum likelihood estimator of θ?

3.15. Suppose that survival time T follows the Weibull distribution with the probability density function

$$f(t; \beta_1, \beta_2) = \frac{\beta_2}{\beta_1}\left(\frac{t}{\beta_1}\right)^{\beta_2 - 1} \exp\left\{-(t/\beta_1)^{\beta_2}\right\}, \quad t > 0, \beta_1 > 0, \beta_2 > 0.$$

Let $\{(t_i, \delta_i) : i = 1, \ldots, n\}$ be a censored random sample of survival times chosen from the population of the same distribution as T.

(a) Find the likelihood function and score function for inference about $\theta = (\beta_1, \beta_2)^\mathsf{T}$.

(b) Consider the transformation $Y = \log T$. Derive the distribution of Y and then repeat (a).

(c) Compare the estimation procedures for θ obtained in (a) and (b).

(Lawless, 2003, Section 5.2)

3.16. Consider the transformation model in Example 3.37 and its extension with covariates.

(a) For the location-scale version, give the likelihood ratio test procedure for testing the hypotheses (i) $H_0 : \mu = 0$ versus $H_a : \mu \neq 0$; and (ii) $H_0 : \sigma = 1$ versus $H_a : \sigma \neq 1$.

(b) Suppose that, along with the censored sample $\{(y_i, \delta_i) : i = 1, \ldots, n\}$ in Example 3.37, there is a covariate vector x_i associated with the survival time T_i for subject $i = 1, \ldots, n$. Suppose that the survival times and the covariates are modeled as

$$\log T_i = \beta_0 + x_i^\mathsf{T} \beta_1 + \sigma \epsilon_i, \quad i = 1, \ldots, n,$$

where $\beta = (\beta_0, \beta_1^{\mathsf{T}})^{\mathsf{T}}$ is the regression parameter, σ is the scale parameter, and the error terms $\{\epsilon_1, \ldots, \epsilon_n\}$ are independent with the density f_0. Give expressions for the likelihood function for $\theta = (\beta^{\mathsf{T}}, \sigma)^{\mathsf{T}}$ and for the asymptotic distribution of the maximum likelihood estimator of θ.

3.17. Suppose $\{Y_1, \ldots, Y_n\}$ is a random sample from a distribution with probability density or mass function

$$f(y; p) = pf_1(y) + (1-p)f_2(y), \quad 0 < p < 1,$$

where $f_1(y)$ and $f_2(y)$ are given probability density or mass functions. Let α denote a constant in $(0, 1)$.

(a) Find the likelihood function for p.

(b) Find the expected Fisher information for p.

(c) Assume that $f_1(y)$ and $f_2(y)$ are the probability density functions of the $N(0, 1)$ and $N(1, 1)$ distributions, respectively. Construct an approximate $(1-\alpha)100\%$ confidence interval for p.

(d) Assume that $f_1(y)$ and $f_2(y)$ are the probability density functions of Exp(1) and Exp(2), respectively. Construct an approximate $(1-\alpha)100\%$ confidence interval for p.

3.18. Suppose $\{Y_1, \ldots, Y_n\}$ is a random sample from the distribution with probability density or mass function

$$f(y; p, \beta_1, \beta_2) = pf_1(y; \beta_1) + (1-p)f_2(y; \beta_2), \quad 0 < p < 1,$$

where $f_1(y; \beta_1)$ and $f_2(y; \beta_2)$ are two given probability density or mass functions with unknown parameters β_1 and β_2. Let $\theta = (p, \beta_1^{\mathsf{T}}, \beta_2^{\mathsf{T}})^{\mathsf{T}}$.

(a) Find the likelihood function for inference about θ.

(b) Find the expected Fisher information matrix for θ.

(c) Assume that $f_1(y; \beta_1)$ and $f_2(y; \beta_2)$ are the probability density functions of $N(\beta_1, 1)$ and $N(\beta_2, 1)$ distributions, respectively. Work out the estimation procedure for θ.

(d) Assume that $f_1(y; \beta_1)$ and $f_2(y; \beta_2)$ are the probability density function of Exp(β_1) and Exp(β_2), respectively. Work out the estimation procedure for θ.

3.19. In Section 3.6.3, survival times in the sample are assumed to be distinct (i.e., no ties). Extend the development in Section 3.6.3 to accommodate settings where the sample includes tied survival times.

3.20. Consider the same setting as in Section 3.6.5, but replace the survival model (3.73) with the accelerated failure time model

$$\log T_i = \beta_0 + \beta_1 X_i(T_i) + \beta_2 Z_i + \epsilon_i,$$

where $\beta = (\beta_0, \beta_1, \beta_2)^{\mathsf{T}}$ is the vector of regression parameters, and ϵ_i is independent of $\{T_i, C_i, Z_i, X_i(t) : t > 0\}$. Suppose that the survivor function of ϵ_i is given by one of the following form:

$$S_0(\epsilon_i) = \exp\{-\exp(\epsilon_i)\},$$
$$S_0(\epsilon_i) = 1 - \Phi(\epsilon_i),$$
$$S_0(\epsilon_i) = \{1 + \exp(\epsilon_i)\}^{-1},$$

corresponding to the Weibull, log-normal, and log-logistic distributions, respectively, where $\Phi(\cdot)$ is the cumulative distribution for the standard normal distribution.

(a) Derive the likelihood function for the observed data.

(b) Discuss estimation of β using the likelihood function in (a).

(c) Discuss estimation of β using the EM algorithm.

4

Likelihood Inference in Misspecified Models

4.1 Model misspecification

Chapters 1 and 2 derive a number of likelihood results under the assumption that the statistical model $\mathcal{F} = \{f(\cdot;\theta);\ \theta \in \Theta\}$ includes the true data-generating mechanism, corresponding to a particular parameter value $\theta_0 \in \Theta$. A more general viewpoint is that the true model does not necessarily belong to the specified statistical model. The parameter θ is then interpreted as a convenient summary of the data, often with a meaningful interpretation in spite of the misspecification. For example, fitting a normal model is helpful when the data can be fruitfully summarized by the mean and the variance, even if the normality assumption does not necessarily hold. The term "working model" is often used in the statistical literature to indicate a misspecified model used for convenience.

Suppose $Y = \{Y_1, \ldots, Y_n\}$ is a sample of independent and identically distributed random variables from a distribution with density function $h(y)$, but inference is based on a working model $f(y;\theta)$, with $\theta \in \Theta$. Maximum likelihood estimation under model misspecification is characterized by the expected log-likelihood

$$\mathrm{E}_h\{\log f(Y_1;\theta)\} = \int \log\{f(t;\theta)\}h(t)\,\mathrm{d}t$$

also called Fraser information in Kent (1982); see Fraser (1965).

Suppose that the Fraser information is finite, $\mathrm{E}_h\{|\log f(Y_1;\theta)|\} < \infty$. Under model misspecification, maximizing the log-likelihood function $\ell(\theta;y) = \Sigma_i \log f(y_i;\theta)$ is asymptotically equivalent to maximizing the Fraser information or to minimizing the Kullback–Leibler divergence between the assumed model f and the true model h because

$$n^{-1}\ell(\theta;Y) \xrightarrow{p} \mathrm{E}_h\{\log f(Y_1;\theta)\} = \mathrm{E}_h\{\log h(Y_1)\} - \mathrm{KL}(\theta),$$

where the last term is the Kullback–Leibler divergence (Exercise 1.18) written as a function of θ,

$$\mathrm{KL}(\theta) = \mathrm{E}_h\left\{\log h(Y_1) - \log f(Y_1;\theta)\right\}. \qquad (4.1)$$

By Jensen's inequality, the Kullback–Leibler divergence is non-negative,

$$\mathrm{KL}(\theta) = -\mathrm{E}_h\left[\log\left\{\frac{f(Y_1;\theta)}{h(Y_1)}\right\}\right] \geq -\log\left[\mathrm{E}_h\left\{\frac{f(Y_1;\theta)}{h(Y_1)}\right\}\right] = 0.$$

The inequality is strict unless there is a $\theta_0 \in \Theta$ such that $f(y;\theta_0)$ coincides with $h(y)$ for all y in $\mathcal{Y}$; i.e., the model is correctly specified, in which case $\mathrm{KL}(\theta) = 0$. Otherwise, the value θ_* that minimizes the Kullback–Leibler divergence (4.1) identifies the model closest to the data-generating mechanism. It has been given a variety of names, including "pseudo-true

DOI: 10.1201/9780429090936-4

value" (Vuong, 1989) and "least-false parameter" (Hjort, 1992). If the model family is correctly specified, then $\theta_* = \theta_0$. Under mild regularity conditions that are essentially equivalent to those of Section 1.7.1 with θ_0 replaced by θ_* and expectations computed with respect to h, the maximizer of the misspecified log-likelihood function $\ell(\theta)$ converges to the least-false parameter θ_*.

Example 4.1 (An incorrect exponential model) Suppose that the working model for Y is the exponential distribution with expected value θ but the true model is indeed log-normal, $Y = \exp(X)$ with $X \sim N(\mu, \sigma^2)$. The Fraser information $\mathrm{E}_h\{\log f(Y;\theta)\} = -\log\theta - \mathrm{E}_h(Y)/\theta$ is maximized at $\theta_* = \mathrm{E}_h(Y) = \exp(\mu + \sigma^2/2)$, showing that the maximum likelihood estimator based on a sample of n independent observations converges to a quantity with a meaningful interpretation even if the model is misspecified. □

Example 4.2 (Un-modelled skewness) Suppose that the working model is $N(\mu, \sigma^2)$ but the true model is the location-scale transform of the skew-normal distribution (2.87), with location and scale parameters ξ and ω, respectively, and shape parameter α (Azzalini, 1985). The working model parameter is $\theta = (\mu, \sigma^2)^\top$. The Fraser information is maximized at $\theta_* = (\mu_*, \sigma_*^2)^\top$, where $\mu_* = \mathrm{E}_h(Y) = \xi + \omega\delta(2/\pi)^{1/2}$ and $\sigma_*^2 = \omega^2(1 - 2\delta^2/\pi)$, with $\delta = \alpha/(1+\alpha^2)^{1/2}$.

Although the maximum misspecified likelihood estimator of θ converges to a quantity with a meaningful interpretation, inference conclusions that ignore the misspecification would be misleading. For example, confidence or prediction intervals based on the estimated normal model would be symmetric. Unless the asymmetry of the true distribution is negligible; i.e., its shape parameter α is close to zero, such symmetric intervals are inefficient and could be highly inaccurate. □

Example 4.3 (Probit or logit) For a binary regression problem, suppose that true model is probit, $\mathrm{pr}(Y = 1 \mid x) = \Phi(x^\top\beta)$, but one fits an incorrect logit model, $\mathrm{pr}(Y = 1 \mid x) = \pi(x^\top\theta)$, with $\pi(u) = \{1 + \exp(-u)\}^{-1}$ for a vector of covariates x of length p. There is no closed-form expression for the least-false parameter θ_*. However, a reasonable approximation to θ_* is obtained by linearizing both the probit and the logit links around zero,

$$\Phi(u) = \frac{1}{2} + \frac{u}{\sqrt{2\pi}} + o(u), \quad \pi(u) = \frac{1}{2} + \frac{u}{4} + o(u), \quad \text{as } u \to 0.$$

Matching the slopes of the probit and logit linearizations gives $\theta_* = 4\beta/(2\pi)^{1/2} = 1.5958\beta$. This rule of thumb is commonly used to convert regression coefficients from probit to logit models. Other approximate conversion rules have also been proposed; Agresti (2002, p. 247) noted that "parameter estimates in logistic regression are about 1.6 to 1.8 times those in probit models". □

Example 4.4 (Linear regression with omitted quadratic term) Consider observations generated according to the linear model,

$$Y = \beta_0 + \beta_1 X + \beta_2 X^2 + \epsilon,$$

where ϵ are errors such that $\mathrm{E}(\epsilon \mid X) = 0$. Furthermore, suppose that X is distributed as $U(0,1)$. For the working model, we specify a simple linear regression,

$$Y = \alpha_0 + \alpha_1 X + \eta,$$

where η are errors with $\mathrm{E}(\eta \mid X) = 0$. The working model, with parameter $\theta = (\alpha_0, \alpha_1)^\top$ thus omits the quadratic term from the true underlying process. The least-false parameter minimizes the expected squared error between the true conditional mean and the assumed linear model,

$$\min_{\alpha_0,\alpha_1} \mathrm{E}[\{\mathrm{E}_h(Y \mid X) - (\alpha_0 + \alpha_1 X)\}^2];$$

this gives the best linear approximation to a quadratic model. The least-false parameter has components $\alpha_{*0} = \mathrm{E}\{\mathrm{E}_h(Y \mid X)\} - \alpha_{*1}\mathrm{E}(X)$ and $\alpha_{*1} = \mathrm{cov}\{\mathrm{E}_h(Y \mid X), X\}/\mathrm{var}(X)$. Under the assumption that $X \sim U(0,1)$, $\alpha_{*0} = \beta_0 - \beta_2/6$ and $\alpha_{*1} = \beta_1 + \beta_2$.

The interpretation of the least-false parameter deviates from what would be expected in a standard simple linear regression model: specifically, the estimated slope $\widehat{\alpha}_1$ cannot be interpreted as the effect on Y of a one-unit change in X unless $\beta_2 = 0$. □

4.2 Sandwich variance

One consequence of misspecification is the failure of the first two Bartlett identities discussed in Section 1.2 for correctly specified likelihoods and later in Section 5.2 for quasi-score functions. Since θ_* maximizes the Fraser information, the expected value of the misspecified score function is zero only at the least-false parameter,

$$\mathrm{E}_h\{U(\theta)\} = 0 \text{ only at } \theta_*.$$

Under misspecification, the expected Fisher information (1.37) is computed with respect to the true density function h,

$$I(\theta) = \mathrm{var}_h\{U(\theta)\}.$$

The second Bartlett identity (1.39) fails: $I(\theta) \neq \mathrm{E}_h\{J(\theta)\}$, unless the model is correct. Adopting the terminology of estimating equations from Chapter 5, $I(\theta)$ is called the variability matrix, whereas the expected observed information

$$K(\theta) = \mathrm{E}_h\{J(\theta)\} = \mathrm{E}_h\{-\partial U(\theta)/\partial\theta^{\mathsf{T}}\}$$

is called the sensitivity matrix.

Example 4.5 (Assuming an incorrect expected value) Consider estimation of the variance of independent normal random variables $\{Y_1, \ldots, Y_n\}$ under the incorrect assumption that their expected value is zero, while in fact $\mathrm{E}_h(Y_i) = \mu_0 \neq 0$. The maximum likelihood estimator $\widehat{\sigma}^2 = \Sigma_i Y_i^2/n$ is not consistent because it converges to $\sigma_*^2 = \sigma_0^2 + \mu_0^2$.

Furthermore, under the incorrect assumption of zero expectation, there is also the failure of the second Bartlett identity, that is $I(\sigma_*^2) \neq K(\sigma_*^2)$. In fact, we have $K(\sigma_*^2) = n/(2\sigma_*^4)$ and $I(\sigma_*^2) = \mathrm{var}(\Sigma_i Y_i^2)/(4\sigma_*^8)$. Using the properties of the non-central chi-squared distribution, we find that $\mathrm{var}(\Sigma_i Y_i^2) = 2\sigma_0^2(n\sigma_0^2 + 2n\mu_0^2) = 2n(\sigma_*^4 - \mu_0^4)$ so that $I(\sigma_*^2) = n/(2\sigma_*^4) - n\mu_0^4/(2\sigma_*^8)$ which differs from $K(\sigma_*^2)$ unless μ_0 is indeed zero. □

The amount of information for θ in the misspecified framework is summarized by the Godambe information

$$G(\theta) = K(\theta)I^{-1}(\theta)K(\theta).$$

Under the identifiability assumption and regularity conditions that parallel those of Section 1.7.2, a Taylor series expansion of $U(\theta)$ based on a sample of size n shows that the maximum likelihood estimator $\widehat{\theta}$ has a limiting normal distribution centred at the least-false parameter θ_*, with asymptotic variance

$$\text{avar}(\widehat{\theta}) = G^{-1}(\theta_*) = K^{-1}(\theta_*)I(\theta_*)K^{-1}(\theta_*), \tag{4.2}$$

often referred to as the "sandwich variance". For a formal derivation of the asymptotic distribution of the maximum misspecified likelihood estimator see Knight (2000, Theorem 5.5). The square roots of the diagonal elements of $G^{-1}(\widehat{\theta})$ are known as robust standard errors or Huber–White standard errors.

Example 4.6 (Working exponential distribution) Consider a sample $\{Y_1, \ldots, Y_n\}$ of independent and identically distributed variables. As a working model suppose that $Y_1 \sim \exp(\theta)$, then $\widehat{\theta} = n/\Sigma_i Y_i$ converges to $\theta_* = \{\text{E}_h(Y_1)\}^{-1}$. The sandwich variance is $\text{avar}(\widehat{\theta}) = I(\theta_*)/K^2(\theta_*) = n^{-1}\text{var}_h(Y_1)/\{\text{E}_h(Y_1)\}^4$, yielding an estimator of the robust standard error, $n^{-1/2}S/\overline{Y}^2$, where S is the sample standard deviation. In contrast, if the exponential distribution were the true data-generating process, the usual model-based estimate of the asymptotic standard error of $\widehat{\theta}$ would be $n^{-1/2}/\overline{Y}$. □

Example 4.7 (Working assumption of normality) Consider estimation of the expected value μ and variance σ^2 for independent and identically distributed random variables $\{Y_1, \ldots, Y_n\}$. Assume the working model is $Y_1 \sim N(\mu, \sigma^2)$, with model parameter $\theta = (\mu, \sigma^2)^{\mathsf{T}}$. The maximum likelihood estimator $\widehat{\theta}$ has components $\widehat{\mu} = \overline{Y}$ and $\widehat{\sigma}^2 = n^{-1}\Sigma_i(Y_i - \overline{Y})^2$, which are consistent for the true values of the population mean and variance, as they converge to $\mu_* = \mu_0$ and $\sigma_*^2 = \sigma_0^2$, respectively.

The sensitivity and variability matrices are

$$K(\theta_0) = n\begin{bmatrix} 1/\sigma_0^2 & 0 \\ 0 & 1/(2\sigma_0^4) \end{bmatrix}, \quad I(\theta_0) = n\begin{bmatrix} 1/\sigma_0^2 & \gamma_1/(2\sigma_0^3) \\ \gamma_1/(2\sigma_0^3) & \gamma_2/(4\sigma_0^4) \end{bmatrix},$$

where $\gamma_1 = \text{E}_h\{(Y_t - \mu_0)^3\}/\sigma_0^3$ and $\gamma_2 = \text{E}_h\{(Y_t - \mu_0)^4\}/\sigma_0^4 - 1$ are the skewness and kurtosis indices, respectively. The second Bartlett identity is satisfied at θ_0 if $\gamma_1 = 0$ and $\gamma_2 = 2$, which is the case when the assumed normal model is indeed correct. The sandwich asymptotic variance of $\widehat{\theta} = (\widehat{\mu}, \widehat{\sigma}^2)^{\mathsf{T}}$ is

$$\text{avar}(\widehat{\theta}) = n^{-1}\begin{bmatrix} \sigma_0^2 & \gamma_1\sigma_0^3 \\ \gamma_1\sigma_0^3 & \gamma_2\sigma_0^4 \end{bmatrix}.$$

Inference for θ thus critically depends on the skewness and kurtosis of true density $h(y)$. □

Example 4.8 (Unmodelled dependence) Consider the linear regression model $Y = X\beta + \epsilon$, where $\epsilon \sim N(0, \sigma^2 R)$, for some correlation matrix R and suppose β is estimated under the incorrect assumption of uncorrelated errors. The maximum likelihood estimator $\widehat{\beta} = (X^{\mathsf{T}}X)^{-1}X^{\mathsf{T}}Y$ is unbiased even if the model is misspecified, that is $\beta_0 = \beta_*$. The sandwich variance of $\widehat{\beta}$ is $\sigma^2(X^{\mathsf{T}}X)^{-1}X^{\mathsf{T}}RX(X^{\mathsf{T}}X)^{-1}$ and collapses to the variance of the ordinary least squares estimator if R is the identity matrix.

As a simple numerical illustration, consider the special case of an intercept-only model with an exchangeable correlation matrix $R = [r_{ij}]$, where $r_{ij} = \alpha$ if $i \neq j$ and $r_{ij} = 1$ if $i = j$, for some correlation parameter $-(n-1)^{-1} < \alpha < 1$. The maximum likelihood estimator of β is the sample mean $\overline{Y}$ with variance $n^{-1}\sigma^2\{1 + \alpha(n-1)\}$. Imprudently neglecting even small levels of dependence between the errors gives an artificial impression of precision of the maximum likelihood estimator. For example, with $\alpha = 0.1$ and $n = 100$, the standard error $n^{-1/2}\sigma$ of the maximum likelihood estimator that wrongly assumes $\alpha = 0$ is $\{1 + 0.1(99)\}^{1/2} = 3.3$ times smaller than the standard error expected under the correct dependence model. □

Example 4.9 (Unobserved grouping factors in counts) Suppose $\{Y_1, \ldots, Y_n\}$ are modelled as independent Poisson random variables with expected value μ but the distribution of the observations depends indeed upon a hidden binary grouping factor, so that Y_i follows a Poisson distribution with expected value μ_0 if $U_i = 0$ and $\mu_0 + \delta$ if $U_i = 1$. This is a simple example of unobserved heterogeneity.

Denoting by π the probability that $U_i = 1$, then the maximum likelihood estimator $\widehat{\mu} = \overline{Y}$ has expected value $\mu_* = \mu_0 + \delta\pi$. It follows that $K(\mu_*) = n/\mu_*$ and $I(\mu_*) = n\{\mu_* + \delta^2\pi(1-\pi)\}/\mu_*^2$ and thus the asymptotic variance of $\widehat{\mu}$ is $\{\mu_* + \delta^2\pi(1-\pi)\}/n$. The unobserved heterogeneity due to the grouping factor inflates the variance of $\widehat{\mu}$ with respect to the quantity μ_0/n expected in a standard Poisson model: an example of overdispersion. □

Example 4.10 (Proportional hazards regression) The proportional hazards regression model, discussed in Example 2.77 and Section 3.6.3, is a semi-parametric model with the hazard function for the survival time t, modelled as a function of covariates x,

$$\lambda(t; x, \beta) = \lambda_0(t) \exp(x^{\mathsf{T}}\beta).$$

As noted there, the partial likelihood function for β, (2.82), can be derived by multiplying successive conditional probabilities at each failure point. In this sense the likelihood function is misspecified, as it treats the factors of (2.82) as if they were independent, when they are not. However, this partial likelihood is special in that the Hessian of the log partial likelihood function, evaluated at the maximum, is a consistent estimate of the asymptotic variance of $\widehat{\beta}$, obviating the need for the sandwich formula. □

The sandwich variance is not limited to misspecified likelihood inference; it applies more generally. As we will see in Chapter 5, the asymptotic variance for a broad class of estimators derived from estimating equations also has this sandwich form. A special case that is used in robust estimation is discussed in the next two examples.

Example 4.11 (M-estimation) M-estimation, an abbreviation for "maximum likelihood-type" estimation, is a generalization of maximum likelihood estimation due to Huber (1964). This involves minimizing an objective function, typically chosen to ensure robustness against violations of distributional assumptions, particularly the presence of outliers. For a set of n independent random variables $\{Y_1, \ldots, Y_n\}$, an M-estimator $\widehat{\theta}$ is defined as the minimizer of the loss function

$$\varphi(\theta) = \sum_{i=1}^{n} \rho(Y_i; \theta).$$

The maximum likelihood estimator has $\rho(Y_i; \theta) = -\log f(Y_i; \theta)$. M-estimation permits flexible choices of ρ, for example providing robustness while avoiding strict distributional assumptions.

If the loss function ρ is differentiable, then $\widehat{\theta}$ is the solution of the estimating equation $\Psi(\theta; Y) = \Sigma_i \Psi(\theta; Y_i) = 0$, with $\Psi(\theta; Y_i) = \partial\rho(Y_i; \theta)/\partial\theta$. Estimating equations are discussed in Chapter 5. In the robust statistics literature Ψ is called the influence curve or the influence function because it measures the impact of an individual data point on the estimated parameter.

Under suitable regularity conditions, M-estimators are asymptotically normal with asymptotic sandwich variance (4.2), where the variability and sensitivity matrices are $I(\theta) = \mathrm{var}_h\{\Psi(\theta; Y)\}$ and $K(\theta) = \mathrm{E}_h\{\partial\Psi(\theta; Y)/\partial\theta^{\mathsf{T}}\}$, respectively. Here $K(\theta)$ does not require the minus sign as M-estimators minimize the loss function ρ rather than maximizing the likelihood function. □

Example 4.12 (Robust regression) M-estimation can be used for a robust approach to regression. Consider the linear regression model $Y_i = x_i^{\mathsf{T}}\beta + \epsilon_i$, where ϵ_i are independent and identically distributed errors with variance σ^2, for $i = 1, \ldots, n$. The M-estimate $\widehat{\beta}$ minimizes the residual loss function

$$\varphi(\beta) = \sum_{i=1}^{n} \rho(y_i - x_i^{\mathsf{T}}\beta).$$

Let $e_i = y_i - x_i^{\mathsf{T}}\beta$ denote the residual for observation $i = 1, \ldots, n$. A reasonable choice for ρ is a non-negative symmetric function that is zero at the origin, $\rho(0) = 0$, and increases with the magnitude of the residual, $|e_i|$. Standard least-squares estimation, corresponding to $\rho(e_i) = e_i^2$, satisfies these properties but is not robust, as the corresponding influence function is unbounded. The influence of an observation on $\widehat{\beta}$ grows linearly with its residual.

Robust fitting is achieved by down-weighting large residuals, as for example in the Huber loss function,

$$\rho_\kappa(e_i) = \begin{cases} \dfrac{e_i^2}{2}, & |e_i| \le \kappa, \\ \kappa|e_i| - \dfrac{\kappa^2}{2}, & |e_i| > \kappa, \end{cases}$$

for a tuning constant κ. Another popular choice for ρ is the Tukey loss function (Beaton and Tukey, 1974),

$$\rho_\kappa(e_i) = \begin{cases} \dfrac{\kappa^2}{6}\left\{1 - \left(1 - \dfrac{e_i^2}{\kappa^2}\right)^3\right\}, & |e_i| \le \kappa, \\ \dfrac{\kappa^2}{6}, & |e_i| > \kappa. \end{cases} \tag{4.3}$$

For a fixed value of κ, the robust M-estimator $\widehat{\beta}_\kappa$ solves the weighted least squares equations

$$\sum_{i=1}^{n} w_i x_i \widehat{e}_i = 0,$$

for weights $w_i = \Psi_\kappa(\widehat{e}_i)/\widehat{e}_i$, with $\Psi_\kappa(\widehat{e}_i) = \rho_\kappa'(\widehat{e}_i)$, where $\widehat{e}_i$ is the residual evaluated at $\widehat{\beta}_\kappa$. In the case of the Huber loss function, $w_i = 1$ if $|\widehat{e}_i| \le \kappa$ and $w_i = \kappa/|\widehat{e}_i|$ if $|\widehat{e}_i| > \kappa$. The Tukey weights are $w_i = \{1 - (\widehat{e}_i/\kappa)^2\}^2$ if $|\widehat{e}_i| \le \kappa$ and $w_i = 0$ otherwise. The Tukey weight function is also called the biweight kernel; see Section 6.2.1. In contrast to the Huber estimator, the Tukey estimator assigns zero weight to any observation whose residual has magnitude exceeding κ.

The mutual dependence among weights, residuals, and coefficients creates a circular problem. This is solved with an iteratively reweighted least squares algorithm using weights calculated in the previous step to find the updated coefficients.

The asymptotic variance of $\widehat{\beta}_\kappa$ has the sandwich form

$$\text{avar}(\widehat{\beta}_\kappa) = \frac{\mathrm{E}\{\Psi_\kappa^2(\epsilon_1)\}}{[\mathrm{E}\{\Psi_\kappa'(\epsilon_1)\}]^2}\left\{\sum_{i=1}^{n} x_i x_i^{\mathsf{T}}\right\}^{-1}.$$

Robust standard errors are computed from the estimated variance of $\widehat{\beta}_\kappa$,

$$\widehat{\text{var}}(\widehat{\beta}_\kappa) = \lambda \frac{(n-p)^{-1}\Sigma_i \Psi_\kappa^2(\widehat{e}_i)}{\{n^{-1}\Sigma_i \Psi_\kappa'(\widehat{e}_i)\}^2}\left\{\sum_{i=1}^{n} x_i x_i^{\mathsf{T}}\right\}^{-1},$$

where $\lambda = 1 + (p/n)(1-a)/a$ with $a = n^{-1}\Sigma_i \Psi_\kappa'(\widehat{e}_i)$ is a correction proposed by Huber (1981) to remove the first-order bias from the variance estimate; see also Street et al. (1988).

The tuning parameter κ is typically selected to ensure high efficiency under normal errors. For Huber–ρ, setting $\kappa = 1.345\sigma$ yields about 95% efficiency while still protecting against outliers. For Tukey's estimator, $\kappa = 4.685\sigma$ achieves the same 95% efficiency under normality. Usually, a robust measure of spread such as $\widehat{\sigma} = \mathrm{MAR}/0.6745$, where MAR is the median absolute residual, is used to estimate σ. □

4.2.1 Sandwich estimators

Inference (Section 4.3) and model selection (Section 4.4) in misspecified models require estimation of the components of the sandwich variance. While the observed Fisher information $J(\widehat{\theta}\,)$ provides a consistent estimator of $K(\theta_*)$, precise and accurate estimation of the variance of the score function, $I(\theta_*)$, is far from trivial.

In the case of n independent observations, it is perhaps natural to consider the "outer-product" estimator

$$\widehat{I}(\widehat{\theta}) = \sum_{i=1}^{n} u(\widehat{\theta}; Y_i) u^{\mathsf{T}}(\widehat{\theta}; Y_i). \tag{4.4}$$

However, the finite-sample performance of the outer-product estimator may be affected by considerable bias and inefficiency when the assumed model is in fact correct.

Example 4.13 (Heteroscedasticity-consistent variance estimation) Suppose the linear regression model $Y = X\beta + \epsilon$ is fitted under the homoscedasticity assumption $\epsilon \sim N(0, \sigma^2 \mathcal{I}_{n\times n})$ when the errors are instead heteroscedastic, $\epsilon \sim N(0, \Omega)$, with $\Omega = \mathrm{diag}(\sigma_1^2, \dots, \sigma_n^2)$. The ordinary least squares estimator $\widehat{\beta} = (X^{\mathsf{T}}X)^{-1}X^{\mathsf{T}}Y$ is unbiased and consistent, but the model-based variance estimator $\widehat{\sigma}^2(X^{\mathsf{T}}X)^{-1}$, where $\widehat{\sigma}^2 = \Sigma_i \widehat{e}_i^2/(n-p)$ and $\widehat{e}_i = Y_i - x_i^{\mathsf{T}}\widehat{\beta}$, is inconsistent. The consistent estimator using the sandwich variance is

$$\widehat{\mathrm{var}}(\widehat{\beta}\,) = (X^{\mathsf{T}}X)^{-1}\widehat{I}(\widehat{\beta}\,)(X^{\mathsf{T}}X)^{-1},$$

where $\widehat{I}(\widehat{\beta}\,) = X^{\mathsf{T}}\mathrm{diag}(\widehat{e}_i^2)X$ is the outer product estimator.

When the assumed homoscedastic model is correct, the outer product estimator $\widehat{I}(\widehat{\beta})$ is affected by downward bias because $\mathrm{E}_h(\widehat{e}_i^2) = \sigma^2(1 - p_{ii})$, where p_{ii} is the ith diagonal element of the projection hat matrix $P = X(X^{\mathsf{T}}X)^{-1}X^{\mathsf{T}}$; see Kauermann and Carroll (2001). Bias correction is obtained by replacing $\widehat{e}_i$ with $\widehat{e}_i/(1-p_{ii})^{1/2}$ and the resulting bias-corrected sandwich variance estimator is called heteroscedasticity-consistent in the econometric literature (MacKinnon and White, 1985). However, the bias-corrected sandwich variance suffers from considerable loss of efficiency if the errors in fact have constant variance; see Exercise 4.7. □

An alternative approach to estimating the variance of $\widehat{\theta}$ is the jackknife estimate, which some authors report to have superior finite-sample properties compared to the sandwich estimator (MacKinnon and White, 1985). The jackknife estimate of variance is

$$\widehat{\mathrm{var}}(\widehat{\theta}\,) = \frac{n-1}{n}\sum_{i=1}^{n}(\widehat{\theta}_{-i} - \widehat{\theta}\,)(\widehat{\theta}_{-i} - \widehat{\theta}\,)^{\mathsf{T}}, \tag{4.5}$$

where $\widehat{\theta}_{-i}$ is the "leave-one-out estimate" obtained by fitting the model to the sample with the ith observation omitted.

Example 4.14 (The jackknife variance estimator in the linear model) In the linear regression model the leave-one-out estimate $\widehat{\beta}_{-i}$ has the closed-form expression

$$\widehat{\beta}_{-i} = \widehat{\beta} - (X^{\mathsf{T}}X)^{-1}x_i\widetilde{e}_i, \tag{4.6}$$

where $\widetilde{e}_i = \widehat{e}_i/(1-p_{ii})$ are the scaled residuals with p_{ii} the diagonal elements of the projection hat matrix; see Exercise 4.3. The jackknife variance estimator (4.5) is

$$\widehat{\text{var}}(\widehat{\beta}) = \frac{n-1}{n}(X^\top X)^{-1}X^\top \text{diag}(\widetilde{e}_i^{\,2})X(X^\top X)^{-1},$$

and differs from the heteroscedasticity-consistent estimator described in Example 4.13 by the factor $(n-1)/n$ and the use of scaled residuals $\widetilde{e}_i$. □

The major drawback of the jackknife variance estimator outside the linear model is the need to compute the leave-one-out estimates $\widehat{\theta}_{-i}$. The computational cost is often reduced with an approximation of $\widehat{\theta}_{-i}$ by the estimator obtained from one iteration of the Newton–Raphson algorithm, starting from the maximum likelihood estimate $\widehat{\theta}$ computed from the full data,

$$\widehat{\theta}_{-i} \doteq \widehat{\theta} - J^{-1}(\widehat{\theta})u(\widehat{\theta}; Y_i), \tag{4.7}$$

where $u(\widehat{\theta}; Y_i) = \partial \log f(Y_i; \theta)/\partial\theta$ is the contribution from observation i to the score function whereas the observed information J is computed with the full data.

The approximate estimates (4.7) are sometimes referred to as one-step estimates and they are asymptotically equivalent to the estimates obtained by iterating the algorithm until convergence (Lehmann and Casella, 1998, p. 454). The jackknife estimate of the variance computed with the one-step estimates is

$$\widehat{\text{var}}(\widehat{\theta}) = \frac{n-1}{n}J^{-1}(\widehat{\theta})\left\{\sum_{i=1}^n u(\widehat{\theta}; Y_i)u^\top(\widehat{\theta}; Y_i)\right\}J^{-1}(\widehat{\theta}),$$

which reduces to the sandwich variance computed using the outer-product estimator (4.4) of $I(\theta_*)$, up to the finite-sample correction factor $(n-1)/n$. Consequently, the computational advantage of one-step estimation may come at the cost of the bias and inefficiency inherent in the outer-product approach, especially when the assumed model is correct.

In certain statistical models, most notably generalized linear models (Example 1.12), the one-step estimates are computed by performing a single iteration of the Fisher Scoring algorithm. This is achieved by replacing the observed information, $J(\widehat{\theta})$, with its expectation, $K(\widehat{\theta})$, in the approximation formula (4.7) for leave-one-out estimation.

It is substantially more difficult to estimate the variance of the score function when data cannot be decomposed into independent replicates because there is no internal replication. For regular time series, $I(\theta_*)$ is often estimated by the heteroscedasticity and autocorrelation-consistent (HAC) estimator

$$\widehat{I}(\widehat{\theta}) = \sum_{i,j} w_{|i-j|}u(\widehat{\theta}; Y_i)u^\top(\widehat{\theta}; Y_j), \tag{4.8}$$

where the weights $\{w_0, \ldots, w_{n-1}\}$ are chosen to ensure consistency of the estimator. Newey and West (1987) considered linearly decaying weights $w_r = 1 - r/(L+1)$ if $r \leq L$ and $w_r = 0$ if $r > L$ for some maximal lag L predetermined or estimated with a nonparametric procedure. Andrews (1991) proposed the kernel estimator obtained by setting w_r equal to a kernel function (Section 6.2.1).

In temporal and spatial processes with good mixing properties, a sample estimate of $I(\theta)$ can be obtained using window subsampling (Heagerty and Lele, 1998; Lumley and Heagerty, 1999). The study region $\mathcal{R}$ is split into subregions sufficiently large to be considered as nearly independent and then $I(\theta_*)$ is estimated with the scaled weighted outer product

$$\widehat{I}(\widehat{\theta}) = \frac{1}{B}\sum_{b=1}^B |\mathcal{R}_b| U_b(\widehat{\theta})U_b^\top(\widehat{\theta}), \quad U_b(\widehat{\theta}) = \sum_{\{i: Y_i \in \mathcal{R}_b\}} u(\widehat{\theta}; Y_i),$$

where $\mathcal{R}_1, \ldots, \mathcal{R}_B$ are B possibly overlapping subregions of $\mathcal{R}$ and $|\mathcal{R}_b|$ is the dimension of region $\mathcal{R}_b$. The quantity $U_b(\theta)$ is the score function restricted to the observations in the region $\mathcal{R}_b$. The performance of window subsampling requires careful selection of the subregions and may become unstable when data are sparsely distributed in the region $\mathcal{R}$.

4.3 Asymptotic distributions

The limiting χ^2 distribution of the likelihood-based statistical summaries described in Section 1.7 does not generally hold under model misspecification. The Wald statistic for misspecified models,

$$w_e(\theta_*) = (\widehat{\theta} - \theta_*)^{\mathsf{T}} K(\theta_*)(\widehat{\theta} - \theta_*)$$

has the asymptotic distribution of the quadratic form AA^{T}, where $A = Z^{\mathsf{T}} G^{-1/2}(\theta_*) K^{1/2}(\theta_*)$ and $Z = (Z_1, \ldots, Z_p)^{\mathsf{T}}$ is multivariate standard normal, from which it follows that

$$w_e(\theta_*) \xrightarrow{d} \sum_{i=1}^{p} \omega_i Z_i^2, \tag{4.9}$$

where $\omega_1, \ldots, \omega_p$ are the eigenvalues of $G^{-1}(\theta_*)K(\theta_*)$. If the model is well-specified, then $G(\theta) = K(\theta)$, all ω_i are equal to one and we recover to the standard χ^2_p distribution (1.50).

Evaluation of the distribution of the weighted sum (4.9) typically requires simulation and is complicated by the difficulty of estimating the eigenvalues of $G^{-1}(\theta_*)K(\theta_*)$. Section 4.2.1 describes problems associated with the estimation of the Godambe information.

It is however straightforward to construct a generalized version of the Wald statistic which is asymptotically distributed as χ^2_p, by simply replacing the Fisher information with the Godambe information giving the generalized Wald statistic

$$\widetilde{w}_e(\theta_*) = (\widehat{\theta} - \theta_*)^{\mathsf{T}} G(\theta_*)(\widehat{\theta} - \theta_*) \xrightarrow{d} \chi^2_p. \tag{4.10}$$

The score statistic under misspecification has the usual form

$$\widetilde{w}_u(\theta_*) = U(\theta_*) I^{-1}(\theta_*) U(\theta_*) \xrightarrow{d} \chi^2_p, \tag{4.11}$$

but, differently from the case where the assumed model is correct, the limiting distribution of $\widetilde{w}_u(\theta_*)$ may not be χ^2_p if $I(\theta_*)$ is replaced by $K(\theta_*)$ or by $J(\theta_*)$.

If θ is composed of a parameter of interest ψ of dimension $q \leq p$ and a nuisance parameter λ of dimension $p - q$, then, as discussed in Section 2.4, inference for ψ is based on the profile log-likelihood function $\ell_p(\psi) = \ell(\psi, \widehat{\lambda}_\psi)$. An equivalent version of the generalized Wald statistic for ψ is

$$\widetilde{w}_e(\psi_*) = (\widehat{\psi} - \psi_*)^{\mathsf{T}} G_p(\widehat{\psi})(\widehat{\psi} - \psi_*) \xrightarrow{d} \chi^2_q, \tag{4.12}$$

where $G_p(\psi) = \{G^{\psi\psi}(\psi, \widehat{\lambda}_\psi)\}^{-1}$ with $G^{\psi\psi} = (G_{\psi\psi} - G_{\psi\lambda} G_{\lambda\lambda}^{-1} G_{\lambda\psi})^{-1}$. The generalized score statistic for ψ is

$$\widetilde{w}_u(\psi_*) = \ell_p'(\psi_*)^{\mathsf{T}} K_p^{-1}(\psi_*) G_p(\psi_*) K_p^{-1}(\psi_*) \ell_p'(\psi_*) \xrightarrow{d} \chi^2_q, \tag{4.13}$$

where $K_p(\psi) = \{K^{\psi\psi}(\psi, \widehat{\lambda}_\psi)\}^{-1}$. See Rotnitzky and Jewell (1990) for the derivation of (4.13). In the case that ψ is scalar, it is common to use the signed square roots of the generalized score and Wald test statistics, and the limiting theory enables the construction of significance functions for ψ, as in Sections 1.3 and 2.4.

Example 4.15 (Working assumption of normality, cont'd) In the setting of Example 4.7, the Wald statistic for θ is $w_e(\theta_*) = n(\widehat{\mu} - \mu_*)^2/\sigma_*^2 + n(\widehat{\sigma}^2 - \sigma_*^2)^2/(2\sigma_*^4)$ and has limiting distribution $\omega_1 Z_1^2 + \omega_2 Z_2^2$, where ω_1 and ω_2 are the eigenvalues of the matrix

$$G^{-1}(\theta_*)K(\theta_*) = \begin{bmatrix} 1 & \gamma_1/(2\sigma_*) \\ \sigma_*\gamma_1 & \gamma_2/2 \end{bmatrix},$$

where $\gamma_1 = \mathrm{E}_h\{(Y_t - \mu_*)^3\}/\sigma_*^3$ and $\gamma_2 = \mathrm{E}_h\{(Y_t - \mu_*)^4\}/\sigma_*^4 - 1$ are the skewness and the kurtosis indices computed at θ_*.

As a simple illustration suppose that the true density function h is symmetric about the expected value μ, so $\gamma_1 = 0$. The limiting distribution of the Wald statistic is that of the linear combination $Z_1^2 + Z_2^2\gamma_2/2$ instead of the familiar χ_2^2. For example, if $\gamma_2 = 4$, the 95-th percentile of the χ_2^2 distribution is 5.99, but this value corresponds only to the 86-th percentile of the correct limiting distribution.

The Godambe information at θ_* is

$$G(\theta_*) = \frac{n}{\sigma_*^2(\gamma_2 - \gamma_1^2)} \begin{bmatrix} \gamma_2 & -\gamma_1/\sigma_* \\ -\gamma_1/\sigma_* & 1/\sigma_*^2 \end{bmatrix}$$

giving the generalized Wald statistic

$$\widetilde{w}_e(\theta_*) = n\{(\widehat{\mu} - \mu_*)^2\gamma_2/\sigma_*^2 - 2(\widehat{\mu} - \mu_*)(\widehat{\sigma}^2 - \sigma_*^2)\gamma_1/\sigma_*^3 + (\widehat{\sigma}^2 - \sigma_*^2)^2/\sigma_*^4\}/(\gamma_2 - \gamma_1^2),$$

where $\widehat{\mu} = \overline{Y}$ and $\widehat{\sigma}^2 = n^{-1}\Sigma_i(Y_i - \overline{Y})^2$. The generalized Wald statistic has the usual χ_2^2 limit distribution.

Finally, consider the case that a single component of θ is of interest. The Wald statistic for μ treating σ^2 as a nuisance parameter is robust to misspecification because $w_e(\mu) = \widetilde{w}_e(\mu_*) = n(\overline{Y} - \mu_*)^2/\sigma^2$ has a limiting χ_1^2 distribution. On the other hand, the Wald statistic for the parameter of interest σ^2 is sensitive to misspecification of the kurtosis because $\widetilde{w}_e(\sigma_*^2) = 2w_e(\sigma_*^2)/\gamma_2$, where $w_e(\sigma_*^2) = n(\widehat{\sigma}^2 - \sigma_*^2)^2/(2\sigma_*^4)$. □

The log-likelihood ratio statistic $w(\theta_*)$ has the same limiting distribution (4.9) because it is asymptotically equivalent to the Wald statistic under model misspecification. When θ is a scalar parameter, dividing the log-likelihood ratio by the scalar quantity $G^{-1}(\theta_*)K(\theta_*)$ is enough to restore the χ_1^2 limit.

Example 4.16 (Equicorrelated Bernoulli observations) The binomial model arises as the sum of independent Bernoulli random variables with the same probability of success. Suppose that the binomial model of size n and parameter π is assumed for the sum $\Sigma_i Y_i$ when the n binary observations $\{Y_1, \ldots, Y_n\}$ are instead equicorrelated, $\mathrm{cor}(Y_i, Y_j) = \rho$, for $i \neq j$. In this model $K(\pi_*) = n/\{\pi_*(1 - \pi_*)\}$, whereas the variance of the score function is

$$I(\pi_*) = \mathrm{var}\left\{\frac{\sum_{i=1}^n Y_i - n\pi_*}{\pi_*(1 - \pi_*)}\right\} = \frac{n\{1 + (n-1)\rho\}}{\pi_*(1 - \pi_*)},$$

which follows from

$$\begin{aligned} \mathrm{var}\left(\sum_{i=1}^n Y_i\right) &= \sum_{i=1}^n \mathrm{var}(Y_i) + 2\sum_{i<j}^n \rho\{\mathrm{var}(Y_i)\mathrm{var}(Y_j)\}^{1/2} \\ &= n\pi_*(1 - \pi_*)\{1 + (n-1)\rho\}. \end{aligned}$$

The Godambe information is $G(\pi_*) = K(\pi_*)/\{1 + (n-1)\rho\}$. The re-scaled log-likelihood ratio statistic $\widetilde{w}(\pi_*) = w(\pi_*)/\{1 + (n-1)\rho\}$ therefore has the usual χ_1^2 limit distribution. □

For a vector parameter of interest, several adjustments of $w(\theta_*)$ have been proposed to avoid the non-standard limit distribution (4.9). A simple method consists in rescaling the log-likelihood ratio by the average of the eigenvalues $\overline{\omega} = \Sigma_i \omega_i / p$ and approximating the rescaled test statistic $w(\theta_*)/\overline{\omega}$ by a χ^2_p limit distribution. However, only the expected value of the rescaled statistic matches that of the χ^2_p distribution.

Often a significant improvement is achieved by the Satterthwaite adjustment $w(\theta_*)\, \nu/(p\overline{\omega})$ using the approximate chi-squared limit distribution with $\nu = (\Sigma_i \omega_i)^2/\Sigma_i \omega_i^2$ degrees-of-freedom. Both the expected value and the variance of the Satterthwaite adjustment coincide with those of the limit distribution.

An alternative to moment-matching adjustments is the vertical adjustment proposed by Chandler and Bate (2007),

$$\frac{(\widehat{\theta} - \theta_*)^\top G(\theta_*)(\widehat{\theta} - \theta_*)}{(\widehat{\theta} - \theta_*)^\top K(\theta_*)(\widehat{\theta} - \theta_*)} w(\theta_*) \xrightarrow{d} \chi^2_p, \tag{4.14}$$

which is, however, not invariant to reparametrization when $p > 1$. Furthermore, simulations in Pace et al. (2011) suggest this adjustment may lead to confidence regions that have a coverage level substantially lower than the nominal coverage level.

Invariance to reparametrization and better coverage is recovered with another vertical adjustment proposed by Pace et al. (2011) based on quadratic forms of the composite scores,

$$\widetilde{w}(\theta_*) = \frac{U^\top(\theta_*) I^{-1}(\theta_*) U(\theta_*)}{U^\top(\theta_*) K^{-1}(\theta_*) U(\theta_*)} w(\theta_*) \xrightarrow{d} \chi^2_p. \tag{4.15}$$

In the case of nuisance parameters, the profile log-likelihood ratio statistic $w(\psi_*) = 2\{\ell_p(\widehat{\psi}) - \ell_p(\psi_*)\}$ has a limiting distribution of the form (4.9), with $\{\omega_1, \ldots, \omega_q\}$ the eigenvalues of $G_p^{-1}(\psi_*) K_p(\psi_*)$. The parametrization-invariant vertical adjustment for $w(\psi_*)$ is

$$\widetilde{w}(\psi_*) = \frac{\widetilde{w}_u(\psi_*)}{\ell_p'(\psi_*)^\top K^{\psi\psi}(\psi_*) \ell_p'(\psi_*)} w(\psi_*). \tag{4.16}$$

The vertical adjustments were proposed for a particular misspecified likelihood called composite likelihood discussed in Section 4.7; see Examples 4.31 and 4.32.

4.4 Model selection

Model selection is also affected by the assumption that the model includes the true data-generating mechanism. The model dimension penalty used in the AIC statistic defined at (1.77) is not valid when the model is misspecified.

Consider a prediction problem where a first set of data $Y = \{Y_1, \ldots, Y_n\}$ is used for estimation of θ with $\widehat{\theta} = \widehat{\theta}(Y)$ and a second independent set of data $Z = \{Z_1, \ldots, Z_n\}$ from the same distribution is used for model selection through minimization of the expected Kullback–Leibler divergence between $f(z; \widehat{\theta})$ and $h(z)$,

$$\mathrm{E}_h\{\mathrm{KL}(\widehat{\theta})\} = \iint \log\left\{\frac{h(z)}{f(z; \widehat{\theta})}\right\} h(z) h(y)\, \mathrm{d}z\, \mathrm{d}y,$$

or, equivalently, maximization of the expected Fraser information

$$\varphi_h(f) = \iint \log f(z; \widehat{\theta}) h(z) h(y)\, \mathrm{d}z\, \mathrm{d}y.$$

Since we do not know the true model h, the expected Fraser information $\varphi_h(f)$ must be replaced with a suitable estimator $\widehat{\varphi}_h(f)$; a natural estimator is the maximized log-likelihood function $\ell(\widehat{\theta}; Z)$. As shown below, this estimator suffers from bias that increases linearly with the model dimension. The AIC statistic and its generalization derived below are first-order corrections of the bias of this estimator.

A stochastic Taylor series expansion of $\ell(\widehat{\theta}; Z)$ about θ_* gives

$$\ell(\widehat{\theta}; Z) = \ell(\theta_*; Z) + (\widehat{\theta} - \theta_*)^{\mathsf{T}} u(\theta_*; Z) - \frac{1}{2}(\widehat{\theta} - \theta_*)^{\mathsf{T}} J(\theta_*; Z)(\widehat{\theta} - \theta_*) + o_p(1), \tag{4.17}$$

where dependence on Z is made explicit, while dependence of the maximum likelihood estimator $\widehat{\theta}$ on Y is implicit. Taking expectation with respect to the true distribution of Z and using the fact that Y and Z are independent and identically distributed gives

$$\mathrm{E}_h\{\ell(\widehat{\theta}; Z)\} = \mathrm{E}_h\{\ell(\theta_*; Y)\} - \frac{1}{2}(\widehat{\theta} - \theta_*)^{\mathsf{T}} K(\theta_*)(\widehat{\theta} - \theta_*) + o_p(1).$$

Taking expectation again, this time with respect to the true distribution of Y, we have

$$\varphi_h(f) = \mathrm{E}_h\{\ell(\theta_*; Y)\} - \frac{1}{2}\mathrm{tr}\{K(\theta_*)V(\theta_*)\} + o(1),$$

where $\mathrm{tr}(\cdot)$ is the matrix trace operator and $V(\theta_*) = \mathrm{E}_h\{(\widehat{\theta} - \theta_*)(\widehat{\theta} - \theta_*)^{\mathsf{T}}\}$ is the asymptotic variance of $\widehat{\theta}$. Since $V(\theta_*)$ is asymptotically equivalent to the inverse of the Godambe information (4.2),

$$\varphi_h(f) = \mathrm{E}_h\{\ell(\theta_*; Y)\} - \frac{1}{2}\mathrm{tr}\{K(\theta_*)G^{-1}(\theta_*)\} + o(1). \tag{4.18}$$

By also expanding $\ell(\theta; Y)$ in a Taylor series around θ_*, as in (4.17), and using the fact that $u(\theta_*; Y) = J(\theta_*; Y)(\widehat{\theta} - \theta_*) + o_p(1)$, we obtain

$$\ell(\widehat{\theta}; Y) = \ell(\theta_*; Y) + \frac{1}{2}(\widehat{\theta} - \theta_*)^{\mathsf{T}} J(\theta_*; Y)(\widehat{\theta} - \theta_*) + o_p(1).$$

Taking expectations with respect to the true distribution of Y and using, again, the approximation of the variance of $\widehat{\theta}$ by the inverse of the Godambe information gives

$$\mathrm{E}_h\{\ell(\widehat{\theta}; Y)\} = \mathrm{E}_h\{\ell(\theta_*; Y)\} + \frac{1}{2}\mathrm{tr}\{K(\theta_*)G^{-1}(\theta_*)\} + o(1). \tag{4.19}$$

Together (4.18) and (4.19) yield

$$\varphi_h(f) = \mathrm{E}_h\{\ell(\widehat{\theta}; Y)\} - \mathrm{tr}\{K(\theta_*)G^{-1}(\theta_*)\} + o(1),$$

showing that the maximized log-likelihood is a biased estimator of $\varphi_h(f)$. A first-order unbiased estimator is

$$\widehat{\varphi}_h(f) = \ell(\widehat{\theta}; Y) - \mathrm{tr}\{K(\widehat{\theta})G^{-1}(\widehat{\theta})\}.$$

The quantity $\mathrm{tr}\{K(\widehat{\theta})G^{-1}(\widehat{\theta})\}$ measures the effective dimension of the model and reduces to $p = \dim(\theta)$ if the model is correctly specified, as in that case the Fisher and Godambe information functions coincide.

Information criteria are traditionally presented on the scale of the log-likelihood ratio statistic, which requires multiplying $\widehat{\varphi}_h(f)$ by minus two. The result is the Takeuchi Information Criterion,

$$\mathrm{TIC} = -2\ell(\widehat{\theta}) + 2\,\mathrm{tr}\{K(\widehat{\theta})G^{-1}(\widehat{\theta})\} = -2\ell(\widehat{\theta}) + 2\,\mathrm{tr}\{I(\widehat{\theta})K^{-1}(\widehat{\theta})\}, \tag{4.20}$$

and reduces to the usual Akaike information criterion, $\mathrm{AIC} = -2\ell(\widehat{\theta}) + 2p$, if the model is correct; see Takeuchi (1976). Model selection with these information criteria chooses the model that minimizes TIC or AIC.

Example 4.17 (Working normal errors) Suppose the linear regression model $Y = X\beta + \epsilon$ is estimated by maximum likelihood under the working assumption that ϵ is a vector of uncorrelated normal errors with variance σ^2. The maximized log-likelihood is $\ell(\widehat{\theta}) = -(n/2)\log\widehat{\sigma}^2$, where $\widehat{\sigma}^2 = n^{-1}\Sigma_i\widehat{e}_i^2$, $\widehat{e}_i = Y_i - x_i^\mathsf{T}\widehat{\beta}$ and $\widehat{\beta} = (X^\mathsf{T}X)^{-1}X^\mathsf{T}Y$. The maximum misspecified likelihood estimators are consistent for the true parameter values, $\beta_* = \beta_0$ and $\sigma_*^2 = \sigma_0^2$. The sensitivity and variability matrices at $\theta_0 = (\beta_0^\mathsf{T}, \sigma_0^2)^\mathsf{T}$ are

$$K(\theta_0) = \begin{bmatrix} X^\mathsf{T}X/\sigma_0^2 & 0 \\ 0 & n/(2\sigma_0^4) \end{bmatrix}, \quad I(\theta_0) = \begin{bmatrix} X^\mathsf{T}X/\sigma_0^2 & X^\mathsf{T}\gamma_1/(2\sigma_0^3) \\ \gamma_1^\mathsf{T}X/(2\sigma_0^3) & n\gamma_2^\mathsf{T}1/(4\sigma_0^4) \end{bmatrix},$$

where γ_1 is the vector of the skewness indices $\gamma_{1,i} = \mathrm{E}_h(\epsilon_i^3)/\sigma_0^3$ and γ_2 is the vector of kurtosis indices $\gamma_{2,i} = \{\mathrm{E}_h(\epsilon_i^4) - \sigma_0^4\}/\sigma_0^4$. The Takeuchi information criterion is

$$\begin{aligned} \mathrm{TIC} &= n\log\widehat{\sigma}^2 + 2\,\mathrm{tr}\{I(\widehat{\theta})K^{-1}(\widehat{\theta})\} \\ &= n\log\widehat{\sigma}^2 + 2p + \overline{\gamma}_2, \end{aligned}$$

where p is the number of columns of X and $\overline{\gamma}_2 = \Sigma_i\gamma_{2,i}/n$ is the average of the kurtosis indices. If the errors are indeed normally distributed, then all the kurtosis indices are equal to two and we obtain the Akaike statistic $\mathrm{AIC} = n\log\widehat{\sigma}^2 + 2(p+1)$. TIC is estimated by replacing $\overline{\gamma}_2$ with a consistent estimate such as $\widehat{\gamma}_2 = (\Sigma_i\widehat{e}_i^{\,4}/n - \widehat{\sigma}^{\,4})/\widehat{\sigma}^{\,4}$. In the context of model selection, a common practice is to estimate γ_2 only once using the residuals obtained from the most complex candidate model. □

Example 4.18 (Overdispersion in Poisson regression) Suppose some count data Y_i are modelled as Poisson variables with expected value $\mu_i(\beta) = \exp(x_i^\mathsf{T}\beta)$ but the true model is overdispersed, that is, $\mathrm{var}(Y_i) = \phi\mu_i(\beta)$, where ϕ is the dispersion parameter, here assumed to be larger than one. The sensitivity matrix is $K(\beta) = \Sigma_i\mu_i(\beta)x_ix_i^\mathsf{T}$ and the variability matrix is $I(\beta) = \phi K(\beta)$. The Takeuchi statistic is $\mathrm{TIC} = -2\ell(\widehat{\beta}) + 2\widehat{\phi}\,p$, where $\widehat{\beta}$ is the maximum likelihood estimate of β and $\widehat{\phi}$ is a consistent estimate of ϕ. A common choice is the method-of-moments estimate $\widehat{\phi} = \Sigma_i\{y_i - \mu_i(\widehat{\beta})\}^2/\{(n-p)\mu_i(\widehat{\beta})\}$. As in the previous example, when evaluating competing models, ϕ is typically estimated only once from the most complex candidate model. The effect of the overdispersion is to inflate the model-dimension penalty of the Akaike statistic. □

Stone (1977) proved that selecting the model that minimizes TIC is asymptotically equivalent to selecting the model that minimizes the cross-validated likelihood ratio. Let $\widehat{\theta}_{-i}$ denote the leave-one-out maximum likelihood estimate omitting observation i. In regular models, $\widehat{\theta}_{-i}$ satisfies the likelihood equations

$$u(\widehat{\theta}_{-i}) - u_i(\widehat{\theta}_{-i}) = 0, \quad i = 1, \ldots, n, \tag{4.21}$$

where $u_i(\theta) = u(\theta; y_i) = \partial\log f(y_i;\theta)/\partial\theta$ is the ith component of the score function. Expanding $u(\widehat{\theta}_{-i})$ around $\widehat{\theta}$ in equation (4.21) gives

$$-J\{\widehat{\theta} + a_i(\widehat{\theta}_{-i} - \widehat{\theta})\}(\widehat{\theta}_{-i} - \widehat{\theta}) - u_i(\widehat{\theta}_{-i}) = 0, \quad |a_i| \le 1. \tag{4.22}$$

Consider the cross-validated log-likelihood

$$\ell_{\mathrm{CV}} = \sum_{i=1}^{n} \log f(y_i; \widehat{\theta}_{-i}), \tag{4.23}$$

and the Taylor series expansion of this about the maximum likelihood estimator

$$\begin{aligned}\ell_{\mathrm{CV}} &= \ell(\widehat{\theta}) + \sum_{i=1}^{n} (\widehat{\theta}_{-i} - \widehat{\theta})^{\mathsf{T}} u_i\{\widehat{\theta} + b_i(\widehat{\theta}_{-i} - \widehat{\theta})\} \\ &= \ell(\widehat{\theta}) - \sum_{i=1}^{n} u_i^{\mathsf{T}}(\widehat{\theta}_{-i}) J^{-1}\{\widehat{\theta} + a_i(\widehat{\theta}_{-i} - \widehat{\theta})\} u_i\{\widehat{\theta} + b_i(\widehat{\theta}_{-i} - \widehat{\theta})\},\end{aligned}$$

where $|b_i| \leq 1$, for $i = 1, \ldots, n$. Since $n^{-1}\Sigma_i u_i(\widehat{\theta}_{-i}) u_i\{\widehat{\theta} + b_i(\widehat{\theta}_{-i} - \widehat{\theta})\}^{\mathsf{T}} = n^{-1} I(\theta_*) + o_p(1)$ and $n^{-1} J\{\widehat{\theta} + a_i(\widehat{\theta}_{-i} - \widehat{\theta})\} = n^{-1} K(\theta_*) + o_p(1)$, because $\mathrm{E}_h\{J(\theta)\} = K(\theta)$, then

$$\ell_{\mathrm{CV}} = \ell(\widehat{\theta}) - \mathrm{tr}\{K^{-1}(\theta_*) I(\theta_*)\} + o_p(1).$$

Finally multiplication by minus two gives the asymptotic equivalence between twice the negative cross-validated log-likelihood and TIC in (4.20).

Example 4.19 (Cross-validation in the linear model) The leave-one-out cross-validated sum of squares is defined as

$$\mathrm{CV} = \frac{1}{n} \sum_{i=1}^{n} (y_i - \widehat{y}_{i,-i})^2,$$

where $\widehat{y}_{i,-i}$ is the prediction of y_i computed from all the data except the ith observation. In the linear regression model we have

$$\mathrm{CV} = \frac{1}{n} \sum_{i=1}^{n} (y_i - x_i^{\mathsf{T}} \widehat{\beta}_{-i})^2 = \frac{1}{n} \sum_{i=1}^{n} \widetilde{e}_i^{\,2},$$

where $\widetilde{e}_i = \widehat{e}_i/(1 - p_{ii})$ are the scaled residuals as in Example 4.14.

The generalized cross-validation criterion (Golub et al., 1979) simplifies calculation of CV by replacing p_{ii} with the average $\Sigma_i p_{ii}/n = p/n$,

$$\mathrm{GCV} = \frac{1}{n} \sum_{i=1}^{n} \frac{\widehat{e}_i^{\,2}}{(1 - p/n)^2}.$$

The logarithm of GCV is asymptotically equivalent to AIC because

$$\begin{aligned} n \log \mathrm{GCV} &= n \log \left\{ \frac{1}{n} \sum_{i=1}^{n} \frac{\widehat{e}_i^{\,2}}{(1 - p/n)^2} \right\} \\ &= n \log \widehat{\sigma}^2 + 2n \log\{(1 - p/n)^{-1}\} \\ &= n \log \widehat{\sigma}^2 + 2n\{p/n + o(p/n)\} \\ &\doteq n \log \widehat{\sigma}^2 + 2p, \end{aligned}$$

for p fixed or for p growing slower than n, as n increases.

The equivalence between CV and AIC requires large n and the p_{ii} to be not too different from each other. In finite samples the difference between CV and AIC may not be negligible. □

4.5 Local misspecification

The previous sections have discussed likelihood inference for globally misspecified models. An alternative view considers situations where the interest is focused on the robustness of statistical conclusions to local deviations from model conditions.

Suppose that we have a "wide" or full model with density or probability function $f(y;\theta,\gamma)$ but inference is performed on a "narrow" model with density $f(y;\theta) = f(y;\theta,\gamma_0)$ corresponding to the particular choice $\gamma = \gamma_0$. The parameter of interest is $\varphi(\theta,\gamma)$. In the local misspecification setting the wide model has density function

$$f(y;\theta_0,\gamma_0 + \delta n^{-1/2}). \tag{4.24}$$

The quantity $\delta n^{-1/2}$ describes the amount of departure from the narrow model. The order $O(n^{-1/2})$ of the departure is chosen so that the squared bias is of the same order of the variance. The choice of the narrow or wide model is thus reframed as the classical bias-variance trade-off. The discussion below follows Hjort and Claeskens (2003) and Claeskens and Hjort (2008, Chapter 5).

The limiting distribution of the estimator $\widehat{\theta}_{\text{narr}}$ of θ under the narrow model is obtained from the first-order Taylor series expansion

$$\widehat{\theta}_{\text{narr}} - \theta_0 = J_{\theta\theta}^{-1}(\widetilde{\theta},\gamma_0)U_\theta(\theta_0,\gamma_0),$$

for some $\widetilde{\theta}$ that lies between θ_0 and $\widehat{\theta}_{\text{narr}}$. The distribution of the score component U_θ is approximately

$$U_\theta(\theta_0,\gamma_0) \dot{\sim} N\{I_{\theta\gamma}(\theta_0,\gamma_0)\delta n^{-1/2}, I_{\theta\theta}(\theta_0,\gamma_0)\}.$$

To see this, note that the expected value of U_θ is

$$\mathrm{E}\{U_\theta(\theta_0,\gamma_0)\} = \sum_{i=1}^{n} \int u_\theta(y_i;\theta_0,\gamma_0) f(y_i;\theta_0,\gamma_0+\delta n^{-1/2})\,\mathrm{d}y_i, \tag{4.25}$$

and the wide model density is related to the narrow model density as

$$f(y_i;\theta_0,\gamma_0+\delta n^{-1/2}) = f(y_i;\theta_0,\gamma_0)\{1 + u_\gamma(y_i;\theta_0,\gamma_0)\delta n^{-1/2}\} + o(\delta n^{-1/2}), \tag{4.26}$$

where $u_\theta(y_i;\theta_0,\gamma_0)$ and $u_\gamma(y_i;\theta_0,\gamma_0)$ are the components of the score for a single observation,

$$u_\theta(y_i;\theta_0,\gamma_0) = \left.\frac{\partial}{\partial\theta}\log f(y_i;\theta,\gamma_0)\right|_{\theta=\theta_0} \quad \text{and} \quad u_\gamma(y_i;\theta_0,\gamma_0) = \left.\frac{\partial}{\partial\gamma}\log f(y_i;\theta_0,\gamma)\right|_{\gamma=\gamma_0}.$$

Substituting the Taylor expansion (4.26) into (4.25) gives $\mathrm{E}\{U_\theta(\theta_0,\gamma_0)\} = I_{\theta\gamma}(\theta_0,\gamma_0)\delta n^{-1/2} + O(1)$, and by a similar argument $\mathrm{E}\left\{U_\theta(\theta_0,\gamma_0)U_\theta^{\mathsf{T}}(\theta_0,\gamma_0)\right\} = I_{\theta\theta}(\theta_0,\gamma_0) + O(n^{1/2})$. It follows that the narrow model estimator is approximately distributed as

$$\widehat{\theta}_{\text{narr}} \dot{\sim} N(\theta_0 + I_{\theta\theta}^{-1} I_{\theta\gamma}\delta n^{-1/2}, I_{\theta\theta}^{-1}),$$

where here and in the rest of the section the information components are implicitly evaluated at the narrow-model values (θ_0,γ_0). The approximate distribution of the maximum likelihood estimator in the wide model is

$$\begin{bmatrix}\widehat{\theta}_{\text{wide}}\\ \widehat{\gamma}_{\text{wide}}\end{bmatrix} \dot{\sim} N\left(\begin{bmatrix}\theta_0\\ \gamma_0+\delta n^{-1/2}\end{bmatrix}, \begin{bmatrix}I^{\theta\theta} & I^{\theta\gamma}\\ I^{\gamma\theta} & I^{\gamma\gamma}\end{bmatrix}\right).$$

Assuming that the parameter of interest $\varphi = \varphi(\theta, \gamma)$ has continuous partial derivatives in a neighbourhood of (θ_0, γ_0), the limiting distributions of the estimators of φ based on the narrow and the wide model are derived from the delta method as

$$\widehat{\varphi}_{\text{narr}} \dot{\sim} N(\varphi_0 + a^{\mathsf{T}}\delta n^{-1/2}, b), \tag{4.27}$$

$$\widehat{\varphi}_{\text{wide}} \dot{\sim} N(\varphi_0, a^{\mathsf{T}} I^{\gamma\gamma} a + b), \tag{4.28}$$

where the "true" value of φ is $\varphi_0 = \varphi(\theta_0, \gamma_0 + \delta n^{-1/2})$, $a = I_{\gamma\theta} I_{\theta\theta}^{-1}(\partial\varphi/\partial\theta) - (\partial\varphi/\partial\gamma)$ and $b = (\partial\varphi/\partial\theta)^{\mathsf{T}} I_{\theta\theta}^{-1}(\partial\varphi/\partial\theta)$, with the partial derivatives evaluated at (θ_0, γ_0). See Corollary 5.1 of Claeskens and Hjort (2008). The limiting mean-squared errors for the narrow and the wide model estimators are then $n^{-1}(a^{\mathsf{T}}\delta)^2 + b$ and $a^{\mathsf{T}} I^{\gamma\gamma} a + b$, respectively. The narrow model is thus preferable to the wide model in mean-squared error sense if δ lies in the region

$$R_\varphi = \{\delta : n^{-1/2}|a^{\mathsf{T}}\delta| \leq (a^{\mathsf{T}} I^{\gamma\gamma} a)^{1/2}\}.$$

This preference region depends on the particular choice of function $\varphi(\theta, \gamma)$. It is also possible to identify a universal preference region that holds for all possible φ. Narrow model estimation is always preferable to wide model estimation if $n^{-1/2}|a^{\mathsf{T}}\delta| \leq (a^{\mathsf{T}} I^{\gamma\gamma} a)^{1/2}$ for all vectors a. Write D for the square-root matrix of $I^{\gamma\gamma}$, so that $I^{\gamma\gamma} = DD^{\mathsf{T}}$. Let $c = Da$, then the narrow model is preferred if $n^{-1/2}|c^{\mathsf{T}} D^{-1}\delta| \leq \|c\|_2$ for all c. The Cauchy–Schwarz inequality implies that $n^{-1/2}|c^{\mathsf{T}} D^{-1}\delta| \leq n^{-1/2}\|c\|_2 \|D^{-1}\delta\|_2$ for all c and thus the narrow model is preferred when δ lies in the region

$$\{\delta : n^{-1/2}\|D^{-1}\delta\|_2 \leq 1\},$$

and this is valid for any φ.

Example 4.20 (Omitted covariates) Suppose data follow the linear model $Y = X\beta + z\gamma + \epsilon$, where ϵ is a vector of uncorrelated normal errors with variance σ^2. As narrow model take the model with covariate z omitted, corresponding to $\gamma = 0$. The information matrix computed at $\gamma = 0$ is

$$I(\sigma^2, \beta, \gamma = 0) = \frac{1}{\sigma^2} \begin{bmatrix} n/(2\sigma^2) & 0 & 0 \\ 0 & X^{\mathsf{T}}X & X^{\mathsf{T}}z \\ 0 & z^{\mathsf{T}}X & z^{\mathsf{T}}z \end{bmatrix}.$$

It follows that $I^{\gamma\gamma} = \sigma^2\{z^{\mathsf{T}}z - z^{\mathsf{T}}X(X^{\mathsf{T}}X)^{-1}X^{\mathsf{T}}z\}^{-1}$ and narrow model estimation is preferred in mean-squared error to wide model estimation if $|\delta| n^{-1/2} \leq (I^{\gamma\gamma})^{1/2}$. Since $\gamma = \gamma_0 + \delta n^{-1/2} = \delta n^{-1/2}$, preference for the narrow model corresponds to $|\gamma| \leq (I^{\gamma\gamma})^{1/2}$. In particular, if z is orthogonal to X, then the narrow model is preferred if $|\gamma| \leq n^{-1/2}\sigma/s_z$, with $s_z^2 = n^{-1}\Sigma_i z_i^2$. □

Example 4.21 (Variance heteroscedasticity) Consider the linear regression model $Y = X\beta + \epsilon$, with $\epsilon \sim N(0, \Omega)$ where $\Omega = \text{diag}(\sigma_1^2, \ldots, \sigma_n^2)$ and $\sigma_i = \sigma(1 + \gamma c_i)$, $i = 1, \ldots, n$. The information matrix at $\gamma = 0$ is

$$I(\sigma^2, \beta, \gamma = 0) = \begin{bmatrix} n/(2\sigma^4) & 0 & \Sigma_i c_i/\sigma^2 \\ 0 & X^{\mathsf{T}}X/\sigma^2 & 0 \\ \Sigma_i c_i/\sigma^2 & 0 & 2\Sigma_i c_i^2 \end{bmatrix}.$$

The simpler homoscedastic model performs better in mean-square error than the heteroscedastic model as long as $|\gamma| \leq (2ns_c^2)^{-1/2}$, where $s_c^2 = \Sigma_i(c_i - \bar{c})^2/n$. The above general result may be tailored to specific choices for the function $\varphi(\beta, \gamma)$. In the special case that

$\varphi = \beta$, or, more generally, that φ is a function of β, then $a = 0$ and the wide model estimator $\widehat{\beta}_{\text{wide}} = (X^{\mathsf{T}}W^{-1}X)^{-1}X^{\mathsf{T}}W^{-1}Y$, where W is the diagonal matrix of elements $1 + \widehat{\gamma}c_i$, is asymptotically equivalent to the least-squares estimator $\widehat{\beta}_{\text{narr}} = (X^{\mathsf{T}}X)^{-1}X^{\mathsf{T}}Y$ obtained under the narrow model. □

The discussion can be extended to situations involving a boundary parameter γ. For example, if $\gamma \geq \gamma_0$, then $\delta \geq 0$ within the local misspecification model as defined in (4.24). This scenario is relevant, for instance, when evaluating the amount of overdispersion a Poisson model can tolerate or the degree of "t-ness" a normal model can handle (Hjort, 1994). The limiting distributions of maximum likelihood estimators and related quantities are notably more complex under such parameter constraints; see Claeskens and Hjort (2008, Section 10.2).

4.6 Indirect likelihood

Indirect inference is a class of methods for estimating a parameter of a complex model using an intermediate estimator derived from a convenient but misspecified model. Indirect inference is often motivated by models with intractable likelihoods but from which simulation is feasible (Gouriéroux et al., 1993; Smith, 1993), although the general concept of indirect inference does not necessarily require simulation.

Consider a sample $\{Y_1, \ldots, Y_n\}$ from a random variable with distribution function $F(\cdot;\theta_0)$, for some $\theta_0 \in \mathbb{R}^p$. The first step in indirect inference is the computation of some intermediate statistic $\widehat{\varphi}$, typically an estimator built from a convenient misspecified model or a robust nonparametric estimator such as the nonparametric maximum likelihood estimator discussed in Section 6.1. The dimension of $\widehat{\varphi}$ is $q \geq p$; otherwise, it is not possible to identify θ. In the second step, a bridge relationship $\varphi = g\{F(\cdot;\theta)\}$ is defined; φ is called an intermediate parameter. The final step is to estimate θ by the indirect estimator $\widehat{\theta}_{\text{ind}} = \theta(\widehat{\varphi})$ which depends on the data through $\widehat{\varphi}$.

Example 4.22 (Method of moments) The method of moments is a basic example of indirect inference with $q = p$. The intermediate statistic is a vector of p sample moments and the intermediate parameter is the vector of the p corresponding population moments. For example, if θ has dimension two and the mean μ and variance σ^2 are both functions of θ, then the intermediate statistic is the vector with the sample mean and the sample variance, $\widehat{\varphi} = (\overline{Y}, S^2)^{\mathsf{T}}$, and the intermediate parameter is $\varphi(\theta) = (\mu(\theta), \sigma^2(\theta))^{\mathsf{T}}$. The indirect estimator $\widehat{\theta}_{\text{ind}}$ is found by solving the moment conditions $\widehat{\varphi} = \varphi(\widehat{\theta})$. □

Example 4.23 (Exponential observations with censoring) Let $\{T_1, \ldots, T_n\}$ be a sample of independent observations from an exponential distribution with mean θ and suppose the data are subject to censoring at a fixed threshold c; the observed data are $Y_i = \min\{T_i, c\}$. See Section 3.6 for some general discussion about survival data.

Consider the intermediate statistic $\widehat{\varphi} = \overline{Y}$ that corresponds to the estimate of θ if the censoring mechanism is ignored. The intermediate statistic consistently estimates the intermediate parameter $\varphi(\theta) = \theta\{1 - \exp(-c/\theta)\}$. The indirect estimate of θ is then obtained by solving $\overline{Y} = \widehat{\theta}\{1 - \exp(-c/\widehat{\theta})\}$. For comparison the maximum likelihood estimator of θ is $\widehat{\theta} = n\overline{Y}/\Sigma_i 1(Y_i < c)$. See Example 1 of Jiang and Turnbull (2004). □

Example 4.24 (Measurement errors in covariates) Suppose we are interested in the relationship between a response variable Y and a covariate X described by a conditional distribution $f(y \mid x; \theta_0)$. In measurement error problems, we observe a contaminated version Z of the covariate X. Furthermore, suppose that Z is a surrogate of X in the sense that Y is independent of Z conditional on X. The likelihood function based on the observed data $\{(y_1, z_1), \ldots, (y_n, z_n)\}$ requires integration over the unobserved covariates

$$L(\theta, \lambda_1, \lambda_2) = \prod_{i=1}^{n} \int f(y_i \mid x; \theta) f(x \mid z_i; \lambda_1) f(z_i; \lambda_2)\, \mathrm{d}x,$$

where λ_1 and λ_2 are the nuisance parameters that appear in the conditional distribution of X given Z and in the marginal distribution of Z.

Jiang and Turnbull (2004) discussed an indirect analysis based on the misspecified model that ignores the measurement error and assumes that Z is the true covariate. For example, suppose that $Y_i = \theta X_i + \epsilon_i$, where ϵ_i are uncorrelated normal errors with zero mean and variance σ_ϵ^2, and that $Z_i = X_i + U_i$, where X_i and U_i are independent normal variables with variances σ_x^2 and σ_u^2, respectively. The naive maximum likelihood estimator that ignores the measurement error, $\widehat{\varphi} = \Sigma_i Y_i Z_i / \Sigma_i Z_i^2$, consistently estimates $\varphi(\theta) = \mathrm{E}(Y_1 Z_1)/\mathrm{E}(Z_1^2) = \theta\{\sigma_x^2/(\sigma_x^2 + \sigma_u^2)\}$, which shows the so-called attenuation phenomenon: the absolute value of the naive slope, $|\widehat{\varphi}\,|$, underestimates the size of true slope, $|\theta|$. The indirect estimator is then simply $\widehat{\theta}_{\text{ind}} = \widehat{\varphi}\,\{(\sigma_x^2 + \sigma_u^2)/\sigma_x^2\}$, which requires knowledge of σ_x^2 and σ_u^2. Sometimes these unknown variances are estimated from an independent validation set where the true covariate X and its surrogate Z are observed directly. □

In the above examples, the indirect estimator was computed directly by solving the intermediate parameter equation. More commonly, and in particular if $q > p$, there is no explicit expression for the indirect estimator as a function of the intermediate parameter. In this case θ can be estimated by minimizing a suitable distance between the intermediate statistic and its expectation computed under the true model, or alternatively by maximizing the indirect likelihood as discussed below.

Suppose that the intermediate statistic, computed with a sample of n observations, has an asymptotic normal distribution centred at $\varphi(\theta)$ with covariance matrix V. Then the maximum indirect likelihood estimator $\widehat{\theta}_{\text{ind}}$ is the maximizer of the indirect log-likelihood function

$$\ell_{\text{ind}}(\theta; \widehat{\varphi}\,) = -\frac{1}{2}\log|\widehat{V}| - \frac{1}{2}\{\widehat{\varphi} - \varphi(\theta)\}^{\mathsf{T}}\, \widehat{V}^{-1}\{\widehat{\varphi} - \varphi(\theta)\},$$

where $\widehat{V}$ is a sample estimate of V. If computation of $\widehat{V}$ is difficult or robustness arguments suggest avoiding it, V may be replaced by a more convenient covariance matrix, at the cost of some loss of efficiency.

More generally, $\widehat{\varphi}$ is defined implicitly as the solution of an estimating equation $\Psi(\varphi; Y) = 0$. Estimating equations are discussed in Chapter 5. The indirect log-likelihood function when φ is derived from a generic estimating equation is

$$\ell_{\text{ind}}(\theta; \widehat{\varphi}\,) = -\frac{1}{2}\log|\widehat{\Sigma}_\Psi(\widehat{\varphi})| - \frac{1}{2}\mathrm{E}_\theta\{\Psi(\widehat{\varphi}; Y)\}^{\mathsf{T}}\widehat{\Sigma}_\Psi^{-1}(\widehat{\varphi})\mathrm{E}_\theta\{\Psi(\widehat{\varphi}; Y)\}, \qquad (4.29)$$

where $\widehat{\Sigma}_\Psi(\widehat{\varphi})$ is a consistent estimator of the variability matrix $\Sigma_\Psi(\varphi) = \mathrm{var}_\theta\{\Psi(\varphi; Y)\}$ for the estimating function $\Psi(\varphi; Y)$ computed at $\widehat{\varphi}$.

In some complex models the required expectations for computing the indirect likelihood are estimated by simulating from the true model.

Example 4.25 (Stochastic volatility model) Consider the discrete-time stochastic volatility model mentioned in Example 3.21. This is a parameter-driven time series, with $\{Y_1, \ldots, Y_n\}$ assumed to follow the model

$$Y_t = \exp(b_t/2)\epsilon_t,$$
$$b_t = \theta_1 b_{t-1} + \theta_2 \eta_t,$$

where ϵ_t and η_t are independent $N(0,1)$ white noise processes, $|\theta_1| < 1$, and $\theta_2 > 0$. For a time series of n observations, the likelihood for $\theta = (\theta_1, \theta_2)^\mathsf{T}$ is the intractable integral (3.34).

An alternative to approximating the likelihood integral is indirect inference with working model a generalized autoregressive model, $Y_t = \sigma_t \epsilon_t$, where

$$\sigma_t^2(\varphi) = \varphi_0 + \sum_{i=1}^{p} \varphi_i Y_{t-i}^2 + \sum_{j=1}^{q} \varphi_{p+j} \sigma_{t-j}^2,$$

and $\epsilon_t \sim N(0,1)$ is white noise. The intermediate parameter φ has dimension $1+p+q$. The log-likelihood function for the working model conditional on the first observation is

$$\ell(\varphi) = -\frac{1}{2}\sum_{t=2}^{n} \log \sigma_t^2(\varphi) - \frac{1}{2}\sum_{t=2}^{n} \frac{y_t^2}{\sigma_t^2(\varphi)}.$$

Numerical maximization of $\ell(\varphi)$ gives the intermediate statistic $\widehat{\varphi}$. The bridge function $\varphi(\theta) = \mathrm{E}_\theta(\widehat{\varphi})$ needed to compute the indirect likelihood can be approximated via simulation from the stochastic volatility model. For a given value of θ, we simulate m time series of size n from the stochastic volatility model, then for each simulated time series compute the intermediate statistics $\widehat{\varphi}(\theta)^{(j)}$, for $j = 1, \ldots, m$. The simulated indirect log-likelihood function is

$$\ell_{\text{ind}}(\theta; \widehat{\varphi}) = -\frac{1}{2}\log|\widehat{V}| - \frac{1}{2}\Big\{\widehat{\varphi} - \frac{1}{m}\sum_{j=1}^{m} \widehat{\varphi}(\theta)^{(j)}\Big\}^\mathsf{T} \widehat{V}^{-1}\Big\{\widehat{\varphi} - \frac{1}{m}\sum_{j=1}^{m} \widehat{\varphi}(\theta)^{(j)}\Big\}.$$

The simulated intermediate statistics $\widehat{\varphi}(\theta)^{(j)}$ can also be used to estimate V. □

Under mild regularity conditions, the indirect likelihood function essentially behaves as a genuine likelihood function. Let $U_{\text{ind}}(\theta) = \partial \ell_{\text{ind}}(\theta)/\partial\theta$ be the indirect score function and $J_{\text{ind}}(\theta) = -\partial U_{\text{ind}}(\theta)/\partial\theta^\mathsf{T}$ the indirect observed information. The first two Bartlett identities,

$$\mathrm{E}\{U_{\text{ind}}(\theta)\} = 0, \ \mathrm{var}\{U_{\text{ind}}(\theta)\} = \mathrm{E}\{J_{\text{ind}}(\theta)\}, \quad \theta \in \Theta,$$

are also valid for the indirect likelihood function as the asymptotic variance of $U_{\text{ind}}(\theta_0)$ is $G(\theta) = [\partial\varphi(\theta)/\partial\theta]^\mathsf{T} V^{-1} [\partial\varphi(\theta)/\partial\theta]$ evaluated at θ_0 and $G^{-1}(\theta_0) J_{\text{ind}}(\theta_0) \xrightarrow{p} \mathcal{I}_{p\times p}$. It follows that the maximum indirect likelihood estimator has a limiting normal distribution centred at θ_0 with asymptotic variance $J_{\text{ind}}^{-1}(\theta_0)$ and the indirect log-likelihood ratio has the usual χ^2 limit distribution,

$$w_{\text{ind}}(\theta_0) = 2\{\ell_{\text{ind}}(\widehat{\theta}_{\text{ind}}) - \ell_{\text{ind}}(\theta_0)\} \xrightarrow{d} \chi_p^2.$$

Furthermore, it can be shown that the maximum indirect likelihood estimator has smallest asymptotic variance among all consistent asymptotically normal estimators constructed from the intermediate statistics $\widehat{\varphi}$; see Jiang and Turnbull (2004).

4.7 Composite likelihoods

4.7.1 Intractable likelihood

Likelihood functions can be intractable in high-dimensional problems or with large data sets. Some common reasons for this are the need for numerical inversion of large correlation matrices, the approximation of high-dimensional integrals, and combinatorial difficulties involved in normalizing constants. The growing availability of vast, well-structured datasets, along with the need to model them effectively, has further exacerbated these computational problems. In such situations, analytical approximations may have low statistical or computational efficiency, and Monte Carlo methods may not be viable because they do not scale well with the sample size. These difficulties have motivated several authors to replace the ordinary likelihood with a particular type of misspecified likelihood called composite likelihood.

Lindsay (1988) coined the term composite likelihood to describe a class of pseudo-likelihood functions constructed by multiplying a set of component likelihoods as if they were independent. Suppose $Y = (Y_1, \dots, Y_d)^\top$ is a random vector with density function $f(y;\theta)$. Let $\{\mathcal{A}_s : s \in S \subset \mathbb{N}\}$ be a set of marginal or conditional measurable events. A composite log-likelihood function for θ is a weighted sum of the log-likelihood functions constructed from the events $\mathcal{A}_s$,

$$c\ell(\theta) = \sum_{s \in S} w_s \log f(y \in \mathcal{A}_s; \theta), \tag{4.30}$$

for a set of non-negative weights $\{w_s : s \in S\}$. Composite likelihood functions are used to replace likelihood functions in situations where inference based on the latter would be difficult or even impossible.

Different types of composite likelihood have been considered in the literature, depending on the assumed model and the reason for the intractability. For example, when the ordinary likelihood function involves an integral whose dimension grows with the sample size, then it is natural to combine likelihoods for subsets of the data whose size make the integrals manageable.

Although composite likelihoods can be constructed with combinations of conditional and marginal densities, this is not often used, and thus composite likelihoods are typically classified as either conditional or marginal.

For approximate inference in spatial processes, Besag (1974) proposed what he called a pseudo-likelihood, which is in our terminology a composite conditional likelihood, constructed as the product of the densities of each observation conditional on its neighbours with neighbourhoods defined relative to some distance. Besag's pseudo-likelihood was discussed in Example 3.27 for spatial autologistic regression models. In that example, the pseudo-likelihood enables likelihood-type inference in a situation where ordinary likelihood inference would require the computation of a normalization constant with 2^d terms, where d is the number of observed spatial locations.

The conditional subsets used to build composite conditional likelihoods depend on the specific application. Commonly used versions are constructed by pooling the pairwise conditional densities $f(y_i \mid y_j; \theta)$ or the full conditional densities $f(y_i \mid y_{-i}; \theta)$, where y_{-i} is the vector of all the observations except y_i. Composite marginal likelihoods include the independence likelihood, which combines univariate densities $f(y_i; \theta)$, the pairwise likelihood, which is based on bivariate densities $f(y_i, y_j; \theta)$, and more complex versions constructed from triplets or larger blocks of data.

Example 4.26 (Stationary Gaussian random field) The difficulty of maximum likelihood estimation for stationary Gaussian random fields was discussed in Example 3.22. The profile likelihood function (3.37) for that class of models has a computational cost of $O(d^3)$ for a vector of d spatial observations, which is prohibitive in many applications. Even when the computational cost is manageable, maximizing the likelihood function is difficult because it is likely to be multimodal.

To avoid the numerical difficulties associated with a full likelihood analysis, Hjort and Omre (1994) considered the pairwise log-likelihood function

$$c\ell(\theta) = \sum_{i<j}^{d} w_{ij} \log f(y_i, y_j; \theta), \tag{4.31}$$

where weights w_{ij} are typically chosen by taking only pairs of nearby observations to include in $c\ell(\theta)$, for example, $w_{ij} = 1$ if $\|x_i - x_j\|_2 \leq r$ and 0 otherwise, for prespecified distance r. Using the notation of Example 3.22, the log-density for a pair of observations is

$$\log f(y_i, y_j; \theta) = -\log \sigma^2 - \frac{1}{2} \log |\Gamma_{ij}(\psi)| - \frac{1}{2\sigma^2} \frac{Q_{ij}(y, \mu, \psi)}{|\Gamma_{ij}(\psi)|},$$

where $\Gamma_{ij}(\psi)$ is the version of $\Gamma(\psi) = \tau \mathcal{I}_{d\times d} + R(\alpha)$ restricted to observations i and j, $|\Gamma_{ij}(\psi)| = (1+\tau)^2 - \rho_{ij}^2(\alpha)$, with $\rho_{ij}(\alpha) = \rho(\|x_i - x_j\|_2; \alpha)$, and

$$Q_{ij}(y, \mu, \psi) = (y_i - \mu)^2(1+\tau) - 2(y_i - \mu)(y_j - \mu)\rho_{ij}(\alpha) + (y_j - \mu)^2(1+\tau).$$

Maximization of (4.31) with respect to μ and σ^2 gives the estimators

$$\widehat{\mu}_{c\ell\,\psi} = \frac{\sum_{i<j}^{d} v_{ij}(\psi)(y_i + y_j)}{2\sum_{i<j}^{d} v_{ij}(\psi)}, \quad \widehat{\sigma}^2_{c\ell\,\psi} = \frac{\sum_{i<j}^{d} w_{ij} Q_{ij}(y, \widehat{\mu}_{c\ell\,\psi}, \psi)/|\Gamma_{ij}(\psi)|}{2\sum_{i<j}^{d} w_{ij}},$$

where $v_{ij}(\psi) = w_{ij}\{1 + \tau + \rho_{ij}(\alpha)\}^{-1}$. Substitution of $\widehat{\mu}_{c\ell\,\psi}$ and $\widehat{\sigma}^2_{c\ell\,\psi}$ into $c\ell(\theta)$ gives the profile pairwise log-likelihood function for ψ,

$$c\ell_p(\psi) = -\log \widehat{\sigma}^2_{c\ell\,\psi} \sum_{i<j}^{d} w_{ij} - \frac{1}{2} \sum_{i<j}^{d} w_{ij} \log |\Gamma_{ij}(\psi)|.$$

Maximizing this numerically to obtain the composite likelihood estimator of ψ requires inverting only 2×2 matrices rather than the $d \times d$ matrix needed with the full likelihood function (3.37).

□

Example 4.27 (Multivariate extremes) As anticipated in Example 1.30, pairwise likelihood inference is widely used in the analysis of multivariate extremes; see Davison et al. (2012). Environmental extremes naturally exhibit spatial or spatiotemporal dependence, which has motivated the development of a variety of modelling approaches.

For d locations $x_1, \dots, x_d$ in space or in space-time, the multivariate extreme value distribution of the component-wise maxima $\{Y_1, \dots, Y_d\}$, with $Y_i = Y(x_i)$, is

$$\text{pr}(Y_1 \leq y_1, \dots, Y_d \leq y_d) = \exp\{-V(y_1, \dots, y_d; \theta)\},$$

for an exponent function $V(\cdot; \theta)$ that describes spatial or spatio-temporal dependence between component-wise maxima. The likelihood function is

$$L(\theta) = \exp\{-V(y_1, \dots, y_d; \theta)\} \sum_{\pi \in \mathcal{P}} \prod_{i=1}^{k} \{-V_{\pi_i}(y_1, \dots, y_d; \theta)\}, \tag{4.32}$$

where $\mathcal{P}$ is the set of all partitions $\pi = \{\pi_1, \ldots, \pi_k\}$ of $\mathcal{D} = \{1, \ldots, d\}$ with $\pi_i \subset \mathcal{D}$, $i = 1, \ldots, k \leq d$, and V_{π_i} is the partial derivative of V with respect to the variables indexed by the set π_i.

The likelihood involves a sum over all possible partitions of a set. The number of terms in the sum is the Bell number, which grows faster than the exponential function. The terms in the sum are combinations of the $2^d - 1$ partial derivatives of the exponent function $V(\cdot; \theta)$, which is also challenging. The combinatorial explosion of terms that appear in the likelihood makes its computation prohibitive even for small d. For example, when $d = 10$, the likelihood sum has $115,975$ terms.

The combinatorial difficulties associated with the computation of the ordinary likelihood function have motivated pairwise likelihood inference for various models for spatial or spatiotemporal extrema as discussed, for example, in Huser and Wadsworth (2022). The bivariate densities used to construct the pairwise likelihood are specified by two-dimensional exponent functions. The maximum pairwise likelihood estimate of θ must be computed numerically, but the computational cost is at most $O(d^2)$ if all pairs are included in the pairwise likelihood.

When d is large it is reasonable to take a subset of all possible pairs, perhaps selecting only observation pairs within a certain distance. Selecting pairs of observations that are close together is not only useful to reduce the computational cost, but may also be necessary to preserve reasonable statistical efficiency of the maximum pairwise likelihood estimator, because those pairs of observations that are too far apart contribute little or nothing to the estimation of the dependence parameters, while increasing the variability of the estimator. □

4.7.2 Composite likelihood inference

Consider n independent observations $\{Y_1, \ldots, Y_n\}$ of a d-dimensional random vector. The maximum composite likelihood estimate $\widehat{\theta}_{c\ell}$ maximizes the composite log-likelihood function,

$$c\ell(\theta; y) = \sum_{i=1}^{n} \sum_{s \in S} w_s \log f(y_i \in \mathcal{A}_s; \theta).$$

In regular problems, $\widehat{\theta}_{c\ell}$ is the solution of the composite score equations,

$$u_{c\ell}(\widehat{\theta}_{c\ell}; y) = 0,$$

where $u_{c\ell}(\theta; y) = \partial c\ell(\theta; y)/\partial\theta$. Maximizing the composite likelihood function is equivalent to finding the parameter value that minimizes the composite Kullback–Leibler divergence

$$\mathrm{KL}_{c\ell}(\theta) = \sum_{s \in S} w_s \mathrm{E}_h\{\log h(Y \in \mathcal{A}_s) - \log f(Y \in \mathcal{A}_s; \theta)\}. \tag{4.33}$$

The composite Kullback-Leibler divergence quantifies the expected mismatch between the assumed parametric sub-models $f(Y \in \mathcal{A}_s; \theta)$ and the underlying true distributions $h(Y \in \mathcal{A}_s)$. Composite likelihood inference thus follows the framework of misspecified likelihoods discussed in Section 4.1, using a modification of the usual Kullback–Leibler divergence from the true model to the proposed model.

If all the component sub-likelihoods are correctly specified, then the composite score function is an unbiased estimating equation,

$$\mathrm{E}_h\{u_{c\ell}(\theta; Y)\} = 0, \quad \forall\theta \in \Theta, \tag{4.34}$$

and the theory of estimating equations discussed in Sections 5.3 and 5.4 can be applied.

Xu and Reid (2011) derived the convergence of the maximum composite likelihood estimator $\widehat{\theta}_{c\ell}$ to the least-false parameter θ_*, which is the value of $\theta \in \Theta$ that minimizes (4.33), using a set of regularity conditions which are essentially those of Wald's classical proof for the consistency of the maximum likelihood estimator (Wald, 1949) but applied to each sub-likelihood of the composite likelihood.

The limiting distribution of $\widehat{\theta}_{c\ell}$ is normal with expected value θ_* and covariance matrix equal to the inverse of Godambe information $G_{c\ell}(\theta_*) = K_{c\ell}(\theta_*)I_{c\ell}^{-1}(\theta_*)K_{c\ell}(\theta_*)$, where $I_{c\ell}(\theta)$ and $K_{c\ell}(\theta)$ are the variability and sensitivity matrices derived from the composite score function $U_{c\ell}(\theta) = u_{c\ell}(\theta; Y)$. If all the component likelihood functions are compatible with the true full model, θ_* is the true parameter value θ_0.

Example 4.28 (Dichotomized equicorrelated normal) Consider a sample of n independent vectors of binary random variables $\{Y_1, \ldots, Y_n\}$, with $Y_i = (Y_{i1}, \ldots, Y_{id})^{\mathsf{T}}$. The vector Y_i is produced by dichotomizing a vector of equicorrelated normal variables, $Y_{ij} = 1(Z_{ij} > \mu)$ where $Z_{ij} \sim N(\mu, 1)$ and $\mathrm{cor}(Z_{ij}, Z_{ik}) = \rho$, for any $j \neq k$. The likelihood function for $\theta = (\mu, \rho)^{\mathsf{T}}$ is a product of n multivariate normal integrals each of size d.

Under the assumed equicorrelation, the dimension of the integrals can be reduced from d to one,

$$\begin{aligned} L(\theta) &= \prod_{i=1}^{n} \mathrm{pr}(Y_{i1} = y_{i1}, \ldots, Y_{id} = y_{id}; \theta) \\ &= \prod_{i=1}^{n} \int_{-\infty}^{\infty} \left[1 - \Phi\left\{\frac{t\sqrt{\rho} - \mu}{\sqrt{1-\rho}}\right\}\right]^{y_{i\bullet}} \left[\Phi\left\{\frac{t\sqrt{\rho} - \mu}{\sqrt{1-\rho}}\right\}\right]^{d - y_{i\bullet}} \phi(t)\, dt, \end{aligned} \tag{4.35}$$

where $y_{i\bullet} = \Sigma_j y_{ij}$. See Tong (1990, p.192) for the derivation of (4.35). The n integrals can be approximated with adaptive Gauss–Hermite quadrature (Liu and Pierce, 1994).

As an alternative to ordinary likelihood inference, Cox and Reid (2004) investigated the pairwise log-likelihood function based on all the pairs of correlated observations,

$$c\ell(\theta) = \sum_{i=1}^{n} \sum_{j<k}^{d} \log \mathrm{pr}(Y_{ij} = y_{ij}, Y_{ik} = y_{ik}; \theta) = \sum_{r \in \{0,1\}} \sum_{s \in \{0,1\}} n_{rs} \log p_{rs}(\theta),$$

where $n_{rs} = \Sigma_i \Sigma_{j<k} 1(Y_{ij} = r, Y_{ik} = s)$ and $p_{rs}(\theta) = \mathrm{pr}(Y_{ij} = r, Y_{ik} = s; \theta)$. For example, $p_{11}(\theta) = \Phi_2(\mu, \mu, \rho)$, where $\Phi_2(a, b, r)$ is the distribution function of a bivariate standard normal variable with correlation r computed at $(a, b)^{\mathsf{T}}$.

In the special case $\mu = 0$, the computations simplify and the pairwise likelihood function becomes

$$c\ell(\rho) = \sum_{i=1}^{n} T_i \log\left\{\frac{p_{11}(\rho)}{1/2 - p_{11}(\rho)}\right\} + n\binom{d}{2} \log p_{11}(\rho),$$

where $T_i = \Sigma_{j<k}(2Y_{ij}Y_{ik} - Y_{ij} - Y_{ik})$ and $p_{11}(\rho) = 1/4 + \arcsin(\rho)/(2\pi)$; see Cox and Wermuth (2002). The variability matrix is

$$I_{c\ell}(\rho) = n\left\{\frac{1}{p_{11}(1 - 2p_{11})}\right\}^2 (p'_{11})^2 \mathrm{var}(T_1),$$

where p'_{11} is the derivative of $p_{11} = p_{11}(\rho)$ with respect to ρ and $\mathrm{var}(T_1)$ is a polynomial of order four in d which involves three and four-dimensional multivariate normal integrals. The sensitivity matrix is

$$K_{c\ell}(\rho) = nd(d-1)\left\{\frac{1}{p_{11}(1 - 2p_{11})}\right\} (p'_{11})^2$$

and the Godambe information is

$$G_{c\ell}(\rho) = nd^2(d-1)^2\frac{(p'_{11})^2}{\mathrm{var}(T_1)} = \frac{n}{4\pi^2(1-\rho^2)}\frac{d^2(d-1)^2}{\mathrm{var}(T_1)}.$$

Since $\mathrm{var}(T_i) = O(d^4)$, the Godambe information is $O(1)$ if n is fixed but $d \to \infty$, and the maximum pairwise likelihood estimator of ρ is inconsistent. However if n diverges with fixed d, the maximum pairwise likelihood estimator is consistent and the loss of efficiency compared to the maximum likelihood estimator is small. The largest loss of efficiency is 3% for $d = 5$, 8% for $d = 8$ and 15% for $d = 10$. □

Example 4.29 (Equicorrelated multivariate normal) Consider the linear mixed-effects model of Example 1.35,

$$Y_{ij} = \mu + b_i + \epsilon_{ij}, \quad i = 1, \dots, n;\ j = 1, \dots, d,$$

where $b_i \sim N(0, \sigma_b^2)$ and $\epsilon_{ij} \sim N(0, \sigma_\epsilon^2)$ are mutually uncorrelated. This model corresponds to n independent observations of a multivariate normal variable of size d whose components have mean μ, variance $\sigma^2 = \sigma_b^2 + \sigma_\epsilon^2$ and correlation between any two components of $\rho = \sigma_b^2/(\sigma_b^2 + \sigma_\epsilon^2)$. This is an exchangeable correlation normal model; see Example 3.2.

The model parameter is $\theta = (\mu, \sigma^2, \rho)^\mathsf{T}$. The log-likelihood for θ has the form (3.2) with $m_i = d$ for all i, $\alpha = \rho$, $X = 1_d$ and $\beta = \mu$. Using the explicit formulas (3.3) for the inverse and the determinant of the exchangeable correlation matrix, we obtain

$$\begin{aligned}\ell(\theta) = &-\frac{nd}{2}\log\sigma^2 - \frac{n(d-1)}{2}\log(1-\rho) - \frac{n}{2}\log\{1+\rho(d-1)\}\\ &-\frac{\mathrm{SS_w}}{2\sigma^2(1-\rho)} - \frac{d\,\mathrm{SS_b} + nd(\bar{y}_{\bullet\bullet}-\mu)^2}{2\sigma^2\{1+\rho(d-1)\}},\end{aligned} \tag{4.36}$$

which depends on the data through the sufficient statistics $\bar{y}_{\bullet\bullet} = (nd)^{-1}\Sigma_{ij}y_{ij}$, $\mathrm{SS_w} = \Sigma_{i,j}(y_{ij} - \bar{y}_{i\bullet})^2$ and $\mathrm{SS_b} = \Sigma_i(\bar{y}_{i\bullet} - \bar{y}_{\bullet\bullet})^2$, with $\bar{y}_{i\bullet} = d^{-1}\Sigma_j y_{ij}$.

Since all observations within a vector are equicorrelated, it is natural to consider the pairwise log-likelihood function

$$c\ell(\theta) = \sum_{i=1}^{n}\sum_{r<s}^{d}\log f(y_{ir}, y_{is}; \theta),$$

which depends on the data through the sufficient statistics of the ordinary likelihood,

$$c\ell(\theta) = -n\binom{d}{2}\log\sigma^2 - \frac{n}{2}\binom{d}{2}\log(1-\rho^2) - \frac{d-1+\rho}{2\sigma^2(1-\rho^2)}\mathrm{SS_w} - \binom{d}{2}\frac{\mathrm{SS_b} + n(\bar{y}_{\bullet\bullet}-\mu)^2}{\sigma^2(1+\rho)}. \tag{4.37}$$

Calculations reported in Pace et al. (2011) show that for this model $U_{c\ell}(\theta) = I_{c\ell}(\theta)K_{c\ell}^{-1}(\theta)U(\theta)$, from which it can be verified that the maximum pairwise likelihood estimator of θ coincides with the maximum likelihood estimator; see Exercise 4.23. □

Maximum composite likelihood estimation in presence of nuisance parameters can result in some unexpected behaviour. It is well-known that the maximum likelihood estimator for the parameter of interest ψ has smaller asymptotic variance when the nuisance parameter λ is known than when it is estimated. The same result does not hold in general for maximum composite likelihood estimators. The composite likelihood function is said to be information unbiased when the second Bartlett identity holds, i.e., $I_{c\ell}(\theta) = K_{c\ell}(\theta)$. Inference from information-unbiased composite likelihood functions behaves in some respects as does

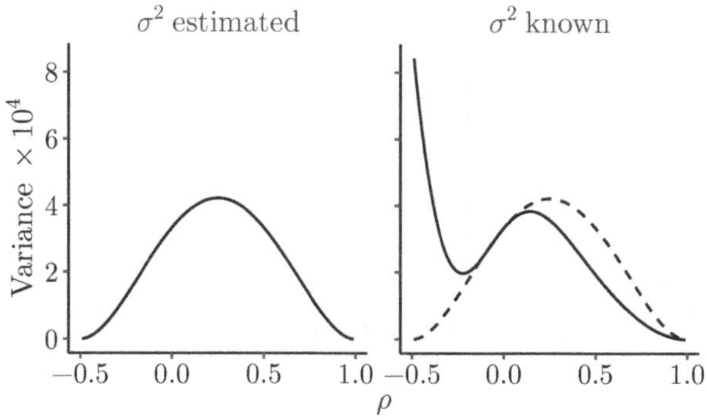

FIGURE 4.1
Asymptotic variance for the maximum pairwise likelihood estimator of ρ in the equicorrelated normal model with $d = 3$ when σ^2 is estimated ($\widehat{\rho}_{c\ell}$, left panel) or known ($\widetilde{\rho}_{c\ell}$, right panel). The variance of $\widehat{\rho}_{c\ell}$ is shown as a reference dashed curve in the graph of the variance of $\widetilde{\rho}_{c\ell}$. Based on Xu et al. (2024).

that from genuine likelihood functions. In particular, estimators of parameters of interest computed from information-unbiased composite likelihood functions are more efficient when the nuisance parameter is known than when it is estimated. In contrast, an information-biased composite likelihood function may produce estimators of the parameters of interest that are less efficient when the nuisance parameters are known (Xu et al., 2024).

Example 4.30 (Equicorrelated multivariate normal, cont'd) Consider the setting of Example 4.29 but suppose that $\mu = 0$. If σ^2 is unknown, then the maximum pairwse likelihood estimator is equal to the maximum likelihood estimator. If instead σ^2 is known, the maximum pairwise likelihood estimator of ρ differs from the maximum likelihood estimator of ρ; see Exercise 4.20. Xu et al. (2024) show that the asymptotic variance of the maximum pairwise likelihood estimator of ρ when σ^2 is fixed is

$$\text{avar}(\widetilde{\rho}_{c\ell}) = \frac{2}{nd(d-1)} \frac{(1-\rho)^2 c(d,\rho)}{(1+\rho^2)^2},$$

where $c(d,\rho) = (1-\rho)^2(3\rho^2 + \rho^2 d^2 + 1) + d\rho(-3\rho^3 + 8\rho^2 - 3\rho + 2)$, whereas the asymptotic variance of the maximum pairwise likelihood estimator of ρ when σ^2 is estimated equals that of the maximum likelihood estimator,

$$\text{avar}(\widehat{\rho}_{c\ell}) = \frac{2}{nd(d-1)}(1-\rho)^2\{1+(d-1)\rho\}^2.$$

Figure 4.1 shows the asymptotic variances of $\widetilde{\rho}_{c\ell}$ and $\widehat{\rho}_{c\ell}$ for $d = 3$: $\widetilde{\rho}_{c\ell}$ is more efficient than $\widehat{\rho}_{c\ell}$ except for negative values of ρ that approach the lower bound -0.5, because the asymptotic variance of $\widetilde{\rho}_{c\ell}$ diverges as ρ approaches that value. This paradox is explained by Henmi and Eguchi (2004, Theorem 1). □

The next two examples demonstrate the application of the vertical adjustments, given in equations (4.15) and (4.16), to composite likelihood ratio statistics. While these adjustments hold generally for any misspecified likelihood, they were originally proposed for composite likelihoods.

Example 4.31 (Equicorrelated multivariate normal, cont'd) Consider again the setting of Example 4.29. It was already noted in Example 4.29 that for this model

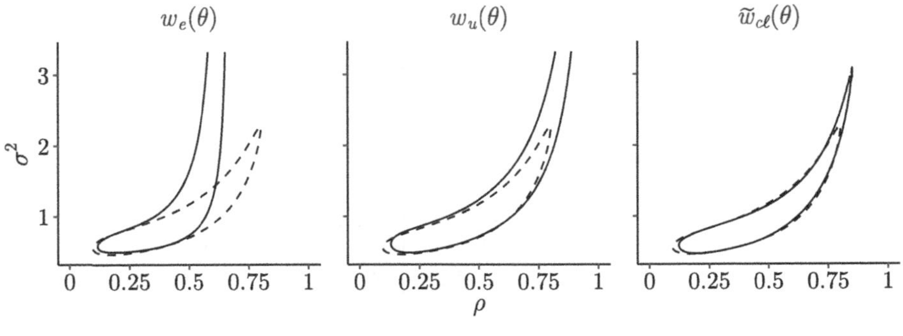

FIGURE 4.2
95% confidence regions for $\theta = (\sigma^2, \rho)^\mathsf{T}$ in the equicorrelated normal model for a simulated set of $n = 5$ observations from a multivariate normal vector of size $d = 30$ with mean $\mu = 0$, variance $\sigma^2 = 1$ and correlation $\rho = 0.5$. The confidence regions shown are based on the Wald statistic $w_e(\theta)$ (left), score statistic $w_u(\theta)$ (center) and the adjusted pairwise log-likelihood ratio statistic $\widetilde{w}_{c\ell}(\theta)$ (right). The dashed curve corresponds to the confidence region based on the log-likelihood ratio statistic $w(\theta)$. Adapted from Pace et al. (2011, Figure 1).

$U_{c\ell}(\theta) = I_{c\ell}(\theta)K_{c\ell}^{-1}(\theta)U(\theta)$, from which it follows not only that the maximum pairwise likelihood estimator of θ coincides with the maximum likelihood estimator (Exercise 4.23), but also that the Wald test statistic and the score statistic are the same, whether they are computed with the full likelihood function or the pairwise likelihood function: $\widetilde{w}_e(\theta) = w_e(\theta)$ and $\widetilde{w}_u(\theta) = w_u(\theta)$. In contrast, the likelihood ratio statistic and the pairwise likelihood ratio statistic differ; see Pace et al. (2011).

Figure 4.2 illustrates the differences between the pairwise likelihood significance functions when μ is known so $\theta = (\sigma^2, \rho)^\mathsf{T}$. Had we considered μ unknown, the following discussion would be largely unchanged, because the constrained estimate of μ for a fixed value of $\theta = (\sigma^2, \rho)^\mathsf{T}$ is the sample mean $\bar{y}$, which does not depend on θ. Figure 4.2 displays the 95% confidence regions for $\theta = (\sigma^2, \rho)^\mathsf{T}$ using simulated data with $n = 5$, $d = 30$, $\mu = 0$, $\sigma^2 = 1$ and $\rho = 0.5$. In each panel of the figure there is a confidence region based on the pairwise likelihood significance function for $w_e(\theta)$, $w_u(\theta)$ and $\widetilde{w}_{c\ell}(\theta)$ compared to that based on the log-likelihood ratio statistic $w(\theta)$, which is taken as the gold standard.

The shape of the Wald confidence region is degenerate and matches that of the log-likelihood ratio statistic confidence region only for small values of σ^2 and ρ. The shape of the score confidence region is better than that based on the Wald statistic but degenerates as both σ^2 and ρ increase. The vertically adjusted pairwse log-likelihood ratio confidence region does not degenerate and agrees well with the log-likelihood ratio confidence region. □

Example 4.32 (Equicorrelated multivariate normal, cont'd) Figure 4.3 displays the 95% confidence intervals for ρ when σ^2 is a nuisance parameter for the simulated data used in Example 4.31 and Figure 4.2. The pairwse likelihood Wald confidence interval appears too narrow compared to the profile log-likelihood ratio confidence interval, but the pairwise likelihood score confidence interval is much too wide. The confidence interval based on the adjusted profile pairwise log-likelihood is very close to that from the profile log-likelihood ratio.

□

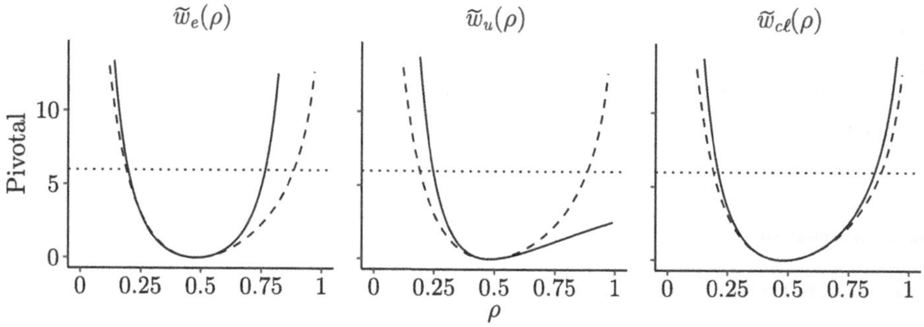

FIGURE 4.3
95% confidence intervals for ρ with σ^2 treated as nuisance parameter in the equicorrelated normal model for the simulated data used for Figure 4.2. The confidence intervals shown are based on the Wald statistic $\widetilde{w}_e(\rho)$ (left), the score statistic $\widetilde{w}_u(\rho)$ (centre) and the adjusted profile pairwise log-likelihood ratio statistic $\widetilde{w}_{c\ell}(\rho)$. The dashed curve corresponds to the confidence interval built on the profile log-likelihood ratio statistic $w_p(\rho)$. Adapted from Pace et al. (2011, Figure 2).

4.7.3 Bayesian composite likelihood

Let $\pi(\theta)$ be the prior distribution for the model parameter $\theta \in \mathbb{R}^p$. A naive use of the composite likelihood in the Bayes formula leads to the composite posterior

$$\pi_{c\ell}(\theta \mid y) = \frac{\exp\{c\ell(\theta)\}\pi(\theta)}{\int \exp\{c\ell(\theta)\}\pi(\theta)\,\mathrm{d}\theta}. \tag{4.38}$$

However, inference based on the above composite posterior may be misleading because the working assumption of independence used in the construction of the composite likelihood gives an incorrect impression of high precision. The failure of the naive composite likelihood Bayesian inference can be understood using the Laplace approximation (see Section 1.4) as discussed in as discussed in Pauli et al. (2011) and Ribatet et al. (2012, Appendix A). For large n, the Laplace approximation gives

$$\begin{aligned}\pi_{c\ell}(\theta \mid y) &\doteq (2\pi)^{-p/2}|J_{c\ell}(\widehat{\theta}_{c\ell})|^{1/2}\exp\left\{-\frac{1}{2}w_{c\ell}(\theta)\right\}\frac{\pi(\theta)}{\pi(\widehat{\theta}_{c\ell})}\\ &\propto \exp\left\{-\frac{1}{2}w_{c\ell}(\theta) + \log\pi(\theta)\right\}.\end{aligned} \tag{4.39}$$

Expanding $w_{c\ell}$ around the maximum composite likelihood estimator $\widehat{\theta}_{c\ell}$ and $h(\theta) = -\log\pi(\theta)$ around its mode θ_π yields

$$\pi_{c\ell}(\theta \mid y) \mathrel{\dot{\propto}} \exp\left\{-\frac{1}{2}(\theta-\widehat{\theta}_{c\ell})^{\mathsf{T}}J_{c\ell}(\widehat{\theta}_{c\ell})(\theta-\widehat{\theta}_{c\ell}) - \frac{1}{2}(\theta-\theta_\pi)^{\mathsf{T}}h''(\theta_\pi)(\theta-\theta_\pi)\right\}.$$

For large n, the influence of the prior vanishes and the composite posterior distribution $\pi_{c\ell}(\theta \mid y)$ converges to the distribution of a normal random variable with expected value θ_* and variance $K_{c\ell}^{-1}(\theta_*)$, but the correct variance of $\widehat{\theta}_{c\ell}$ is $G_{c\ell}^{-1}(\theta_*)$. Bayesian inference based on $\pi_{c\ell}(\theta \mid y)$ is appropriate only if the individual likelihoods that form the composite likelihood are truly independent, otherwise the precision of the inferential conclusion will be overstated.

Expression (4.39) shows that asymptotically correct Bayesian inference can be obtained by replacing the composite log-likelihood ratio statistic $w_{c\ell}(\theta)$ with the adjusted statistic $\widetilde{w}_{c\ell}(\theta)$ given in (4.15). With this replacement, $\pi_{c\ell}(\theta \mid y)$ has the same limiting normal approximation as the density of the maximum composite likelihood estimator.

4.8 Bibliographic notes

The study of statistical methods when the assumed model is incorrect began with the seminal papers of Cox (1961) and Cox (1962) on testing separate families of hypotheses. The profound impact of this work on the statistical and econometric literature was celebrated in the Royal Statistical Society retrospective read paper Cox (2013). The results discussed in Cox's papers were formalized in Kent (1986) and motivated subsequent influential work such as Huber (1967) and White (1982). White (1994) is a classical reference for likelihood inference under misspecification.

In the econometric literature the maximizer of a misspecified likelihood is called pseudo- or quasi-maximum likelihood estimator; see for example Gouriéroux et al. (1984) and Bollerslev and Wooldridge (1992). The term pseudo-maximum likelihood estimator was also used in Gong and Samaniego (1981) to denote the maximizer of the parameter of interest in a pseudo-likelihood in which the nuisance parameters are replaced by estimates.

One of the earlier proofs that the maximizer of the misspecified likelihood converges to the least-false parameter is due to Huber (1967). Parallel work on Bayesian inference when the model is incorrect began with Berk (1966).

There is a large body of literature in econometrics on specification tests, used to determine if a model is correctly specified or if it is sufficiently close to the correct model that its bias, at least for the parameter of interest, is negligible. The best-known specification test is due to Hausman (1978).

Vuong (1989) developed a likelihood ratio test to assess the null hypothesis that two competing models are equally close to the true data-generating process against the alternative hypothesis that one model is closer. Viraswami and Reid (1996; 1998) obtained higher-order asymptotic results for likelihood-based tests under model misspecification.

Lin and Wei (1989) derived the asymptotic distribution of the maximum partial likelihood estimator and associated test statistics when the Cox proportional hazard model is misspecified. Heagerty and Kurland (2001) evaluated the asymptotic bias due to random effects misspecification in generalized linear mixed models. Battey and Reid (2024) studied situations in which the maximum likelihood estimator of the parameter of interest is consistent under arbitrary misspecification of the nuisance part of the model.

M-estimation was introduced in Huber (1964) and Huber (1967), although the work of Godambe (1960) on estimating equations may be considered a precursor; see Stefanski and Boos (2002). Estimating equations are the topic of Chapter 5. Theoretical properties of the influence curve were described in Hampel (1974). Asymptotic properties of M-estimators were detailed in Serfling (1980, Chapter 7) and van der Vaart (1998, Chapter 5). Monographs on robust statistics that contain detailed discussion of M-estimators include Huber and Ronchetti (2009), and Maronna et al. (2019).

In econometrics, several variants of the sandwich estimator have been developed to account for different forms of dependence. The heteroscedasticity and autocorrelation consistent (HAC) estimator remains the standard for time series applications. Key references include Newey and West (1987), Andrews (1991), and Newey and West (1994). Cluster-robust covariance estimators were developed in generalized estimating equations (Liang and Zeger, 1986) and introduced in econometrics by Arellano (1987) for panel data and grouped structures; see also the review Abadie et al. (2023) or the monograph on micro-econometrics by Cameron and Trivedi (2005).

The R package `sandwich` (Zeileis, 2006; Zeileis et al., 2020) provides convenient object-oriented implementations of various forms of the sandwich variance.

Monographs on model selection include Linhart and Zucchini (1986) and Burnham and Anderson (2002). Claeskens and Hjort (2008, Chapter 2) provided a rigorous account of the derivation and properties of various information criteria, including their behaviour under model misspecification. The consequences of model selection for subsequent statistical inference are often overlooked. One approach to address the variability introduced by model selection is model averaging, which involves conducting inference on a parameter defined as a weighted average of estimators obtained from a set of candidate models; see Claeskens and Hjort (2008, Chapter 7).

The local misspecification analysis in Section 4.5 is taken from Claeskens and Hjort (2008). Local misspecification is related to the concept of local asymptotic normality of Le Cam (1960); see also Cox and Hinkley (1974, Section 4.8) and van der Vaart (1998, Chs. 6, 7). Dæhlen et al. (2024) developed accurate approximations for the mean square error under potential model misspecification. These approximations are used in computation of the focused information criterion (Claeskens and Hjort, 2003), which ranks models based on the quality of their estimates for a parameter of interest.

Indirect inference has been developed mainly in econometrics, building on Gouriéroux et al. (1993) where the approach was motivated by applications in dynamic financial models with an intractable likelihood function, and Smith (1993), who developed estimators of nonlinear time-series models using simulated vector autoregression; see also Gouriéroux and Monfort (1996, Chapter 4). However, the use of simulation to implement indirect inference in implicit statistical models, bypassing explicit likelihood evaluation, can be traced back to Diggle and Gratton (1984).

The method of simulated moments (McFadden, 1989) is a special case of indirect inference. Heggland and Frigessi (2004) discussed estimating functions in indirect inference. Fearnhead and Prangle (2012) highlighted the relation between approximate Bayesian computation (Tavaré et al., 1997) and indirect inference; see also Drovandi (2018).

Another form of indirect inference based on simulations is provided by the synthetic likelihood proposed by Wood (2010) and applied to noisy nonlinear ecological dynamic systems.

The definition of composite likelihood first appeared in Lindsay (1988), who also set out the general theory of estimating equations from composite likelihood, and introduced the term information unbiased. This was a formalization of the pseudo-likelihood developed in Besag (1974) for spatial modelling. The advent of increasingly large datasets due to the lower cost of collecting data compared to the computational cost of analysis has attracted new interest in Besag's pseudo-likelihood; see Khare et al. (2015), Lee and Hastie (2015), and Barber and Drton (2015).

General overviews of composite likelihood, with many references to the literature, are Varin (2008) and Varin et al. (2011). Bevilacqua and Gaetan (2015) compared different types of composite likelihoods for inference in geostatistical models and also give conditions for consistency under increasing-domain asymptotics. Composite likelihood inference in genetics is discussed in Larribe and Fearnhead (2011).

Full conditional composite likelihoods were considered by Molenberghs and Verbeke (2005) for statistical analysis of longitudinal studies. Those authors used the term pseudo-likelihood. Pairwise likelihood is discussed in Cox and Reid (2004), and the term pseudo-likelihood was also used there, in the context of increasing dimension and increasing sample size. Composite likelihood has since become the preferred term, as pseudo-likelihood is also invoked in other contexts.

Since composite likelihood methods typically use only a partial specification of the data distribution, there is the possibility that the model parameters are not identifiable, as illustrated in Yi (2017a). When the sub-likelihoods used in the composite likelihood are not consistent with a joint distribution, the maximum composite likelihood estimator might not have a meaningful interpretation; see Xu and Reid (2011) and Yi (2017a).

Bayesian inference based on composite likelihood functions is discussed in Pauli et al. (2011), Ribatet et al. (2012), and Miller (2021). Kleijn and van der Vaart (2012) provided a detailed treatment of the asymptotic properties of the posterior distribution when the likelihood function is misspecified. Gao and Song (2010) derived a composite likelihood version of the Bayesian information criterion.

Composite likelihood estimators in Markov chains are studied in Hjort and Varin (2008). Stoltenberg and Hjort (2021) discussed pairwise likelihood inference for binary sequences modelled with clipped Ornstein–Uhlenbeck processes.

Alfonzetti et al. (2025) combined stochastic approximation methods with composite likelihoods for inference in large-scale applications where the computational cost of a direct application of composite likelihood inference would otherwise be too demanding.

4.9 Further results and exercises

4.1. Suppose that the working model is the Laplace distribution with location α and scale β but the true model is normal with mean μ and variance σ^2. Obtain the Fraser information and show that it is maximized at the least-false parameters $\alpha_* = \mu$ and $\beta_* = \sigma(2/\pi)^{1/2}$.

4.2. Explain how the regularity conditions of Section 1.7.2 should be modified to prove that the maximum misspecified likelihood estimator has a limiting normal distribution centred at θ_* with sandwich variance (4.2).

4.3. Verify the expression (4.6) for the leave-one-out estimator $\widehat{\beta}_{-i}$ in the linear regression model using the matrix inversion lemma also known as the Sherman–Morrison–Woodbury formula.

4.4. Suppose we have n independent observations from the logistic regression model

$$\Pr(Y_i = 1 \mid x_i, z_i) = \frac{\exp(x_i^\top \beta + z_i^\top \gamma)}{1 + \exp(x_i^\top \beta + z_i^\top \gamma)}, \quad i = 1, \ldots, n.$$

Consider the misspecified model in which z_i is omitted.

(a) Derive the Godambe information for β.

(b) Derive Wald, Rao and likelihood ratio test statistics for assessing β.

4.5. Let $\{Y_1, \ldots, Y_n\}$ be a time series generated from the linear regression model $Y_t = x_t^\top \beta + \epsilon_t$ with autocorrelated errors $\epsilon_t = \phi\epsilon_{t-1} + \eta_t$, where $|\phi| < 1$ and $\eta_t \sim N(0, \sigma^2)$ is a white noise process. Assume that $\epsilon_0 = 0$. Consider the working model in which the errors are assumed uncorrelated ($\phi = 0$).

(a) Derive the Godambe information matrix for $\theta = (\beta^\top, \sigma^2)^\top$.

(b) Derive Wald, Rao and likelihood ratio test statistics for assessing β.

4.6. Consider the linear regression model with heteroscedastic errors $Y_i = \beta_0 + \beta_1 x_i + \epsilon_i$, where ϵ_i are independent $N(0, \sigma_i^2)$ with $\sigma_i^2 = \sigma^2(1 + \gamma x_i^2)$, for $i = 1, \ldots, n$. Simulate 1,000 datasets of length $n = 50$ with parameters $\beta_0 = 1, \beta_1 = 1, \sigma^2 = 3, \gamma = 2$ and covariate x_i simulated from $U(-1, 1)$. Fit each dataset with a linear regression model with homoscedastic errors, corresponding to $\gamma = 0$, and study the performance of the sandwich variance estimator of $\widehat{\beta}_1$ that uses the heteroscedasticity and autocorrelation-consistent estimator (4.8) of the variability matrix.

4.7. Consider the simple linear regression $Y_i = \beta_0 + \beta_1 x_i + \epsilon_i$ and show that the asymptotic relative efficiency of the bias-corrected sandwich variance for the slope β_1 is equal to the inverse of the kurtosis index of the covariate x.

(a) For a covariate with an approximately normal distribution, verify that the sandwich estimator of variance is three times more variable than the model-based counterpart.

(b) If the covariate follows a Laplace distribution, verify that the sandwich estimator of variance is six times more variable. *(Kauermann and Carroll, 2001)*

4.8. Suppose Y follows a negative binomial distribution with expected value $\mathrm{E}(Y \mid x) = \mu(x)$ and variance $\mathrm{var}(Y \mid x) = \mu(x) + \kappa\mu^2(x)$, where $\mu(x) = \exp(\beta_0 + \beta_1 x_1 + \beta_2 x_2)$, $\kappa \geq 0$ and the covariate vector $x = (x_1, x_2)^\top$ is drawn from a bivariate standard normal distribution with correlation 0.5. This is an example of negative binomial regression, which generalizes the Poisson log-linear regression model corresponding to $\kappa = 0$.

In a simulation study with varying sample size, and a selection of values of κ and β, compare the full and one-step jackknife methods in their accuracy in estimating the standard errors of the estimates of the vector of regression coefficients $\beta = (\beta_0, \beta_1, \beta_2)^\top$.

4.9. Starting from the limiting distributions of $\widehat{\theta}_{\mathrm{narr}}$, $\widehat{\theta}_{\mathrm{wide}}$ and $\widehat{\gamma}_{\mathrm{wide}}$ in Section 4.5, use the delta method to derive the limiting distributions given in (4.27) and (4.28).

4.10. Suppose $\{Y_1, \ldots, Y_n\}$ is a random sample from the Weibull distribution,

$$f(y; \theta, \gamma) = \frac{\gamma}{\theta}\left(\frac{y}{\theta}\right)^{\gamma-1} \exp\left\{-\left(\frac{y}{\theta}\right)^{\gamma}\right\}, \quad y > 0.$$

The exponential distribution corresponds to $\gamma = 1$.

(a) Use results about local misspecification in Section 4.5 to show how much γ must differ from one to make the estimator of $\varphi = \varphi(\theta, \gamma)$ based on the Weibull distribution better in terms of mean-squared error than the estimator based on the exponential distribution.

(b) Derive the TIC criterion (4.20) to discriminate between the two models. How does model selection based on TIC compare with results based on local misspecification? *(Claeskens and Hjort, 2008)*

4.11. Suppose we have n observations from the zero-truncated Poisson distribution,

$$\mathrm{pr}(Y = y; \theta) = \frac{\exp(-\theta)\theta^y}{\{1 - \exp(-\theta)\}y!}, \quad y = 1, 2, \ldots$$

Discuss indirect inference for θ using as intermediate statistic the naive estimator $\widehat{\varphi}(\theta) = \overline{Y}$. This is equivalent to assuming a working Poisson distribution for Y. *(Jiang and Turnbull, 2004)*

4.12. Suppose $Y_0, Y_1, \ldots$ is a Markov chain with states $\{0, 1\}$ and transition matrix

$$P = \begin{bmatrix} 1-\alpha & \alpha \\ \beta & 1-\beta \end{bmatrix}, \quad \alpha, \beta \in [0,1].$$

Also suppose the chain starts out in its equilibrium distribution and is observed until time n. Consider inference for $\theta = (\alpha, \beta)^{\mathsf{T}}$ based on the ordinary log-likelihood function

$$\ell(\theta) = \sum_{t=2}^{n} \log \mathrm{pr}(Y_t = y_t \mid Y_{t-1} = y_{t-1}; \theta)$$

and the pairwise log-likelihood function

$$c\ell(\theta) = \sum_{t=2}^{n} \log \mathrm{pr}(Y_t = y_t, Y_{t-1} = y_{t-1}; \theta).$$

Show that while the estimators of θ that maximize $\ell(\theta)$ and $c\ell(\theta)$ are different, they are asymptotically equivalent. *(Hjort and Varin, 2008)*

4.13. Suppose $\{Y_1, \ldots, Y_n\}$ are independent and identically distributed random vectors of length $d = 4$ from a multivariate normal distribution with mean zero and covariance matrix

$$\Sigma = \begin{bmatrix} 1 & \rho & 2\rho & 2\rho \\ \rho & 1 & 2\rho & 2\rho \\ 2\rho & 2\rho & 1 & \rho \\ 2\rho & 2\rho & \rho & 1 \end{bmatrix}.$$

Consider estimation of ρ when it is assumed only that the correlation between the first two components of Y_i is the same as the correlation between the second two components.

(a) Show that the maximum likelihood estimator based on the equicorrelated normal model that wrongly assumes the covariance $\Sigma = (1-\rho)\mathcal{I}_{d\times d} + \rho 1_d 1_d^{\mathsf{T}}$ is inconsistent for ρ.

(b) Show that a consistent estimator of ρ is the maximizer of the pairwise likelihood function,

$$c\ell(\rho) = \sum_{i=1}^{n} \{\log f(y_{i1}, y_{i2}; \rho) + \log f(y_{i3}, y_{i4}; \rho)\}.$$

(Xu and Reid, 2011)

4.14. Suppose $\{Y_1, \ldots, Y_n\}$ are observations from the autoregressive model of order one,

$$Y_t = \phi Y_{t-1} + \epsilon_t,$$

where $\epsilon_t \sim N(0, \sigma^2)$ is a white noise process. The model parameter is $\theta = (\phi, \sigma^2)^{\mathsf{T}}$. Consider the pairwise conditional log-likelihood function of order m, composed from pairs of observations that are separated up to distance m,

$$c\ell_m(\theta) = \sum_{i=1}^{m} \sum_{t=i+1}^{n} \log f(y_t \mid y_{t-i}; \theta). \tag{4.40}$$

(a) Show that the maximum pairwise likelihood estimator is fully efficient for $m = 1$. *(Davis and Yau, 2011)*

(b) Explain why as m increases, the maximizer $\widehat{\theta}_m$ of $c\ell_m(\theta)$ loses efficiency and in the limiting case, as $m \to \infty$, the maximizer of $c\ell_m(\theta)$ is not consistent.

4.15. Suppose $\{Y_1, \ldots, Y_n\}$ are observations from the moving average model

$$Y_t = \epsilon_t + \omega\epsilon_{t-1},$$

where $\epsilon_t \sim N(0, \sigma^2)$ is a white noise process. The model parameter is $\theta = (\omega, \sigma^2)^\top$. Suppose θ is estimated with the pairwise conditional log-likelihood function (4.40) with $m = 1$.

(a) Verify that the maximum pairwise likelihood estimator of θ is the method-of-moments estimator.

(b) Derive the asymptotic variance of the maximum pairwise likelihood estimator of ω.

(c) Plot the ratio of the asymptotic variance of the pairwise maximum likelihood estimator of ω to the asymptotic variance of the ordinary maximum likelihood estimator of ω.

4.16. Consider a sample of n independent bivariate random vectors $\{(Y_{11}, Y_{21}), \ldots, (Y_{1n}, Y_{2n})\}$. Instead of specifying a bivariate joint model for the pairs of observations, assume that Y_{1i} given $Y_{2i} = y_{2i}$ has an exponential distribution with expected value θy_{2i}, and symmetrically also Y_{2i} given $Y_{1i} = y_{1i}$ has an exponential distribution with expected value θy_{1i}, for some $\theta > 0$. Consider the pairwise conditional log-likelihood function

$$c\ell(\theta) = \sum_{i=1}^{n} \{\log f(y_{1i} \mid y_{2i}; \theta) + \log f(y_{2i} \mid y_{1i}; \theta)\}.$$

Using the Hammersley–Clifford theorem show that the joint distribution of (Y_1, Y_2) associated with the assumed exponential pairwise conditional distribution should have the form

$$f(y_1, y_2; \theta) = \frac{1}{C} \exp(-\theta y_1 y_2),$$

for some $C = C(\theta)$. Verify that this is not a valid density function. What conclusion can be drawn about inference for θ based on the conditional pairwise log-likelihood function $c\ell(\theta)$?
(Yi, 2017a)

4.17. Consider a vector $Y = (Y_1, Y_2, Y_3)^\top$ of positive continuous random variables with joint density

$$f(y_1, y_2, y_3; \theta) = \theta_1 \exp\left(-\frac{\theta_2}{\theta_1} y_1 - \frac{\theta_1}{\theta_2} y_2 - \theta_1 y_3\right), \quad y_1, y_2, y_3 > 0,$$

where $\theta = (\theta_1, \theta_2)^\top$, with $\theta_1 > 0$ and $\theta_2 > 0$.

(a) Obtain the univariate marginal densities $f(y_1; \theta)$, $f(y_2; \theta)$ and $f(y_3; \theta)$.

(b) Verify that a composite likelihood based on the marginal densities of just Y_1 and Y_2 cannot identify θ but only the ratio θ_1/θ_2.

(c) Obtain the bivariate densities $f(y_1, y_2; \theta)$, $f(y_1, y_3; \theta)$ and $f(y_2, y_3; \theta)$;

(d) Verify that a pairwise likelihood based on just the pair (Y_1, Y_2) can identify only the ratio θ_1/θ_2, whereas the full vector θ is identified using the distribution of the pairs (Y_1, Y_3) or (Y_2, Y_3), or the full triple (Y_1, Y_2, Y_3). *(Yi, 2017a)*

4.18. Suppose $Y = (Y_1, Y_2, Y_3)^\top$ is a vector of uncorrelated normal variables with mean $\mu > 0$ and unit variance. For a sample of n independent observations of Y, consider the pairwise log-likelihood function

$$c\ell(\mu) = \sum_{i=1}^{n} \{\log f(y_{i1}, y_{i2}; \mu) + \log f(y_{i2}, y_{i3}; \mu) + \log f(y_{i1}, y_{i3}; \mu)\},$$

where the first two terms in the product are correctly specified as the densities of a bivariate normal distribution with mean vector $(\mu, \mu)^\top$ and identity covariance matrix, but the density of (Y_{i1}, Y_{i3}) is incorrectly specified as

$$f(y_{i1}, y_{i3}; \mu) = \frac{1}{\mu} \exp\left(-\frac{y_{i1}}{\mu}\right) \frac{1}{\sqrt{2\pi}} \exp\left\{-\frac{(y_{i3}-\mu)^2}{2}\right\}.$$

(a) Show that there is no joint density for Y that is compatible with the pairwise likelihood.

(b) Derive the pairwise score $U_{c\ell}(\mu)$ and show that its unique real root converges to μ.

(Xu and Reid, 2011).

4.19. Suppose that $Y_1 \sim N(\mu_1, \sigma_1^2)$, $Y_2 \sim N(\mu_2, \sigma_2^2)$ and $\epsilon \sim N(0,1)$ are independent random variables and define $Y_3 = Y_1 + Y_2 + \beta Y_1 Y_2 + \epsilon$, for some $\beta \neq 0$.

(a) Verify that the full conditionals $f(y_1 \mid y_2, y_3)$, $f(y_2 \mid y_1, y_3)$, $f(y_3 \mid y_1, y_2)$ are normal but the joint distribution $f(y_1, y_2, y_3)$ is not multivariate normal.

(b) Compute the maximum likelihood estimator based on the misspecified multivariate normal model for the joint data.

(c) Show that the maximum composite likelihood estimator $\widehat{\beta}_{c\ell}$ of β based only on the full conditional $f(y_3 \mid y_1, y_2)$ is $\widehat{\beta}_{c\ell} = \Sigma_i\{y_{i1}y_{i2}(y_{i3} - y_{i1} - y_{i2})\}/\Sigma_i y_{i1}^2 y_{i2}^2$. Verify that $\widehat{\beta}_{c\ell}$ is consistent for β.

(Andrei and Kendziorski, 2009; Xu and Reid, 2011)

4.20. Consider a random vector $Y = (Y_1, \ldots, Y_d)^\top$ that follows the canonical exponential family of distributions with density given in (1.16), canonical parameter $\varphi(\theta)$, and associated sufficient statistic $s(Y)$. Mardia et al. (2009) define the canonical exponential family as closed if, for any subset $B \subset \{1, \ldots, d\}$, the marginal distribution of Y_B, the subvector of Y made with the Y_i with $i \in B$, follows a canonical exponential family with sufficient statistic $s(Y_B)$. Mardia et al. (2009, Theorem 2) states that maximum pairwise marginal and maximum pairwise conditional likelihood estimators coincide with maximum likelihood estimator in closed exponential family models in which only first-order interactions are present.

Suppose that Y follows a multivariate normal distribution with zero mean vector and covariance matrix Σ.

(a) Show that if Σ is unrestricted then the assumptions of Theorem 2 of Mardia et al. (2009) apply and thus the maximum pairwise likelihood estimators are identical to the maximum likelihood estimator.

(b) Show that if Σ is an equicovariance matrix, $\Sigma = \sigma^2\{(1-\rho)\mathcal{I}_{d\times d} + \rho 1_d 1_d^\top\}$, then the assumptions of Theorem 2 of Mardia et al. (2009) do not apply, but the maximum pairwise likelihood estimators are still identical to the maximum likelihood estimator.

(c) Show that if Σ is an equicorrelation matrix, i.e. $\sigma^2 = 1$ and $\Sigma = (1-\rho)\mathcal{I}_{d\times d} + \rho 1_d 1_d^\top$, then the model is no longer a canonical exponential family and the maximum pairwise likelihood estimators are not identical to the maximum likelihood estimator.

(Mardia et al., 2009)

4.21. Show that the vertically adjusted composite likelihood ratio statistic (4.14) is not parametrization-invariant when $p > 1$, whereas the alternative rescaled version (4.15) is invariant. *(Pace et al., 2011)*

4.22. Some insight on the efficiency of composite likelihood can be obtained from the theory of optimal estimating functions in Section 5.5.

(a) Show that $K_{c\ell}(\theta) = \text{cov}_h\{U_{c\ell}(\theta), U(\theta)\}$. Explain why this result implies that the maximum composite likelihood estimator cannot be more efficient than the maximum likelihood estimator in the sense that the matrix $\Delta(\theta) = I(\theta) - G_{c\ell}(\theta)$ is positive semi-definite.

(b) Show that the maximum composite likelihood estimator is fully efficient, $\Delta(\theta) = 0$, if and only if

$$U(\theta) = K_{c\ell}(\theta)I_{c\ell}^{-1}(\theta)U_{c\ell}(\theta) + b(\theta),$$

for a vector $b(\theta)$ that does not depend on Y. If $b(\theta) = 0$, then the maximum composite likelihood estimator coincides with the maximum likelihood estimator.

(Xu et al., 2024)

4.23. Consider the equicorrelated multivariate normal model of Examples 4.29–4.32.

(a) Check the expressions for log-likelihood function (4.36) and pairwise log-likelihood function (4.37).

(b) Derive $U(\theta)$, $U_{c\ell}(\theta)$, $I_{c\ell}(\theta)$ and $K_{c\ell}(\theta)$.

(c) Show that $U_{c\ell}(\theta) = I_{c\ell}(\theta)K_{c\ell}^{-1}(\theta)U(\theta)$, which proves that the maximum pairwise likelihood estimator coincides with the maximum likelihood estimator in this model (see (b) in the previous exercise).

(d) Check that the Wald statistic and the score statistic do not change whether they are computed with full likelihood or pairwise likelihood, that is $\widetilde{w}_e(\theta) = w_e(\theta)$ and $\widetilde{w}_u(\theta) = w_u(\theta)$ as claimed in Example 4.31. *(Pace et al., 2011).*

5

Quasi-likelihood and Estimating Functions

5.1 Introduction

The validity of likelihood inference hinges on the correct specification of the underlying distribution. As illustrated in Chapter 4, adjustments to classical likelihood inference methods are required if the model is misspecified. Extensions to the likelihood function, including composite likelihood functions and quasi-likelihood functions, can be developed to accommodate model misspecification. The maximization of these functions to obtain point estimators also falls into the framework of estimating functions. An alternative to changing the inference function is to use more flexible models, specifying for example equations that provide estimators of relevant parameters without specifying the complete probability structure. This leads naturally to a theory of estimating functions, the focus of this chapter.

An estimating function has some of the properties of a likelihood-based score function, and in particular defines an estimating equation that links the parameter and the data. We begin with quasi-likelihood functions and their role in inference, and then introduce the general framework and basic theory of estimating functions. As applications, we examine several estimating equations that have been commonly used in the literature.

In Sections 5.2, 5.7, and 5.8, and in Example 5.5 of Section 5.4, we distinguish the response variable from the covariates, and denote the regression parameter by θ rather than β, as is more common elsewhere. In other sections, however, we simply use the generic notation Y (or y) to represent relevant random variables (or their realizations) and denote the associated parameter by θ.

5.2 Quasi-likelihood

Consistency and asymptotic normality of the maximum likelihood estimator, under some regularity conditions, are useful for constructing approximate confidence intervals and computing approximate p-values, as outlined in Sections 1.3 and 2.5. The derivation of these properties relies on the first two Bartlett identities (1.36) and (1.38) of Section 1.2,

$$\mathrm{E}_\theta\left\{\frac{\partial \ell(\theta;Y)}{\partial\theta}\right\}=0 \tag{5.1}$$

and

$$\mathrm{E}_\theta\left\{-\frac{\partial^2 \ell(\theta;Y)}{\partial\theta\,\partial\theta^{\mathsf{T}}}\right\}=\mathrm{E}_\theta\left\{\frac{\partial \ell(\theta;Y)}{\partial\theta}\,\frac{\partial \ell(\theta;Y)}{\partial\theta^{\mathsf{T}}}\right\}, \tag{5.2}$$

DOI: 10.1201/9780429090936-5

where $\ell(\theta; y)$ is the log-likelihood function and E_θ denotes expectation under the model with probability density or mass function $f(y;\theta)$. Identity (5.1) is useful for establishing the consistency of the maximum likelihood estimator (see Section 1.7) and (5.2) serves to simplify the expression for the asymptotic covariance matrix of the maximum likelihood estimators.

To illustrate the utility of these identities, suppose that $\{Y_1, \ldots, Y_n\}$ are independently distributed, each with a probability density or mass function of the generalized linear model form introduced in Example 1.12,

$$f(y_i; \theta_i, \phi) = \exp\left\{\frac{y_i\theta_i - b(\theta_i)}{a(\phi)} + h(y_i, \phi)\right\}, \tag{5.3}$$

where $a(\cdot), b(\cdot)$ and $h(\cdot,\cdot)$ are known functions, and ϕ is a dispersion parameter. Here, θ_i is a scalar canonical parameter that may vary with subject index i to reflect differences in covariate information across observations.

Using (5.1) and (5.2) we have

$$\mathrm{E}_{\theta_i}(Y_i) = b'(\theta_i), \quad \mathrm{var}_{\theta_i}(Y_i) = a(\phi)\, b''(\theta_i),$$

and

$$\frac{\partial \log f(y_i; \theta_i)}{\partial \theta_i} = \frac{\partial \mathrm{E}_{\theta_i}(Y_i)}{\partial \theta_i}\left\{\frac{y_i - \mathrm{E}_{\theta_i}(Y_i)}{\mathrm{var}_{\theta_i}(Y_i)}\right\}.$$

Thus the score equation based on the sample $\{y_1, \ldots, y_n\}$ is

$$\sum_{i=1}^{n}\left\{\frac{\partial \mathrm{E}_{\theta_i}(Y_i)}{\partial \theta_i}\frac{y_i - \mathrm{E}_{\theta_i}(Y_i)}{\mathrm{var}_{\theta_i}(Y_i)}\right\} = 0, \tag{5.4}$$

which shows that the maximum likelihood estimator in this model depends on just the first two moments of Y_i and the sensitivity of the expected value of Y_i to a change in the canonical parameter.

Expression (5.4) is useful in the context of generalized linear regression, where only certain conditional moment structures of the response variable, given covariates, are assumed. Consider the scenario where the response Y_i is assumed to depend on a vector of associated covariates X_i, but we model only the conditional expected value and variance by analogy with the generalized linear model (1.21):

$$g(\mu_i) = g\{\mathrm{E}(Y_i \mid X_i = x_i)\} = x_i^{\mathsf{T}}\theta \tag{5.5}$$

and

$$\mathrm{var}(Y_i \mid X_i = x_i) = a(\phi)V(\mu_i) \tag{5.6}$$

for specified functions $g(\cdot)$ and $V(\cdot)$. To highlight the main idea, we treat the dispersion parameter ϕ as known and focus the discussion on the regression parameter θ, which is usually of primary interest. The intercept is not explicitly included for simplicity, as in (3.1) and in many other models.

Given only the moment assumptions (5.5) and (5.6), a natural question arises: How can we estimate θ? The likelihood method is not directly applicable due to the lack of a specified distribution for the random variables. However, the structure of (5.4) motivates the function

$$\Psi(\theta; y_i, x_i) = \frac{\partial \mu_i}{\partial \theta}\{\mathrm{var}(Y_i \mid X_i = x_i)\}^{-1}(y_i - \mu_i), \tag{5.7}$$

which, based on (5.6) and the fact that $a(\phi)$ does not involve θ, can be equivalently written as

$$\Psi(\theta; y_i, x_i) = \frac{\partial \mu_i}{\partial \theta}\{V(\mu_i)\}^{-1}(y_i - \mu_i).$$

Thus for a sample $\{\{y_i, x_i\} : i = 1, \ldots, n\}$, mimicking the score equation (5.4), we estimate θ as the solution of

$$\Psi(\theta; y, x) = \sum_{i=1}^{n} \Psi(\theta; y_i, x_i) = 0. \tag{5.8}$$

The resulting estimator $\widehat{\theta}$ is the quasi-likelihood estimator, and the function $\Psi(\theta; y, x)$ is referred to as the quasi-score function. If the distribution of Y_i does indeed follow the generalized linear model (5.3), then (5.8) is the usual score equation, and is optimal among regular unbiased estimating functions; see Section 5.3. Modelling only the conditional expected value and variance of Y_i through (5.5) and (5.6), (5.8) can yield a consistent estimator for θ and $\sqrt{n}(\widehat{\theta} - \theta)$ is asymptotically normally distributed with expected value zero, under some regularity conditions (McCullagh, 1983).

Quasi-score functions resemble ordinary likelihood score functions. In particular, the first two Bartlett identities (5.1) and (5.2) also hold for quasi-score functions:

$$\mathrm{E}\{\Psi(\theta; Y_i, X_i)\} = 0,$$
$$\mathrm{E}\{\Psi(\theta; Y_i, X_i)\Psi^{\mathsf{T}}(\theta; Y_i, X_i)\} = -\mathrm{E}\left\{\frac{\partial \Psi(\theta; Y_i, X_i)}{\partial \theta^{\mathsf{T}}}\right\}.$$

Consequently, the statistical properties of quasi-likelihood score functions are similar to some key properties of ordinary likelihood score functions. Using the quasi-likelihood estimator $\widehat{\theta}$ and the quasi-score function $\Psi(\theta; y, x)$, hypothesis tests and confidence intervals for θ can be obtained using the quasi-Wald-type statistic or the quasi-score statistic, as summarized in McCullagh (1991, Section 11.6).

However, there are two major differences between the likelihood method and the quasi-likelihood approach. First, the quasi-likelihood approach does not provide an immediate pathway to conduct a test similar to the usual likelihood ratio test, because there is no function analogous to the log-likelihood function. One solution is to replicate the relationship between the score and the log-likelihood functions by constructing a quasi-log-likelihood function using the quasi-score functions. As the likelihood score function is the derivative with respect to θ of the log-likelihood function, we integrate the quasi-score function to define

$$Q(\theta; y_i, x_i) = \int_{y_i}^{\mu_i} \frac{y_i - t}{a(\phi)V(t)}\,\mathrm{d}t,$$

where the denominator in the integrand is related to the conditional variance in (5.6). This then leads to the quasi-log-likelihood function for the sample,

$$Q(\theta; y, x) = \sum_{i=1}^{n} Q(\theta; y_i, x_i).$$

A second distinction between the likelihood and quasi-likelihood methods pertains to the estimation of the nuisance parameter, ϕ. While the likelihood method allows for the simultaneous estimation of both θ and ϕ, the quasi-likelihood method treats the estimation of ϕ separately from θ. In cases where, for example, only the mean and variance are specified, the method of moments is commonly used to estimate ϕ, coupled with the use of the quasi-score equation (5.8) for estimation of θ.

5.3 Theory for estimating equations

The quasi-likelihood method provides a useful basis for analysing generalized linear models, requiring only the modelling of the conditional expected value and variance of the response variable, given covariates. This approach fits within the broader framework of estimating equations, which offer robustness and flexibility under model misspecification, as discussed in Chapter 4. Unlike likelihood methods, estimating equations avoid exact distributional assumptions, requiring only partial specification and producing results less sensitive to errors in the assumptions. This robustness makes them particularly valuable in complex or high-dimensional settings where likelihood methods are difficult to implement.

In this section, we present the basic theory of inference based on estimating equations without distinguishing between response and covariates, in contrast to Section 5.2, where such a distinction is explicitly made. Suppose Y is a random variable or vector defined on the sample space $\mathcal{Y}$. Let $f(y;\theta)$ denote the probability density or mass function of Y, where $\theta \in \Theta$, and Θ is an open subset of $\mathbb{R}^p$ with a positive integer p.

An estimating function for θ is a function that maps $\Theta \times \mathcal{Y}$ to $\mathbb{R}^p$. In cases where $p > 1$, an estimating function for θ is a p-dimensional column vector. This function can be denoted by $\Psi(\theta; y)$, explicitly indicating its dependence on data through the variable y, or simply by $\Psi(\theta)$ with the dependence on the variable suppressed in the notation.

An estimating function $\Psi(\theta; y)$ is called unbiased if

$$\mathrm{E}_\theta\{\Psi(\theta; Y)\} = 0, \quad \theta \in \Theta.$$

The expectation is evaluated with respect to $f(y;\theta)$.

To obtain meaningful estimators with desirable statistical properties, such as consistency and asymptotic normality, we often confine our attention to unbiased estimating functions. Let $u(\theta; y) = \partial \log f(y;\theta)/\partial\theta$ denote the score function, analogously to Section 1.2, except that we consider a single observation here rather than a sample. The following regularity conditions are commonly imposed on estimating functions and the underlying model $f(y;\theta)$:

R1: $\Psi(\theta; y)$ and $u(\theta; y)$ are unbiased.

R2: $\Psi(\theta; Y)$ and $u(\theta; Y)$ have finite second moments.

R3: The matrix $\mathrm{E}\{\Psi(\theta; Y)u^{\mathsf{T}}(\theta; Y)\}$ is non-singular for all $\theta \in \Theta$.

R4: $\Psi(\theta; y)$ is differentiable with respect to θ. The integrals $\int u(\theta; y)f(y;\theta)\,dy$ and $\int \Psi(\theta; y)f(y;\theta)\,dy$ are differentiable with respect to θ, and the operations of integration and differentiation for these integrals, as well as $\int f(y;\theta)\,dy$, can be interchanged. In the case of discrete random variables, the integrals are replaced by summations.

A function $\Psi(\theta; y)$ satisfying R1—R3 is called a regular estimating function. We define the sensitivity matrix and the variability matrix as

$$\Gamma_\Psi(\theta) = \mathrm{E}\left\{-\frac{\partial\Psi(\theta; Y)}{\partial\theta^{\mathsf{T}}}\right\} \quad \text{and} \quad \Sigma_\Psi(\theta) = \mathrm{E}\{\Psi(\theta; Y)\Psi^{\mathsf{T}}(\theta; Y)\}, \tag{5.9}$$

respectively. Define the Fisher information matrix as

$$i(\theta) = \mathrm{E}\{u(\theta; Y)u^{\mathsf{T}}(\theta; Y)\}, \tag{5.10}$$

analogously to Section 1.2, except that the expectation is taken with respect to $f(y;\theta)$ of a single observation rather than that of a sample.

An estimating function $\Psi(\theta; y)$ is said to be smooth if it satisfies the requirements in Assumption R4. Let $\mathcal{G}$ denote the class of all estimating functions for θ that are both regular and smooth. If $\Psi(\theta; y) \in \mathcal{G}$, then, differentiating $\mathrm{E}_\theta\{\Psi(\theta; Y)\} = 0$ and using the identity $\partial f(y; \theta)/\partial\theta = u(\theta; y) f(y; \theta)$,

$$\Gamma_\Psi(\theta) = \mathrm{E}\{\Psi(\theta; Y) u^\mathsf{T}(\theta; Y)\},$$

or equivalently,

$$\mathrm{E}\left\{-\frac{\partial\Psi(\theta; Y)}{\partial\theta^\mathsf{T}}\right\} = \mathrm{cov}\{\Psi(\theta; Y), u(\theta; Y)\}. \tag{5.11}$$

This relationship is sometimes referred to as the extended Bartlett identity.

When $\Psi(\theta; y)$ is regular, the Godambe information matrix

$$G_\Psi(\theta) = \Gamma_\Psi(\theta)\Sigma_\Psi^{-1}(\theta)\Gamma_\Psi^\mathsf{T}(\theta),$$

is positive definite. This matrix reflects the information carried by an estimating function $\Psi(\theta; y)$ about the parameter θ. Its inverse gives the asymptotic variance of the estimator derived from the estimating function $\Psi(\theta; y)$; see Theorem 5.3 in Section 5.4.

Theorem 5.1 For any $\Psi(\theta; y) \in \mathcal{G}$, $i(\theta) - G_\Psi(\theta)$ is non-negative definite for all $\theta \in \Theta$. □

For square matrices A and B of the same dimension, if $B - A$ is non-negative definite, then we say A is less than or equal to B in the Loewner order, and write $A \preccurlyeq B$ or $B \succcurlyeq A$. Thus, Theorem 5.1 states that $G_\Psi(\theta) \preccurlyeq i(\theta)$.

If our attention is on the entire class $\mathcal{G}$, this theorem identifies the Fisher information matrix $i(\theta)$ as the attainable upper bound of the Godambe information matrix $G_\Psi(\theta)$ for all $\Psi(\theta; y) \in \mathcal{G}$. In applications, however, exploring the entire class $\mathcal{G}$ may be hindered by the unavailability of likelihood score functions. In such cases, we consider a subclass of $\mathcal{G}$ that excludes these functions. The upper bound for the Godambe information matrices associated with the estimating functions in this subclass may become unattainable. Examining those Godambe information matrices that are "closer" to the Fisher information matrix $i(\theta)$ becomes informative, as discussed in Section 5.5. A geometric perspective of estimating functions is often helpful in this regard; see Appendix 5.C.

Definition 5.2 Suppose $\Psi(\theta; y)$ and $\Psi^*(\theta; y)$ are estimating functions in $\mathcal{G}$.

(a) The standardized estimating function of $\Psi(\theta; y)$ is

$$\Psi_{\mathrm{std}}(\theta; y) = \Gamma_\Psi^{-1}(\theta)\Psi(\theta; y),$$

and the projected estimating function of $\Psi(\theta; y)$ is

$$\Psi_{\mathrm{proj}}(\theta; y) = \Gamma_\Psi(\theta)\Sigma_\Psi^{-1}(\theta)\Psi(\theta; y). \tag{5.12}$$

(b) The projection of $\Psi^*(\theta; y)$ onto $\Psi(\theta; y)$ is

$$\mathrm{proj}\{\Psi^*(\theta; y) \mid \Psi(\theta; y)\} = \mathrm{cov}\{\Psi^*(\theta; Y), \Psi(\theta; Y)\}[\mathrm{var}\{\Psi(\theta; Y)\}]^{-1}\Psi(\theta; y),$$

or equivalently,

$$\mathrm{E}\{\Psi^*(\theta; Y)\Psi^\mathsf{T}(\theta; Y)\}[\mathrm{E}\{\Psi(\theta; Y)\Psi^\mathsf{T}(\theta; Y)\}]^{-1}\Psi(\theta; y).$$

□

The function $\Psi^*(\theta; y) - \text{proj}\{\Psi^*(\theta; Y) \mid \Psi(\theta; Y)\}$ is orthogonal to $\Psi(\theta; y)$, as

$$\mathrm{E}\left([\Psi^*(\theta; Y) - \text{proj}\{\Psi^*(\theta; Y) \mid \Psi(\theta; Y)\}]\, \Psi^{\mathsf{T}}(\theta; Y)\right) = 0.$$

The projected estimating function of $\Psi(\theta; y)$ is the projection of the likelihood score function $u(\theta; y)$ onto $\Psi(\theta; y)$, given by

$$\Psi_{\text{proj}}(\theta; y) = \text{proj}\{u(\theta; y) \mid \Psi(\theta; y)\}. \tag{5.13}$$

The covariance matrix of $\Psi_{\text{proj}}(\theta; y)$ is the Godambe information matrix,

$$G_\Psi(\theta) = \mathrm{E}\{\Psi_{\text{proj}}(\theta; Y)\Psi_{\text{proj}}^{\mathsf{T}}(\theta; Y)\}; \tag{5.14}$$

a comparison to (5.10) highlights its analogy to the Fisher information.

5.4 Estimators derived from estimating equations

Section 5.3 outlines the fundamentals of estimating functions, with a primary focus on regular estimating functions. The use of such functions is motivated by their utility for inference about model parameters, particularly in the asymptotic sense. The resulting estimators often possess desirable statistical properties—such as consistency and asymptotic normality, akin to estimators derived from the likelihood method. This section presents these properties.

Theorem 5.3 Suppose $\{Y_1, \ldots, Y_n\}$ are i.i.d. with probability density or mass function $f(y; \theta)$. Assume that the estimating function $\Psi(\theta; y)$ is unbiased, and the estimating equation

$$\sum_{i=1}^{n} \Psi(\theta; Y_i) = 0$$

has a unique solution $\widehat{\theta}$. Then, under the regularity conditions in Section 5.3 and those in Appendix 5.A, as $n \to \infty$,

$$\widehat{\theta} \xrightarrow{p} \theta \quad \text{and} \quad \sqrt{n}(\widehat{\theta} - \theta) \xrightarrow{d} N\{0, G_\Psi^{-1}(\theta)\}.$$

□

In contrast to the asymptotic distribution (2.48) in Section 2.5, which is derived from the likelihood function and requires correct model specification, Theorem 5.3 underscores the utility of regular estimating functions. For models lacking a likelihood function to derive maximum likelihood estimators, regular estimating functions provide a practical alternative for inferring model parameters, constructing confidence intervals, and conducting hypothesis tests. Theorem 5.3 follows from van der Vaart (1998, Lemma 5.10 and Section 5.3); see the bibliographic notes.

While the regularity conditions required by Theorem 5.3 are not unique, assumption R1 in Section 5.3 is usually imposed. However the existence and uniqueness of a solution for estimating equations are not guaranteed without some further conditions specific to the problem considered. Unlike likelihood score equations, which may have multiple roots but allow the identification of the maximum likelihood estimator through the likelihood function, estimating equations present difficulties in choosing a suitable estimator among multiple roots. Heyde and Morton (1998) and Small et al. (2000) proposed criteria to guide the selection of a consistent estimator in this case.

Example 5.4 (Nonlinear moving average model, cont'd) In continuation of Example 3.18, we employ the estimating equation approach to estimate the parameter $\theta = (\alpha, \beta, \sigma^2)^{\mathsf{T}}$ in the nonlinear model (3.27), with observations indexed here by i instead of t. For $i = 1, \dots, n$, we define

$$\Psi(\theta; y_i) = (\Psi_1(\theta; y_i), \Psi_2(\theta; y_i), \Psi_3(\theta; y_i))^{\mathsf{T}},$$

where

$$\begin{aligned}
\Psi_1(\theta; y_i) &= y_i^2 - \{(1+\alpha^2)\sigma^2 + \beta^2\sigma^4\}; \\
\Psi_2(\theta; y_i) &= y_i y_{i-1} - \alpha\sigma^2; \\
\Psi_3(\theta; y_i) &= y_i y_{i-1} y_{i-2} - \alpha\beta\sigma^4,
\end{aligned}$$

with $\Psi_2(\theta; y_1) = \Psi_3(\theta; y_1) = \Psi_3(\theta; y_2) = 0$. Here the dependence on y_{i-1} or y_{i-2} is suppressed in $\Psi_2(\cdot;\cdot)$ and $\Psi_3(\cdot;\cdot)$.

These functions are designed to capture key features of model (3.27). It is easily seen that $\Psi(\theta; y_i)$ is an unbiased estimating function for θ. Writing $\widehat{\theta} = (\widehat{\alpha}, \widehat{\beta}, \widehat{\sigma}^2)^{\mathsf{T}}$ for the solution of $\sum_{i=1}^{n} \Psi_i(\theta) = 0$, and

$$a_n = \frac{1}{n-2}\sum_{i=3}^{n} y_i y_{i-1} y_{i-2}, \quad b_n = \frac{1}{n-1}\sum_{i=2}^{n} y_i y_{i-1}, \quad c_n = \frac{1}{n}\sum_{i=1}^{n} y_i^2,$$

then

$$\widehat{\sigma}^2 = \frac{b_n}{\widehat{\alpha}}, \quad \widehat{\beta} = \frac{\widehat{\alpha} a_n}{b_n^2}, \quad \widehat{\alpha}^2 + \left(\frac{a_n^2}{b_n^3} - \frac{c_n}{b_n}\right)\widehat{\alpha} + 1 = 0. \tag{5.15}$$

This quadratic equation in (5.15) may have zero, one, or two solutions. There will be no solution if the discriminant

$$\left[\frac{\{(n-2)^{-1}\sum_{i=3}^{n} y_i y_{i-1} y_{i-2}\}^2}{\{(n-1)^{-1}\sum_{i=2}^{n} y_i y_{i-1}\}^3} - \frac{n^{-1}\sum_{i=1}^{n} y_i^2}{(n-1)^{-1}\sum_{i=2}^{n} y_i y_{i-1}}\right]^2 - 4$$

is negative, but if the discriminant is positive there may be two solutions. □

Example 5.5 (Generalized estimating equation) In the discussion of longitudinal data in Section 3.2, we distinguish between response and covariate variables. Section 3.2.1 considers the conditional mean of Y_{ij} given the subject-level covariates X_i. In contrast, here we define $\mu_{ij} = \mathrm{E}(Y_{ij} \mid X_{ij} = x_{ij})$ by conditioning on subject-time-specific covariates X_{ij}, for $i = 1, \dots, n$ and $j = 1, \dots, m_i$. Suppose that marginal regression models are used to describe the dependence of μ_{ij} on the time-dependent covariate vector X_{ij},

$$g(\mu_{ij}) = x_{ij}^{\mathsf{T}}\theta,$$

where θ is the vector of regression parameters, and $g(\cdot)$ is the link function. An estimating function for θ can be developed as follows.

With $\mu_i = (\mu_{i1}, \dots, \mu_{im_i})^{\mathsf{T}}$, $V_i = \mathrm{var}(Y_i \mid X_i)$, and $D_i = \partial\mu_i^{\mathsf{T}}/\partial\theta$, we follow (5.7) and define the estimating function

$$\Psi(\theta; y_i, x_i) = D_i V_i^{-1}(y_i - \mu_i).$$

Then solving $\Sigma_i \Psi(\theta; y_i, x_i) = 0$ yields an estimator of θ. The solution of this generalized estimating equation, sometimes called GEE, is a consistent estimator of θ under mild regularity conditions, including that $\Psi(\theta; y_i, x_i)$ is unbiased.

A sufficient condition for $\Psi(\theta; y_i, x_i)$ to be unbiased is

$$E(Y_{ij} \mid X_i) = E(Y_{ij} \mid X_{ij}), \quad j = 1, \ldots, m_i, \tag{5.16}$$

which is justified as follows. Let D_{ijk} denote the (j, k) component of D_i, and let V_i^{jk} denote the (j, k) element of the inverse matrix V_i^{-1}. Let $\Psi_l(\theta; y_i, x_i)$ denote the lth element of $\Psi(\theta; y_i, x_i)$, for any $l = 1, \ldots, \dim(\theta)$. Then

$$\begin{aligned}
E\{\Psi_l(\theta; Y_i, X_i)\} &= E\left\{\sum_{k=1}^{m_i}\sum_{j=1}^{m_i} D_{ilj} V_i^{jk}(Y_{ik} - \mu_{ik})\right\} \\
&= \sum_{k=1}^{m_i}\sum_{j=1}^{m_i} E\left[E\left\{D_{ilj} V_i^{jk}(Y_{ik} - \mu_{ik}) \mid X_i\right\}\right] \\
&= \sum_{k=1}^{m_i}\sum_{j=1}^{m_i} E\left[D_{ilj} V_i^{jk}\left\{E(Y_{ik} \mid X_i) - \mu_{ik}\right\}\right] \\
&= \sum_{k=1}^{m_i}\sum_{j=1}^{m_i} E\left[D_{ilj} V_i^{jk}\left\{E(Y_{ik} \mid X_{ik}) - \mu_{ik}\right\}\right] \\
&= 0,
\end{aligned}$$

where the second last step requires (5.16) unless V_i is an identity matrix, and the last step follows from the definition of μ_{ik}. □

5.5 Optimality

As discussed in the comments following Theorem 5.1 in Section 5.3, while score functions are optimal in the class $\mathcal{G}$ of smooth regular estimating functions, different applications may lead us to consider a subset of $\mathcal{G}$, denoted by $\mathcal{G}^*$. It is interesting to explore optimal estimating functions within any "reasonable" class of unbiased estimating functions.

We first establish criteria for the optimality of an estimating function in $\mathcal{G}^*$ when θ is a scalar. For an estimating function $\Psi^*(\theta; y)$ to be considered optimal, it should tightly concentrate around zero, its expected value. At the same time, the expectation $E_\theta\{\Psi^*(\theta; Y)\}$ should be highly sensitive to changes in θ, so that even small variations in θ should lead to substantial changes in the expectation.

To achieve these objectives, we simultaneously impose two requirements: (i) minimizing the variability of $\Psi^*(\theta; Y)$, and (ii) maximizing the absolute value of $E_\theta\{\partial\Psi^*(\theta; Y)/\partial\theta\}$. Both requirements can be combined by minimizing

$$\frac{E_\theta[\{\Psi^*(\theta; Y)\}^2]}{[E_\theta\{\partial\Psi^*(\theta; Y)/\partial\theta\}]^2}$$

for $\Psi(\theta; y) \in \mathcal{G}^*, \theta \in \Theta$, which, by Definition 5.2, is equivalent to minimizing the variance of $\Psi^*_{\text{std}}(\theta; Y)$, the standardized version of $\Psi^*(\theta; Y)$.

Definition 5.6 An unbiased estimating function for a scalar parameter θ, $\Psi^*(\theta; y) \in \mathcal{G}^*$, is optimal in $\mathcal{G}^*$ if for any $\theta \in \Theta$ and any other $\Psi(\theta; y) \in \mathcal{G}^*$,

$$\frac{E_\theta[\{\Psi^*(\theta; Y)\}^2]}{[E_\theta\{\partial\Psi^*(\theta; Y)/\partial\theta\}]^2} \le \frac{E_\theta[\{\Psi(\theta; Y)\}^2]}{[E_\theta\{\partial\Psi(\theta; Y)/\partial\theta\}]^2}.$$

□

While Definition 5.6 characterizes the nature of estimating functions regardless of data information or sample size, Theorem 5.3 relates it to the asymptotic variance of the resulting estimators. Consequently, Definition 5.6 is also relevant from the perspective of large-sample theory.

Using Definition 5.2, $G^{-1}_{\Psi_{\rm std}}(\theta) = \text{var}_\theta\{\Psi_{\rm std}(\theta; Y)\}$ for any $\Psi(\theta; y) \in \mathcal{G}^*$, implying that a standardized optimal estimating function attains the minimum variance among all standardized estimating functions in $\mathcal{G}^*$. Furthermore, under some regularity conditions, optimal estimating functions are unique up to a constant multiple; see Exercise 5.3.

The following theorem reveals the relationship between the optimal estimating function and the score function $u(\theta; y)$. This is particularly useful for describing the nature of optimal estimating functions when the class $\mathcal{G}^*$ of unbiased estimating functions does not contain the likelihood score function.

Theorem 5.7 The estimating function $\Psi^*(\theta; y) \in \mathcal{G}^*$ is an optimal estimating function for a scalar parameter θ if and only if $\Psi^*(\theta; Y)$ has maximal absolute correlation with the score function $u(\theta; Y)$, i.e.,

$$|\text{cor}\{\Psi(\theta; Y), u(\theta; Y)\}| \le |\text{cor}\{\Psi^*(\theta; Y), u(\theta; Y)\}|$$

for any $\Psi(\theta; y) \in \mathcal{G}^*$. □

We now extend this discussion to the vector parameter θ. The following definitions are equivalent in the scalar case but reflect different properties when θ is vector–valued.

Definition 5.8 Suppose $\Psi^*(\theta; y) \in \mathcal{G}^*$ and $\theta \in \Theta$.

(a) Matrix optimality: $\Psi^*(\theta; y)$ is M-optimal if $G_{\Psi^*}(\theta) - G_\Psi(\theta)$ is non-negative definite for all $\Psi(\theta; y) \in \mathcal{G}^*$ and all $\theta \in \Theta$.

(b) Trace optimality: $\Psi^*(\theta; y)$ is T-optimal if $\text{trace}\{G_{\Psi^*}(\theta)\} \ge \text{trace}\{G_\Psi(\theta)\}$ for all $\Psi(\theta; y) \in \mathcal{G}^*$ and all $\theta \in \Theta$.

(c) Determinant optimality: $\Psi^*(\theta; y)$ is D-optimal if $\det\{G_{\Psi^*}(\theta)\} \ge \det\{G_\Psi(\theta)\}$ for all $\Psi(\theta; y) \in \mathcal{G}^*$ and all $\theta \in \Theta$.

□

According to Theorem 5.3, these measures can be employed to assess the asymptotic relative efficiency of estimators derived from different estimating functions. T-optimality and D-optimality are often more useful than M-optimality, as they involve scalar summaries, which facilitate more intuitive comparisons. T-optimality quantifies the total variability in estimating all components of θ, while M-optimality and D-optimality also account for potential correlations among the estimators of these components. The determinant of the covariance matrix, known as the generalized variance, is commonly used to assess the overall spread within a distribution. These three measures are interconnected; see Exercise 5.6.

In applications, identifying the optimal estimating function for a general class $\mathcal{G}^*$ can be difficult: we often restrict attention to a more accessible class and formulate an optimal estimating function within this restricted class. In particular, classes formed by the linear combination of convenient functions are frequently used.

Suppose $\{b_j(\theta; y) : j = 1, \ldots, m\}$ is a collection of unbiased estimating functions for θ, each of dimension p, that are easy to construct and partially reflect the model structure for the distribution $f(y; \theta)$. Here, m is a given positive integer, which may exceed the dimension p of θ. Define the vector

$$b(\theta; y) = (b_1^{\mathsf{T}}(\theta; y), \ldots, b_m^{\mathsf{T}}(\theta; y))^{\mathsf{T}},$$

and define the class of estimating functions generated by linear combinations of the components of $b(\theta; y)$,

$$\mathcal{L} = \{\Psi(\theta; y) : \Psi(\theta; y) = C^{\mathsf{T}}(\theta)b(\theta; y), \text{with an } mp \times p \text{ matrix } C(\theta) \text{ of constants}\}.$$

An optimal estimating function in $\mathcal{L}$ is given in the following theorem.

Theorem 5.9 Define

$$\Gamma_b(\theta) = \mathrm{E}_\theta \left\{ \frac{\partial b^{\mathsf{T}}(\theta; Y)}{\partial \theta} \right\} \quad \text{and} \quad \Sigma_b(\theta) = \mathrm{E}_\theta\{b(\theta; Y)b^{\mathsf{T}}(\theta; Y)\}.$$

If $\Sigma_b(\theta)$ is non-singular, then the estimating function

$$\Psi^*(\theta; y) = \Gamma_b(\theta)\Sigma_b^{-1}(\theta)b(\theta; y) \tag{5.17}$$

is optimal in the class $\mathcal{L}$ of linear combinations of the components of $b(\theta; y)$. □

The optimal estimating function $\Psi^*(\theta; y)$ in (5.17) satisfies properties similar to those of the likelihood score functions in (5.2), except that a single observation is considered here rather than a sample. In particular,

$$\begin{aligned} \mathrm{E}_\theta \left\{ \frac{\partial \Psi^{\mathsf{T}}(\theta; Y)}{\partial \theta} \right\} &= -\mathrm{E}_\theta\{\Psi^*(\theta; Y)\Psi^{\mathsf{T}}(\theta; Y)\}, \quad \text{for any } \Psi(\theta; y) \in \mathcal{L}, \\ \mathrm{E}_\theta \left\{ \frac{\partial \Psi^{*\mathsf{T}}(\theta; Y)}{\partial \theta} \right\} &= -\mathrm{E}_\theta\{\Psi^*(\theta; Y)\Psi^{*\mathsf{T}}(\theta; Y)\}. \end{aligned} \tag{5.18}$$

Furthermore, the optimal estimating function $\Psi^*(\theta; y)$ in (5.17) resembles the projected estimating function of $b(\theta; y)$, if we overlook the implicit requirement in Definition 5.2 that the estimating function should have the same dimension as θ.

Theorem 5.9 provides a strategy for constructing useful estimating functions. The theorem below provides a dual version of Theorem 5.9 for a conditional formulation; see Exercise 5.6.

Theorem 5.10 Suppose that for $j = 1, \ldots, m$, $b_j(\theta; y)$ is an estimating function of θ and $\mathcal{A}_j$ is a set of random variables satisfying

$$\mathrm{E}_\theta\{b_j(\theta; Y) \mid \mathcal{A}_j\} = 0,$$

and

$$\mathrm{E}_\theta\{b_j(\theta; Y)b_k(\theta; Y) \mid \mathcal{A}_j, \mathcal{A}_k\} = 0, \quad j \neq k.$$

Let $b(\theta; y) = (b_1^{\mathsf{T}}(\theta; y), \ldots, b_m^{\mathsf{T}}(\theta; y))^{\mathsf{T}}$. Then the estimating function

$$\Psi^*(\theta; y) = \sum_{j=1}^{m} \mathrm{E}_\theta \left\{ \left. \frac{\partial b_j(\theta; Y)}{\partial \theta} \right| \mathcal{A}_j \right\} [\mathrm{var}_\theta \{b_j(\theta; Y) \mid \mathcal{A}_j\}]^{-1} b_j(\theta; y)$$

is optimal in the class of linear combinations of the components of $b(\theta; y)$, where the inverse matrix is assumed to exist. □

Theorem 5.10 is useful in situations such as time series, where observations are not independent: conditioning on the history allows one to isolate the incremental information contributed by the current observation. See Exercise 5.6.

5.6 Generalized method of moments

While Theorem 5.9 aims to construct optimal estimating functions within the class $\mathcal{L}$, it also provides a way to create a broader class of unbiased estimating functions. In particular, the formulation in (5.17) can be relaxed by replacing $\Sigma_b^{-1}(\theta)$ with a positive definite matrix of the same dimension, denoted by W, which is free of the data but may depend on parameter θ. We define

$$\Psi_W^*(\theta; y) = \Gamma_b(\theta) W b(\theta; y). \tag{5.19}$$

The function $\Psi_W^*(\theta; y)$ has expected value zero and the same dimension as θ, and thus serves as an unbiased estimating function. It yields a consistent estimator for θ as the solution to

$$\sum_{i=1}^{n} \Psi_W^*(\theta; y_i) = 0, \tag{5.20}$$

under the regularity conditions of Theorem 5.3.

This extended perspective establishes a connection between the estimating equation method and the generalized method of moments, an extension of the method of moments introduced by Hansen (1982). The method of moments estimates the model parameters by matching the sample moments of the data with the population moments implied by the assumed model. When the number of equations exceeds the number of unknown parameters, resulting in an over-identified model, the generalized method of moments is particularly useful.

Suppose $S(\theta; y)$ is a $q \times 1$ vector of functions such that

$$\mathrm{E}_{\theta_0}\{S(\theta_0; Y)\} = 0,$$

where q is a positive integer larger than the dimension p of θ, and θ_0 is the true value of θ. The resulting estimating equation $\Sigma_i S(\theta; y_i) = 0$ based on a sample $\{y_1, \ldots, y_n\}$ of observations from i.i.d. random variables $\{Y_1, \ldots, Y_n\}$ is over-determined, so we consider instead a quadratic form

$$Q_n(\theta) = \left\{\frac{1}{n}\sum_{i=1}^{n} S(\theta; y_i)\right\}^{\mathsf{T}} W_n \left\{\frac{1}{n}\sum_{i=1}^{n} S(\theta; y_i)\right\},$$

where W_n is a $q \times q$ symmetric, positive semi-definite, weight matrix. The generalized method of moments estimate is then defined as

$$\widehat{\theta} = \operatorname{argmin}_{\theta \in \Theta} Q_n(\theta).$$

In regular settings, $\widehat{\theta}$ solves

$$\frac{\partial Q_n(\theta)}{\partial \theta} = 0. \tag{5.21}$$

By the formulation of $Q_n(\theta)$,

$$\begin{aligned}
\frac{\partial Q_n(\theta)}{\partial \theta} &= 2\left\{\frac{1}{n}\sum_{i=1}^{n} \frac{\partial S^{\mathsf{T}}(\theta; y_i)}{\partial \theta}\right\} W_n \left\{\frac{1}{n}\sum_{i=1}^{n} S(\theta; y_i)\right\} \\
&= 2\left[\mathrm{E}_\theta\left\{\frac{\partial S^{\mathsf{T}}(\theta; Y_i)}{\partial \theta}\right\} + o_p(1)\right] W_n \left\{\frac{1}{n}\sum_{i=1}^{n} S(\theta; y_i)\right\} \\
&= \frac{2}{n}\sum_{i=1}^{n} \mathrm{E}_\theta\left\{\frac{\partial S^{\mathsf{T}}(\theta; Y_i)}{\partial \theta}\right\} W_n S(\theta; y_i) + 2W_n \left\{\frac{1}{n}\sum_{i=1}^{n} S(\theta; y_i)\right\} o_p(1).
\end{aligned} \tag{5.22}$$

Under mild conditions on the weight matrix W_n, such as the condition that $W_n\{n^{-1}\Sigma_i S(\theta; Y_i)\}$ converges in probability to a matrix, the second term in (5.22) converges in probability to zero by the weak law of large numbers applied to $\Sigma_i S(\theta; Y_i)$. The solution of (5.21) is asymptotically equivalent to the solution of

$$\sum_{i=1}^{n} \mathrm{E}_\theta\left\{\frac{\partial S^\top(\theta; y_i)}{\partial \theta}\right\} W_n S(\theta; y_i) = 0; \tag{5.23}$$

an equation that resembles (5.20).

The GMM approach distinguishes itself from the estimating equations approach, as suggested in Theorem 5.3, by allowing the dimension of the function $S(\theta; y)$ to be larger than that of the parameter θ.

As discussed in Section 5.4, solving estimating equations for unknown parameters may lead to multiple roots, and choosing the appropriate root is not always obvious. However, the GMM estimates are obtained through a minimization procedure, eliminating concerns about multiple solutions. In numerical implementation, where several local minimizers may be obtained, one can compare the resulting values of $Q_n(\theta)$ and identify the solution that minimizes $Q_n(\theta)$.

5.7 Missing data

Likelihood-based methods for handling missing data, as outlined in Section 3.5, require complete specification of the distributions for both the variables of interest and the missing data indicators. Here, we instead describe a two-step approach that modifies easily constructed but potentially biased estimating functions to achieve unbiasedness. In the first step, we assume an ideal scenario where measurements for all associated variables are available, allowing us to construct unbiased estimating functions. In the second step, we introduce modifications to address the effects of missing data. This two-step approach avoids full distribution specification, focusing instead on modelling marginal features such as the expected value and variance, which leads to more robust inference.

In this section, we describe three strategies for adjusting the estimating equation to accommodate data with missing values, where variables are differentiated by their roles. Specifically, let Y denote the response variable, and let $\{X, Z\}$ represent the associated covariates. Assume that the conditional expected value and variance of Y given $\{X, Z\}$ are modeled as

$$\mathrm{E}(Y \mid X = x, Z = z) = \mu(x, z; \theta), \tag{5.24}$$

and

$$\mathrm{var}(Y \mid X = x, Z = z) = k\{\mu(x, z; \theta); \phi\}, \tag{5.25}$$

respectively, where $\mu(\cdot)$ and $k(\cdot)$ are known functions, and θ is the vector of regression parameters of interest. Here ϕ is a dispersion parameter, treated as a nuisance parameter and assumed to be known. For estimation involving both parameters of interest and nuisance parameters, see Section 5.9.

By setting $b(\theta; y, x, z) = y - \mu(x, z; \theta)$ and $m = 1$, and applying Theorem 5.9, an optimal estimating function for θ is

$$\Psi(\theta; y, x, z) = A(x, z)\{y - \mu(x, z; \theta)\}, \tag{5.26}$$

where

$$A(x, z) = \left(\frac{\partial \mu(x, z; \theta)}{\partial \theta}\right) \{\text{var}(Y \mid X = x, Z = z)\}^{-1}.$$

Given a sample $\{\{y_i, x_i, z_i\} : i = 1, \ldots, n\}$ of independent observations of $\{Y, X, Z\}$, we obtain an estimate as the solution to

$$\sum_{i=1}^{n} \Psi(\theta; y_i, x_i, z_i) = 0.$$

The resulting estimator, under standard regularity conditions, is consistent and asymptotically normal (after an appropriate transformation), as described in Theorem 5.3.

This formulation assumes complete availability of measurements $\{y_i, x_i, z_i\}$ for $i = 1, \ldots, n$. However, in practice, data are often incomplete. For example, when measurements of X are very costly compared to those of Z, measurements for Z may be obtained for all subjects, while measurements of X are obtained only for a subset. It is necessary to introduce a missing data indicator and incorporate the missingness effects into the estimation procedure. In the following discussion, we consider the setting in which X is subject to missingness, while Y and Z are fully observed.

5.7.1 Inverse probability weighted estimating equation

For any subject, let $R = 1$ if X is observed, and 0 otherwise. Assume a missing at random mechanism

$$\text{pr}(R = 1 \mid Y, X, Z) = \text{pr}(R = 1 \mid Y, Z),$$

denoted by $\pi(Y, Z)$. In the presence of missing data, there may be a temptation to conduct a complete-case analysis using the estimating equation

$$\sum_{i=1}^{n} r_i \Psi(\theta; y_i, x_i, z_i) = 0 \tag{5.27}$$

for θ, where r_i is the realized value of the missing data indicator R_i for subject i. However, this approach may not lead to a consistent estimator, as the estimating functions are no longer unbiased, as shown below,

$$\begin{aligned} \text{E}\{R\Psi(\theta; Y, X, Z)\} &= \text{E}[\text{E}\{R\Psi(\theta; Y, X, Z) \mid Y, X, Z\}] \\ &= \text{E}[\text{E}(R \mid Y, X, Z) A(X, Z)\{Y - \mu(X, Z; \theta)\}] \\ &= \text{E}[\pi(Y, Z) A(X, Z)\{Y - \mu(X, Z; \theta)\}] \qquad (5.28) \\ &\neq 0. \end{aligned}$$

The dependence of the missingness probability $\pi(y, z)$ on y and z implies that the estimating function (5.26) is not even conditionally unbiased, given X and Z.

A quick remedy is to re-weight the estimating function in (5.27) to remove the missingness probability $\pi(y, z)$ from (5.28). Different weights are assigned to reflect distinct contributions from subjects with varying missingness probabilities. This adjustment accounts for the non-random nature of the subsample with complete observations, accommodating its possible lack of representativeness of the underlying population.

Define

$$\Psi_{\text{IPW}}(\theta; r, y, x, z) = \frac{r}{\pi(y, z)} \Psi(\theta; y, x, z),$$

and $\widehat{\theta}$ the solution of

$$\sum_{i=1}^{n} \Psi_{\text{IPW}}(\theta; r_i, y_i, x_i, z_i) = 0. \tag{5.29}$$

This approach is often referred to as the inverse probability weighted generalized estimating equation method.

From the derivation of (5.28), it is clear that the modified estimating function $\Psi_{\text{IPW}}(\theta; r, y; x, z)$ is unbiased, so from Theorem 5.3, $\widehat{\theta}$ is consistent and $\sqrt{n}(\widehat{\theta} - \theta)$ has an asymptotic normal distribution with expected value zero, subject to certain regularity conditions, including some assumptions on the missingness process; see Robins et al. (1995).

5.7.2 Augmented inverse probability weighted estimating equation

The estimator $\widehat{\theta}$ from (5.29) relies only on complete-case observations and discards information from subjects with incomplete data. By attaching the inverse of the probability $\pi(y, z)$ as a weight, the initial estimating function $\Psi(\theta; y, x, z)$ is adjusted to account for selection bias, making complete cases resemble a random sample. However, since (5.29) is based on a reduced sample size, it has higher variability than methods that include both complete and incomplete cases. The estimating function $\Psi_{\text{IPW}}(\theta; r, y; x, z)$ can be augmented to include incomplete cases,

$$\Psi_{\text{A}}(\theta; r, y, x, z) = \Psi_{\text{IPW}}(\theta; r, y, x, z) + \frac{\pi(y, z) - r}{\pi(y, z)} B(y, z), \tag{5.30}$$

where $B(y, z)$ is a function of y and z that may involve the parameter θ but is free of x. The augmented estimator $\widehat{\theta}_{\text{A}}$ solves

$$\sum_{i=1}^{n} \Psi_{\text{A}}(\theta; r_i, y_i, x_i, z_i) = 0. \tag{5.31}$$

As $\text{E}(R \mid Y, Z) = \pi(Y, Z)$, the augmented estimating function $\Psi_{\text{A}}(\theta; r, y, x, z)$ is also unbiased. The efficiency of the estimator $\widehat{\theta}_{\text{A}}$ depends on the choice of $A(x, z)$ in (5.26) and $B(y, z)$ in (5.30), with the goal of minimizing the variance of the resulting estimator. Optimal forms for $A(X, Z)$ and $B(Y, Z)$ are difficult to identify in applications. As noted by Tsiatis (2006, pp. 174–179), even the optimal choice of $A(X, Z)$ in the presence of missing data may differ from that for the full data. However, the augmented estimator $\widehat{\theta}_{\text{A}}$ generally exhibits higher efficiency than $\widehat{\theta}$ derived from (5.29), as discussed by many authors, including Robins et al. (1994) and Chen et al. (2010).

5.7.3 Doubly robust method

While there is no unique choice for $B(y, z)$ in (5.30), a common choice is

$$B_{\text{E}}(y, z) = \text{E}\{\Psi(\theta; Y, X, Z) \mid Y = y, Z = z\}, \tag{5.32}$$

where the subscript E highlights the expectation taken with respect to the conditional distribution of X given $\{Y, Z\}$. Let

$$\Psi_{\text{AE}}(\theta; r, y, x, z) = \Psi_{\text{IPW}}(\theta; r, y, x, z) + \frac{\pi(y, z) - r}{\pi(y, z)} B_{\text{E}}(y, z) \tag{5.33}$$

denote the resulting estimating function, which is alternatively expressed as

$$\Psi_{\text{AE}}(\theta; r, y, x, z) = \Psi(\theta; y, x, z) + \frac{r - \pi(y, z)}{\pi(y, z)} \{\Psi(\theta; y, x, z) - B_{\text{E}}(y, z)\}.$$

The construction of $\Psi_{\rm AE}(\theta; r, y, x, z)$ entails both the missingness probability $\pi(y, z)$ and the conditional expectation $B_{\rm E}(y, z)$, which require models for the conditional distributions of X and R, given $\{Y, Z\}$. However, the unbiasedness of $\Psi_{\rm AE}(\theta; r, y, x, z)$ does not require both models to be correctly specified simultaneously. If either the model for the conditional distribution of R given $\{Y, Z\}$ or the model for the conditional distribution of X given $\{Y, Z\}$ is correct, then $\Psi_{\rm AE}(\theta; r, y; x, z)$ remains unbiased. This property is known as double robustness.

To illustrate, let $\pi(y, z)$ and $\pi^*(y, z)$ denote the missingness probabilities under the true and working models, respectively, and let $B_{\rm E}(y, z)$ and $B^*_{\rm E}(y, z)$ denote the corresponding conditional expectations in (5.32). By the unbiasedness of $\Psi(\theta; r, y, x, z)$ derived from (5.24),

$$\mathrm{E}\{\Psi_{\rm AE}(\theta; R, Y, X, Z)\} = \mathrm{E}\left[\frac{R - \pi(Y, Z)}{\pi(Y, Z)}\{\Psi(\theta; Y, X, Z) - B_{\rm E}(Y, Z)\}\right], \tag{5.34}$$

which can be assessed through different iterative expectations, considering either the conditional distribution of X given $\{R, Y, Z\}$ or the conditional distribution of R given $\{X, Y, Z\}$.

In other words, (5.34) can be expressed as

$$\begin{aligned}
&\mathrm{E}\left(\mathrm{E}\left[\frac{R - \pi(Y, Z)}{\pi(Y, Z)}\{\Psi(\theta; Y, X, Z) - B_{\rm E}(Y, Z)\} \mid R, Y, Z\right]\right)\\
&= \mathrm{E}\left(\frac{R - \pi(Y, Z)}{\pi(Y, Z)}\mathrm{E}[\{\Psi(\theta; Y, X, Z) - B_{\rm E}(Y, Z)\} \mid R, Y, Z]\right)\\
&= \mathrm{E}\left(\frac{R - \pi(Y, Z)}{\pi(Y, Z)}\mathrm{E}[\{\Psi(\theta; Y, X, Z) - B_{\rm E}(Y, Z)\} \mid Y, Z]\right),
\end{aligned} \tag{5.35}$$

or alternatively,

$$\begin{aligned}
&\mathrm{E}\left(\mathrm{E}\left[\frac{R - \pi(Y, Z)}{\pi(Y, Z)}\{\Psi(\theta; Y, X, Z) - B_{\rm E}(Y, Z)\} \mid X, Y, Z\right]\right)\\
&= \mathrm{E}\left[\frac{\mathrm{E}(R \mid X, Y, Z) - \pi(Y, Z)}{\pi(Y, Z)}\{\Psi(\theta; Y, X, Z) - B_{\rm E}(Y, Z)\}\right].
\end{aligned} \tag{5.36}$$

If the covariate model for the conditional distribution of X given $\{Y, Z\}$ is correctly specified, then $\mathrm{E}[\{\Psi(\theta; Y, X, Z) - B_{\rm E}(Y, Z)\} \mid Y, Z] = 0$ in (5.35). In this case, (5.34) equals zero regardless of whether $\pi(Y, Z)$ is correctly specified or replaced by $\pi^*(Y, Z)$ based on a potentially misspecified working model.

On the other hand, if in (5.36) the missing data model is correctly specified as $\pi(Y, Z)$, then $\mathrm{E}(R|X, Y, Z) - \pi(Y, Z) = 0$, so that (5.34) is zero regardless of whether $B_{\rm E}(Y, Z)$ is correctly specified or replaced by $B^*_{\rm E}(Y, Z)$, based on a potentially misspecified working model.

5.8 Error-contaminated data

We now turn our attention to scenarios where data are error-contaminated, rather than missing. Our objective is to formulate unbiased estimating equations in terms of the observed data, enabling us to leverage Theorem 5.3 for inference. We outline two strategies to accomplish this goal, using the notation of Section 5.7.

5.8.1 Subtraction

There may be a natural choice of a biased working estimating function that provides a starting point. Devising a method to modify this working function to create an unbiased estimating function can be useful. Let $\Psi_{\mathrm{B}}(\theta; y, x, z)$ denote a biased estimating function for θ. The most direct method to address the bias in $\Psi_{\mathrm{B}}(\theta; y, x, z)$ is to subtract its expected value

$$\Psi_{\mathrm{S}}(\theta; y, x, z) = \Psi_{\mathrm{B}}(\theta; y, x, z) - \mathrm{E}\{\Psi_{\mathrm{B}}(\theta; Y, X, Z)\}.$$

Alternatively, to more explicitly link to a model specification that conditions on the covariates$X$ and Z, we may define

$$\Psi_{\mathrm{S}}(\theta; y, x, z) = \Psi_{\mathrm{B}}(\theta; y, x, z) - \mathrm{E}\{\Psi_{\mathrm{B}}(\theta; Y, X, Z) \mid X, Z\}.$$

By construction, either version of $\Psi_{\mathrm{S}}(\theta; y, x, z)$ is an unbiased estimating function for θ, and under the usual regularity conditions leads to a consistent and asymptotically normal estimator by Theorem 5.3. This approach is referred to as the subtraction correction method. The following example provides an illustration.

Example 5.11 (Binary data with misclassification) Using our notation for longitudinal data in Section 3.2.1, $Y_i = (Y_{i1}, \ldots, Y_{im})^{\mathsf{T}}$ and $X_i = (X_{i1}^{\mathsf{T}}, \ldots, X_{im}^{\mathsf{T}})^{\mathsf{T}}$ are the response and covariate variables, respectively. Suppose Y_{ij} is 0 or 1, with expected value $\mu_{ij} = \mathrm{E}(Y_{ij}|X_i = x_i)$ of logistic form

$$\operatorname{logit} \mu_{ij} = x_{ij}^{\mathsf{T}}\theta. \tag{5.37}$$

If the components in Y_i are independent, the score equation $\Sigma_{i=1}^{n}\Psi(\theta; y_i, x_i) = 0$, where

$$\Psi(\theta; y_i, x_i) = \sum_{j=1}^{m} x_{ij}\left\{y_{ij} - \frac{\exp(x_{ij}^{\mathsf{T}}\theta)}{1 + \exp(x_{ij}^{\mathsf{T}}\theta)}\right\}, \tag{5.38}$$

gives the maximum likelihood estimator of θ.

If the independence assumption is violated, the resulting estimator is not the maximum likelihood estimator of θ, but $\Psi(\theta; y_i, x_i)$ is an unbiased estimating function using only the model assumption (5.37), and can be used to analyze data with responses subject to misclassification.

Let Y_{ij}^* denote an observed measurement of Y_{ij}, linked to Y_{ij} with known misclassification probabilities

$$p_1 = \mathrm{pr}(Y_{ij}^* = 1 \mid Y_{ij} = 0), \quad p_0 = \mathrm{pr}(Y_{ij}^* = 0 \mid Y_{ij} = 1).$$

The observations are $\{\{y_i^*, x_i\} : i = 1, \ldots, n\}$, where $y_i^* = (y_{i1}^*, \ldots, y_{im}^*)^{\mathsf{T}}$. The obvious estimating function

$$\Psi(\theta; y_i^*, x_i) = \sum_{j=1}^{m} x_{ij}\left\{y_{ij}^* - \frac{\exp(x_{ij}^{\mathsf{T}}\theta)}{1 + \exp(x_{ij}^{\mathsf{T}}\theta)}\right\} \tag{5.39}$$

is biased, with expected value

$$\mathrm{E}\{\Psi(\theta; Y_i^*, x_i)\} = \sum_{j=1}^{m}\left[\frac{x_{ij}\{(1-p_0)\exp(x_{ij}^{\mathsf{T}}\theta) + p_1\}}{1 + \exp(x_{ij}^{\mathsf{T}}\theta)} - \frac{\exp(x_{ij}^{\mathsf{T}}\theta)}{1 + \exp(x_{ij}^{\mathsf{T}}\theta)}\right], \tag{5.40}$$

where for a reason similar to that in Example 5.5, we assume $\mathrm{pr}(Y_{ij}^* = y_i^* \mid Y_i = y_i, X_i = x_i) = \mathrm{pr}(Y_{ij}^* = y^* \mid Y_{ij} = y_{ij})$; see Exercise 5.9.

Although the estimating function $\Psi(\theta; y_i^*, x_i)$ is naturally motivated by the available formulation (5.38) constructed under the ideal setting with precisely measured data, the resulting estimator is generally inconsistent. However, using the subtraction correction method, one can readily construct an unbiased estimating function $\Psi_{\mathrm{S}}(\theta; y_i, x_i) = \Psi(\theta; y_i^*, x_i) - \mathrm{E}\{\Psi(\theta; Y_i^*, X_i)\}$ using (5.40). □

5.8.2 Expected estimating equations

We now consider the expected estimating equation method when Y and Z are observed precisely, but we observe a surrogate X^* of the true measurement of X. Our objective is to derive an unbiased estimating function solely in terms of the observed variables Y, Z, and X^*.

Assuming our ideal estimating function $\Psi(\theta; Y, X, Z)$ is unbiased, let

$$\Psi_{\mathrm{E}}(\theta; Y, X^*, Z) = \mathrm{E}\left\{\Psi(\theta; Y, X, Z) \mid Y, X^*, Z\right\}.$$

This estimating function is also unbiased, as shown below,

$$\begin{aligned}
\mathrm{E}_{(Y,X,X^*,Z)}\left\{\Psi_{\mathrm{E}}(\theta; Y, X^*, Z)\right\} &= \mathrm{E}_{(Y,X,X^*,Z)}\left[\mathrm{E}\left\{\Psi(\theta; Y, X, Z) \mid Y, X^*, Z\right\}\right] \\
&= \mathrm{E}_{(Y,X^*,Z)}\left(\mathrm{E}_{X|(Y,X^*,Z)}\left[\mathrm{E}\left\{\Psi(\theta; Y, X, Z) \mid Y, X^*, Z\right\}\right]\right) \\
&= \mathrm{E}_{(Y,X^*,Z)}\left[\mathrm{E}_{X|(Y,X^*,Z)}\left\{\Psi(\theta; Y, X, Z)\right\}\right] \\
&= \mathrm{E}_{(Y,X,X^*,Z)}\left\{\Psi(\theta; Y, X, Z)\right\} \\
&= \mathrm{E}_{(Y,X,Z)}\left\{\Psi(\theta; Y, X, Z)\right\} \\
&= 0.
\end{aligned}$$

The expected estimating equation method shares certain similarities to the EM algorithm discussed in Section 3.5.2, particularly in the need for the conditional expectation of the unobserved quantities given the observed variables. However, these two methods are distinct. While the EM algorithm requires specification of a likelihood function, the expected estimating equation approach is applicable to general estimating functions. The EM algorithm requires the conditional expectation of the log-likelihood function for the complete data $\{Y, X, X^*, Z\}$, whereas the expected estimating equation approach requires the expectations of estimating functions that involve $\{Y, X, Z\}$ only, but not X^*. Furthermore, when implementing the EM algorithm, the parameter in the conditional distribution of the missing data, given the observed data, differs from the parameter in the log-likelihood function for the complete data. In contrast, the expected estimating equation method has the same parameter in both the conditional distribution and the estimating function.

The expected estimating equation approach can be applied not only to data with measurement error but also to data with missing observations, as illustrated in the following example.

Example 5.12 (Internal validation study) Consider a main study with an internal validation study comprising data from n subjects. The subjects in the internal validation sample have measurements of $\{Y_i, X_i, Z_i\}$, whereas the main study subjects have measurements of $\{Y_i, Z_i\}$ only, with measurements of X_i missing. Let $\mathcal{V}$ and $\mathcal{M}$ denote the subject index sets for the validation sample and the main study, respectively, where $\mathcal{V}$ is a subset of $\mathcal{M}$.

Suppose the expected value and variance of Y_i are given by (5.24) and (5.25), respectively. An unbiased estimating function for the model parameter is given by (5.26). However, as the main study contributes no measurements on X_i, the estimating function (5.26) cannot be used for the main study subjects.

To address this, we utilize the expected estimating equation method. The contribution from the main study subjects is $\mathrm{E}\{\Psi(\theta; Y, X, Z)|Y, Z\}$ rather than $\Psi(\theta; Y, X, Z)$ due to the unavailability of X. For subjects in the internal validation sample, their contributions to the estimating functions are given by $\Psi(\theta; Y, X, Z)$ directly. The estimating equations are

$$\sum_{j\in\mathcal{V}} \Psi(\theta; Y_j, X_j, Z_j) + \sum_{i\in\mathcal{M}\setminus\mathcal{V}} \mathrm{E}\{\Psi(\theta; Y_i, X_i, Z_i) \mid Y_i, Z_i\} = 0; \tag{5.41}$$

see Exercise 5.10. □

5.9 Nuisance parameters

The preceding sections focus on unbiased estimating functions for parameters of interest. However, when additional nuisance parameters are present, such as those arising from modeling the missing data or measurement error processes as in Sections 5.7 and 5.8, it is necessary to construct corresponding estimating functions, typically unbiased, for those nuisance parameters. These supplementary functions are then coupled with those for the parameters of interest.

5.9.1 Joint estimation and projected estimating function

Using notation similar to that in Section 2.1, we write $\theta = (\lambda^\mathsf{T}, \psi^\mathsf{T})^\mathsf{T}$, but here we put the nuisance parameters λ first, followed by ψ, the parameters of interest. Consider two unbiased estimating functions, $\Psi_\lambda(\theta; y)$ and $\Psi_\psi(\theta; y)$. The function $\Psi_\lambda(\theta; y)$ is used to estimate λ assuming ψ is known, while $\Psi_\psi(\theta; y)$ is used to estimate ψ assuming λ is known. Define

$$\Psi(\theta; y) = (\Psi_\lambda^\mathsf{T}(\theta; y), \Psi_\psi^\mathsf{T}(\theta; y))^\mathsf{T}.$$

Let $\widehat{\theta} = (\widehat{\lambda}^\mathsf{T}, \widehat{\psi}^\mathsf{T})^\mathsf{T}$ be the solution of

$$\sum_{i=1}^{n} \Psi(\theta; y_i) = 0. \tag{5.42}$$

Assume the conditions in Theorem 5.3, so that $\widehat{\theta}$ is consistent and $\sqrt{n}(\widehat{\theta}-\theta)$ has an asymptotic normal distribution with expected value zero.

Recalling the definition of the projected estimating function in (5.12), solving (5.42) is equivalent to solving

$$\sum_{i=1}^{n} \Psi_{\mathrm{proj}}(\theta; y_i) = 0. \tag{5.43}$$

We partition the projected estimating function as $\Psi_{\mathrm{proj}}(\theta; y) = \left(\Psi_{\mathrm{proj},\lambda}^\mathsf{T}(\theta; y), \Psi_{\mathrm{proj},\psi}^\mathsf{T}(\theta; y)\right)^\mathsf{T}$, and similarly partition its covariance matrix as

$$\begin{aligned}
\Omega_{\lambda\lambda} &= \mathrm{E}\left\{\Psi_{\mathrm{proj},\lambda}(\theta; Y)\Psi_{\mathrm{proj},\lambda}^\mathsf{T}(\theta; Y)\right\}, \\
\Omega_{\psi\lambda} &= \mathrm{E}\left\{\Psi_{\mathrm{proj},\psi}(\theta; Y)\Psi_{\mathrm{proj},\lambda}^\mathsf{T}(\theta; Y)\right\} = \Omega_{\lambda\psi}^\mathsf{T}, \\
\Omega_{\psi\psi} &= \mathrm{E}\left\{\Psi_{\mathrm{proj},\psi}(\theta; Y)\Psi_{\mathrm{proj},\psi}^\mathsf{T}(\theta; Y)\right\},
\end{aligned}$$

and let

$$\Sigma_{\psi\psi} = \Omega_{\psi\psi} - \Omega_{\psi\lambda}\Omega_{\lambda\lambda}^{-1}\Omega_{\lambda\psi}.$$

Theorem 5.13 Assume the regularity conditions in Theorem 5.3. Let $\widehat{\theta} = (\widehat{\lambda}^\mathsf{T}, \widehat{\psi}^\mathsf{T})^\mathsf{T}$ denote the estimator of θ determined by the solution to (5.42), or equivalently, to (5.43). Then as $n \to \infty$,

(a) $$\sqrt{n}(\widehat{\psi} - \psi) \xrightarrow{d} N(0, \Sigma_{\psi\psi}^{-1}),$$

(b) $$\mathrm{avar}(\widehat{\psi}) = \left(\mathrm{var}\left[\Psi_{\mathrm{proj},\psi}(\theta; Y) - \mathrm{proj}\{\Psi_{\mathrm{proj},\psi}(\theta; Y) \mid \Psi_{\mathrm{proj},\lambda}(\theta; Y)\}\right]\right)^{-1}.$$

Consequently,

$$\text{avar}(\widehat{\psi}) \succcurlyeq \left[\text{var}\left\{\Psi_{\text{proj},\psi}(\theta; Y)\right\}\right]^{-1},$$

with equality if and only if $\text{E}\{\Psi_{\text{proj},\psi}(\theta; Y)\Psi^{\mathsf{T}}_{\text{proj},\lambda}(\theta; Y)\} = 0$, where $\text{avar}(\widehat{\psi}) = \Sigma^{-1}_{\psi\psi}$ is the asymptotic covariance matrix of $\sqrt{n}(\widehat{\psi} - \psi)$. □

5.9.2 Profile estimation

One way to address nuisance parameters is to imitate the profile log-likelihood function which was introduced in Section 2.4.

We first estimate λ for fixed ψ by solving for λ in the equation

$$\sum_{i=1}^{n} \Psi_\lambda(\lambda, \psi; y_i) = 0, \tag{5.44}$$

to obtain $\widehat{\lambda}_\psi$, and then solve for ψ in

$$\sum_{i=1}^{n} \Psi_\psi(\widehat{\lambda}_\psi, \psi; y_i) = 0. \tag{5.45}$$

We refer to $\Psi_\psi(\widehat{\lambda}_\psi, \psi; y)$ as a profile estimating function of ψ.

As in Section 2.4, where the profile score function $\partial\ell_{\text{p}}(\psi)/\partial\psi$ does not have zero expected value, the profile estimating function $\Psi_\psi(\widehat{\lambda}_\psi, \psi; y)$ is also not an unbiased estimating function for ψ. However, just as solving the profile score equation gives the same solution $\widehat{\psi}$ as solving the system of paired equations in (2.30), the solution to equation (5.45) for ψ is identical to the component corresponding to ψ in the solution obtained from simultaneously solving for both λ and ψ in equation (5.42).

This profile approach prompts us to examine how estimation of ψ using (5.45) may vary with the value of the nuisance parameter λ. In particular, if the true value λ_0 of λ is known, we would estimate ψ by solving

$$\sum_{i=1}^{n} \Psi_\psi(\lambda_0, \psi; y_i) = 0 \tag{5.46}$$

for ψ.

Theorem 5.14 Assume the regularity conditions of Theorem 5.3. Let $\widetilde{\psi}$ denote the estimator of ψ determined by solving (5.46). Then

(a)
$$\sqrt{n}(\widetilde{\psi} - \psi) \xrightarrow{d} N(0, \widetilde{\Sigma}^{-1}_{\psi\psi}),$$

where $\widetilde{\Sigma}_{\psi\psi} = \text{var}[\text{proj}\{\Psi_{\text{proj},\psi}(\lambda_0, \psi; Y) \mid \Psi_\psi(\lambda_0, \psi; Y)\}]$,

(b)
$$\text{avar}(\widetilde{\psi}) \succcurlyeq \left[\text{var}\{\Psi_{\text{proj},\psi}(\lambda_0, \psi; Y)\}\right]^{-1},$$

with equality if $\Psi_{\psi,\text{proj}}(\lambda_0, \psi; y) = \Psi_{\text{proj},\psi}(\lambda_0, \psi; y)$, where $\text{avar}(\widetilde{\psi})$ denotes the asymptotic covariance matrix of $\sqrt{n}(\widetilde{\psi} - \psi)$. □

Here $\Psi_{\psi,\text{proj}}(\lambda, \psi; y)$ is the projected estimating function of $\Psi_\psi(\lambda, \psi; y)$, and $\Psi_{\text{proj},\psi}(\lambda, \psi; y)$ is the ψ-component of the projected estimating function $\Psi_{\text{proj}}(\lambda, \psi; y)$.

It is evident that the two matrices $\text{avar}(\widetilde{\psi})$ and $\text{avar}(\widehat{\psi})$ are both greater than $[\text{var}\{\Psi_{\text{proj},\psi}(\lambda_0, \psi_0; Y)\}]^{-1}$ in the Loewner order, where λ_0 and ψ_0 are the true values of λ and ψ. However, they do not have an order relationship except in special cases.

Theorem 5.15 Let $\widetilde{\psi}$ and $\widehat{\psi}$ denote the estimators of ψ described in Theorems 5.13 and 5.14, respectively. Assume the regularity conditions in Theorem 5.3. Then

(a) If $\Psi(\theta; y) = \Psi_{\text{proj}}(\theta; y)$, then $\text{avar}(\widehat{\psi}) \succcurlyeq \text{avar}(\widetilde{\psi})$.

(b) If the two components of $\Psi_{\text{proj}}(\theta; y)$ are orthogonal, i.e., $\text{E}\{\Psi_{\text{proj},\psi}(\theta; Y)\Psi^{\mathsf{T}}_{\text{proj},\lambda}(\theta; Y)\} = 0$, then

$$\text{avar}(\widehat{\psi}) \preccurlyeq \text{avar}(\widetilde{\psi}),$$

with equality if and only if $\text{E}\{\Psi_{\psi}(\theta; Y)u^{\mathsf{T}}_{\lambda}(\theta; Y)\} = 0$.

□

In Theorem 5.15 (a), the estimator $\widehat{\psi}$ of ψ obtained from the joint estimation procedure with λ treated as unknown can be less efficient than the estimator $\widetilde{\psi}$ obtained from the profile estimation procedure with λ set to its true value. This pattern holds because the former procedure involves additional variation in the estimation of the nuisance parameter λ. However, if $\Psi(\theta; y)$ differs from its projected estimating function $\Psi_{\text{proj}}(\theta; y)$, this phenomenon is generally untrue. When the two components of the projected function $\Psi_{\text{proj}}(\theta; y)$ are orthogonal, $\widetilde{\psi}$ can be less efficient than $\widehat{\psi}$, as in Theorem 5.15 (b). Furthermore, when $\Psi_{\psi}(\theta; y)$ is orthogonal to the score function $u_{\lambda}(\theta; y)$ for λ, the two estimators $\widehat{\psi}$ and $\widetilde{\psi}$ have the same asymptotic covariance.

This comparison highlights that if λ must be estimated, neglecting the variability induced by estimating λ in the first stage and simply replacing λ with its estimate for (5.42) often results in an invalid variance estimate for the estimator of ψ, leading to distortion in the asymptotic covariance matrix and its estimate. Overlooking the variability from the estimation of λ can yield a larger variance estimate for the ψ estimator than accounting for the variation caused by the first stage estimation of λ. This counterintuitive paradox does not occur with likelihood methods but may arise when using estimating functions.

5.10 Bibliographic notes

Estimating functions provide a broad framework for statistical inference, proving particularly useful when one cannot or chooses not to specify a likelihood function. The formal groundwork traces back to seminal works by Godambe (1960) and Durbin (1960), although the term "estimating function" itself finds roots in Kimball (1946). Estimating function theory has been expanded and refined by numerous authors, including Godambe and Thompson (1974, 1984), Godambe and Heyde (1987), and Jørgensen and Knudsen (2004). Reviews of estimating function theory can be found in Desmond (1997) and Jacod and Sørensen (2018).

The estimating function framework encompasses various extensions of likelihood, including quasi-likelihood (Wedderburn, 1974; McCullagh, 1983), pseudo-likelihood (Gouriéroux et al., 1984), and composite likelihood (Besag, 1974; Lindsay, 1988; Varin, 2008; Cox and Reid, 2004; Lindsay et al., 2011; Varin et al., 2011; Xu and Reid, 2011; Yi, 2017a). The quasi-likelihood function, introduced in Section 5.2, has been used in a wide variety of problems. The initial concept of this method dates back at least to Williams (1959, Section 4.5), with the term "quasi-likelihood" coined by Wedderburn (1974). Further development of the theory quasi-likelihood inference was undertaken in McCullagh (1983), Firth (1987), McCullagh and Nelder (1989, Section 9.3), and McCullagh (1991, Section 11.7), among others. Extensions of the quasi-likelihood method include local quasi-likelihood (Tibshirani and Hastie, 1987;

Firth et al., 1991) and extended quasi-likelihood (Pregibon, 1984; Nelder and Pregibon, 1987; Godambe and Thompson, 1989). Overviews of quasi-likelihood inference are provided by McCullagh (1986, 1991) and Firth (1993b).

Research on estimating functions can be divided into two main categories: the study of estimating functions themselves and the analysis of estimators derived from them. The former focuses on studying properties from a finite sample viewpoint, as summarized in works such as Godambe (1991) and Heyde (1997). The latter concentrates on developing properties from an asymptotic perspective. Despite this distinction, these categories are closely linked because optimal estimating functions often yield efficient estimators. While most estimating functions only allow asymptotic properties of the resulting estimators to be established, the derived approximations can still be useful in finite samples. Both aspects are discussed in this chapter, with the properties of estimators presented in Sections 5.4 and 5.9, and characteristics of estimating functions in other sections.

This chapter draws on material from Heyde (1997), Mukhopadhyay (2004), Tsiatis (2006), and Yi (2017b). The definition of the projected estimating function in Definition 5.2 differs from that of Heyde (1997, Chapter 1) who used the term "projected" for what we call the standardized estimating function. Further discussion of $\Psi_{\text{std}}(\theta; y)$ and $\Psi_{\text{proj}}(\theta; y)$ may be found in Godambe (1991), Desmond (1997), and Henmi and Eguchi (2004).

In terms of estimating function properties, Theorem 5.1 was obtained by Godambe (1960) for scalar parameters and extended to multiple parameters by Bhapkar (1972). Theorem 5.10 was established by Godambe (1985). Discussions on multiple roots of estimating functions are available in Hanfelt and Liang (1995), Heyde (1997, Chapter 13), and Heyde and Morton (1998).

For estimators derived from unbiased estimating functions, Theorem 5.3 provides theoretical support for inference on model parameters. Consistency in Theorem 5.3 (a) follows van der Vaart (1998, Lemma 5.10), while asymptotic normality in Theorem 5.3 (b) is discussed in van der Vaart (1998, Section 5.3), along with related regularity conditions. Regularity conditions ensuring Theorem 5.3 are not unique; sufficient conditions, though not necessarily minimal, are often imposed in various settings. Yuan and Jennrich (1998) established conditions that ensure the existence, consistency, and asymptotic normality of the estimators derived from estimating functions. Further discussions on these conditions are available in Godambe and Thompson (1989), Newey and McFadden (1994), Heyde (1997), and Jørgensen and Knudsen (2004), among others. Consistency of M-estimators under weaker conditions, in particular suitable for regression models conditional on the observed covariates, is established in Bücher et al. (2025).

In applying Theorem 5.3 for data analysis, the unbiasedness of estimating functions is often essential. Yanagimoto and Yamamoto (1991) provided examples illustrating the importance of unbiasedness in estimating equations and its relevance to conditional likelihood inference within exponential families. For situations where estimating functions are biased, Yi and Reid (2010) and Yi (2017b) discussed the issues and introduced strategies for using easily constructed but biased estimating functions. Sections 5.7 and 5.8 outline methods to derive unbiased estimating functions from biased ones caused by data quality issues, such as missing observations or measurement errors.

Estimating functions derived using Theorem 5.9 have been extended in various directions and appear in different forms, including the generalized estimating equations (GEE) method (Liang and Zeger, 1986; Hardin and Hilbe, 2013), GEE2 (Prentice, 1988; Liang et al., 1992), and quadratic inference functions (Qu et al., 2000). However, GEE estimators are not necessarily consistent with an arbitrary working correlation structure when there are time-dependent covariates, as discussed in Example 5.5, which is based on the discussions by Pepe and Anderson (1994) and Lai and Small (2007), among others. The sufficient condition

(5.16), if holding in the error-free context, can be distorted in the presence of covariate measurement error, as noted by Yi et al. (2012).

The concept of estimating equations is relevant to both the method of moments (Pearson, 1936) and the generalized method of moments (Hansen, 1982), with extensions provided by authors such as Hansen (1985), Chamberlain (1987), and Kuersteiner (2002). Comprehensive discussions are available in Hall (2005) and Jesus and Chandler (2011). Hansen (2001) provides insights into the interplay between the econometric and statistical literature. Section 5.6 outlines the connection between unbiased estimating functions and the generalized method of moments.

Section 5.7 discusses extended versions of estimating function (5.26) that incorporate inverse probability weighting to address missing data effects, applicable to data types such as longitudinal data (Robins et al., 1995) and longitudinal data arising in clusters (Yi and Cook, 2002). The doubly robust approach in Section 5.7.3 can be found in Bang and Robins (2005, Chapter 7) and Tsiatis (2006, Chapter 7).

The methods in Section 5.8, while also addressing missing data, focus on measurement error or misclassification. The subtraction correction method in Section 5.8.1 is useful in settings where $E\{\Psi_B(\theta; Y, X, Z)\}$ has a manageable form, as explored by Yan and Yi (2016) and Yi et al. (2018). An approximation to $E\{\Psi_B(\theta; Y, X, Z)\}$ was suggested by McCullagh and Tibshirani (1990). The expected estimating equations approach in Section 5.8.2 is commonly used for data with missing observations or measurement error; for applications, see, e.g., Pepe (1992), Pepe et al. (1994), Wang and Pepe (2000), Wang et al. (2008), Yi et al. (2010), and Wang and Song (2013). Additional applications of estimating functions are documented in Yi (2017b).

Section 5.9 discusses the use of estimating equations when nuisance parameters are present. Interestingly, and somewhat counterintuitively, ignoring the variability induced by estimating λ can sometimes result in a larger variance estimate for the estimator of ψ than when this variability is accounted for. This paradox, observed by authors such as Robins et al. (1994) and Ning et al. (2017), typically does not occur with likelihood-based methods (Pierce, 1982). Newey and McFadden (1994, Section 6) provided a detailed discussion of this phenomenon, and Henmi and Eguchi (2004) offered a geometric explanation, along with a theoretical justification summarized in Theorems 5.13 and 5.14.

This chapter primarily centres on constructing estimating functions using marginal features of the underlying distribution, with a brief discussion on conditional estimating functions highlighted by Theorem 5.10. Applications of this theorem are available in the literature, as seen in conditional least squares estimation methods by Klimko and Nelson (1978) and Godambe (1985).

5.11 Further results and exercises

5.1. In Example 5.4, show that the estimating function $\Psi(\theta; y_i)$ is unbiased and the resulting estimate of θ is given by (5.15).

5.2. Consider the estimating functions in Sections 5.3 and 5.5.

(a) Verify (5.11).

(b) Verify (5.14).

(c) Verify the two identities in (5.18).

5.3. Suppose θ is an unknown scalar parameter, and let $\mathcal{G}^*$ be a subset of $\mathcal{G}$.

(a) Suppose $\Psi_1^*(\theta; y) \in \mathcal{G}$ and $\Psi_2^*(\theta; y) \in \mathcal{G}$ are both optimal estimating functions. Let

$$\Psi_+^*(\theta; y) = \Psi_1^*(\theta; y) + \Psi_2^*(\theta; y)$$
$$\Psi_-^*(\theta; y) = \Psi_1^*(\theta; y) - \Psi_2^*(\theta; y).$$

Compare the variances of $\Psi_+^*(\theta; y)$ and $\Psi_-^*(\theta; y)$ with those of $\Psi_1^*(\theta; y)$ and $\Psi_2^*(\theta; y)$.

(b) If $\Psi_1^*(\theta; y)$ and $\Psi_2^*(\theta; y)$ are both optimal in $\mathcal{G}^*$, show that their standardized versions satisfy

$$\Psi_{1,\text{std}}^*(\theta; Y) = \Psi_{2,\text{std}}^*(\theta; Y)$$

almost everywhere in probability for all $\theta \in \Theta$, under regularity conditions to be stated.

(c) Can the results in (a) and (b) be extended to the case where θ is a vector?

(Godambe, 1976; Kale, 1987)

5.4. Can Theorem 5.7 be generalized to the case where θ is a vector?

5.5. This question compares definitions of optimality for estimating functions $\Psi^*(\theta; y) \in \mathcal{G}$.

(a) If $\Psi^*(\theta; y)$ is T-optimal, and there exists an M-optimal estimating function, show that $\Psi^*(\theta; y)$ is M-optimal.

(b) If $\Psi^*(\theta; y)$ is D-optimal, and there exists an M-optimal estimating function, show that $\Psi^*(\theta; y)$ is M-optimal.

5.6. Suppose the $b_j(\theta; y)$ are elementary estimating functions for $j = 1, \ldots, m$, and for each j, there is a set of conditioning variables $\mathcal{A}_j = \mathcal{A}_j(\theta; y)$ such that

$$\mathrm{E}\{b_j(\theta; Y) \mid \mathcal{A}_j\} = 0,$$

and

$$\mathrm{E}\{b_j(\theta; Y) b_k(\theta; Y) \mid \mathcal{A}_j, \mathcal{A}_k\} = 0, \ j \neq k.$$

(a) Let $C_j(\mathcal{A}_j, \theta)$ denote any function of $\mathcal{A}_j$ and θ, and let $\Psi(\theta; y)$ denote a linear combination of the elementary functions constructed using $C_j(\mathcal{A}_j, \theta)$,

$$\Psi(\theta; y) = \sum_{j=1}^{m} C_j(\mathcal{A}_j, \theta) b_j(\theta; y).$$

Show that

$$\mathrm{E}\{\Psi(\theta; Y)\Psi^{\mathsf{T}}(\theta; Y)\} = -\mathrm{E}\left\{\frac{\partial \Psi(\theta; Y)}{\partial \theta}\right\}.$$

(b) Prove Theorem 5.10. *(Godambe, 1985)*

5.7. Suppose $\{Y_1, Y_2, \ldots\}$ is an autoregressive process of order 1,

$$Y_j = \theta Y_{j-1} + \epsilon_j, \quad j = 1, 2, \ldots,$$

where $Y_0 = 0$ and ϵ_j are independent increments with the expected value zero and common variance σ^2.

Let $b_j(\theta; y_j, y_{j-1}) = y_j - \theta y_{j-1}$ for $j = 1, 2, \ldots,$. Show that for a given m,

$$\Psi^*(\theta; y_1, \ldots, y_m) = -\sum_{j=1}^{m} \frac{y_{j-1}(y_j - \theta y_{j-1})}{\sigma^2}$$

is an optimal estimating function for θ in the class of estimating functions of the form

$$\Psi(\theta;\, y_1, \ldots, y_m) = \sum_{j=1}^{m} a_j(y_0, \ldots, y_{j-1})\, b_j(\theta;\, y_j, y_{j-1}),$$

where $a_j(y_0, \ldots, y_{j-1})$ are arbitrary functions of the past observations. Discuss the role of σ^2.

5.8. Consider the missing data structure introduced in Section 5.7.

(a) Develop an asymptotic distribution for the estimator $\widehat{\theta}$ from (5.29). What conditions do you need?

(b) Develop an asymptotic distribution for the estimator $\widehat{\theta}_{\mathrm{A}}$ from (5.31). What conditions do you need?

(c) Compare the asymptotic relative efficiency of $\widehat{\theta}$ and $\widehat{\theta}_{\mathrm{A}}$.

(d) Develop an asymptotic distribution for the estimator derived from the doubly robust estimating function (5.33). What conditions do you need?

5.9. Consider the structure of error-contaminated data introduced in Section 5.8.1.

(a) Verify (5.38).

(b) Verify (5.39).

(c) Verify (5.40).

5.10. In Section 5.8.2 we assumed that Y and Z are observed precisely, but that only a surrogate X^* of the true measurement X is available.

(a) Let $\widehat{\theta}$ denote the estimator of θ from (5.41). Develop an asymptotic distribution for $\widehat{\theta}$. What conditions do you need?

(b) Let $\widetilde{\theta}$ denote the estimator of θ obtained from solving $\Sigma_{j\in\mathcal{V}}\Psi(\theta; Y_j, X_j, Z_j) = 0$ for θ. Develop an asymptotic distribution for $\widetilde{\theta}$. What conditions do you need?

(c) Compare the asymptotic relative efficiency of $\widehat{\theta}$ and $\widetilde{\theta}$.

5.11. Suppose h and g are two random column vectors of the same dimension with expectation zero, and $\mathrm{E}(gg^{\mathsf{T}})$ is invertible. Show that

$$\mathrm{E}(hg^{\mathsf{T}})\{\mathrm{E}(gg^{\mathsf{T}})\}^{-1}\mathrm{E}(gh^{\mathsf{T}}) \preccurlyeq \mathrm{E}(hh^{\mathsf{T}}).$$

Consequently, $\mathrm{E}\{\mathrm{proj}(h \mid g)h^{\mathsf{T}}\} \preccurlyeq \mathrm{E}(hh^{\mathsf{T}})$.

5.12. Suppose $\Psi(\theta; y)$ and $\Psi^*(\theta; y)$ are unbiased estimating functions, and $\mathrm{E}\{\Psi(\theta; y)\Psi^{\mathsf{T}}(\theta; y)\}$ is invertible.

(a) Show that

$$\begin{aligned}&\mathrm{E}\left(\mathrm{proj}\{\Psi^*(\theta;Y) \mid \Psi(\theta;Y)\}[\mathrm{proj}\{\Psi^*(\theta;Y) \mid \Psi(\theta;Y)\}]^{\mathsf{T}}\right)\\ &\quad= \mathrm{E}\left[\mathrm{proj}\{\Psi^*(\theta;y) \mid \Psi(\theta;Y)\}\Psi^{*\mathsf{T}}(\theta;Y)\right]\end{aligned}$$

(b) Verify that $\mathrm{var}\{\Psi^*(\theta;Y)\} \succcurlyeq \mathrm{var}\left[\Psi^*(\theta;Y) - \mathrm{proj}\{\Psi^*(\theta;Y) \mid \Psi(\theta;Y)\}\right]$.

5.13. In Section 5.9, under the model parameter decomposition $\theta = \left(\lambda^{\mathsf{T}}, \psi^{\mathsf{T}}\right)^{\mathsf{T}}$, consider the corresponding decompositions $u(\theta; y) = (u_\lambda^{\mathsf{T}}(\theta; y), u_\psi^{\mathsf{T}}(\theta; y))^{\mathsf{T}}$, $\Psi(\theta; y) = (\Psi_\lambda^{\mathsf{T}}(\theta; y), \Psi_\psi^{\mathsf{T}}(\theta; y))^{\mathsf{T}}$, and $\Psi_{\text{proj}}(\theta; y) = (\Psi_{\text{proj},\lambda}^{\mathsf{T}}(\theta; y), \Psi_{\text{proj},\psi}^{\mathsf{T}}(\theta; y))^{\mathsf{T}}$ for the score function $u(\theta; y)$, unbiased estimating function $\Psi(\theta; y)$, and its projection function, respectively. Let

$$\Psi_{\psi,\text{proj}}(\lambda_0, \psi; y) = \left[-\mathrm{E}\left\{\frac{\partial \Psi_\psi(\lambda_0, \psi; Y)}{\partial \psi^{\mathsf{T}}}\right\}\right] [\mathrm{E}\{\Psi_\psi(\lambda_0, \psi; Y)\Psi_\psi^{\mathsf{T}}(\lambda_0, \psi; Y)\}]^{-1}\Psi_\psi(\lambda_0, \psi; y),$$

where λ_0 denotes the true value of the parameter λ.

(a) Show that $\Psi_{\text{proj},\psi}(\theta; y) = \text{proj}\{u_\psi(\theta; y) \mid \Psi(\theta; y)\}$.

(b) Show that $\Psi_{\psi,\text{proj}}(\lambda_0, \psi; y) = \text{proj}\{\Psi_{\text{proj},\psi}(\lambda_0, \psi; y) \mid \Psi_\psi(\lambda_0, \psi; y)\}$.

(c) Show that $\text{var}(Z) = \text{var}\{\text{proj}(Z \mid W)\} + \text{var}\{Z - \text{proj}(Z \mid W)\}$, where $Z = \Psi_{\text{proj},\psi}(\lambda_0, \psi; Y)$ and $W = \Psi_\psi(\lambda_0, \psi; Y)$.

Appendix 5.A Additional regularity conditions

Consistency of the root of the equation $\Psi_n(\theta) = \sum_{i=1}^n \Psi(\theta; Y_i) = 0$ in Theorem 5.3, denoted by $\widehat{\theta}$, can be established under suitable regularity conditions on the function $\Psi_n(\theta)$. Identifying conditions that are both necessary and sufficient is generally difficult, and hence sufficient conditions are commonly imposed. The consistency arguments here follow van der Vaart (1998, Section 5.3), which provides conditions that are sufficient but not necessary.

When θ is scalar, van der Vaart (1998, Lemma 5.10) states that if $\Psi_n(\theta) \xrightarrow{p} \Psi_0(\theta)$ as $n \to \infty$ for every θ, and either (i) $\Psi_n(\theta)$ is continuous with a unique root $\widehat{\theta}$, or (ii) $\Psi_n(\theta)$ is non-decreasing with $\Psi_n(\widehat{\theta}) = o_p(1)$, then $\widehat{\theta} \xrightarrow{p} \theta_0$ as $n \to \infty$, where θ_0 satisfies, for every $\epsilon > 0$,

$$\Psi_0(\theta_0 - \epsilon) < 0 < \Psi_0(\theta_0 + \epsilon).$$

Here, $\Psi_0(\theta)$ is a deterministic function of θ.

For vector-valued parameters θ, stronger conditions on the convergence of $\Psi_n(\theta)$ and on the structure of $\Psi_0(\theta)$ are required; in particular, van der Vaart (1998, Theorem 5.9) shows that if

$$\sup_\theta \|\Psi_n(\theta) - \Psi_0(\theta)\| \xrightarrow{p} 0 \quad \text{and} \quad \inf_{\theta: d(\theta, \theta_0) < \epsilon} \|\Psi_0(\theta)\| > 0 = \|\Psi_0(\theta_0)\|$$

for some θ_0 and every $\epsilon > 0$, then for any sequence $\widehat{\theta}_n$ satisfying $\Psi_n(\widehat{\theta}_n) = o_p(1)$,

$$\widehat{\theta}_n \xrightarrow{p} \theta_0 \quad \text{as} \quad n \to \infty,$$

where $d(\cdot, \cdot)$ denotes a distance on the parameter space and $\|\cdot\|$ is the associated norm. Thus, consistency of $\widehat{\theta}$ follows.

Once consistency is established, conditions on $\Psi_n(\theta)$ and $f(y; \theta)$ sufficient to derive convergent Taylor series expansions are typically imposed to derive the asymptotic distribution of $\widehat{\theta}$; see van der Vaart (1998, Section 5.3).

Appendix 5.B Proofs of theorems

Proof of Theorem 5.1

Define $b(\theta; y) = (u^\mathsf{T}(\theta; y)\ \Psi^\mathsf{T}(\theta; y))^\mathsf{T}$. Let $\mathcal{I}_{p\times p}$ denote the $p \times p$ identity matrix, and $0_{p\times p}$ the $p \times p$ zero matrix. Then by (5.11),

$$\mathrm{E}\left\{b(\theta;Y)b^\mathsf{T}(\theta;Y)\right\} = \begin{bmatrix} \mathrm{E}\left\{u(\theta;Y)u^\mathsf{T}(\theta;Y)\right\} & \mathrm{E}\left\{u(\theta;Y)\Psi^\mathsf{T}(\theta;Y)\right\} \\ \mathrm{E}\left\{\Psi(\theta;Y)u^\mathsf{T}(\theta;Y)\right\} & \mathrm{E}\left\{\Psi(\theta;Y)\Psi^\mathsf{T}(\theta;Y)\right\} \end{bmatrix} = \begin{bmatrix} i(\theta) & \Gamma_\Psi^\mathsf{T}(\theta) \\ \Gamma_\Psi(\theta) & \Sigma_\Psi(\theta) \end{bmatrix},$$

whose determinant is non-negative.

Left-multiplying and right-multiplying this matrix by the nonsingular matrix

$$M = \begin{bmatrix} \mathcal{I}_{p\times p} & -\Gamma_\Psi^\mathsf{T}(\theta)\Sigma_\Psi^{-1}(\theta) \\ 0_{p\times p} & \mathcal{I}_{p\times p} \end{bmatrix}.$$

and its transpose, respectively, we obtain that

$$M\mathrm{E}\{b(\theta;Y)b^\mathsf{T}(\theta;Y)\}M^\mathsf{T} = \begin{bmatrix} i(\theta) - \Gamma_\Psi^\mathsf{T}(\theta)\Sigma_\Psi^{-1}(\theta)\Gamma_\Psi(\theta) & 0 \\ 0 & \Sigma_\Psi(\theta) \end{bmatrix}.$$

Since $M\mathrm{E}\{b(\theta;Y)b^\mathsf{T}(\theta;Y)\}M^\mathsf{T}$ is non-negative definite and block diagonal, each diagonal block must then be non-negative definite. In particular,

$$i(\theta) - \Gamma_\Psi^\mathsf{T}(\theta)\Sigma_\Psi^{-1}(\theta)\Gamma_\Psi(\theta) \succcurlyeq 0,$$

which, by the definition of the Godambe information matrix $G_\Psi(\theta) = \Gamma_\Psi(\theta)\Sigma_\Psi^{-1}(\theta)\Gamma_\Psi^\mathsf{T}(\theta)$, is equivalent to $G_\Psi(\theta) \preccurlyeq i(\theta)$ for all $\theta \in \Theta$.

□

Proof of Theorem 5.7

By (5.11),

$$\mathrm{E}_\theta\left\{\frac{\partial\Psi(\theta;Y)}{\partial\theta}\right\} = -\mathrm{E}_\theta\{\Psi(\theta;Y)u(\theta;Y)\}.$$

Since $u(\theta; y)$ is unbiased,

$$\mathrm{cor}_\theta\{\Psi(\theta;Y), u(\theta;Y)\} = \frac{\mathrm{E}_\theta\{\Psi(\theta;Y)u(\theta;Y)\}}{\sqrt{\mathrm{var}_\theta\{\Psi(\theta;Y)\}}\sqrt{\mathrm{var}_\theta\{u(\theta;Y)\}}}.$$

Thus it suffices to show that

$$\left|\frac{\mathrm{E}_\theta\{\Psi(\theta;Y)u(\theta;Y)\}}{\sqrt{\mathrm{var}_\theta\{\Psi(\theta;Y)\}}\sqrt{\mathrm{var}_\theta\{u(\theta;Y)\}}}\right| \le \left|\frac{\mathrm{E}_\theta\{\Psi^*(\theta;Y)u(\theta;Y)\}}{\sqrt{\mathrm{var}_\theta\{\Psi^*(\theta;Y)\}}\sqrt{\mathrm{var}_\theta\{u(\theta;Y)\}}}\right|,$$

which is equivalent to showing that

$$\left|\frac{-\mathrm{E}_\theta\left\{\dfrac{\partial\Psi(\theta;Y)}{\partial\theta}\right\}}{\sqrt{\mathrm{E}_\theta[\{\Psi(\theta;Y)\}^2]}}\right| \le \left|\frac{-\mathrm{E}_\theta\left\{\dfrac{\partial\Psi^*(\theta;Y)}{\partial\theta}\right\}}{\sqrt{\mathrm{E}_\theta[\{\Psi^*(\theta;Y)\}^2]}}\right|,$$

or

$$\frac{\left[\mathrm{E}_\theta\left\{\dfrac{\partial\Psi(\theta;Y)}{\partial\theta}\right\}\right]^2}{\mathrm{E}_\theta[\{\Psi(\theta;Y)\}^2]} \le \frac{\left[\mathrm{E}_\theta\left\{\dfrac{\partial\Psi^*(\theta;Y)}{\partial\theta}\right\}\right]^2}{\mathrm{E}_\theta[\{\Psi^*(\theta;Y)\}^2]}.$$

This is indeed the definition for $\Psi^*(y;\theta)$ to be the optimal estimating function for θ, and thus the result follows.

□

Proof of Theorem 5.9

Consider the residual from projecting $\Psi^*(\theta;y)$ on $\Psi(\theta;y)$, denoted by $k(\theta;y)=\Psi^*(\theta;y)-\text{proj}\{\Psi^*(\theta;y)\mid\Psi(\theta;y)\}$. By Definition 5.2 and the unbiasedness of $\Psi^*(\theta;y)$,

$$\begin{aligned}k(\theta;y)&=\Psi^*(\theta;y)-\mathrm{E}\{\Psi^*(\theta;Y)\Psi^{\mathsf{T}}(\theta;Y)\}[\mathrm{E}\{\Psi(\theta;Y)\Psi^{\mathsf{T}}(\theta;Y)\}]^{-1}\Psi(\theta;y)\\&=\Psi^*(\theta;y)-\Gamma_\Psi\Sigma_\Psi^{-1}\Psi(\theta;y),\end{aligned}$$

where the second step follows from (5.9) and (5.18).

By (5.18),

$$\begin{aligned}G_{\Psi^*}&=\mathrm{E}\left\{-\frac{\partial\Psi^{*\mathsf{T}}(\theta;Y)}{\partial\theta}\right\}\left[\mathrm{E}\left\{\Psi^*(\theta;Y)\Psi^{*\mathsf{T}}(\theta;Y)\right\}\right]^{-1}\left[\mathrm{E}\left\{-\frac{\partial\Psi^*(\theta;Y)}{\partial\theta^{\mathsf{T}}}\right\}\right]\\&=\mathrm{E}\{\Psi^*(\theta;Y)\Psi^{*\mathsf{T}}(\theta;Y)\}.\end{aligned}\tag{5.47}$$

Then, by the unbiasedness of $\Psi^*(\theta;y)$ and $\Psi(\theta;y)$, the covariance matrix of $k(\theta;Y)$ is

$$\begin{aligned}\mathrm{E}\{k(\theta;Y)k^{\mathsf{T}}(\theta;Y)\}&=\mathrm{E}[\{\Psi^*(\theta;Y)-\Gamma_\Psi\Sigma_\Psi^{-1}\Psi(\theta;Y)\}\{\Psi^{*\mathsf{T}}(\theta;Y)-\Psi^{\mathsf{T}}(\theta;Y)\Sigma_\Psi^{-1}\Gamma_\Psi^{\mathsf{T}}\}]\\&=\mathrm{E}\{\Psi^*(\theta;Y)\Psi^{*\mathsf{T}}(\theta;Y)\}-\Gamma_\Psi\Sigma_\Psi^{-1}\mathrm{E}\{\Psi(\theta;Y)\Psi^{*\mathsf{T}}(\theta;Y)\}\\&\quad-\mathrm{E}\{\Psi^*(\theta;Y)\Psi^{\mathsf{T}}(\theta;Y)\}\Sigma_\Psi^{-1}\Gamma_\Psi^{\mathsf{T}}+\mathrm{E}\{\Gamma_\Psi\Sigma_\Psi^{-1}\Psi(\theta;Y)\Psi^{\mathsf{T}}(\theta;Y)\Sigma_\Psi^{-1}\Gamma_\Psi^{\mathsf{T}}\}\\&=\mathrm{E}\{\Psi^*(\theta;Y)\Psi^{*\mathsf{T}}(\theta;Y)\}-\Gamma_\Psi\Sigma_\Psi^{-1}\Gamma_\Psi^{\mathsf{T}}\\&=G_{\Psi^*}-G_\Psi,\end{aligned}$$

where the third step follows from (5.9) and (5.18), and the last step from (5.47). Therefore, $G_{\Psi^*}-G_\Psi$ is non-negative definite for any $\Psi(\theta;y)\in\mathcal{L}$, suggesting that $\Psi^*(\theta;y)$ is optimal in $\mathcal{L}$. □

Proof of Theorem 5.13

(a) By Theorem 5.3 and the identity (5.14), it is immediate that the asymptotic covariance matrix of $\sqrt{n}(\widehat{\theta}-\theta)$ is

$$\text{avar}(\widehat{\theta})=[\mathrm{E}\{\Psi_{\text{proj}}(\theta;Y)\Psi_{\text{proj}}^{\mathsf{T}}(\theta;Y)\}]^{-1}.$$

Examining this matrix by using the block matrices corresponding to the decomposition $\Psi_{\text{proj}}(\theta;y)=(\Psi_{\text{proj},\lambda}^{\mathsf{T}}(\theta;y),\Psi_{\text{proj},\psi}^{\mathsf{T}}(\theta;y))^{\mathsf{T}}$, we obtain

$$\begin{aligned}\text{avar}(\widehat{\theta})&=[\mathrm{E}\{\Psi_{\text{proj}}(\theta;Y)\Psi_{\text{proj}}^{\mathsf{T}}(\theta;Y)\}]^{-1}\\&=\begin{bmatrix}\mathrm{E}\{\Psi_{\text{proj},\lambda}(\theta;Y)\Psi_{\text{proj},\lambda}^{\mathsf{T}}(\theta;Y)\}&\mathrm{E}\{\Psi_{\text{proj},\lambda}(\theta;Y)\Psi_{\text{proj},\psi}^{\mathsf{T}}(\theta;Y)\}\\\mathrm{E}\{\Psi_{\text{proj},\psi}(\theta;Y)\Psi_{\text{proj},\lambda}^{\mathsf{T}}(\theta;Y)\}&\mathrm{E}\{\Psi_{\text{proj},\psi}(\theta;Y)\Psi_{\text{proj},\psi}^{\mathsf{T}}(\theta;Y)\}\end{bmatrix}^{-1}\\&=\begin{bmatrix}\Omega_{\lambda\lambda}&\Omega_{\lambda\psi}\\\Omega_{\psi\lambda}&\Omega_{\psi\psi}\end{bmatrix}^{-1}=\begin{bmatrix}\Omega^{\lambda\lambda}&\Omega^{\lambda\psi}\\\Omega^{\psi\lambda}&\Omega^{\psi\psi}\end{bmatrix},\end{aligned}$$

say, where

$$\Omega^{\psi\psi}=(\Omega_{\psi\psi}-\Omega_{\psi\lambda}\Omega_{\lambda\lambda}^{-1}\Omega_{\lambda\psi})^{-1}\quad\text{and}\quad\Omega^{\lambda\lambda}=\Omega_{\lambda\lambda}^{-1}+\Omega_{\lambda\lambda}^{-1}\Omega_{\lambda\psi}\Omega^{\psi\psi}\Omega_{\psi\lambda}\Omega_{\lambda\lambda}^{-1}.$$

Thus, the asymptotic covariance matrix of $\sqrt{n}(\widehat{\psi} - \psi)$ is

$$\operatorname{avar}(\widehat{\psi}) = (\Omega_{\psi\psi} - \Omega_{\psi\lambda}\Omega_{\lambda\lambda}^{-1}\Omega_{\lambda\psi})^{-1}. \tag{5.48}$$

(b) Let $Q_{\mathrm{P}} = \Psi_{\mathrm{proj},\psi}(\theta; Y) - \mathrm{proj}\{\Psi_{\mathrm{proj},\psi}(\theta; Y) \mid \Psi_{\mathrm{proj},\lambda}(\theta; Y)\}$. For ease of exposition, we simply write $\Psi_{\mathrm{proj},\psi}(\theta; Y)$ as $\Psi_{\mathrm{P}\psi}$ and $\Psi_{\mathrm{proj},\lambda}(\theta; Y)$ as $\Psi_{\mathrm{P}\lambda}$. Then

$$\mathrm{E}(Q_{\mathrm{P}}Q_{\mathrm{P}}^{\mathsf{T}}) = \mathrm{E}[\{\Psi_{\mathrm{P}\psi} - \mathrm{proj}(\Psi_{\mathrm{P}\psi} \mid \Psi_{\mathrm{P}\lambda})\}\,\{\Psi_{\mathrm{P}\psi} - \mathrm{proj}(\Psi_{\mathrm{P}\psi} \mid \Psi_{\mathrm{P}\lambda})\}^{\mathsf{T}}],$$

and since

$$\Psi_{\mathrm{P}\psi} - \mathrm{proj}(\Psi_{\mathrm{P}\psi} \mid \Psi_{\mathrm{P}\lambda}) = \Psi_{\mathrm{P}\psi} - \mathrm{E}(\Psi_{\mathrm{P}\psi}\Psi_{\mathrm{P}\lambda}^{\mathsf{T}})\{\mathrm{E}(\Psi_{\mathrm{P}\lambda}\Psi_{\mathrm{P}\lambda}^{\mathsf{T}})\}^{-1}\Psi_{\mathrm{P}\lambda},$$

we have

$$\begin{aligned}
\mathrm{E}(Q_{\mathrm{P}}Q_{\mathrm{P}}^{\mathsf{T}}) &= \mathrm{E}(\Psi_{\mathrm{P}\psi}\Psi_{\mathrm{P}\psi}^{\mathsf{T}}) - \mathrm{E}(\Psi_{\mathrm{P}\psi}\Psi_{\mathrm{P}\lambda}^{\mathsf{T}})\{\mathrm{E}(\Psi_{\mathrm{P}\lambda}\Psi_{\mathrm{P}\lambda}^{\mathsf{T}})\}^{-1}\{\mathrm{E}(\Psi_{\mathrm{P}\lambda}\Psi_{\mathrm{P}\psi}^{\mathsf{T}})\}^{\mathsf{T}} \\
&\quad - \mathrm{E}(\Psi_{\mathrm{P}\psi}\Psi_{\mathrm{P}\lambda}^{\mathsf{T}})\{\mathrm{E}(\Psi_{\mathrm{P}\lambda}\Psi_{\mathrm{P}\lambda}^{\mathsf{T}})\}^{-1}\mathrm{E}(\Psi_{\mathrm{P}\lambda}\Psi_{\mathrm{P}\psi}^{\mathsf{T}}) \\
&\quad + \mathrm{E}(\Psi_{\mathrm{P}\psi}\Psi_{\mathrm{P}\lambda}^{\mathsf{T}})\{\mathrm{E}(\Psi_{\mathrm{P}\lambda}\Psi_{\mathrm{P}\lambda}^{\mathsf{T}})\}^{-1}\mathrm{E}(\Psi_{\mathrm{P}\lambda}\Psi_{\mathrm{P}\psi}^{\mathsf{T}}) \\
&= \mathrm{E}(\Psi_{\mathrm{P}\psi}\Psi_{\mathrm{P}\psi}^{\mathsf{T}}) - \mathrm{E}(\Psi_{\mathrm{P}\psi}\Psi_{\mathrm{P}\lambda}^{\mathsf{T}})\{\mathrm{E}(\Psi_{\mathrm{P}\lambda}\Psi_{\mathrm{P}\lambda}^{\mathsf{T}})\}^{-1}\mathrm{E}(\Psi_{\mathrm{P}\lambda}\Psi_{\mathrm{P}\psi}^{\mathsf{T}}) \\
&= \Omega_{\psi\psi} - \Omega_{\psi\lambda}\Omega_{\lambda\lambda}^{-1}\Omega_{\lambda\psi}.
\end{aligned}$$

Noting that $\mathrm{E}(Q_{\mathrm{P}}) = 0$, we conclude that

$$\operatorname{avar}(\widehat{\psi}) = [\operatorname{var}\{\Psi_{\mathrm{P}\psi} - \mathrm{proj}(\Psi_{\mathrm{P}\psi} \mid \Psi_{\mathrm{P}\lambda})\}]^{-1}.$$

Finally, (5.48) implies that $\operatorname{avar}(\hat{\psi}) \succcurlyeq \Omega_{\psi\psi}^{-1}$, with equality if and only if $\Omega_{\psi\lambda}\Omega_{\lambda\lambda}^{-1}\Omega_{\lambda\psi} = 0$, which is equivalent to $\Omega_{\psi\lambda} = 0$. This completes the proof.

□

Proof of Theorem 5.14

(a) By Theorem 5.3 and (5.14), $\sqrt{n}(\widetilde{\psi} - \psi) \xrightarrow{d} N(0, \widetilde{\Sigma}_{\psi\psi}^{-1})$ as $n \to \infty$, where $\widetilde{\Sigma}_{\psi\psi} = \operatorname{var}\{\Psi_{\psi,\mathrm{proj}}(\lambda_0, \psi; Y)\}$. By Exercise 5.13 (b),

$$\widetilde{\Sigma}_{\psi\psi} = \operatorname{var}[\mathrm{proj}\{\Psi_{\mathrm{proj},\psi}(\lambda_0, \psi; Y) \mid \Psi_{\psi}(\lambda_0, \psi; Y)\}].$$

(b) Let $Z = \Psi_{\mathrm{proj},\psi}(\lambda_0, \psi; Y)$ and $W = \Psi_{\psi}(\lambda_0, \psi; Y)$. Then part (a) implies that $\widetilde{\Sigma}_{\psi\psi} = \operatorname{var}\{\mathrm{proj}(Z \mid W)\}$. Applying Exercise 5.13 (c) leads to

$$\operatorname{avar}(\widetilde{\psi}) \succeq [\operatorname{var}\{\Psi_{\mathrm{proj},\psi}(\lambda_0, \psi; Y)\}]^{-1},$$

with equality if and only if $\operatorname{var}\{Z - \mathrm{proj}(Z \mid W)\} = 0$, which is equivalent to $Z = \mathrm{proj}(Z \mid W)$ almost surely. By Exercise 5.13 (b), this means $\Psi_{\mathrm{proj},\psi}(\lambda_0, \psi; y) = \Psi_{\psi,\mathrm{proj}}(\lambda_0, \psi; y)$, which completes the proof.

□

Proof of Theorem 5.15

(a) If the estimating function $\Psi(\theta; y)$ in (5.42) coincides with the projected estimating function $\Psi_{\text{proj}}(\theta; y)$ in (5.43), then by Theorem 5.3 and the identity (5.14), the asymptotic covariance matrix of $\sqrt{n}(\widetilde{\psi} - \psi)$ equals the inverse of $\mathrm{E}\{\Psi_{\text{proj},\psi}(\lambda_0, \psi; Y)\Psi^{\mathsf{T}}_{\text{proj},\psi}(\lambda_0, \psi; Y)\}$. By Theorem 5.13 (b), we obtain that

$$\operatorname{avar}(\widetilde{\psi}) \preccurlyeq \operatorname{avar}(\widehat{\psi}).$$

(b) By the assumption that $\mathrm{E}(\Psi_{\text{proj},\psi}\Psi^{\mathsf{T}}_{\text{proj},\lambda}) = 0$ and Theorem 5.13 (b), we obtain that

$$\operatorname{avar}(\widehat{\psi}) = \left[\operatorname{var}\{\Psi_{\text{proj},\psi}(\lambda_0, \psi; Y)\}\right]^{-1}.$$

Then applying Theorem 5.14 (b) completes the proof.

□

Appendix 5.C Geometric view

Estimating functions and their projected functions can be understood from a geometric viewpoint. Below, we outline some basics, with further details available in Tsiatis (2006, Chapter 2).

Definition 5.16 (Linear space) A class $\mathcal{B}$ of estimating functions of a parameter θ is called a linear space if for any $\Psi_1(\theta; y), \Psi_2(\theta; y) \in \mathcal{B}$ and any constants c_1 and c_2,

$$c_1\Psi_1(\theta; y) + c_2\Psi_2(\theta; y) \in \mathcal{B}.$$

□

Definition 5.17 (Orthogonal estimating functions) Let $\mathcal{U}$ be the class of estimating functions of θ that are unbiased and have finite second moments. For $\Psi_1(\theta; y), \Psi_2(\theta; y) \in \mathcal{U}$, define their covariance inner product by

$$\langle \Psi_1(\theta; Y), \Psi_2(\theta; Y)\rangle = \mathrm{E}\{\Psi_1^{\mathsf{T}}(\theta; Y)\Psi_2(\theta; Y)\}.$$

If $\langle \Psi_1(\theta; Y), \Psi_2(\theta; Y)\rangle = 0$, then $\Psi_1(\theta; y)$ and $\Psi_2(\theta; y)$ are said to be orthogonal. Define the length of $\Psi(\theta; y) \in \mathcal{U}$ to be

$$\|\Psi(\theta; Y)\| = \langle \Psi(\theta; Y), \Psi(\theta; Y)\rangle.$$

□

Theorem 5.18 (Pythagorean theorem) For $\Psi_1(\theta; y), \Psi_2(\theta; y) \in \mathcal{U}$, if $\Psi_1(\theta; y)$ and $\Psi_2(\theta; y)$ are orthogonal, then

$$\|\Psi_1(\theta; Y) + \Psi_2(\theta; Y)\| = \|\Psi_1(\theta; Y)\| + \|\Psi_2(\theta; Y)\|.$$

□

Theorem 5.19 Suppose that $\mathcal{H}$ is a linear subspace of $\mathcal{U}$ that is closed (i.e., contains all its "limit" points). Then for any $\Psi(\theta; y) \in \mathcal{U}$, there exists a unique $\Psi^*(\theta; y) \in \mathcal{H}$ such that

(a) $\Psi^*(\theta; y)$ is the element in $\mathcal{H}$ that is closest to $\Psi(\theta; y)$, i.e.,

$$\|\Psi(\theta; Y) - \Psi^*(\theta; Y)\| \leq \|\Psi(\theta; Y) - k(\theta; Y)\| \text{ for any } k(\theta; y) \in \mathcal{H};$$

(b) $\Psi(\theta; y) - \Psi^*(\theta; y)$ is orthogonal to the subspace $\mathcal{H}$, i.e.,

$$\langle \Psi(\theta; Y) - \Psi^*(\theta; Y), k(\theta; Y) \rangle = 0 \text{ for every } k(\theta; y) \in \mathcal{H}.$$

□

Consequently, $\Psi^*(\theta; y)$ is defined to be the projection of $\Psi(\theta; y)$ onto $\mathcal{H}$, denoted by

$$\text{proj}\{\Psi(\theta; y) \mid \mathcal{H}\}.$$

Definition 5.20 Suppose $\Psi(\theta; Y)$ is a $p \times 1$ random vector with expected value zero. For a positive integer r, define the r-dimensional linear space generated by $\Psi(\theta; y)$ as

$$\mathcal{L}_{\Psi}^{(r)} = \{C\Psi(\theta; y) : C \in \mathbb{R}^{r \times p}\}.$$

□

Definition 5.21 Suppose that $\Psi_1(\theta; Y)$ is a $p \times 1$ random vector and $\Psi_2(\theta; Y)$ is a $q \times 1$ random vector, both with expected value zero, and that $\mathbb{E}\{\Psi_2(\theta; Y)\Psi_2^{\top}(\theta; Y)\}$ is invertible. Define

$$\text{proj}\{\Psi_1(\theta; y) \mid \Psi_2(\theta; y)\} = \mathbb{E}\{\Psi_1(\theta; Y)\Psi_2^{\top}(\theta; Y)\}\left[\mathbb{E}\{\Psi_2(\theta; Y)\Psi_2^{\top}(\theta; Y)\}\right]^{-1}\Psi_2(\theta; y)$$

to be the projection of $\Psi_1(\theta; y)$ onto the linear space $\mathcal{L}_{\Psi_2}^{(p)} = \{C\Psi_2(\theta; y) : C \in \mathbb{R}^{p \times q}\}$. We refer to this simply as the projection of $\Psi_1(\theta; y)$ onto $\Psi_2(\theta; y)$, as in Definition 5.2. □

6

Nonparametric Likelihood

6.1 Nonparametric maximum likelihood estimation

In Chapter 4 we discussed likelihood inference when the assumed model is incorrect. While in some cases likelihood inference remains approximately valid and useful even if it is based on a working model, there are situations where using an incorrect parametric model is unacceptable because it produces estimates that may be excessively inaccurate or even meaningless. Nonparametric methods offer a flexible approach by making fewer assumptions about the functional form of the model or the underlying data distribution. They are considered when specifying a reasonable parametric model is difficult, or when there are concerns about the robustness of inferences to deviations from model assumptions.

Consider a sample of n independent and identically distributed observations $\{y_1, \ldots, y_n\}$ of a random variable with distribution function $F_0(y) = \mathrm{pr}(Y \leq y)$. The empirical distribution function for the sample is

$$\widehat{F}_n(y) = \frac{1}{n}\sum_{i=1}^{n} 1(y_i \leq y), \quad y \in \mathbb{R}, \tag{6.1}$$

where $1(E)$ is the indicator function of the event E. The nonparametric likelihood function of F is defined as

$$L(F) = \prod_{i=1}^{n} \left\{F(y_i) - F(y_i^-)\right\}, \tag{6.2}$$

where $F(y^-) = \lim_{t \to y^-} \mathrm{pr}(Y \leq t)$, so that $\mathrm{pr}(Y = y) = F(y) - F(y^-)$. The nonparametric maximum likelihood estimate of F is $\widehat{F}_n$. A proof of this result is given in Owen (2001, p. 9). Let $z_1, \ldots, z_m$ be $m \leq n$ distinct values in the sample, let $n_j = \Sigma_i 1(y_i = z_j)$, and let $p_j = F(z_j) - F(z_j^-)$ and $\widehat{p}_j = n_j/n$. Suppose that $p_j > 0$ for all j, otherwise $L(F) = 0 < L(\widehat{F}_n)$. Also suppose that $p_j \neq \widehat{p}_j$ for at least one j. Using the fact that $\log x \leq x - 1$, we have

$$\begin{aligned}\log\left\{\frac{L(F)}{L(\widehat{F}_n)}\right\} &= \sum_{j=1}^{m} n_j \log\left(\frac{p_j}{\widehat{p}_j}\right) = n\sum_{j=1}^{m} \widehat{p}_j \log\left(\frac{p_j}{\widehat{p}_j}\right) \\ &\leq n\sum_{j=1}^{m} \widehat{p}_j\left(\frac{p_j}{\widehat{p}_j} - 1\right) = 0.\end{aligned}$$

This result extends to a sample of independent and identically distributed random vectors; see Owen (2001, Chapter 3).

In a typical application, interest focuses on some functional $\theta = T(F)$. Denoting the true value of θ by $\theta_0 = T(F_0)$, the nonparametric maximum likelihood estimator of θ is $\widehat{\theta} = T(\widehat{F}_n)$. Without further constraints, nonparametric maximum likelihood estimators

DOI: 10.1201/9780429090936-6

are classical estimators based on the method of moments. The sample mean $\widehat{\theta} = \overline{Y}$ is the nonparametric maximum likelihood estimator of the expected value $\theta_0 = \mathrm{E}(Y)$ and the sample proportion $\widehat{p}_A = n^{-1}\Sigma_i 1(Y_i \in A)$ is the nonparametric maximum likelihood estimator of the population proportion $p_A = \mathrm{pr}(Y \in A)$, for some $A \subset \mathbb{R}$. More elaborate cases of nonparametric maximum likelihood estimators are summarized in the following examples.

Example 6.1 (The product-limit estimator) Consider the problem of estimating the survivor function $S(t) = \mathrm{pr}(T > t)$ of a non-negative random variable T which may be subject to right-censoring. The observed time is $Y = \min(T, c)$ where c is a censoring time known in advance; see Section 3.6.1. We begin by assuming that the distribution of T is discrete. Suppose that the event could occur at pre-specified failure times $0 = t_0 < t_1 < t_2 < \cdots < t_m < t_{m+1} = \infty$. Define the discrete hazard functions

$$\lambda_1 = \mathrm{pr}(T = t_1), \quad \lambda_i = \mathrm{pr}(T = t_i \mid T > t_{i-1}),\ i > 1.$$

For $t \in [t_i, t_{i+1})$, the survivor function is

$$\begin{aligned} S(t) &= \mathrm{pr}(T > t_i) \\ &= \mathrm{pr}(T > t_i \mid T > t_{i-1})\mathrm{pr}(T > t_{i-1} \mid T > t_{i-2}) \cdots \mathrm{pr}(T > t_1) \\ &= \prod_{\{i:t_i \le t\}} (1 - \lambda_i). \end{aligned}$$

Consider a sample of n units. Denote by d_i the number of units that fail at time t_i and with r_i the number of units at risk at time t_i, that is, units that either did not have an event before time t_i or were not censored before or at that time; that is

$$d_i = \sum_{j=1}^{n} \delta_j 1(Y_j = t_i), \quad r_i = \sum_{j=1}^{n} 1(Y_j \ge t_i),$$

where δ_j is equal to one if unit j was not censored and to zero otherwise.

The log-likelihood function for the vector of hazard constants $\lambda = (\lambda_1, \ldots, \lambda_m)^{\mathsf{T}}$ is

$$\ell(\lambda) = \sum_{i=1}^{m} \{d_i \log \lambda_i + (r_i - d_i) \log(1 - \lambda_i)\}. \tag{6.3}$$

Maximizing $\ell(\lambda)$ over λ_i we obtain $\widehat{\lambda}_i = d_i / r_i$, and the nonparametric maximum likelihood estimate of the survivor function is

$$\widehat{S}_n(t) = \prod_{\{i:t_i \le t\}} \left(1 - \frac{d_i}{r_i}\right), \text{ if } t \ge t_1 \quad \text{and} \quad \widehat{S}_n(t) = 1, \text{ if } t < t_1.$$

Kaplan and Meier (1958) showed that the same estimator arises also in the more general case that the distribution of T is not discrete and the t_i are not known in advance. When there is no censoring, $1 - \widehat{S}_n(t)$ coincides with the empirical distribution function (Exercise 6.1).

An approximation to the variance of $\widehat{S}_n(t)$ is derived using the fact that the $\widehat{\lambda}_i$ are asymptotically independent because the observed information matrix is diagonal,

$$-\left.\frac{\partial^2 \ell(\lambda)}{\partial \lambda_i^2}\right|_{\lambda = \widehat{\lambda}} = \frac{r_i}{\widehat{\lambda}_i(1 - \widehat{\lambda}_i)}, \quad -\frac{\partial^2 \ell(\lambda)}{\partial \lambda_i \partial \lambda_j} = 0, \quad i \ne j.$$

It then follows that the variance of $\log \widehat{S}_n(t)$ is asymptotically

$$\mathrm{var}\{\log \widehat{S}_n(t)\} = \mathrm{var}\left\{\textstyle\sum_{\{i:t_i \le t\}} \log(1 - \widehat{\lambda}_i)\right\} \doteq \sum_{\{i:t_i \le t\}} \mathrm{var}\{\log(1 - \widehat{\lambda}_i)\}.$$

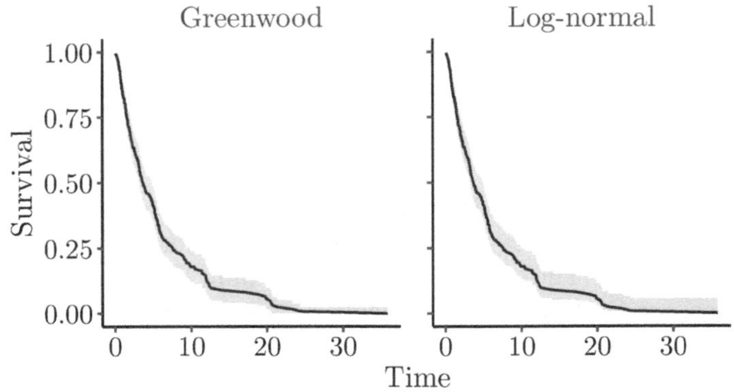

FIGURE 6.1
Kaplan-Meier estimate of the survival curve for a sample of 250 simulated times from a log-normal distribution with $\mu = 1$ and $\sigma = 1$; 25% of the simulated times were randomly chosen as censored times. In the left panel, a 95% approximate confidence band is computed using Greenwood's formula (6.5), truncated at zero to avoid negative lower bounds. In the right panel, a 95% approximate normal confidence band is computed on the log scale as in (6.4) and reported on the original scale.

Using the delta method,

$$\widehat{\text{var}}\{\log \widehat{S}_n(t)\} \doteq \sum_{\{i:t_i \leq t\}} \frac{1}{(1-\widehat{\lambda}_i)^2} \frac{\widehat{\lambda}_i(1-\widehat{\lambda}_i)}{r_i} = \sum_{\{i:t_i \leq t\}} \frac{d_i}{r_i(r_i - d_i)}. \tag{6.4}$$

The uncertainty around $\widehat{S}(t)$ is often estimated by Greenwood's formula

$$\widehat{\text{var}}\{\widehat{S}_n(t)\} = \widehat{S}_n(t)^2 \sum_{\{i:t_i \leq t\}} \frac{d_i}{r_i(r_i - d_i)}, \tag{6.5}$$

also derived by the delta method, as $\text{var}\{\log \widehat{S}_n(t)\} \doteq \text{var}\{\widehat{S}_n(t)\}/\{\widehat{S}_n(t)\}^2$.

It is preferable to compute approximate confidence intervals for $S(t)$ by taking the exponential of the approximate confidence intervals for $\log S(t)$ instead of using Greenwood's formula because the distribution of $\widehat{S}_n(t)$ is more symmetric on the log scale (Davison, 2003, p. 197). Furthermore, confidence intervals based on Greenwood's formula may fall outside $[0, 1]$, in which case they must be truncated. This is illustrated in Figure 6.1, where the pointwise confidence intervals at large values of t appear artificially narrow (left panel) compared to those calculated on the logarithmic scale (right panel). □

Example 6.2 (Nonparametric empirical Bayes) Suppose we observe a sample of n independent counts $\{Y_1, \ldots, Y_n\}$ that are assumed to follow the Poisson distribution, where Y_i has mean θ_i. Bayesian inference proceeds by assuming $\theta_i \sim \pi(\theta)$, for some prior density function $\pi(\cdot)$ with non-negative support. A typical objective is estimation of the posterior means

$$\text{E}(\theta_i \mid y_i) = \frac{\int \pi(\theta) \exp(-\theta)\theta^{(y_i+1)}/y_i!\, d\theta}{\int \pi(\theta) \exp(-\theta)\theta^{y_i}/y_i!\, d\theta}, \quad i = 1, \ldots, n.$$

Robbins (1956) proposed a clever approach to estimate the posterior means without assumptions on the prior density. The posterior means are rewritten as

$$\begin{aligned}\mathrm{E}(\theta_i \mid y_i) &= (y_i+1)\frac{\int \pi(\theta)\exp(-\theta)\theta^{(y_i+1)}/(y_i+1)!\,\mathrm{d}\theta}{\int \pi(\theta)\exp(-\theta)\theta^{y_i}/y_i!\,\mathrm{d}\theta} \\ &= (y_i+1)\frac{f(y_i+1)}{f(y_i)}. \qquad (6.6)\end{aligned}$$

From this, a nonparametric estimate of $\mathrm{E}(\theta_i \mid y_i)$ is

$$\widehat{\mathrm{E}}(\theta_i \mid y_i) = (y_i+1)\frac{\sum_{j=1}^{n} 1(Y_j = y_i+1)}{\sum_{j=1}^{n} 1(Y_j = y_i)}.$$

For this problem, the most commonly used parametric empirical Bayes estimator assumes that the prior distribution for θ_i is a gamma distribution with shape α and scale β; this is a conjugate prior. The posterior distribution for θ_i given y_i is again gamma, with shape $\alpha + y_i$ and scale $\beta/(1+\beta)$. The corresponding posterior expectation is a weighted average of the observed count y_i and the prior expected count,

$$\mathrm{E}(\theta_i \mid y_i) = y_i w + \mathrm{E}(\theta_i)(1-w), \qquad (6.7)$$

with $\mathrm{E}(\theta_i) = \alpha\beta$ and $w = \beta/(1+\beta)$. The marginal distribution of Y_i is negative binomial, with size α and probability w,

$$f(y_i;\alpha,w) = \binom{y_i+\alpha-1}{y_i} w^{y_i}(1-w)^{\alpha}, \quad y_i = 0,1,\ldots.$$

The parametric empirical Bayes estimate of the posterior expected value is obtained by replacing α and β in (6.7) with the maximum likelihood estimates $\widehat{\alpha}$ and $\widehat{\beta}$ obtained by maximizing the negative-binomial likelihood function from the sample $\{y_1,\ldots,y_n\}$.

Figure 6.2 compares Robbins' nonparametric estimate and the parametric gamma-Poisson empirical Bayes estimate using simulated counts. The two top panels of the figure display the estimates with 10,000 counts simulated from the gamma-Poisson model with the shape and scale parameters both equal to 0.5. As expected, the parametric empirical Bayes estimator outperforms Robbins' method, which appears unstable, especially for those values of y that are less frequently observed. Conversely, in the two bottom panels, θ_i are simulated from a uniform distribution in the interval $[0,2]$. In this case, the nonparametric Robbins method performs better than the parametric gamma-Poisson model, in particular for larger values of y where the effect of the grossly misspecified parametric prior model becomes more evident. □

Example 6.3 (Generalized linear mixed models) Let $\{Y_1,\ldots,Y_n\}$ be a sample of vector responses, with $Y_i = (Y_{i1},\ldots,Y_{im_i})^{\mathsf{T}}$, for $i = 1,\ldots,n$. Denote by x_{ij} the vector of covariates associated with Y_{ij}. Given a random intercept b_i, the conditional expectation $\mu_{ij} = \mathrm{E}(Y_{ij} \mid x_{ij}, b_i)$ is assumed to follow the generalized linear model (1.12),

$$g(\mu_{ij}) = x_{ij}^{\mathsf{T}}\beta + b_i. \qquad (6.8)$$

Model (6.8) is an example of a generalized linear mixed model, introduced in Example 1.13 and discussed in more detail in Section 3.2. Usually $\{b_1,\ldots,b_n\}$ are assumed to be uncorrelated normal random variables with zero mean and variance σ^2. One potential concern is the sensitivity of the conclusions to the distributional assumption for the random effects, as there is no data to test this assumption.

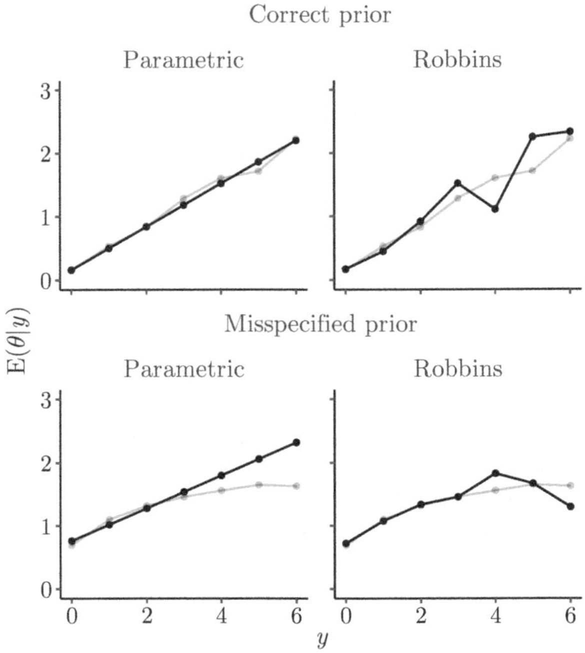

FIGURE 6.2
Comparison between the posterior expectations $\mathrm{E}(\theta_i \mid y_i)$ estimated using a fully parametric empirical Bayesian method which assumes a gamma prior and the nonparametric Robbins' method. The comparison is based on 10,000 simulated counts where the θ_i are generated from either a gamma$(0.5, 0.5)$ distribution (correct prior, top row) or a uniform$(0, 2)$ distribution (misspecified prior, bottom row). The light gray dots denote the true θ_i and black dots represent the estimates.

The likelihood for this generalized linear mixed model is the product of n one-dimensional integrals (3.6) that can be approximated by the Laplace method, Example 3.7, or more accurately with the adaptive Gaussian quadrature method of Liu and Pierce (1994), which combines Laplace approximation and classical Gauss–Hermite quadrature. Rewrite model (6.8) as

$$g(\mu_{ij}) = x_{ij}^{\mathsf{T}}\beta + \sigma\epsilon_i, \tag{6.9}$$

where ϵ_i is a standard normal random variable. The density or probability function of Y_i given ϵ_i and the associated covariates is

$$f(y_i \mid x_i, \epsilon_i; \theta) = \prod_{j=1}^{m_i} f(y_{ij} \mid x_{ij}, \epsilon_i; \theta),$$

where $\theta = (\beta^{\mathsf{T}}, \sigma^2)^{\mathsf{T}}$. The generic quadrature approximation to the likelihood function is

$$L(\theta) = \prod_{i=1}^{n} \int_{-\infty}^{\infty} f(y_i \mid x_i, \epsilon_i; \theta)\phi(\epsilon_i)\,\mathrm{d}\epsilon_i \approx \prod_{i=1}^{n}\sum_{r=1}^{k} w_r f(y_i \mid x_i, \delta_r; \theta), \tag{6.10}$$

for a vector of quadrature nodes $\delta = (\delta_1, \ldots, \delta_k)^{\mathsf{T}}$ and a vector of associated weights $w = (w_1, \ldots, w_k)^{\mathsf{T}}$. The quadrature approximation (6.10) is the exact likelihood for a finite mixture of exponential family densities. The linear predictor for Y_{ij} in the r-th class of the mixture is

$$g(\mu_{ijr}) = x_{ij}^{\mathsf{T}}\beta + \sigma\delta_r, \quad r = 1, \ldots, k. \tag{6.11}$$

Aitkin (1996) suggested avoiding distributional assumptions for the unobserved random effects by replacing the generalized linear mixed model (6.9) with the finite mixture approximation (6.11) treating the quadrature nodes δ and weights w as unknown parameters. Nonparametric maximum likelihood estimation is a standard approach for estimation of finite mixture models; see Example 3.33. The parameter σ is absorbed in the variance of the nodes vector δ so that the mixture model is rewritten as

$$g(\mu_{ijr}) = x_{ij}^{\mathsf{T}}\beta + \alpha_r, \quad r = 1, \ldots, k. \tag{6.12}$$

The parameter α_r is the intercept for the r-th component of the mixture. Model identifiability requires that either the intercept β_0 or one of α_r is omitted, or there is a constraint on the vector α, such as $\Sigma_r\alpha_r = 0$. Estimation conveniently proceeds with an EM algorithm; see Section 3.5.2. Given the current estimates of the vector of weights w, the vector α is estimated together with β by including in the linear predictor a covariate that is a factor variable with k levels and then maximizing a weighted likelihood of a generalized linear model. Given the current estimates of β and α, the vector of weights w is estimated by nonparametric maximum likelihood with the EM iterations detailed in Example 3.33. □

6.2 Nonparametric density estimation

6.2.1 Kernel estimators

Consider a sample of independent variables $\{Y_1, \ldots, Y_n\}$ from a continuous univariate distribution with unknown density $f(y)$. As seen in Section 6.1, the nonparametric maximum likelihood estimator of F is $\widehat{F}_n$. However, we cannot obtain the nonparametric maximum likelihood estimate of $f(y)$ by differentiating $\widehat{F}_n$, as it is a step function whose derivative is zero or undefined.

A nonparametric estimator of the density function at a given point y_0 can instead be obtained by approximating the derivative of $F(y)$ at y_0 with a finite central difference,

$$\widehat{f}_h(y_0) = \frac{1}{2h}\left\{\widehat{F}_n(y_0 + h) - \widehat{F}_n(y_0 - h)\right\} = \frac{1}{2nh}\sum_{i=1}^{n} 1(y_0 - h < Y_i \leq y_0 + h), \tag{6.13}$$

for a suitable bandwidth parameter $h > 0$. The bias is $\mathrm{E}\{\widehat{f}_h(y_0)\} - f(y_0) = (2h)^{-1}\{F(y_0 + h) - F(y_0 - h)\} - f(y_0)$ and thus goes to zero if $h \to 0$ as $n \to \infty$. The variance of $\widehat{f}_h(y_0)$ is

$$\begin{aligned}
\mathrm{var}\{\widehat{f}_h(y_0)\} &= \mathrm{var}\left\{\frac{1}{2nh}\sum_{i=1}^{n} 1(y_0 - h < Y_i \leq y_0 + h)\right\} \\
&= \frac{1}{4nh^2}\mathrm{var}\{1(y_0 - h < Y_1 \leq y_0 + h)\} \\
&= \frac{1}{4nh^2}\{F(y_0 + h) - F(y_0 - h)\}\left[1 - \{F(y_0 + h) - F(y_0 - h)\}\right] \\
&= \frac{f(y_0)}{2nh} + o\{(nh)^{-1}\},
\end{aligned}$$

which converges to zero if $nh \to \infty$ as $n \to \infty$. The mean-squared error of $\widehat{f}_h(y_0) \to 0$ if the bandwidth is chosen so that the bias and variance of $\widehat{f}_h(y_0)$ both tend to zero as n increases, i.e., $\widehat{f}_h(y_0)$ is mean-squared-error consistent.

The tuning parameter h is typically selected to maximize the rate of convergence of the mean-squared error of $\widehat{f}_h(y_0)$ to zero. If $F(y_0 \pm h)$ is smooth enough in a neighbourhood of $h = 0$ to admit four derivatives, then by Taylor expansion

$$F(y_0 \pm h) = F(y_0) \pm f(y_0)h + \frac{1}{2}f'(y_0)h^2 \pm \frac{1}{3!}f''(y_0)h^3 + o(h^4),$$

from which it follows that

$$\frac{1}{2h}\{F(y_0 + h) - F(y_0 - h)\} = f(y_0) + \frac{1}{6}f''(y_0)h^2 + o(h^2),$$

so that the mean square error is

$$\begin{aligned}\text{MSE}\{\widehat{f}_h(y_0)\} &= [\text{E}\{\widehat{f}_h(y_0)\} - f(y_0)]^2 + \text{var}\{\widehat{f}_h(y_0)\}\\ &= \left\{\frac{1}{6}f''(y_0)h^2\right\}^2 + o(h^4) + \frac{f(y_0)}{2nh} + o\{(nh)^{-1}\}.\end{aligned} \tag{6.14}$$

The squared bias is proportional to h^4, while the variance is inversely proportional to nh. Hence, the optimal bandwidth sequence occurs when the squared bias and the variance converge to zero at the same rate, that is when $h^4 \approx (nh)^{-1}$, i.e., $\widehat{h} = O(n^{-1/5})$. The corresponding mean-squared error of $\widehat{f}_h(y_0)$ is $O(n^{-4/5})$.

The nonparametric density estimator (6.13) has the disadvantage of being discontinuous with jumps at $y_i \pm h$, as illustrated in the top left graph of Figure 6.4. Smooth nonparametric density estimation can be achieved by replacing the indicator function in (6.13) with a function that downweights distant observations,

$$\widehat{f}_h(y_0) = \frac{1}{n}\sum_{i=1}^{n} K_h(Y_i - y_0), \tag{6.15}$$

for a weight function that takes the form $K_h(t) = K(t/h)/h$, where $K(\cdot)$ is a kernel function and h is the bandwidth. Kernel functions are symmetric density functions with zero mean and unit variance. Commonly used kernel functions defined in $[-1, 1]$ are the triangular kernel $K(u) = (1 - |u|)$, the Epanechnikov kernel $K(u) = (3/4)(1 - u^2)$, the cosine kernel $K(u) = \pi/4\cos(u\pi/2)$ and the biweight kernel $K(u) = (15/16)(1 - u^2)^2$. The biweight kernel is also known as the Tukey weight function and is used in robust estimation; see Example 4.12. Another commonly used kernel function is the Gaussian density, which unlike the previous kernels has unbounded support. The density estimator (6.13) corresponds to the rectangular kernel $K(u) = 1/2$. Figure 6.3 illustrates the kernel functions.

Example 6.4 (Comparison of kernels) Figure 6.4 shows nonparametric density estimates with a sample of 50 observations from a standard normal distribution. In each panel of the figure a different kernel is used while the bandwidth is estimated for all kernels with the Sheather–Jones method discussed in Section 6.2.2. The rectangular kernel, as already observed, is discontinuous with jumps at $y_i \pm h$. The other kernels produce smooth density estimates with small differences, confirming that the choice of a kernel other than the rectangular one is unlikely to be important.

□

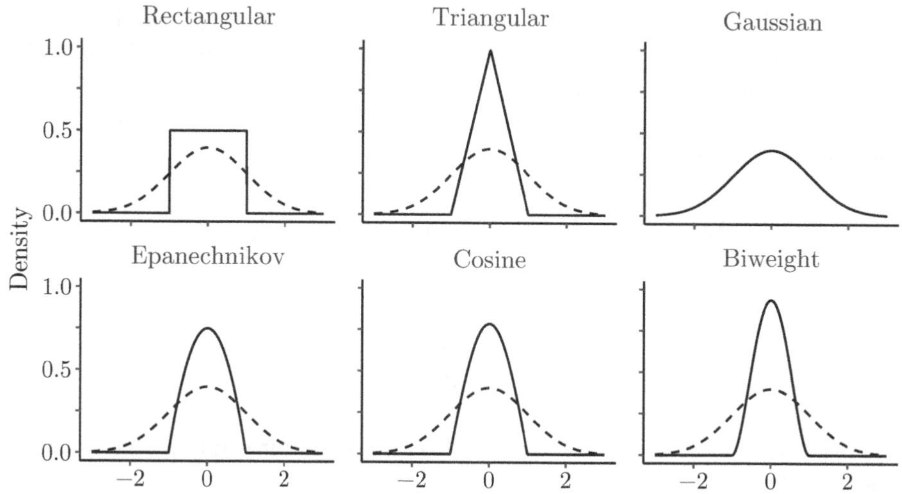

FIGURE 6.3
Some widely used kernel functions. The Gaussian kernel is plotted as a reference curve in the graphs of the other kernels.

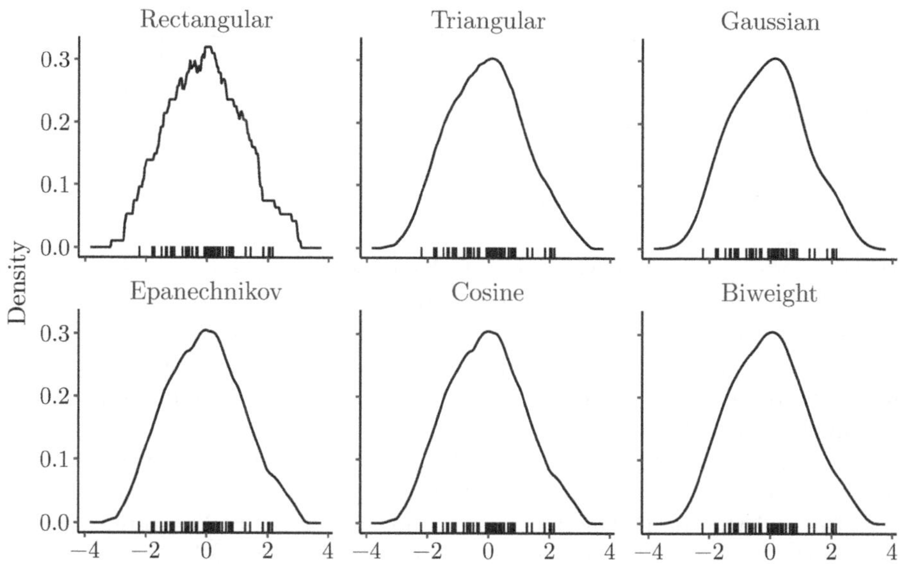

FIGURE 6.4
Nonparametric density estimation using various kernels for a sample of 50 observations drawn from a standard normal distribution.

6.2.2 Bandwidth estimation

If the kernel function has finite fourth moment, then the Taylor expansion used to derive the MSE for the rectangular kernel (6.14) generalizes to any kernel, as

$$\begin{aligned}\text{MSE}\{\widehat{f}_h(y_0)\} &= [\text{E}\{\widehat{f}_h(y_0)\} - f(y_0)]^2 + \text{var}\{\widehat{f}_h(y_0)\} \\ &= \left\{\frac{1}{2}f''(y_0)h^2\sigma_K^2\right\}^2 + o(h^4) + \frac{f(y_0)}{nh}\|K\|_2^2 + o\{(nh)^{-1}\}, \end{aligned} \tag{6.16}$$

where $\sigma_K^2 = \int y^2 K(y)\,dy$ and $\|g\|_2^2 = \int g^2(y)\,dy$ is the squared L^2 norm of $g(\cdot)$. When K is the rectangular kernel, $\sigma_K^2 = 1/3$, $\|K\|_2^2 = 1/2$, and we retrieve (6.14). If instead K is the Gaussian kernel, then $\sigma_K^4 = 1$ and $\|K\|_2^2 = \pi^{-1/2}/2$.

The optimal bandwidth sequence is $\widehat{h} = c_{\text{opt}} n^{-1/5}$. It seems reasonable to choose c_{opt} to minimize the asymptotic mean-squared error,

$$\text{AMSE}\{\widehat{f}_h(y_0)\} = \frac{1}{4}\{f''(y_0)\}^2 h^4 \sigma_K^4 + \frac{f(y_0)}{nh}\|K\|_2^2,$$

which leads to the optimal choice

$$c_{\text{opt}}(y_0) = \left[\frac{\|K\|_2^2 f(y_0)}{\sigma_K^4\{f''(y_0)\}^2}\right]^{1/5}.$$

However, computation of c_{opt} is not feasible because it requires knowledge of the density f and its second derivative.

Calculating a distinct bandwidth at several values of y, such as each observation y_i, is impractical in most applications. Usually either a common bandwidth is determined for all observations or different bandwidths are used for specific observation intervals. A global bandwidth can be chosen by minimizing the asymptotic mean integrated squared error,

$$\text{AMISE}(\widehat{f}_h) = \int_{-\infty}^{\infty} \text{AMSE}\{\widehat{f}_h(y)\}dy = \frac{1}{4}h^4\sigma_K^4\|f''\|_2^2 + \frac{1}{nh}\|K\|_2^2. \tag{6.17}$$

The constant in the optimal global bandwidth that minimizes this is

$$c_{\text{opt}} = \left(\frac{\|K\|_2^2}{\sigma_K^4\|f''\|_2^2}\right)^{1/5}, \tag{6.18}$$

which still depends on the unknown density f. A number of proposals have been formulated to circumvent the need to know f to compute $\widehat{h}$. A simple option is to assume, in the computation of the bandwidth, that f is a normal density with variance σ^2. Then $\|f''\|_2^2 = \{3/(8\pi^{1/2})\}\sigma^{-5}$ (see Exercise 6.4) and $c_{\text{opt}} = (4/3)^{1/5}\sigma = 1.06\sigma$, which requires only estimation of the standard deviation of the data. Scott (1992) suggested the rule of thumb $c_{\text{opt}} = 1.06s$ where s is the sample standard deviation.

However, Silverman (1986) warned against the risk of over-smoothing when f is not normal, recommending instead the rule of thumb $c_{\text{opt}} = 0.9s_{\text{rob}}$, where s_{rob} is a robust estimate of σ and the multiplier 0.9 is an ad hoc adjustment. The robust estimate of σ is $s_{\text{rob}} = \min\{s, \text{IQR}/1.34\}$, where IQR is the empirical interquartile range and $1.34 = \Phi^{-1}(0.75) - \Phi^{-1}(0.25)$ is the interquartile range for the standard normal distribution.

Cross-validation is an alternative approach to bandwidth estimation that does not rely on a reference normal distribution. Consider the integrated squared error

$$\text{ISE}(\widehat{f}_h) = \int_{-\infty}^{\infty}\{\widehat{f}_h(y) - f(y)\}^2 dy = \|\widehat{f}_h\|_2^2 - 2\int \widehat{f}_h(y) f(y)\,dy + \|f\|_2^2.$$

Since the last term of ISE does not depend on the bandwidth h, the unbiased cross-validation method consists in estimating with cross-validation the second term and then minimizing the unbiased cross-validation objective function

$$\text{UCV}(h) = \|\widehat{f}_h\|_2^2 - \frac{2}{n}\sum_{i=1}^{n}\widehat{f}_h^{-i}(y_i),$$

where $\widehat{f}_h^{-i}(y_i)$ is a cross-validation estimate of f at y_i, obtained by excluding the fold containing y_i from the set of observations used for estimation.

The unbiased cross-validation estimate of the bandwidth is very variable and tends to undersmooth, producing artificial fluctuations in the density estimate, as shown in Figure 6.5. A more stable bandwidth estimate is achieved using the biased cross-validation method, which minimizes an estimate of AMISE (6.17) rather than ISE. The AMISE expression (6.17) contains only the unknown term $\|f''\|_2^2$ which naturally suggests estimating it using $\|\widehat{f}_h''\|_2^2$. However, this estimator is biased since

$$\mathrm{E}(\|\widehat{f}_h''\|_2^2) = \|f''\|_2^2 + \frac{\|K''\|_2^2}{nh^5} + O(h^2),$$

which indicates that a more accurate estimate of $\|f''\|_2^2$ is $\|\widehat{f}_h''\|_2^2 - \|K''\|_2^2/(nh^5)$; see Scott and Terrell (1987). The cross-validation estimator of AMISE is the biased cross-validation objective function

$$\begin{aligned}\mathrm{BCV}(h) &= \frac{1}{4}h^4\sigma_K^4\left(\|\widehat{f}_h{''}\|_2^2 - \frac{\|K''\|_2^2}{nh^5}\right) + \frac{1}{nh}\|K\|_2^2, \\ &= \frac{\sigma_K^4}{2n^2h}\sum_{i<j}\phi(c_{ij}) + \frac{1}{nh}\|K\|_2^2, \qquad (6.19)\end{aligned}$$

where $\phi(c) = \int K''(w)K''(w+c)\,dw$ and $c_{ij} = h^{-1}(y_i - y_j)$; see Exercise 6.6. The biased cross-validation estimate of h is the minimizer of $\mathrm{BCV}(h)$. It is called "biased" because it targets the asymptotic MISE rather than the exact MISE. The connection to cross-validation is the exclusion of the diagonal terms ($i = j$) in the summation of the $\phi(c_{ij})$ terms, which implements a leave-one-out approach: the curvature at each point is estimated using only the other observations.

Another approach to bandwidth estimation is the method proposed by Sheather and Jones (1991), which replaces the unknown quantity $\|f''\|_2^2$ in (6.18) with the estimate $\|\widehat{f}_{g(h)}''\|_2^2$, derived using a pilot bandwidth $g(h)$. Subsequently the bandwidth is estimated as the solution of the fixed-point equation

$$h = \left(\frac{\|K\|_2^2}{n\sigma_K^4\|\widehat{f}_{g(h)}''\|_2^2}\right)^{1/5}.$$

In the above expression, the bandwidth $g(h)$ used in estimating $\|f''\|_2^2$ differs from the bandwidth h used in estimating f because the optimal bandwidth naturally depends on the degree of smoothing of the function to be estimated. In fact, Hall and Marron (1987) proved that the estimate of f'' is inconsistent if the optimal bandwidth for estimating f is also used to estimate f''. Some authors report that the Sheather–Jones method is generally preferable to rules of thumb or cross-validation bandwidth estimators; see, for example, Jones et al. (1996) or Venables and Ripley (2002, Section 5.6).

Example 6.5 (Estimation of a gamma density) Figure 6.5 compares bandwidth selection in density estimation with a Gaussian kernel, from a sample of 100 observations simulated from a gamma distribution with shape and scale both equal to 2. Silverman's rule of thumb produces a gamma density reconstruction that is too flat around the mode of the distribution and has some artifacts on the right tail. The bandwidth estimate with Silverman's method is 0.9. Scott's rule of thumb leads to a bandwidth estimate of 1.06 and performs only slightly better than Silverman's method.

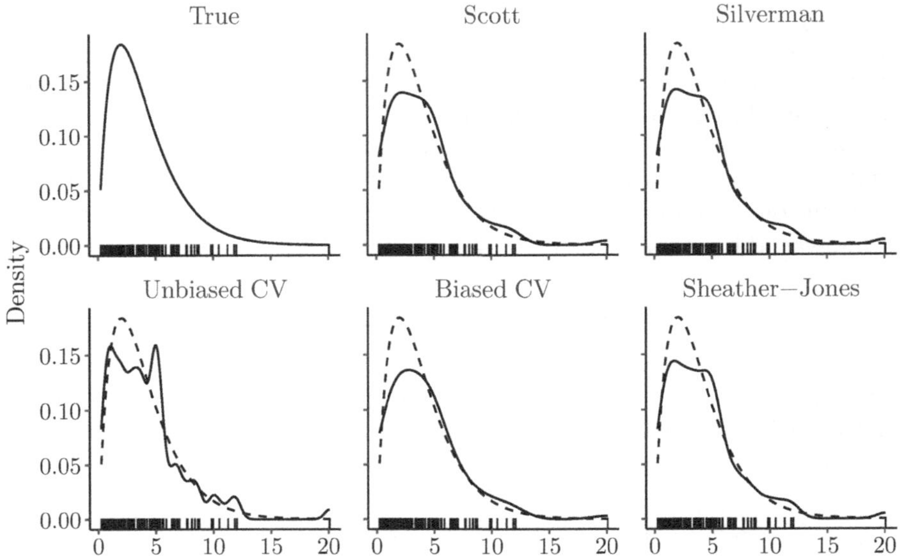

FIGURE 6.5
Nonparametric density estimation for a random sample of 100 observations from a gamma distribution with shape and scale both equal to 2. The top left panel shows the true gamma density, while the other panels show nonparametric density estimates with a Gaussian kernel computed with different bandwidth estimates and the true gamma density drawn dashed for reference.

The excessive variability of the unbiased cross-validation method is evident, leading to a very irregular density reconstruction. With the simulated data, the unbiased cross-validation bandwidth estimate is equal to only 0.45. The biased cross-validation is much more stable and produces a reasonable estimate of the gamma density. The biased cross-validation estimate of the bandwidth is 1.33. The Sheather–Jones plugin method provides a bandwidth estimate of 0.79, thus producing a density estimate similar to that obtained with the Silverman method. □

Example 6.6 (Estimation of a multi-modal density) Consider a random sample of 1000 observations from the normal mixture density

$$\frac{1}{2}N(0,1) + \frac{1}{10}\sum_{j=0}^{4} N(j/2-1, 1/10).$$

Marron and Wand (1992) referred to this mixture density as "the claw"; Wasserman (2006) called it the "Bart Simpson" density. Figure 6.6 illustrates that the Scott and Silverman rules of thumb, along with the biased cross-validation method, oversmooth the density estimate. Bandwidth estimates with these methods range from 0.2 to 0.25. In contrast, unbiased cross-validation and the Sheather–Jones plug-in method successfully capture the multimodal shape of the density function. The bandwidth estimate of the unbiased cross-validation is 0.05 and that of the Sheather–Jones method is 0.06.

□

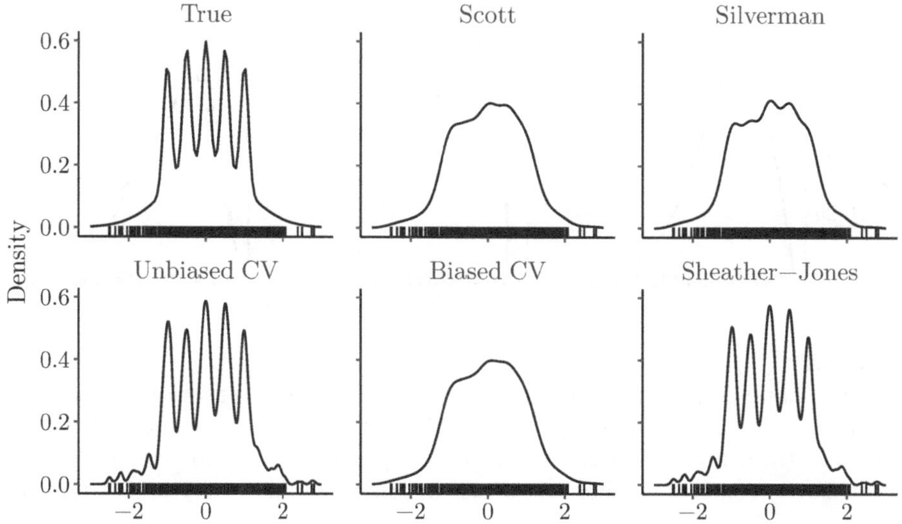

FIGURE 6.6
Nonparametric density estimation for a random sample of 1000 observations from a normal mixture. The top left panel shows the true density function, while the other panels show nonparametric density estimates with a Gaussian kernel computed with different bandwidth estimates.

6.2.3 Shape-constrained density estimators

While highly flexible, nonparametric techniques like kernel density estimation cannot guarantee desirable qualitative properties in the estimated density, such as unimodality or specific curvature. Furthermore, a significant practical challenge in kernel density estimation lies in bandwidth selection, which influences the smoothness of the resulting estimate. Shape-constrained density estimators address these limitations by incorporating prior knowledge about the form of the true underlying density, thereby circumventing the need for a subjective or data-driven bandwidth choice.

An important class of shape-constrained estimators assumes that the density $f : \mathbb{R}^d \to [0, \infty)$ is log-concave: $f(y) = \exp\{g(y)\}$ for some concave function $g : \mathbb{R}^d \to [-\infty, \infty)$, with the convention that $\log 0 = -\infty$. Many common univariate parametric distributions have log-concave densities, including the Gaussian, gamma, beta and Weibull distributions with shape parameters greater than one, and Gumbel, logistic and Laplace families.

Univariate log-concave densities are always unimodal and possess light tails because they are sub-exponential distributions, for which $\text{pr}(|Y| \geq y) = O\{\exp(-cy)\}$ for large y and $c > 0$. The exponential distribution, with its linearly decreasing log-density, is the borderline example of log-concave distribution with light tails. Heavy-tailed distributions like the Cauchy, Pareto, and log-normal are not log-concave.

Interest in log-concave distributions stems from their useful properties. Given a sample $\{Y_1, \ldots, Y_n\}$ of independent and identically distributed random variables with a log-concave density f_0, a unique $\widehat{g}$ maximizes the log-likelihood function

$$\ell(g) = \frac{1}{n}\sum_{i=1}^{n} \log f(Y_i) = \frac{1}{n}\sum_{i=1}^{n} g(Y_i),$$

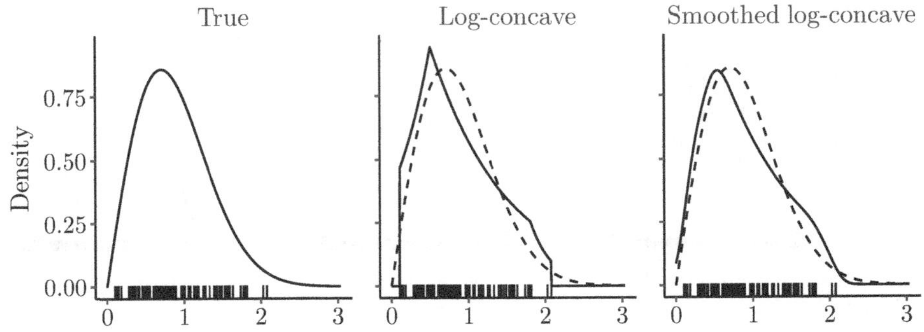

FIGURE 6.7
Log-concave shape-constrained density estimates for 100 observations simulated from a Weibull distribution with shape equal to 2. The left panel shows the true Weibull density function, while the central and right panels present the log-concave and smoothed log-concave density estimates, respectively. For reference, the true Weibull density is shown in the dashed line in the two right panels.

over the space of concave functions such that $\int \exp\{g(y)\}\,\mathrm{d}y = 1$. The nonparametric maximum likelihood estimator $\widehat{g}$ is continuous and piecewise linear on the range of observations, with knots at some of the observations and $\widehat{g} = -\infty$ elsewhere; see, for example, Walther (2009). The shape constraint on the density ensures that the likelihood function is bounded and converts nonparametric maximum likelihood estimation into a finite-dimensional optimization problem, which can be solved efficiently using suitable convex optimization algorithms without having to select tuning parameters.

Example 6.7 (Estimation of a Weibull density) Figure 6.7 shows the log-concave shape-constrained density estimate for a random sample of 100 observations from a Weibull distribution with shape equal to 2. The density estimate obtained with the log-concave shape constraint correctly identifies the essential aspects of the unimodal and right-skewed distribution, but the density estimate is not smooth because it is discontinuous at the minimum and maximum observed values and presents several ridges.

To achieve a more visually appealing density estimate, Dümbgen and Rufibach (2009) suggested a convolution of the initial estimate with a Gaussian density with zero mean and variance γ. This process yields a smooth density estimate that preserves the log-concavity. Moreover, the properties of log-concave distributions allow for an explicit, data-driven estimation of the convolution Gaussian variance γ, thereby ensuring the estimation process remains fully automatic; see Dümbgen and Rufibach (2011).

□

6.3 Local likelihood

6.3.1 Local regression

Local likelihood is a nonparametric approach to estimation that approximates a global function by fitting localized parametric models to weighted subsets of data. Building on the idea of scatterplot smoothing through local adaptation (Tibshirani and Hastie, 1987), this approach constructs a distinct likelihood function for each point of interest, allowing for adaptive and flexible estimation without global parametric assumptions.

Consider n pairs of independent observations $\{(x_1, y_1), \ldots, (x_n, y_n)\}$. The conditional density of Y given $\{X = x\}$ is $f(y; \theta)$, where $\theta = \theta(x)$ is a smooth function of x. Directly estimating a unique parameter θ_i for each observation x_i by maximizing a global likelihood function $L(\theta_1, \ldots, \theta_n)$ would lead to overfitting. Instead, local likelihood methods focus on estimating $\theta(x)$ at specific points by considering only the observations in a neighbourhood around a target point x_0 within the range $[x_{(1)}, x_{(n)}]$ of the observed x values.

Suppose that $\theta(\cdot)$ can be differentiated at least $d+1$ times in a neighbourhood of x_0. A Taylor series expansion of $\theta(x_i)$ in a neighbourhood of x_0 yields the local polynomial approximation

$$\theta(x_i) \doteq \theta(x_0) + \theta'(x_0)(x_i - x_0) + \cdots + \frac{\theta^{(d)}(x_0)}{d!}(x_i - x_0)^d = \sum_{j=0}^{d} \beta_j (x_i - x_0)^j, \tag{6.20}$$

say, where $\beta_j = \theta^{(j)}(x_0)/j!$ for $j = 0, \ldots, d$. The dependence of β_j on x_0 is omitted for notational simplicity. Write $\beta = (\beta_0, \ldots, \beta_d)^{\mathsf{T}}$ and let $\ell_i(\beta; x_0, d)$ denote $\log f\{y_i; \theta(x_i)\}$, with $\theta(x_i)$ replaced by the local polynomial approximation (6.20). The maximum local likelihood estimate $\widehat{\beta}$ is the maximizer of the kernel-weighted log-likelihood function

$$\ell_h(\beta; x_0, d) = \sum_{i=1}^{n} K_h(x_i - x_0)\ell_i(\beta; x_0, d). \tag{6.21}$$

The local estimate of $\theta(x_0)$ is the intercept of the fitted local polynomial, $\widehat{\theta}(x_0) = \widehat{\beta}_0$.

Example 6.8 (Locally weighted least squares regression) Consider the nonparametric regression model

$$Y = \mu(x) + \epsilon, \tag{6.22}$$

where $\mu(x)$ is a smooth function of x and the errors ϵ are assumed to be uncorrelated normal variables with zero mean and variance σ^2. Given n observations from this model, the log-local likelihood function $\ell_h(\beta; x_0, d)$ has the form (6.21) with

$$\ell_i(\beta; x_0, d) = -\frac{1}{2}\log\sigma^2 - \frac{1}{2\sigma^2}\left\{y_i - \sum_{j=0}^{d} \beta_j (x_i - x_0)^j\right\}^2.$$

Maximizing $\ell_h(\beta; x_0, d)$ over β is equivalent to minimizing the local residual sum of squares

$$\sum_{i=1}^{n} K_h(x_i - x_0)\left\{y_i - \sum_{j=0}^{d} \beta_j (x_i - x_0)^j\right\}^2. \tag{6.23}$$

The solution to this problem is generally known as the locally weighted least squares regression estimate. The case $d = 0$ leads to the Nadaraya–Watson estimate

$$\widehat{\mu}_h(x_0) = \frac{\sum_{i=1}^{n} K_h(x_i - x_0) y_i}{\sum_{i=1}^{n} K_h(x_i - x_0)}, \tag{6.24}$$

which is a weighted average of observations in the smoothing window.

The Nadaraya–Watson estimator is rarely a good choice, as it can have large bias near a boundary. Increasing d reduces the bias but increases the variance. Local linear regression corresponds to $d = 1$ and can be written as a combination of the local constant Nadaraya–Watson estimator plus a correction for local slope of the data and skewness of the x_i (Loader, 1999b),

$$\widehat{\mu}_h(x_0) = \frac{\sum_{i=1}^n K_h(x_i - x_0)y_i}{\sum_{i=1}^n K_h(x_i - x_0)} + (x_0 - \overline{x}_h)\frac{\sum_{i=1}^n K_h(x_i - x_0)(x_i - \overline{x}_h)y_i}{\sum_{i=1}^n K_h(x_i - x_0)(x_i - \overline{x}_h)^2},$$

where $\overline{x}_h = \Sigma_i K_h(x_i - x_0)x_i / \Sigma_i K_h(x_i - x_0)$. The correction term reduces the bias near a boundary. When x_0 is not near the boundary, $x_0 \approx \overline{x}_h$ and there is little difference between the local constant and local linear fits. If the true mean function $\mu(x)$ has high curvature around x_0, then large values of d should be considered to reduce the estimation bias.

Smoothing scatterplots by local curve fits is mostly associated with the locally estimated scatterplot smoothing (lowess) algorithm developed in Cleveland (1979). In its original implementation, this method uses a tricube kernel weight $K(u) = 70/81(1 - |u|^3)^3$ and robust local linear fits obtained by replacing the local residual sum of squares (6.23) with the M-estimating function

$$\sum_{i=1}^n K_h(x_i - x_0)\rho_\kappa\{y_i - \beta_0 - \beta_1(x_i - x_0)\},$$

where $\rho_\kappa(\cdot)$ is the Tukey biweight function (4.3). See also Example 4.12. Locally estimated scatterplot smoothing (loess) generalizes lowess to multiple covariates and needs various modifications to handle the increased computational cost (Cleveland and Devlin, 1988). □

Example 6.9 (Local logistic regression) Consider a sample $\{Y_1, \ldots, Y_n\}$ of independent Bernoulli variables with success probability

$$\mu(x) = \text{pr}(Y = 1 \mid x) = \frac{\exp\{\theta(x)\}}{1 + \exp\{\theta(x)\}},$$

for some smooth function $\theta(x) \in \mathbb{R}$. The local log-likelihood $\ell_h(\beta; x_0, d)$ is the kernel-weighted combination of the individual log-likelihoods

$$\ell_i(\beta; x_0, d) = y_i\eta_i(x_0, d) - \log[1 + \exp\{\eta_i(x_0, d)\}], \quad \eta_i(x_0, d) = \sum_{j=0}^d \beta_j(x_i - x_0)^j.$$

The local estimate of the success probability at x_0 is $\widehat{\mu}(x_0) = \exp(\widehat{\beta}_0)/\{1 + \exp(\widehat{\beta}_0)\}$, where $\widehat{\beta}$ maximizes $\ell_h(\beta; x_0, d)$. □

As observed in Section 6.2 for nonparametric density estimation, the choice of kernel is not usually important. Non-rectangular kernels, which descend smoothly to zero as the distance from x_0 increases, are preferable, and different kernel functions produce virtually equivalent results. In contrast, both the degree d of the polynomial and the bandwidth h of the kernel K_h are critical, as they impact the bias-variance trade-off. For practical application $d = 1$ or 2 are the default in most software. A constant order ($d = 0$) is often severely biased, rendering it generally unsuitable, while orders of $d = 3$ or greater entail increased computational burden.

For bandwidth selection, the rules of thumb, cross-validation, and plugin methods discussed in Section 6.2 can be adapted to the local likelihood fitting. However, there is no consensus that these methods perform adequately and robustly. Software implementation more often uses a default bandwidth that holds constant the proportion of nearest neighbours, thus effectively adapting to different covariate densities.

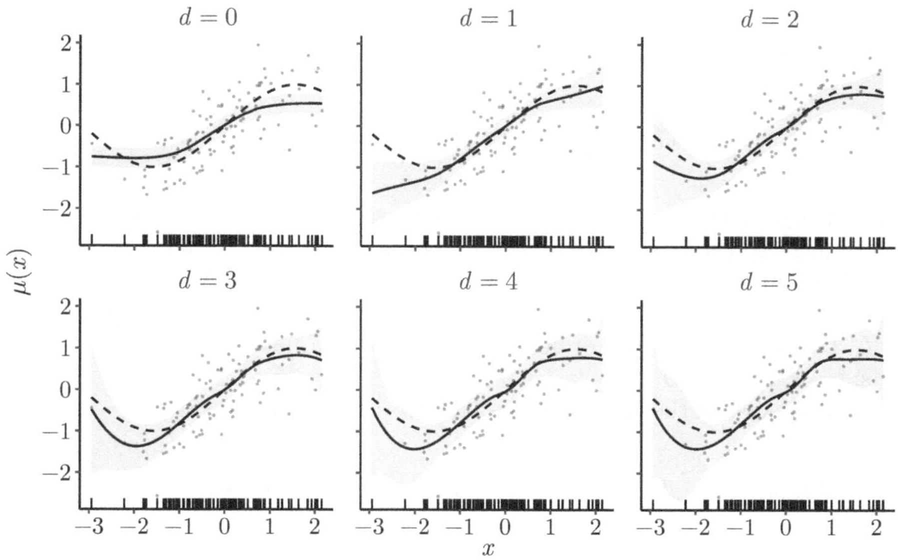

FIGURE 6.8
Local likelihood fitting of simulated data with $d \in \{0, 1, \ldots, 5\}$. The shaded area is a 95% simultaneous confidence band. The dashed line in each plot is the true mean $\mu(x)$ used in the simulations.

Approximate pointwise confidence intervals can be obtained by normal approximation, although some care is needed. The point estimate $\widehat{\beta}$ is biased due to the approximation error in the Taylor series expansion, and its variance typically takes the sandwich form (4.2). While both bias and variance can be estimated using additional terms, these estimates can be unstable as they depend on higher-order derivatives. To improve stability, Fan et al. (1998) proposed a locally weighted averaging method to obtain smoother estimates of both the bias and the variance in a neighbourhood of x_0. Implementing this approach requires the choice of a pilot bandwidth.

Pointwise intervals only reflect the estimation uncertainty at individual points. In practice, when evaluating the global fit, simultaneous confidence bands are often preferred as they guarantee coverage for the entire curve.

Example 6.10 (Locally weighted least squares, cont'd) Figure 6.8 shows the local fit with the Gaussian kernel, for a sample of 100 observations simulated from the model $Y = \sin x + 0.5\epsilon$, where x and ϵ are independently drawn from a standard normal distribution. The shaded gray areas around the fit represent simultaneous 95% confidence bands computed with the method discussed in Sun and Loader (1994) and implemented in the `scb` function of the `R` (R Core Team, 2025) package `locfit` (Loader, 2024). The fits with $d = 0$ and $d = 1$ result in oversmoothing, failing to capture the curvature of the sinusoidal wave, while the fits with $d = 4$ and $d = 5$ have large variance at the outer range of x. The best reconstruction of the signal for this single replicate is obtained with $d = 2$. In all panels, the bandwidth was selected so that the local neighbourhood contains 70% of the observations.

□

6.3.2 Local density estimation

Local likelihood can be used also for density estimation through an extended definition of the log-likelihood function to include non-negative integrable functions. Given a sample of independent observations $\{Y_1, \ldots, Y_n\}$, define an extended log-likelihood function for a given non-negative function $f(\cdot)$ as

$$\ell(f) = \sum_{i=1}^{n} \log f(y_i) - n\left\{\int_{\mathcal{Y}} f(u)\,du - 1\right\}. \tag{6.25}$$

If f is a density function, then the penalty term is zero and the extended log-likelihood function reduces to the standard log-likelihood function. Maximizing (6.25) over the set of non-negative functions is justified because the inequality $\mathrm{E}_f\{\ell(f_1)\} \leq \mathrm{E}_f\{\ell(f)\}$ holds for any non-negative function f_1 (not necessarily a density function), with equality if and only if $f_1 = f$. This is an extension of Jensen's inequality (1.86) for parametric log-likelihood functions; see Exercise 6.9.

The local version of the extended log-likelihood function at y_0 is

$$\ell_h(f; y_0) = \sum_{i=1}^{n} K_h(y_i - y_0) \log f(y_i) - n\left\{\int_{\mathcal{Y}} K_h(u - y_0) f(u)\,du - 1\right\}.$$

As in the regression problems discussed above, $\log f(y)$ can be approximated by a local polynomial of order d in a neighbourhood of y_0,

$$\log f(y) \approx g(y; y_0, d), \quad g(y; y_0, d) = \sum_{j=0}^{d} \beta_j (y - y_0)^j.$$

This local polynomial approximation yields the local log-likelihood function

$$\ell_h(\beta; y_0, d) = \sum_{i=1}^{n} K_h(y_i - y_0) g(y_i; y_0, d) - nC(\beta; y_0, d),$$

where $C(\beta; y_0, d) = \int_{\mathcal{Y}} K_h(u - y_0) \exp\{g(u; y_0, d)\}\,du$. The local estimate $\widehat{\beta}$ solves the system of local equations

$$\frac{1}{n}\sum_{i=1}^{n} K_h(y_i - y_0)\,(y_i - y_0)^j = \int_{\mathcal{Y}} K_h(u - y_0) \exp\{g(u; y_0, d)\}(u - y_0)^j\,du, \tag{6.26}$$

for $j = 0, \ldots, d$. The local likelihood estimate matches local sample moments with local population moments. The local likelihood density estimate at y_0 is $\widehat{f}_h(y_0) = \exp(\widehat{\beta}_0)$. When $d = 0$, there is just one equation in (6.26) and the local likelihood density estimate is

$$\widehat{f}_h(y_0) = \frac{\sum_{i=1}^{n} K_h(y_i - y_0)}{n \int K_h(u - y_0)\,du},$$

which coincides with the kernel density estimate (6.15) discussed in Section 6.2 if the sample space $\mathcal{Y}$ is unbounded.

6.4 Sieve likelihood

In Sections 6.2 and 6.3, we examined nonparametric estimators built using local parametric approximations. This section discusses the method of sieves (Grenander, 1981), an alternative nonparametric estimation approach that instead estimates the function of interest globally across its entire domain. The sieve estimator approximates the infinite-dimensional parameter space with a sequence of increasingly rich finite-dimensional spaces, called sieves. Estimation then proceeds by maximizing the likelihood function over the sieves.

The sieve maximum likelihood estimator provides a regularized approach to nonparametric maximum likelihood estimation by transforming an intractable infinite-dimensional optimization into a sequence of tractable, constrained optimizations over progressively larger and more flexible approximating subspaces.

Let $\{Y_1, \ldots, Y_n\}$ be a sample of independent random variables drawn from an unknown distribution F_0, and suppose the parameter $\theta_0 = T(F_0)$ belongs to an infinite-dimensional parameter space Θ. Direct maximization of the log-likelihood function, $\ell(\theta)$ over Θ is often ill-posed or yields inconsistent estimators in infinite-dimensional settings; see also Section 2.9.

A sequence of sieves $\{\Theta_d : d = 1, 2, \ldots\}$ is constructed such that each Θ_d is a manageable, typically finite-dimensional or otherwise suitably constrained, subspace of Θ. The sieves are chosen so that they are nested, $\Theta_1 \subset \Theta_2 \subset \cdots \subset \Theta$, and their union is dense in the full parameter space Θ under a suitable metric. This ensures that for any $\theta \in \Theta$ (including θ_0), there exists a sequence of elements $\{\theta_d \in \Theta_d : d = 1, 2, \ldots\}$ such that $\theta_d \to \theta$ as $d \to \infty$. Consequently, any parameter in Θ can be approximated arbitrarily closely by elements within the sieve spaces.

Once the sieves are defined, sieve estimation involves two steps: first, specifying a suitable order $d = d_n$ that depends on the sample size n, and then maximizing the sieve likelihood over Θ_d to compute the sieve maximum likelihood estimator $\widehat{\theta}_d$ of θ.

Example 6.11 (Sieve density estimation) Consider a sample of independent variables $\{Y_1, \ldots, Y_n\}$ from a population with a common density f_0. The nonparametric likelihood function for f is

$$L(f) = \prod_{i=1}^{n} f(y_i), \quad f \in \Theta,$$

where $\Theta = \{f : f(y) \geq 0,\ y \in \mathcal{Y} \text{ and } \int_{\mathcal{Y}} f(y)\,dy = 1\}$ is the class of valid density functions.

The simplest version of the sieve estimator for this problem corresponds to the classical histogram estimator, using the sieve corresponding to the class of piecewise constant density functions. For each $d > 0$, define bins of width d^{-1},

$$B_k = \left[\frac{(k-1)}{d}, \frac{k}{d}\right), \quad k \in \mathbb{Z}.$$

The sieve of order d is defined as

$$\Theta_d = \left\{f : f(y) = \omega_k 1(y \in B_k),\ \omega_k \geq 0,\ k \in \mathbb{Z} \text{ and } \textstyle\sum_{k \in \mathbb{Z}} \omega_k = d\right\}.$$

The sieve maximum likelihood estimator is

$$\widehat{f}_d = \arg\max_{f \in \Theta_d} L(f).$$

For each d, $\widehat{f}_d$ solves

$$\max_{\omega} \sum_{k=-\infty}^{\infty} n_k \log \omega_k \quad \text{subject to} \quad \sum_{k=-\infty}^{\infty} \omega_k = d,$$

where ω is the infinite vector of elements ω_k and $n_k \geq 0$ is the number of points that fall in bin k, with $k \in \mathbb{Z}$. The solution of this constrained optimization problem is the stationary point of the Lagrangian

$$\mathcal{L}(\omega) = \sum_{k=-\infty}^{\infty} n_k \log \omega_k - \lambda \left(\sum_{k=-\infty}^{\infty} \omega_k - d \right).$$

Differentiating $\mathcal{L}(\omega)$, we obtain the histogram estimator

$$\widehat{f}_d(y) = \frac{dn_k}{n} 1(y \in B_k).$$

If d diverges sufficiently slowly with n, the sequence of increasingly dense histogram estimators converges to f_0 in the sense that

$$\int_{\mathcal{Y}} |\widehat{f}_d(y) - f_0(y)| \, dy \xrightarrow{a.s.} 0, \quad \text{as } n \to \infty.$$

A more elaborate and efficient estimator is given by the convolution sieve. For example, a Gaussian convolution density yields the sieve

$$\Theta_d = \left\{ f : f(y) = \int_{-\infty}^{\infty} \frac{d}{\sqrt{2\pi}} \exp\left\{ -\frac{d^2}{2}(y-z)^2 \right\} \mathrm{d}G(z), \ G \text{ arbitrary cdf} \right\}. \tag{6.27}$$

The estimator obtained by maximizing the likelihood over Θ_d is consistent if d diverges sufficiently slowly. This sieve estimator is similar to the kernel density estimator with a Gaussian kernel, which is obtained when $G(z)$ in (6.27) is replaced by the empirical distribution function $\widehat{F}_n(z)$. See Geman and Hwang (1982) for more details and extensions. □

Example 6.12 (Sieve density estimation, cont'd) Another approach to density estimation is to maximize the regularized log-likelihood function

$$\ell_\phi(f) = \sum_{i=1}^{n} \log f(y_i) - \lambda\phi(f), \quad f \in \Theta,$$

for some penalty function ϕ, with Θ the space of valid density functions as in Example 6.11. The sieve maximum likelihood estimator solves

$$\widehat{f}_d = \arg\max_{f \in \Theta} \prod_{i=1}^{n} f(y_i) \quad \text{subject to } \phi(f) \leq d,$$

which corresponds to maximizing the likelihood function over the sieve

$$\Theta_d = \{f : f \in \Theta \text{ and } \phi(f) \leq d\}.$$

□

In the two previous examples, the sieves were constructed by imposing a sequence of constraints on the infinite parametric space Θ. An alternative approach constructs sieves algebraically, approximating Θ with a sequence of parametric models of increasing complexity. In many cases of interest, Θ can be represented as the space spanned by an infinite set of bases $\{\varphi_1, \varphi_2, \dots\}$,

$$\theta = \sum_{j=1}^{\infty} \beta_j \varphi_j, \quad \beta \in \mathbb{R}^\infty. \tag{6.28}$$

The corresponding sieve spaces Θ_d are then formed by truncating the expansion to a finite number of terms, such that any $\theta_d \in \Theta_d$ takes the form

$$\theta_d = \sum_{j=1}^{m(d)} \beta_j \varphi_j, \quad d = 1, 2, \ldots,$$

for some $m(d) \geq d$ which depends on the chosen set of basis functions. This version of the method of sieves is known in the econometric literature as series estimation (Newey, 1997; Hansen, 2014).

Example 6.13 (Series regression) Consider a nonparametric version of the generalized linear model discussed in Example 1.12 in which the conditional expectation $\mu(x) = \mathrm{E}(Y \mid x)$ is modelled as a smooth function of x, $g\{\mu(x)\} = \theta(x)$, for a link function $g(\cdot)$. We write

$$\theta(x) = \sum_{j=1}^{\infty} \beta_j \varphi_j(x), \tag{6.29}$$

for some $\beta \in \mathbb{R}^\infty$. There are various ways to specify the coefficients β and the bases φ in (6.29). If $\theta(x)$ is analytic, then it admits a representation as the power series centred at the origin, with coefficients $\beta_j = \theta^{(j-1)}(0)/(j-1)!$ and basis functions $\varphi_j(x) = x^{j-1}$. The sieve Θ_d is the space spanned by the first d power functions, which are used to approximate $\theta(x)$ by the finite sum

$$\theta_d(x) = \sum_{j=1}^{d} \beta_j x^{j-1}.$$

If $\theta(x)$ is periodic of period 1, the sieve can be constructed using trigonometric polynomials with basis functions

$$\varphi_0(x) = 1,\ \varphi_{2j-1}(x) = \cos(2j\pi x),\ \varphi_{2j}(x) = \sin(2j\pi x), \quad j \geq 1,\ x \in [0, 1]. \tag{6.30}$$

The corresponding sieve approximation is the truncated Fourier series

$$\theta_d(x) = \beta_0 + \sum_{j=1}^{d} \{\beta_{2j-1} \cos(2j\pi x) + \beta_{2j} \sin(2j\pi x)\}.$$

The dimension of the sieve Θ_d using trigonometric polynomials is $1 + 2d$.

The vector of sieve coefficients β is estimated by maximizing the sieve log-likelihood function of a selected order d, that is the log-likelihood for a parametric generalized linear model with linear predictor given by $\theta_d(x)$.

Power and trigonometric polynomials can also be combined, leading to a two-dimensional sieve; see Example 6.14. Other commonly employed basis functions are spline functions, Hermite polynomials and wavelets. Splines are discussed in Section 6.5. □

In sieve estimation, the crucial decision is selecting the number of terms to include in the truncated series. Achieving an optimal balance between bias and variance necessitates carefully choosing the growth rate of the sieve order d relative to the sample size n. A larger d reduces the bias from series truncation, but it also increases the variance, as more parameters must be estimated from limited data. The choice of the sieve order is often based on cross-validation or information criteria computed with the sieve log-likelihood, such as AIC or BIC.

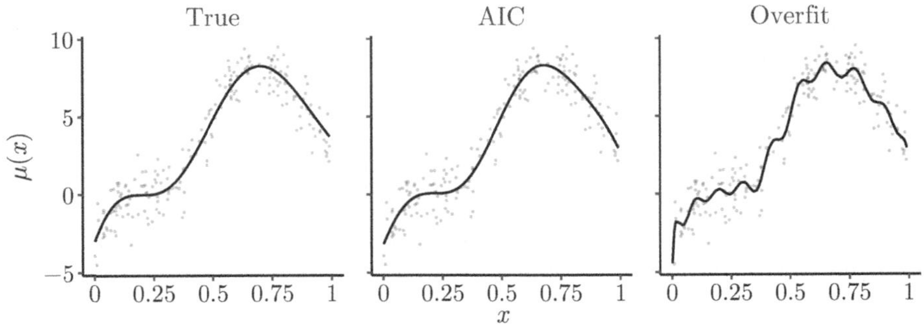

FIGURE 6.9
Sieve estimation using a combination of polynomial and trigonometric terms. The left panel displays 200 simulated observations from the model with mean $\mu(x)$ (solid line). The central panel shows the fitted model selected by AIC, which combines a polynomial of degree 2 ($d_1 = 3$) and a trigonometric series of order 2 ($d_2 = 2$). The right panel shows an overfitted model using a polynomial of degree 5 and a trigonometric series of order 10.

Example 6.14 (Polynomial-trigonometric regression) We simulated a sample of 200 independent observations from the model

$$Y = \mu(x) + \epsilon, \quad \mu(x) = 400 \exp\{-5(x - 0.1)^2\}(x - 0.2)^3,$$

where x is drawn from a $U(0, 1)$ distribution and ϵ from a standard normal random distribution. The approximation of $\mu(x)$ is based on a two-dimensional sieve that combines truncated power and Fourier series,

$$\mu(x) \approx \mu_{\mathrm{P}}(x; d_1) + \mu_{\mathrm{F}}(x; d_2),$$

with components

$$\mu_{\mathrm{P}}(x; d_1) = \sum_{j=1}^{d_1} \alpha_j x^{j-1}, \quad \mu_{\mathrm{F}}(x; d_2) = \sum_{j=1}^{d_2} \{\beta_{2j-1} \cos(2j\pi x) + \beta_{2j} \sin(2j\pi x)\}.$$

The intercept term is omitted in the trigonometric component as it appears in the polynomial component. The values of AIC and BIC both suggest $d_1 = 3$ ans $d_2 = 2$, with

$$\begin{aligned}
\widehat{\mu}_{\mathrm{P}}(x) &= -5.42 + 41.96x - 36.08x^2, \\
\widehat{\mu}_{\mathrm{F}}(x) &= 1.87 \sin(2\pi x) - 2.41 \cos(2\pi x) + 0.32 \sin(4\pi x) + 0.88 \cos(4\pi x).
\end{aligned}$$

Figure 6.9 shows that this model provides a highly accurate reconstruction of $\mu(x)$. For comparison, a model with a polynomial of degree five and a trigonometric series of order ten is also plotted; this model clearly overfits the data.

□

6.5 Smoothing splines

6.5.1 Spline basis functions

A spline of degree k with knots at $\tau_1 < \cdots < \tau_m$ is a piecewise polynomial of degree k that is continuous and has continuous derivatives of orders $1, \ldots, k-1$ at its knots. Any such spline can be represented as a linear combination of a set of basis functions, whose specific form depends on the chosen degree and knot sequence.

As in Section 6.3.1, n pairs of independent observations $\{(y_1, x_1), \ldots, (y_n, x_n)\}$ are assumed to follow the conditional density $f\{y; \theta(x)\}$ for some smooth function $\theta(x) \in \Theta$, but now $\theta(x)$ is approximated by a linear combination of d spline basis functions,

$$\theta_d(x) = \sum_{j=1}^{d} \beta_j \varphi_j(x). \tag{6.31}$$

Assume $a \le x_i \le b$ for all $i = 1, \ldots, n$, and define the interior knots $a < \tau_1 < \cdots < \tau_m < b$ and the boundary knots $\tau_0 = a$ and $\tau_{m+1} = b$. One choice of basis functions is the truncated power spline basis of degree k, formed by combining the standard polynomial basis of degree k with one truncated power function for each internal knot,

$$\varphi_r(x) = \begin{cases} x^r, & r = 0, \ldots, k, \\ (x - \tau_{r-k})_+^k, & r = k+1, \ldots, k+m, \end{cases}$$

where $(x)_+ = \max\{0, x\}$ is the positive part of x. Cubic splines, of degree $k = 3$, are predominantly used in applications. However, truncated power basis functions are highly correlated, leading to numerical instability in computations and sensitivity to small changes in data. To address these limitations, B-splines are usually preferred. These are defined recursively, so that each basis function is non-negative, non-zero only in a local region, and overlaps only with near neighbours. Natural splines use the B-spline basis but with linear boundary constraints; the resulting fit is more stable in the boundary regions $x < \tau_1$ and $x > \tau_m$.

6.5.2 Penalized likelihood

Smoothing splines avoid the specification of the number and exact locations of the knots by placing a knot at each unique observed data point (or on a large number of pre-defined knots). Overfitting is controlled by incorporating a penalty term that constrains the function's roughness, most commonly by penalizing its integrated squared second derivative.

Our discussion will focus on smoothing cubic splines, although the ideas generalize to any polynomial degree. The smoothing spline estimate is the maximizer of the penalized log-likelihood function

$$Q_\lambda(\theta) = \ell(\theta) - \lambda \int_a^b \{\theta''(u)\}^2 \, \mathrm{d}u, \quad \theta(x) \in \Theta,$$

where $\ell(\theta) = \Sigma_i \log f\{y_i; \theta(x_i)\}$ and $\theta''(u)$ denotes the second derivative of $\theta(u)$. The integral term represents a roughness penalty: as $\theta''(u)$ measures the curvature of the function at point u, its square integrated over $[a, b]$ quantifies the total variability of the estimate. The non-negative tuning parameter λ controls the fit-smoothness trade-off: the larger λ, the smoother the resulting estimate $\widehat{\theta}(x)$.

In the nonlinear regression model $Y = \mu(x) + \epsilon$, with $x \in [a, b]$, the smoothing spline estimate is the minimizer of the penalized residual sum of squares

$$\mathrm{RSS}_\lambda(\mu) = \sum_{i=1}^{n} \{y_i - \mu(x_i)\}^2 + \lambda \int_a^b \{\mu''(u)\}^2 \, \mathrm{d}u.$$

It is shown below that the unique minimizer of $\text{RSS}_\lambda(\mu)$ is the natural cubic spline with knots $\tau_i = x_{(i)}$ for $i = 1, \dots, n$ at the n observations. The boundaries are set as $\tau_0 = a$ and $\tau_{n+1} = b$. Let $\widehat{\mu}$ be the minimizer of $\text{RSS}_\lambda(\mu)$. To demonstrate that the minimizer is necessarily a natural cubic spline, we compare $\widehat{\mu}$ to a natural cubic spline $\widetilde{\mu}$ that matches $\widehat{\mu}$ at every observation point x_i. Because $\widetilde{\mu}(x_i) = \widehat{\mu}(x_i)$, for $i = 1, \dots, n$, then the first term of $\text{RSS}_\lambda(\mu)$ is identical at $\widehat{\mu}$ and $\widetilde{\mu}$,

$$\sum_{i=1}^{n} \{y_i - \widetilde{\mu}(x_i)\}^2 = \sum_{i=1}^{n} \{y_i - \widehat{\mu}(x_i)\}^2.$$

It therefore suffices to show that the penalty term is minimized by $\widetilde{\mu}$,

$$\int_a^b \{\widetilde{\mu}''(u)\}^2 \, du \le \int_a^b \{\widehat{\mu}''(u)\}^2 \, du. \tag{6.32}$$

To prove the above inequality, write $\delta(u) = \widetilde{\mu}(u) - \widehat{\mu}(u)$. We have

$$\begin{aligned} \int_a^b \{\widehat{\mu}''(u)\}^2 \, du &= \int_a^b \{\widetilde{\mu}''(u) - \delta''(u)\}^2 \, du \\ &= \int_a^b \{\widetilde{\mu}''(u)\}^2 \, du + \int_a^b \{\delta''(u)\}^2 \, du - 2 \int_a^b \widetilde{\mu}''(u)\delta''(u) \, du. \end{aligned} \tag{6.33}$$

Inequality (6.32) is proved if the third term in (6.33) is zero. Integration by parts yields

$$\int_a^b \widetilde{\mu}''(u)\delta''(u) \, du = \widetilde{\mu}''(u)\delta'(u)\Big|_a^b - \int_a^b \widetilde{\mu}^{(3)}(u)\delta'(u) \, du = -\int_a^b \widetilde{\mu}^{(3)}(u)\delta'(u) \, du$$

because the second-order derivative of a natural cubic spline is zero at the boundaries. We then decompose the remaining integral into the intervals defined by the knots τ_i,

$$\begin{aligned} -\int_a^b \widetilde{\mu}^{(3)}(u)\delta'(u) \, du &= -\sum_{i=0}^{n} \int_{\tau_i}^{\tau_{i+1}} \widetilde{\mu}^{(3)}(u)\delta'(u) \, du \\ &= -\sum_{i=0}^{n} \left\{ \widetilde{\mu}^{(3)}(u)\delta(u)\Big|_{\tau_i}^{\tau_{i+1}} - \int_{\tau_i}^{\tau_{i+1}} \widetilde{\mu}^{(4)}(u)\delta(u) \, du \right\} \end{aligned}$$

Since the fourth-order derivative of a natural cubic spline is null everywhere and the third-order derivative is zero for $x \le \tau_1$ and $x \ge \tau_n$ and constant in each interval (τ_i, τ_{i+1}), then

$$-\int_a^b \widetilde{\mu}^{(3)}(u)\delta'(u) \, du = -\sum_{i=1}^{n-1} c_i\{\delta(\tau_{i+1}) - \delta(\tau_i)\}$$

for some constants c_i. By construction $\delta(x_i) = 0$ for all i and thus also $\delta(\tau_i) = 0$ because each observation is a knot and thus the above integral is null which proves inequality (6.32). Finally, the minimizer $\widehat{\mu}$ is unique because $\text{RSS}_\lambda(\mu)$ is a convex function.

This result is a special case of the representer theorem (Kimeldorf and Wahba, 1970). See also Green and Silverman (1993, Section 2.2). The consequence of this is that the penalized least-squares problem is a function of a vector of n basis coefficients $\beta = (\beta_1, \dots, \beta_n)^\top$:

$$\text{RSS}_\lambda(\beta) = \sum_{i=1}^{n} \left\{ y_i - \sum_{j=1}^{n} \beta_j \varphi_j(x_i) \right\}^2 + \lambda \sum_{r=1}^{n} \sum_{s=1}^{n} \beta_r \beta_s \int_a^b \varphi_r''(u)\varphi_s''(u) \, du,$$

where $\{\varphi_1(x), \ldots, \varphi_n(x)\}$ are the natural spline basis functions. The matrix form of the penalized residual sum of squares is

$$\text{RSS}_\lambda(\beta) = \|y - D\beta\|_2^2 + \lambda\beta^\top\Omega\beta, \tag{6.34}$$

where $y \in \mathbb{R}^n$ is the vector of observed responses, D is the $n \times n$ basis design matrix with entries $[D]_{ij} = \varphi_j(x_i)$ and the $n \times n$ penalty matrix Ω has entries

$$[\Omega]_{rs} = \int_a^b \varphi_r''(u)\varphi_s''(u)\,\mathrm{d}u, \quad r, s = 1, \ldots, n. \tag{6.35}$$

The penalty matrix Ω is positive semi-definite and has rank $n-2$; see, for example, Silverman (1985). If $\{\varphi_1, \ldots, \varphi_n\}$ is a set of B-spline bases, then the integral (6.35) is analytic (Exercise 6.11).

The quadratic penalty term $\lambda\beta^\top\Omega\beta$ in (6.34) is a form of Tikhonov regularization, which is a generalization of ridge regression. This and other forms of regularized likelihood inference are discussed in Chapter 7. The smoothing spline estimate of β is

$$\widehat{\beta}_\lambda = (D^\top D + \lambda\Omega)^{-1}D^\top y \tag{6.36}$$

and the fitted values are

$$\widehat{\mu}_\lambda = D(D^\top D + \lambda\Omega)^{-1}D^\top y = A_\lambda y, \tag{6.37}$$

where A_λ is called the smoother matrix. The number of degrees of freedom of the fit is $\text{tr}(A_\lambda)$, a real number between 0 ($\lambda \to \infty$) and n ($\lambda = 0$), which may not be an integer, in contrast to regression splines with a pre-specified number of knots.

Selecting the penalty coefficient λ for smoothing splines is much simpler than choosing the number and locations of the knots for a regression spline. A standard approach is to minimize the leave-one-out cross validation criterion

$$\text{CV}(\lambda) = \frac{1}{n}\sum_{i=1}^n \left\{\frac{y_i - \widehat{\mu}_{\lambda,i}}{1 - [A_\lambda]_{ii}}\right\}^2,$$

where $\widehat{\mu}_{\lambda,i}$ is the ith element of $\widehat{\mu}_\lambda$ and $[A_\lambda]_{ii}$ is the ith element in the diagonal of the smoother matrix A_λ. Often $\text{CV}(\lambda)$ is replaced by the generalized cross validation criterion

$$\text{GCV}(\lambda) = \frac{1}{n}\sum_{i=1}^n \left\{\frac{y_i - \widehat{\mu}_{\lambda,i}}{1 - n^{-1}\text{tr}(A_\lambda)}\right\}^2,$$

which is simpler to compute, more stable, and invariant to orthogonal transformation of the residual vector. Information criteria are also used for selection of λ, with the number of parameters replaced by the degrees of freedom of the fit. The AIC for selection of λ is

$$\text{AIC}(\lambda) = \log \text{RSS}_0(\widehat{\beta}_\lambda) + 2n^{-1}\text{tr}(A_\lambda),$$

where RSS_0 is the unpenalized residual sum of squares. $\text{AIC}(\lambda)$ may be interpreted as a version of the Takeuchi Information Criterion (4.20) discussed in Chapter 4. In Section 4.4 it is shown that $\text{AIC}(\lambda)$ is asymptotically equivalent to $\text{CV}(\lambda)$. However, in finite samples $\text{AIC}(\lambda)$ has been observed to select too small values of λ, thereby producing undersmoothing. Hurvich et al. (1998) developed a corrected version of AIC which is an approximately unbiased estimate of the expected Kullback–Leibler divergence:

$$\text{AIC}_c(\lambda) = \log \text{RSS}_0(\widehat{\beta}_\lambda) + \frac{2\text{tr}(A_\lambda) + 2}{n - \text{tr}(A_\lambda) - 2}.$$

Computing the coefficients for the interpolating natural cubic spline with knots at each observation has a computational cost of $O(n^3)$. Furthermore, using a basis dimension equal to the number of observations is statistically inefficient. Rank-reduction techniques are thus utilized to achieve a sparser representation of the smooth function by selecting $d \ll n$ knots to reduce the computational burden and improve the statistical efficiency. Asymptotic arguments suggest using $d = O(n^{1/5})$ knots (Wood, 2025).

In more general models, the smoothing spline basis coefficients $\widehat{\beta}_\lambda$ maximize the penalized log-likelihood function

$$Q_\lambda(\beta) = \ell_d(\beta) - \lambda\beta^\mathsf{T}\Omega\beta, \tag{6.38}$$

where $\ell_d(\beta)$ is the log-likelihood with $\theta(x_i)$ approximated by $\theta_d(x_i)$, the linear combination of d spline basis functions (6.31),

$$\ell_d(\beta) = \sum_{i=1}^{n} \log f\{y_i; \theta_d(x_i)\}. \tag{6.39}$$

When $\theta(x)$ represents the linear predictor of a generalized linear model, maximizing Q_λ can be efficiently performed by solving a sequence of weighted least squares problems.

Example 6.15 (Nonparametric generalized linear models) Consider the nonparametric version of the generalized linear model discussed in Example 6.13 in which the conditional expectation $\mu(x) = \mathrm{E}(Y \mid x)$ is modelled as $g\{\mu(x)\} = \theta(x)$, for a link function $g(\cdot)$. The penalized log-likelihood function (6.38) is

$$Q_\lambda(\beta) = \frac{1}{a(\phi)} \sum_{i=1}^{n} [y_i\theta_d(x_i) - b\{\theta_d(x_i)\}] - \lambda\beta^\mathsf{T}\Omega\beta,$$

where ϕ is a dispersion parameter and $a(\cdot)$ and $b(\cdot)$ are known functions. The conditional mean and variance are $\mu(x) = b'\{\theta(x)\}$ and $\mathrm{var}(Y \mid x) = a(\phi)V\{\mu(x)\}$, with $V\{\mu(x)\} = b''\{\theta(x)\}$.

Maximization of $Q_\lambda(\beta)$ uses Fisher scoring, which is a Newton-Raphson algorithm with the observed information matrix replaced by its expectation. In generalized linear models, Fisher scoring simplifies to iteratively reweighted least squares. At each iteration r, the algorithm updates the coefficient estimates by setting

$$\widehat{\beta}_\lambda^{(r+1)} = \{D^\mathsf{T}W^{(r)}D + 2a(\phi)\lambda\Omega\}^{-1}D^\mathsf{T}W^{(r)}z^{(r)}, \quad r = 1, 2, \ldots, \tag{6.40}$$

where the spline design matrix D and the penalty matrix Ω are given in (6.34). The new components in (6.40) are the weight matrix

$$W^{(r)} = \mathrm{diag}(w_1^{(r)}, \ldots, w_n^{(r)}), \quad w_i^{(r)} = \frac{1}{\{g'(\widehat{\mu}_i^{(r)})\}^2 V(\widehat{\mu}_i^{(r)})}, \qquad i = 1, \ldots, n,$$

and the vector of working responses $z^{(r)} = (z_1^{(r)}, \ldots, z_n^{(r)})^\mathsf{T}$ with

$$z_i^{(r)} = g(\widehat{\mu}_i^{(r)}) + (y_i - \widehat{\mu}_i^{(r)})g'(\widehat{\mu}_i^{(r)}), \quad i = 1, \ldots, n,$$

where $\widehat{\mu}_i^{(r)}$ is the estimate of μ_i computed at $\widehat{\beta}^{(r)}$. The algorithm is initialized by setting $\widehat{\mu}_i^{(0)} = y_i$. For models that necessitate strictly positive fitted means (e.g., with a log link function), $\widehat{\mu}_i^{(0)} = \max\{\epsilon, y_i\}$ for some $\epsilon > 0$ is used to prevent issues from zero observations. See Green (1987). □

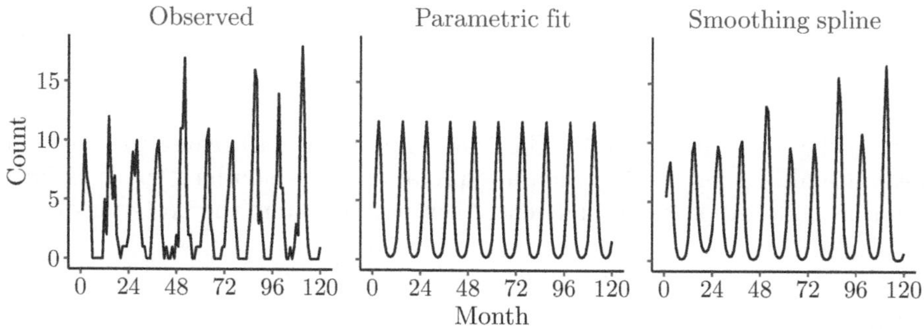

FIGURE 6.10
Comparison between a parametric fit of a Poisson log-linear model (central panel) and penalized cubic spline model (right panel) to a simulated series of 10 years of monthly counts (left panel).

Example 6.16 (Poisson nonparametric regression) Consider a time series of ten years of monthly count data Y_t simulated from a Poisson distribution with expected value

$$\mu_t = \exp\{0.5 + 2\sin(2t\pi/12)\}, \quad t = 1, \dots, 120.$$

The simulated counts are displayed in the left panel of Figure 6.10. We fit a nonparametric Poisson model to the counts, $Y(t) \sim \text{Poisson}(\mu_t)$, using a cubic spline $\mu_t = \exp\{\theta_d(t)\}$. This is an example of a nonparametric generalized linear model discussed in Example 6.15, and can be estimated using the iteratively reweighted least squares algorithm (6.40) with weights and working responses

$$w_t^{(r)} = \widehat{\mu}_t^{(r)}, \quad z_t^{(r)} = \log(\widehat{\mu}_t^{(r)}) + (y_t - \widehat{\mu}_t^{(r)})/\widehat{\mu}_t^{(r)}, \quad t = 1, \dots, n,$$

for $\widehat{\mu}_t^{(r)} = \exp\{\widehat{\theta}_d^{(r)}(t)\}$.

The right-hand panel of Figure 6.10 shows the smoothing spline fit using $d = 30$ knots placed evenly throughout the ten years of monthly data, i.e., three knots per year. The smoothing penalty λ was selected using AIC. The fitted penalized cubic spline gives an accurate reconstruction of the expected number of counts. For comparison, the centre panel of Figure 6.10 shows the parametric fit of the Poisson model used to generate the data, $Y_t \sim \text{Poisson}(\mu_t)$ with $\mu_t = \exp\{\beta_0 + \beta_1 \sin(2t\pi/12)\}$. The maximum likelihood estimates of the regression parameters are $\widehat{\beta}_0 = 0.49$ and $\widehat{\beta}_1 = 1.97$. Here, the nonparametric fit is attractive because it adapts to variations in the seasonal peaks, offering a more realistic summary of the observed pattern in the series. □

Construction of confidence intervals with smoothing splines is complicated by the bias in $\widehat{\beta}_\lambda$, which arises from the roughness penalty. In the special case where Y follows a normal distribution with expected value $\mu(x)$ and variance σ^2, $\widehat{\beta}_\lambda$ has the form (6.36), and its covariance matrix is

$$\text{var}(\widehat{\beta}_\lambda) = (D^\top D + \lambda\Omega)^{-1} D^\top D (D^\top D + \lambda\Omega)^{-1}\sigma^2,$$

but confidence intervals centred at $\widehat{\beta}_\lambda$ typically yield inaccurate coverage probabilities because of the bias in $\widehat{\beta}_\lambda$ (Wood, 2025).

Building on Wahba (1983), various Bayesian approximate interval estimates have been proposed for smoothing splines; they can give better coverage properties compared to naive confidence intervals. The smoothing penalty corresponds to assuming the Gaussian prior distribution on β,

$$\beta \sim N(0, \sigma^2\Omega^-/\lambda), \tag{6.41}$$

where Ω^- is the pseudoinverse of Ω. This is an improper prior because the precision matrix Ω is not full rank. Under this Bayesian interpretation, the smoothing spline estimator $\widehat{\beta}_\lambda$ is the posterior mode. Approximate inference then proceeds from the posterior distribution of β,

$$\beta \mid y \sim N\{\widehat{\beta}_\lambda, (D^\top D + \lambda\Omega)^{-1}\sigma^2\}, \tag{6.42}$$

with σ^2 replaced by its unbiased estimate $\widehat{\sigma}^2 = \|y - \widehat{y}_\lambda\|_2^2/\mathrm{tr}(I - A_\lambda)$. An approximate confidence interval around the fitted line is $\widehat{\mu}_i \pm z_\alpha[A_\lambda]_{ii}^{1/2}\widehat{\sigma}$, where $\widehat{\mu}_i = \mu(x_i; \widehat{\beta}_\lambda)$ and $[A_\lambda]_{ii}$ is the ith element in the diagonal of the smoother matrix in (6.37). This interval depends on the tuning parameter λ, which can be estimated by an empirical Bayes approach. This is equivalent to restricted maximum likelihood estimation (Exercise 6.10). Alternatively, λ can be estimated with cross-validation or selected using AIC. The empirical Bayes interval has been reported to exhibit good coverage averaged over the x_i's (Wood, 2025).

Outside the Gaussian likelihood, the posterior distribution of β does not have a closed form expression. For a sufficiently large sample, Monte Carlo simulation methods can be avoided using the approximation (1.67)

$$\pi_\lambda(\beta \mid y) \mathrel{\dot\sim} N\{\widehat{\beta}_\lambda, (\widehat{J}_\lambda + \lambda\Omega)^{-1}\}, \tag{6.43}$$

where $\widehat{J}_\lambda$ is the observed Fisher information in $\ell_d(\beta)$ defined in (6.39) computed at $\widehat{\beta}_\lambda$. The Laplace approximation in (1.69) and (1.71) may also be used. Empirical Bayes estimation of λ now requires approximating, either numerically or using Laplace's method, the integral

$$L(\lambda) = \int \pi_\lambda(\beta) L_d(\beta)\, \mathrm{d}\beta,$$

where $L_d(\beta) = \exp\{\ell_d(\beta)\}$. This empirical Bayes approach can be interpreted as arising from a Gaussian random effects model.

An alternative approach to selecting $\widehat{\lambda}$ is the Akaike information criterion, with a correction for the degrees of freedom of the smooth fit:

$$\mathrm{AIC}(\lambda) = -2\ell_d(\widehat{\beta}_\lambda) + 2\mathrm{tr}\{(\widehat{J}_\lambda + \lambda\Omega)^{-1}\widehat{J}_\lambda\}.$$

Cross-validation provides another estimate of λ.

6.6 Empirical likelihood

Owen (1988, 1990) introduced the empirical likelihood ratio as the nonparametric analogue of the likelihood ratio statistic discussed in Chapters 1 and 2. The empirical likelihood ratio shares the main theoretical properties of its parametric counterpart for important inferential problems without the need to specify a parametric model.

Empirical likelihood is constructed in a manner distinct from the nonparametric approaches detailed in preceding sections. While methods like local likelihood (Section 6.3) and sieve likelihood (Section 6.4) involve a parametric approximation of an unknown function through smoothing or expansion, empirical likelihood operates by directly assigning probabilities to each observed data point. This nonparametric likelihood is then maximized under constraints derived from the problem, typically involving population moment conditions.

Consider a sample $\{Y_1, \ldots, Y_n\}$ of independent observations from some distribution F that belongs to a family of distributions $\mathcal{F}$. In Section 6.1 it was shown that the nonparametric likelihood $L(F)$ is maximized by the empirical cumulative distribution function $\widehat{F}_n$. By analogy with the parametric setting, hypothesis tests and confidence intervals can be based on the nonparametric likelihood ratio statistic,

$$\widehat{\Lambda}_n(F) = \frac{L(F)}{L(\widehat{F}_n)}, \quad F \in \mathcal{F}.$$

For the likelihood $L(F)$ to be non-zero, the distribution F must place positive probability mass on each observed value y_i. This implies that a distribution F with a high nonparametric likelihood ratio must concentrate its mass on the discrete set of observations $\{y_1, \ldots, y_n\}$. To see this more formally, assume $\widehat{\Lambda}_n(F) \geq 1 - \epsilon$ for some small $\epsilon \in (0, 1)$. By the arithmetic-geometric mean inequality, the average probability mass at the observed points is bounded below by the geometric mean of the individual probabilities

$$\begin{aligned}\frac{1}{n}\sum_{i=1}^{n} \operatorname{pr}(Y_i = y_i) &\geq \left\{\prod_{i=1}^{n} \operatorname{pr}(Y_i = y_i)\right\}^{1/n} \\ &= L(F)^{1/n} = \left\{\widehat{\Lambda}_n(F) L(\widehat{F}_n)\right\}^{1/n} = \frac{1}{n}\widehat{\Lambda}_n(F)^{1/n}.\end{aligned}$$

The total probability mass on the observations is

$$\sum_{i=1}^{n} \operatorname{pr}(Y_i = y_i) \geq \widehat{\Lambda}_n(F)^{1/n} \geq (1-\epsilon)^{1/n} \geq 1 - O(n^{-1}).$$

Hence, the total mass assigned by F to the observed data points approaches unity. Consequently, $L(F)$ behaves, at least asymptotically, like the likelihood of a multinomial distribution with probabilities $\pi = (\pi_1, \ldots, \pi_n)$ assigned to the data points $y = (y_1, \ldots, y_n)$. The nonparametric likelihood ratio written as a function of π is

$$\widehat{\Lambda}_n(\pi) = \prod_{i=1}^{n} n\pi_i, \quad \pi \in \mathcal{S}_{n-1},$$

where $\mathcal{S}_{n-1} = \{c \in \mathbb{R}^n : (c_1 + \cdots + c_n) = 1,\ c_i \geq 0\}$ is the probability simplex.

Suppose the parameter of interest $\theta = T(F) \in \mathbb{R}^p$ is the root of some unbiased estimating equation $\Psi(\theta; Y)$ (Section 5.3). Inference for θ can be based on the profile nonparametric likelihood ratio

$$\widehat{\Lambda}_n(\theta) = \max_{\pi \in \mathcal{S}_{n-1}} \{\widehat{\Lambda}_n(\pi) : \textstyle\sum_i \pi_i \Psi(\theta; Y_i) = 0\}.$$

Maximizing $\Lambda_n(\pi)$ corresponds to finding the stationary point of the Lagrangian

$$\mathcal{L}(\pi, \tau, \gamma) = \sum_{i=1}^{n} \log(n\pi_i) - n\tau^{\mathsf{T}} \sum_{i=1}^{n} \pi_i \Psi(\theta; Y_i) + \gamma\left(\sum_{i=1}^{n} \pi_i - 1\right), \tag{6.44}$$

where τ and γ are the Lagrange multipliers. Taking partial derivatives with respect to γ and π_i gives

$$\widehat{\gamma} = -n, \quad \widehat{\pi}_i(\tau) = \frac{1}{n\{1 + \tau^{\mathsf{T}}\Psi(\theta; Y_i)\}}, \quad i = 1, \ldots, n. \tag{6.45}$$

Writing $h(\tau) = \mathcal{L}\{\widehat{\pi}(\tau), \tau, \widehat{\gamma}\}$, the estimate of τ satisfies

$$\frac{\partial h(\tau)}{\partial \tau} = -\sum_{i=1}^{n} \frac{\Psi(\theta; Y_i)}{1 + \tau^{\mathsf{T}} \Psi(\theta; Y_i)} = 0. \tag{6.46}$$

Since $0 \le \pi_i \le 1$ for each i, $\widehat{\tau}$ must lie in the closed convex region

$$\mathcal{D}_\theta = \{1 + \tau^{\mathsf{T}} \Psi(\theta; Y_i) \ge 1/n,\ i = 1, \ldots, n\}. \tag{6.47}$$

Moreover, the Hessian of $h(\tau)$ is

$$\frac{\partial^2 h(\tau)}{\partial \tau \partial \tau^{\mathsf{T}}} = \sum_{i=1}^{n} \frac{\Psi(\theta; Y_i)\Psi(\theta; Y_i)^{\mathsf{T}}}{\{1 + \tau^{\mathsf{T}} \Psi(\theta; Y_i)\}^2},$$

which is positive definite for $\tau \in D_\theta$ provided that $\Sigma_i \Psi(\theta; Y_i)\Psi(\theta; Y_i)^{\mathsf{T}}$ is positive definite (Qin and Lawless, 1994).

Solution of (6.46) corresponds to minimizing a convex function over a convex domain, under the n inequality constraints (6.47), which may be difficult for large n. Owen (2001) reformulated computation of $\widehat{\tau}$ as an unconstrained minimization; see Exercise 6.15. Owen (2013) showed that the resulting objective function is self-concordant, thereby ensuring guaranteed convergence for Newton-type methods.

The empirical log-likelihood ratio is defined as

$$\widehat{w}_n(\theta) = -2 \log \widehat{\Lambda}_n(\theta) = 2 \sum_{i=1}^{n} \log\{1 + \widehat{\tau}^{\mathsf{T}} \Psi(\theta; Y_i)\}. \tag{6.48}$$

Under mild conditions, the empirical log-likelihood ratio has a limiting chi-squared distribution. This is the important empirical likelihood theorem (Owen, 2001, Theorem 3.4) which generalises Wilks' theorem; see, for example, (1.51). More specifically, if $\mathrm{E}_{\theta_0}\{\Psi(\theta_0; Y)\} = 0$ and $\mathrm{var}_{\theta_0}\{\Psi(\theta_0; Y)\}$ is finite and has rank $q > 0$, then

$$\widehat{w}_n(\theta_0) \xrightarrow{d} \chi^2_q,$$

under the null hypothesis $H_0 : \theta = \theta_0$.

Example 6.17 (Empirical likelihood inference for a mean) Consider a sample of n independent and identically distributed observations from some unknown distribution F_0 with mean $\mathrm{E}(Y_1) = \mu_0$ and finite variance. Further, suppose that $Y_{(1)} < \mu_0 < Y_{(n)}$. The estimating equation for μ, $\Psi(\mu; Y) = Y - \mu$, yields the nonparametric likelihood ratio statistic

$$\widehat{\Lambda}_n(\mu) = \max_{\pi \in \mathcal{S}_{n-1}} \left\{\widehat{\Lambda}_n(\pi) : \textstyle\sum_{i=1}^{n} \pi_i Y_i = \mu\right\}.$$

The corresponding empirical log-likelihood ratio for μ is

$$\widehat{w}_n(\mu) = 2 \sum_{i=1}^{n} \log\{1 + \widehat{\tau}(Y_i - \mu)\}.$$

Computation of $\widehat{w}_n(\mu)$ requires one-dimensional minimization of a convex function with respect to τ. Since π_i must lie between 0 and 1 for all i, the search for $\widehat{\tau}$ is restricted to the interval

$$\left(\frac{1 - n^{-1}}{\mu - Y_{(n)}}, \frac{1 - n^{-1}}{\mu - Y_{(1)}}\right),$$

and can be efficiently performed with the Brent algorithm (Brent, 1973).

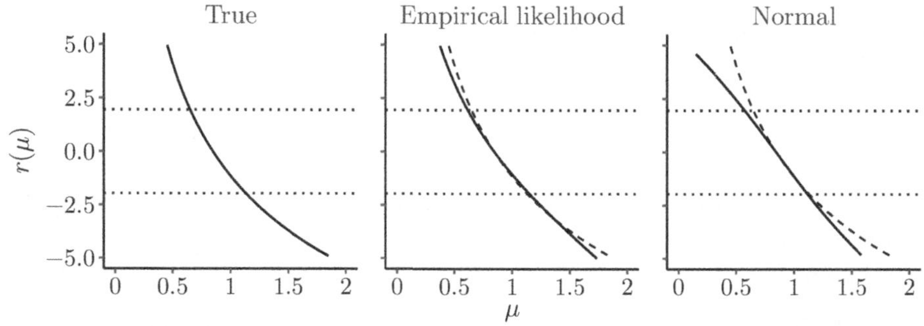

FIGURE 6.11
Comparison between empirical and parametric likelihood inference for the mean of a sample of 50 observations drawn from an exponential distribution with $\mu = 1$. The left panel shows the parametric likelihood root statistic $r(\mu)$ computed with the correct exponential distribution. The central panel is the empirical likelihood ratio root statistic. The right panel is the profile likelihood root when an incorrect normal model is assumed for the observations. The curve based on the correct model in the left-most panel is shown as the reference dashed curve in the two right panels. The horizontal dotted lines identify an approximate 95% confidence interval for μ.

For illustration consider a sample of 50 observations from an exponential distribution with expected value equal to one. The sample mean and standard deviation are $\overline{y} = 0.85$ and $s = 0.97$, respectively. The approximate 95% empirical likelihood confidence interval $\{\mu : \widehat{w}_n(\mu) \leq 3.84\}$ is $[0.62, 1.15]$. The parametric likelihood ratio statistic using the correct exponential distribution is $w(\mu) = 100\{\overline{y}/\mu - 1 + \log(\mu/\overline{y})\}$, and the 95% confidence interval based on the χ^2_1 approximation is $[0.65, 1.13]$, in reasonable agreement with the empirical likelihood confidence interval. This illustrates how empirical likelihood adapts to the skewness of the data without an explicit model assumption. In contrast, the 95% confidence interval under the incorrect assumption that the observations are normally distributed is $[0.58, 1.12]$, which is symmetric around $\overline{y}$ and thus fails to capture the asymmetry of the true distribution.

Figure 6.11 shows how the square root of $\widehat{w}_n(\mu)$ closely approximates the likelihood root statistic derived under the correct exponential distribution compared to the profile likelihood root for the normal model. □

Example 6.18 (Empirical likelihood inference for a quantile) Consider a sample of n independent and identically distributed observations $\{Y_1, \dots, Y_n\}$ from a distribution F. The parameter of interest θ_p is the quantile of order $p \in (0,1)$ defined by $\mathrm{E}\{\Psi(\theta_p; Y)\} = 0$ with $\Psi(\theta_p; Y) = 1(Y \leq \theta_p) - p$. We assume that θ_p lies in the range of the observations, $\theta_p \in [Y_{(1)}, Y_{(n)}]$. The Lagrange multiplier τ solves

$$\sum_{i=1}^{n} \frac{1(Y_i \leq \theta_p) - p}{1 + \tau\{1(Y_i \leq \theta_p) - p\}} = 0, \tag{6.49}$$

which simplifies to

$$n\widehat{p}\frac{1-p}{1+\tau(1-p)} + n(1-\widehat{p})\frac{-p}{1-\tau p} = 0,$$

where $\widehat{p} = n^{-1}\Sigma_i 1(Y_i \leq \theta_p)$. Provided $0 < p < 1$, the solution is $\widehat{\tau} = (\widehat{p} - p)/\{p(1-p)\}$. Substituting $\widehat{\tau}$ into (6.45) gives

$$\widehat{\pi}_i = \frac{1}{n[1 + \widehat{\tau}\{1(Y_i \leq \theta_p) - p\}]},$$

which is either $p/(n\widehat{p})$ or $(1-p)/\{n(1-\widehat{p})\}$, according as $Y_i \leq \theta_p$, or $Y_i > \theta_p$. The corresponding empirical likelihood ratio is

$$\widehat{w}_n(\theta_p) = 2n\left\{\widehat{p}\log\left(\frac{\widehat{p}}{p}\right) + (1-\widehat{p})\log\left(\frac{1-\widehat{p}}{1-p}\right)\right\},$$

which coincides with the parametric likelihood ratio statistic for a binomial model; see Example 1.21. For large n, the empirical likelihood ratio $\widehat{w}_n(\theta_p)$ is approximately distributed as a χ_1^2 random variable. □

We sketch the proof of the empirical likelihood theorem for inference on a mean vector. Suppose that $\{Y_1, \ldots, Y_n\}$ is a sample of independent random vectors with a common distribution F_0 defined on $\mathbb{R}^d$, for some $d \geq 1$. The mean vector is $\mu_0 = \int y \, \mathrm{d}F_0(y)$ and the variance $\Sigma_0 = \int (y-\mu_0)(y-\mu_0)^\mathsf{T} \, \mathrm{d}F_0(y)$. Suppose that $\mathrm{rank}(\Sigma_0) = q \leq d$. The empirical likelihood ratio for the estimating equation $\Psi(\mu; Y) = Y - \mu$ is

$$\begin{aligned}
\widehat{w}_n(\mu_0) &= 2\sum_{i=1}^{n} \log\{1 + \widehat{\tau}^\mathsf{T}(Y_i - \mu_0)\} \\
&\doteq 2\sum_{i=1}^{n}\left\{\widehat{\tau}^\mathsf{T}(Y_i-\mu_0) - \frac{1}{2}\widehat{\tau}^\mathsf{T}(Y_i-\mu_0)(Y_i-\mu_0)^\mathsf{T}\widehat{\tau}\right\} \\
&= n\left\{2\widehat{\tau}^\mathsf{T}(\overline{Y}-\mu_0) - \widehat{\tau}^\mathsf{T}S\widehat{\tau}\right\}, \qquad (6.50)
\end{aligned}$$

where the expansion $\log(1+x) = x - x^2/2 + o(x^2)$ is used. The next step is to derive a convenient approximation for $\widehat{\tau}$ using $(1+x)^{-1} = 1 - x + o(x)$ for $|x| < 1$,

$$\begin{aligned}
0 &= \frac{1}{n}\sum_{i=1}^{n}\frac{Y_i - \mu_0}{1 + \widehat{\tau}^\mathsf{T}(Y_i - \mu_0)} \\
&\doteq \frac{1}{n}\sum_{i=1}^{n}\{(Y_i-\mu_0) - (Y_i-\mu_0)(Y_i-\mu_0)^\mathsf{T}\widehat{\tau}\}.
\end{aligned}$$

It follows that $\widehat{\tau} \doteq S^-(\overline{Y} - \mu_0)$, where $S = n^{-1}\Sigma_i(Y_i-\mu_0)(Y_i-\mu_0)^\mathsf{T}$ and S^- denotes a generalized inverse of S. Using this in (6.50) yields

$$\widehat{w}_n(\mu_0) \doteq n(\overline{Y}-\mu_0)^\mathsf{T}S^-(\overline{Y}-\mu_0) \xrightarrow{d} \chi_q^2.$$

Example 6.19 (Empirical likelihood for parametric regression) Empirical likelihood inference is applied to regression problems by choosing an appropriate estimating equation. The simplest case is the parametric regression model

$$Y = \mu(x;\beta) + \epsilon,$$

where $\mu(x;\beta)$ is a regression function, $\beta \in \mathbb{R}^p$, and ϵ is an error with $\mathrm{E}(\epsilon \mid x) = 0$. Empirical likelihood inference is based on the estimating equations

$$\Psi(\beta; Y, x) = \frac{\partial\mu(x;\beta)}{\partial\beta}\{Y - \mu(x;\beta)\},$$

or, in case of heterogeneous errors $\mathrm{var}(Y \mid x) = \sigma^2(x)$, the weighted estimating equations

$$\Psi(\beta; Y, x) = \frac{\partial\mu(x;\beta)}{\partial\beta}\frac{\{Y - \mu(x;\beta)\}}{\sigma^2(x)}.$$

In some applications, heterogeneity is modelled in terms of some covariates, and two sets of estimating equations can be used for inference on the conditional expectation and the conditional variance of Y given x; see Owen (2001, Section 4.5).

The extension to generalized linear regression models (Example 1.12) uses the quasi-likelihood equations

$$\Psi(\beta; Y, x) = \frac{\partial \mu(x;\beta)}{\partial \beta} \frac{\{Y - \mu(x;\beta)\}}{V\{\mu(x;\beta)\}},$$

where $V(\cdot)$ is the variance function (McCullagh and Nelder, 1989). A general discussion of quasi-likelihood is provided in Section 5.2.

Letting β_0 be the true parameter value for which $\mathrm{E}_0\{\Psi(\beta_0; Y, x)\} = 0$, under the mild regularity assumptions of the empirical likelihood theorem for estimating equations we still have that the empirical likelihood ratio $\widehat{w}_n(\beta_0)$ has a limiting χ^2_q distribution, where q is the rank of the covariance matrix of $\Psi(\beta; Y, x)$. □

Example 6.20 (Empirical likelihoood for nonparametric regression) Under suitable regularity conditions, the chi-squared limit holds also in nonparametric and semi-parametric regression models. Consider the nonparametric regression model

$$Y = \mu(x) + \epsilon,$$

where $\mu(x) = \mathrm{E}(Y \mid x)$ is a smooth function of $x \in \mathbb{R}$. The conditional expectation $\mu(x)$ at a given point x_0 can be estimated with a variety of methods discussed in the previous sections of this chapter, including local polynomials, kernel smoothing, or penalized splines. Empirical likelihood inference can be based on smoothed estimating equations

$$\Psi_h(\mu; Y, x, x_0) = K_h(x - x_0)\{Y - \mu(x_0)\},$$

for a kernel function $K_h(t) = K(t/h)/h$. For large n and $n^{2/5}h^2 \to 0$, it can be shown that the approximate distribution of the empirical likelihood ratio $\widehat{w}_n\{\mu(x_0)\}$ is χ^2_1.

As observed by Chen and Van Keilegom (2009), the empirical log-likelihood ratio can be used to compute confidence intervals about $\widehat{\mu}_h(x)$ without estimating the variance of the errors, because studentization is implicit in the optimization required by the computation of the empirical likelihood ratio.

The discussion extends to semi-parametric modelling. For example, consider the partial linear regression model

$$Y = x^{\mathsf{T}}\beta + g(z) + \epsilon,$$

where x is a covariate vector of length p, β is the corresponding vector of regression coefficients, z is a one-dimensional covariate and $g(\cdot)$ is a continuous but unknown nuisance function. The errors are assumed to be independent with $\mathrm{E}(\epsilon \mid x, z) = 0$ and $\mathrm{var}(\epsilon \mid x, z) = \sigma^2$. For a given β, the nuisance component can be estimated by kernel smoothing. With n observations $\{(Y_1, x_1, z_1), \dots, (Y_n, x_n, z_n)\}$, the kernel smoothing estimator of $g(z)$ is

$$\widehat{g}_h(z) = \frac{\sum_{i=1}^n K_h(z_i - z)(Y_i - x_i^{\mathsf{T}}\beta)}{\sum_{i=1}^n K_h(z_i - z)}.$$

Empirical likelihood inference for β proceeds with the estimating function

$$\Psi(\beta; Y, x, z) = \left\{x + \frac{\partial \widehat{g}_h(z)}{\partial \beta}\right\}\{Y - x^{\mathsf{T}}\beta - \widehat{g}_h(z)\}.$$

Under regularity conditions stated in Wang and Jing (2003), $\widehat{w}_n(\beta_0)$ has a limiting χ^2_p distribution. See Chen and Van Keilegom (2009) for more details and discussion. □

In problems with $\theta = (\psi^\mathsf{T}, \lambda^\mathsf{T})^\mathsf{T}$, where $\psi \in \mathbb{R}^q$ is the parameter of interest and λ is the nuisance parameter, a profile empirical likelihood ratio can be defined, similar to the profile likelihood in Section 2.4, by maximizing the nonparametric likelihood with respect to the nuisance parameter λ

$$\widehat{\Lambda}_{np}(\psi) = \max_{\lambda,\, \pi \in \mathcal{S}_{n-1}} \left\{ \widehat{\Lambda}_n(\pi) : \; \textstyle\sum_{i=1}^n \pi_i \Psi(\psi, \lambda; Y_i) = 0 \right\}.$$

Under mild conditions, the profile empirical likelihood ratio has a limiting χ^2_q distribution; see Owen (2001, Section 3.5).

6.7 Bootstrap likelihood

Another version of nonparametric likelihood inference is provided by the bootstrap as suggested in Davison et al. (1992) and refined in Davison et al. (1995). To motivate the method, consider $\{Y_1, \ldots, Y_n\}$ independent observations from a distribution indexed by $\theta = (\psi^\mathsf{T}, \lambda^\mathsf{T})^\mathsf{T}$, where ψ is the parameter of interest and λ the nuisance parameter. As in Section 2.2, suppose there are statistics $S = s(Y)$ and $T = t(Y)$ that identify a marginal likelihood for ψ:

$$f(y; \psi, \lambda) \propto f_m\{s(y); \psi\} f_c\{t(y) \mid s(y); \psi, \lambda\}.$$

The statistic S is typically an estimator of the parameter of interest ψ. Inference for ψ can be based on the marginal likelihood function $L_m(\psi) \propto f_m\{s(y); \psi\}$ (see Section 2.2) assuming there is little or no information on ψ lost by omitting the conditional density of Y given S. The bootstrap likelihood is a nonparametric analogue of the above marginal likelihood in the sense that bootstrap calculations are used to nonparametrically estimate the density of the statistic S, which is then used as a marginal likelihood function for ψ.

The basic version of the bootstrap likelihood method requires nested bootstrap calculations. The statistic S is assumed to be a function of the empirical distribution function, $S = s(F_n)$. Let $s(y)$ be the value of S computed at the observed data $y = (y_1, \ldots, y_n)^\mathsf{T}$. At the first stage, B_1 bootstrap samples $\{y_1^{*\,b}, \ldots, y_n^{*\,b}\}$ are drawn by resampling with replacement from y, $b = 1, \ldots, B_1$. For each bootstrap sample, the statistic S is computed, giving B_1 replicates, $s_1^*, \ldots, s_{B_1}^*$. Each bootstrap sample is considered to constitute a distinct population indexed by parameter s_b^*, for $b = 1, \ldots, B_1$. Then, B_2 second-level bootstrap samples are drawn by sampling from each of the first-level samples in order to calculate the estimates $s_{b1}^{**}, \ldots, s_{bB_2}^{**}$ of s_b^*, for $b = 1, \ldots, B_1$. Kernel density estimation (Section 6.2) is used to approximate the density of s^{**} at s_b^*,

$$\widehat{f}(s_b^*) = \frac{1}{B_2} \sum_{j=1}^{B_2} K_h(s_{bj}^{**} - s_b^*),$$

for some kernel function $K_h(\cdot)$. The final step consists of some form of nonparametric smoothing of the B_1 density estimates $\widehat{f}(s_b^*), b = 1, \ldots, B_1$, to obtain the bootstrap likelihood for ψ. For this last step of the algorithm, Davison et al. (1992) suggest locally weighted quadratic smoothing (Section 6.3) applied to the logarithm of $\widehat{f}(s_i^*)$ because the log-likelihood is approximately quadratic around the maximum likelihood estimator.

Davison et al. (1992) showed that if the statistic S used to build the marginal likelihood is the root of a monotone estimating equation, standardized so that the leading term is of order one, then the empirical and bootstrap likelihoods agree to $O(n^{-1/2})$. Many commonly

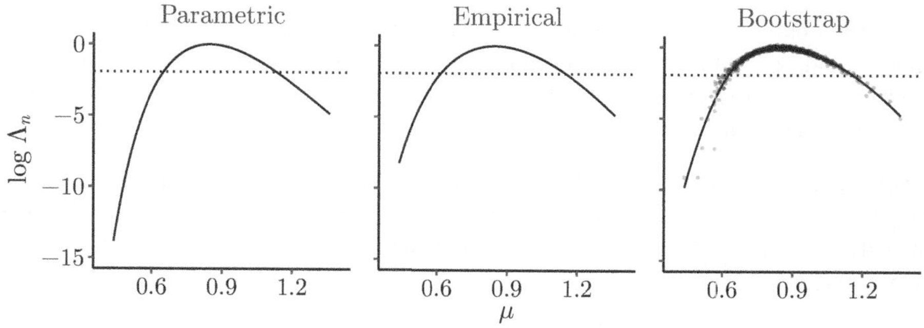

FIGURE 6.12
Comparison of the empirical and bootstrap likelihood functions for the expected value, constructed from a sample of 50 observations from an exponential distribution with $\mu = 1$. The left panel shows the log-likelihood ratio statistic $\log \Lambda_n$ computed with the correct exponential distribution. The central panel and right panel are nonparametric estimates of $\log \Lambda_n$ computed with empirical likelihood and bootstrap likelihood, respectively. The horizontal dotted line identifies the 95% confidence interval for μ. The dots in the right panel are the bootstrap points.

used estimators are indeed determined from a monotone estimating equation, including the sample mean, maximum likelihood estimators that solve the score equations, and many robust estimators.

Example 6.21 (Bootstrap likelihood for a mean, cont'd) As a simple illustration of bootstrap likelihood inference, consider again the sample of 50 observations from the exponential distribution of Example 6.17, with expected value $\mu = 1$. The sufficient statistic is $s(y) = \overline{y} = 0.85$. Figure 6.12 compares the log-likelihood ratio statistic $\log \Lambda_n = \ell(\mu) - \ell(\widehat{\mu})$ based on the true exponential distribution with nonparametric estimates of $\log \Lambda_n$ based on the empirical likelihood and the bootstrap likelihood. The latter was computed using $B_1 = 1000$ bootstrap samples at the first stage, $B_2 = 500$ at the second stage, and approximating the density of s^{**} with a Gaussian kernel density estimate. The bandwidth was estimated using the Sheather–Jones method discussed in Section 6.2.2, and the final smoothing was local weighted quadratic smoothing.

As expected, the empirical and bootstrap likelihood functions closely agree: the 95% confidence interval for μ based on the χ^2 approximation is $[0.62, 1.16]$ for the bootstrap likelihood and $[0.62, 1.15]$ for the empirical likelihood, whereas the parametric likelihood confidence interval is $[0.65, 1.13]$. □

Davison et al. (1995) suggested some variants of the bootstrap likelihood algorithm to reduce its variability, and proposed a saddlepoint approximation to avoid the second level of sampling. When the statistic is too complex to apply the saddlepoint approximation, the variability can be reduced by using importance sampling combined with the bootstrap recycling method of Newton and Geyer (1994).

6.8 Nonparametric Bayesian methods

There is a large and growing literature on Bayesian inference in nonparametric settings. When the parameter is infinite-dimensional, the choice of prior distribution has a potentially large impact on the posterior, even asymptotically, in contrast to the Bayesian theory in Section 1.4. This means that careful attention to the choice of prior is needed in nonparametric problems. Two particular classes of priors that are widely studied are Dirichlet process priors and Gaussian process priors.

Example 6.22 (Density estimation) Suppose that for each i, Y_i has a distribution with density $f(\cdot;\theta_i)$, and we model the parameters θ_i as independent and identically distributed from a distribution P, say. A Dirichlet process prior for P with parameters M and P_0 can be represented as

$$P = \sum_{j=1}^{\infty} p_j \delta_{\nu_j}, \quad p_j = V_j \prod_{k<j}(1 - V_k),$$

where δ_ν is a point mass at ν, $\nu_j \sim P_0$, independently over j, and $V_j \sim \text{Beta}(1, M)$. The resulting model-prior combination is called the Dirichlet process mixture model.

We might instead assume that each Y_i has a density function $f(\cdot)$, and use a Dirichlet prior for the unknown function $f(\cdot)$, but this has the drawback that the posterior densities are discrete, with support on the observations (Ferguson, 1974). In contrast, the posterior density estimate based on the Dirichlet process mixture has a close resemblance to the kernel density estimate of Section 6.2.1, with the modelled density $f(\cdot;\theta_i)$ playing the role of the kernel function. □

Example 6.23 (Nonparametric regression, continued) As in Example 6.20, suppose

$$Y_i = \mu(x_i) + \epsilon_i, \quad i = 1, \ldots, n, \tag{6.51}$$

where the observations are independent, $\mu(x_i) = \text{E}(Y_i \mid x_i)$, and we make the simpler assumption here that $\text{var}(\epsilon_i) = \sigma^2$. A Bayesian approach to inference about the unknown function $\mu(\cdot)$ requires a prior on (smooth) functions. In this model x_i could be scalar and indexed by time, or a spatial variable in $\mathbb{R}^2$ or $\mathbb{R}^3$, or a higher-dimensional feature vector. A Gaussian process prior for a function is determined by its mean function $m(x)$ and a covariance function $\gamma(x, x')$ and is defined by the property that all finite-dimensional distributions are jointly multivariate normal. For example, under a Gaussian process prior the pair of random variables $\mu(x_i)$ and $\mu(x_{i'})$ have covariance function $\text{cov}\{\mu(x_i), \mu(x_{i'})\} = \gamma(x_i, x_{i'})$. A popular stationary model for $\gamma(\cdot)$ is the exponential covariance function $\gamma(x, x'; \alpha) = \sigma^2 \exp\{-||x - x'||/\alpha\}$, where α is a parameter of the prior. Other stationary covariance models for Gaussian process priors are those used in geostatistics; see Section 3.4. The Brownian motion kernel $\gamma(x, x') = \sigma^2 \min(x, x')$ gives a nonstationary Gaussian process with sample paths that are continuous but nowhere differentiable. If the errors in (6.51) are assumed to be distributed as standard normal, then the posterior distribution of $\mu(\cdot)$ given the data is also a Gaussian process. □

Of particular interest in relation to the theory developed in earlier chapters is the asymptotic behaviour of posterior distributions in the nonparametric setting. Two properties of the posterior that are typically studied are posterior consistency and posterior concentration rates. The discussion here draws heavily on Rousseau (2016).

A posterior distribution is consistent if it concentrates around the true value of the parameter; "around" here is interpreted in the context of a loss function or measure of distance in the parameter space. More formally, for $\epsilon > 0$, consistency requires

$$\pi\{d(\theta, \theta_0) < \epsilon \mid Y\} \longrightarrow 1, \quad n \to \infty, \tag{6.52}$$

where the convergence is with respect to the sampling distribution of $Y = \{y_1, \ldots, y_n\}$ under the model with parameter θ_0. The role of the loss function $d(\cdot, \cdot)$ is to describe neighbourhoods of θ_0, typically based on a measure of distance or divergence between the density functions $f(y_i \mid \theta)$ and $f(y_i \mid \theta_0)$, such as Hellinger distance or Kullback–Leibler divergence.

The rate of concentration in (6.52) is useful for studying both the frequentist risk of Bayes estimators and the frequentist coverage of posterior credible regions. Posterior concentration at rate ϵ_n is defined in Rousseau (2016) by

$$\mathrm{E}_{\theta_0}[\pi\{d(\theta, \theta_0) < M\epsilon_n \mid Y\}] \longrightarrow 1, \quad n \to \infty, \tag{6.53}$$

where again the expectation is with respect to the sampling distribution $f(y \mid \theta_0)$. As in the finite-dimensional setting, a posterior credible set C_α is defined by

$$\pi(\theta \in C_\alpha \mid Y) \geq 1 - \alpha. \tag{6.54}$$

In nonparametric problems such sets would typically be obtained using the loss function as

$$C_\alpha = \{\theta : d(\theta, \widehat{\theta}_{\mathrm{B}}) \leq c_\alpha\}, \tag{6.55}$$

where $\widehat{\theta}_{\mathrm{B}}$ is a Bayes estimator of θ, such as the posterior mean or median, or maximum a-posteriori (MAP) estimator. If (6.53) holds, then

$$\int_\Theta P_\theta(\theta \in C_\alpha)\pi(\theta)\, d\theta \geq 1 - \alpha;$$

where $P_\theta(A) = \int_A f(y \mid \theta)\, dy$ is the model probability for the set A in the sample space of $Y = \{Y_1, \ldots, Y_n\}$. This is weaker than frequentist coverage of the credible set, as it is averaged over the prior.

Establishing posterior consistency relies on a strategy first developed by Schwartz (1965), under the conditions (i) that the prior puts enough probability in a neighbourhood of the true parameter, and (ii) that a statistical test can be constructed for testing $H_0 : \theta = \theta_0$ against $H_1 : d(\theta, \theta_0) \geq \epsilon$.

Establishing posterior concentration bounds at rate ϵ_n also requires two similar conditions, but with a stronger constraint on the neighbourhood probability and the size and power of the test. The detailed conditions and various results that have been obtained for density estimation and nonparametric regression are reviewed in Rousseau (2016, Sections 3.2, 3.3).

Spokoiny and Panov (2025) described a different approach to studying the posterior distribution and the Bernstein–Von Mises theorem, which they call non-asymptotic, as it gives finite-sample bounds. The parameter space can be of arbitrarily high dimension, but there are constraints on what they call the effective dimension of the parameter.

Broadly speaking, studying properties of Bayesian posterior distributions for infinite-dimensional parameters depends both on the smoothness of the function that serves as the parameter, such as the density or regression function, and on properties of the prior. Often smoothness of a function is imposed by modelling the function as a linear combination of basis functions with the coefficients treated as a sequence of unknown parameters. For example, Spokoiny and Panov (2025) modelled the log-density function $f(y)$ as

$$\log f(y; \theta) = \sum_{j=1}^{p} \varphi_j(y)\theta_j - \kappa(\theta),$$

where $\{\varphi_1(y), \ldots, \varphi_p(y)\}$ is a set of known basis functions and $\theta = \{\theta_1, \ldots, \theta_p\}$ are parameters, where $1 \leq p \leq \infty$, and consider priors for θ that are either truncated or smoothed; for example concentrated on vectors θ for which $\Sigma_{j=1}^{\infty} j^{2s}\theta_j^2 \leq w$ for some choice of s, w.

There is a close relationship between nonparametric models and models with very high-dimensional parameters; the latter were discussed briefly in Section 2.9.

Example 6.24 (The sequence model) The model

$$Y_i = \theta_i + \epsilon_i, \quad \epsilon_i \sim N(0,1), \quad i = 1, \ldots, p,$$

is often called the sequence model and is a version of nonparametric regression, since the expected values are arbitrary. Instead of imposing smoothness in the context of the nonparametric regression model of Example 6.23, it may be appropriate to assume that only a small number of the θ_i are non-zero. In this setting a three-stage prior is often proposed (Castillo and van der Vaart, 2012). The first prior $\pi_p(\cdot)$ is for the number of non-zero values $|S|$, the second is typically uniform on the set of non-zero values $S \subset \mathbb{N}$, and the third prior is a proper prior $g_S(\cdot)$ on the non-zero values θ_S. The full prior is then

$$\pi(S, \theta) \propto \pi_p(|S|) \binom{p}{|S|}^{-1} \prod_{j \in S} g_S(\theta_j),$$

with zero prior probability on components of θ not in S. The MAP estimator from such a class of priors can often be interpreted as a penalized maximum likelihood estimator. Castillo et al. (2015) suggested using Laplace priors for the components θ_j to form $g_S(\theta_S)$, and a class of priors on the sparsity of the form

$$\pi_p(s) \propto c^{-s} p^{-as}, \quad s = 0, \ldots, p.$$

Posterior concentration of the posterior for θ in the sequence model is studied in Castillo and van der Vaart (2012) and extended to high-dimensional regression in Castillo et al. (2015). □

6.9 Nonparametric prediction

Section 1.5 describes prediction intervals based on parametric models; here we describe a nonparametric method. This could in principle be based on the posterior predictive distribution, but we will confine attention to non-Bayesian methods. Suppose $\{Y_1, \ldots, Y_n\}$ are independent, identically distributed observations from a continuous distribution taking values on a space $\mathcal{Y}$. A prediction set from a sample of size n for a new observation, Y_{n+1}, is a set $C_n \subset \mathcal{Y}$,

$$C_n = C_n(y_1, \ldots, y_n), \tag{6.56}$$

and we would usually require that for some (small) value of α

$$\text{pr}(Y_{n+1} \in C_n) \geq 1 - \alpha, \tag{6.57}$$

and further that C_n be in some sense as small as possible. For $\mathcal{Y} = \mathbb{R}$, C_n will be an interval, either one-sided or two-sided. One way to construct a one-sided interval is to note that as $\{Y_1, \ldots, Y_{n+1}\}$ are assumed to be independent and identically distributed, the rank of Y_{n+1} in this set follows a discrete uniform distribution on the set $\{1, \ldots, n+1\}$. Writing the order statistic for the observations as $y_{(1)} < \cdots < y_{(n)}$, the set $C_{n,r} = \{y \leq y_{(r)}\}$ satisfies $\text{pr}(Y_{n+1} \in C_{n,r}) = r/(n+1)$, so for $1 - \alpha = r/(n+1)$ we have achieved the desired coverage. Defining $r = \lceil (1-\alpha)(n+1) \rceil$ for $0 < \alpha < 1$, it can be shown that if there are no ties among $y_1, \ldots, y_n$, that

$$1 - \alpha \leq \text{pr}(Y_{n+1} \in C_{n,r}) \leq 1 - \alpha + \frac{1}{n+1}. \tag{6.58}$$

In the notation above the dependence on the original sample is implicit, but it is important that in (6.58) the probability refers to the joint distribution of $(Y_1, \dots, Y_{n+1})$. There are situations where it is of more interest to ensure conditional coverage,

$$\mathrm{pr}(Y_{n+1} \in C_n \mid y_1, \dots, y_n) \geq 1 - \alpha. \tag{6.59}$$

In the case of one-sided or two-sided interval sets C_n, these are usually called tolerance intervals; see the bibliographic notes. For a similar approach to prediction in a regression setting, suppose $\{(X_1, Y_1), \dots, (X_n, Y_n)\}$ is an i.i.d. sample from some distribution F on $\mathbb{R}^{p+1}$, where X_i is a p-dimensional covariate vector and $Y_i \in \mathbb{R}$. A prediction set for a new observation Y_{n+1} given X_{n+1}, analogous to (6.56) with a coverage guarantee of the form (6.57) is

$$C_n(X_{n+1}) = C_n\{X_{n+1}; (X_1, Y_1), \dots, (X_n, Y_n)\}, \quad \mathrm{pr}\{Y_{n+1} \in C_n(X_{n+1})\} \geq 1 - \alpha, \tag{6.60}$$

where now the probability is over the joint distribution of $\{(X_1, Y_1), \dots, (X_{n+1}, Y_{n+1})\}$. A construction similar to the order statistic version above is to define some score $S(X, Y)$ and order the observed values $s_i = S(x_i, y_i)$, $i = 1, \dots, n$. Then, as above, if there are no ties among the scores, we can define $C_{n,r}(X) = \{y : S(X, y) \leq s_{(r)}\}$, and by the i.i.d. assumption,

$$\mathrm{pr}\{Y_{n+1} \in C_{n,r}(X_{n+1})\} = r/(n+1), \tag{6.61}$$

and as above

$$1 - \alpha \leq \mathrm{pr}\{Y_{n+1} \in C_{n,r}(X_{n+1})\} \leq 1 - \alpha + \frac{1}{n+1}. \tag{6.62}$$

The scores are typically based on some regression model fit to the observed data, giving fitted values $\widehat{y}_i = \widehat{f}(x_i)$, say, where $f(\cdot)$ might be for example a smoothing spline, or a kernel regression estimator. Scores are designed to be large when the model fits poorly, two examples are $s_i = |y_i - \widehat{f}(x_i)|$ and $s_i = |y_i - \widehat{f}(x_i)|/\widehat{\sigma}$. However the collection of scores $s_1 < \cdots < s_{n+1}$ are not i.i.d., as $\widehat{f}(x_{n+1})$ depends on $(x_1, y_1), \dots, (x_n, y_n)$, so the argument above cannot be applied directly.

A widely-used solution splits the data into two sets; a training set $\mathcal{D}_1$ of size n_1 on which to fit the model, and a calibration set $\mathcal{D}_2$ of size $n_2 = n - n_1$ to define the ordered scores. The estimated model from the training set is $\widehat{f}(x)$, and the scores $s_i, i = 1, \dots, n_2$ are evaluated for the calibration set. Then with $C_{n_2}(X) = \{y : S(X, y) \leq s_{(r)}\}$, where $r = \lceil (1 - \alpha)(n_2 + 1) \rceil$, we have analogously to (6.62),

$$1 - \alpha \leq \mathrm{pr}\{Y_{n+1} \in C_{n,r}(X_{n+1}) \mid \mathcal{D}_1\} \leq 1 - \alpha + \frac{1}{n_2 + 1}. \tag{6.63}$$

Generalizations of this procedure to different regression models and other score functions is in principle straightforward. The coverage guarantee is valid for any n and any method of fitting a model to the training data. However, the coverage does not provide any information about the size of the prediction set, which will depend on the choice of the score function.

The coverage guarantee is over the joint distribution of the calibration data set and the new point X_{n+1} to be predicted. Prediction sets with guaranteed coverage conditional on X_{n+1} are more relevant to specific prediction problems, satisfying

$$\mathrm{pr}\{Y_{n+1} \in C_{n_2}(X_{n+1}) \mid \mathcal{D}_1, X_{n+1}\} \geq 1 - \alpha. \tag{6.64}$$

Lei and Wasserman (2014) showed that it is not possible to attain coverage guarantees with prediction sets of finite size, and develop a method that ensures approximate guarantees of the form (6.63) as $n \to \infty$.

The ordering of the scores is sometimes called conformalization, and this method of constructing prediction sets is called conformal prediction. Adjustments are available for dealing with ties in the ordered scores. At the time of writing research on conformal prediction is developing rapidly. It has gained particular favour as a method of uncertainty quantification in machine learning; see the bibliographic notes. Extensions to quantile regression, multivariate predictors, classification, optimal choice of scores, are just a few of the topics being investigated.

6.10 Bibliographic notes

The first derivation of the nonparametric maximum likelihood estimator of F as $\widehat{F}_n$ is attributed to Kiefer and Wolfowitz (1956). Classical references on the theory of nonparametric maximum likelihood estimation of mixing distributions include Laird (1978) and Lindsay (1995). The EM algorithm for nonparametric maximum likelihood estimation of finite mixtures was derived in Aitkin and Rubin (1985). A detailed overview of mixture models is McLachlan et al. (2019). The product-limit estimator (6.3) is also called the Kaplan–Meier estimate after Kaplan and Meier (1958); Efron (1967) discussed its self-consistency.

Empirical Bayes estimation was introduced in Robbins (1956), with the Robbins formula (6.6) also derived by Good and Toulmin (1956). The concept of "borrowing strength" in the estimation of a vector of parameters was notably demonstrated by the seminal James–Stein shrinkage estimator (James and Stein, 1961). The principles of empirical Bayes were significantly advanced by Efron and Morris (1972). Nonparametric empirical Bayes estimation has been applied to genomics (Efron et al., 2001) and to estimation of false discovery rates (Efron, 2014).

Kernel density estimators were introduced in Parzen (1962) and Rosenblatt (1956), and are also known as Parzen–Rosenblatt window estimators. Classical references on kernel density estimation include Silverman (1986), Scott (1992), and Wand and Jones (1994). There seems to be little consensus on the choice of the bandwidth; see Jones et al. (1996), Loader (1999b) or Heidenreich et al. (2013). See also Köhler et al. (2014) for bandwidth selection in kernel regression.

An early example of shape-constrained density estimation is in Grenander (1956), which considers maximum likelihood estimation of a decreasing density on the non-negative reals. While the properties of the Grenander estimator are appealing, the class of decreasing densities is quite restrictive, and its generalization to multivariate settings is not particularly natural (Samworth, 2018). Dümbgen and Rufibach (2009) investigated the properties and uniform consistency of the nonparametric maximum likelihood estimator of a log-concave density. Wang (2019) reviewed algorithms for computing the nonparametric maximum likelihood estimate of a univariate log-concave density. Overviews of shape-constrained estimation include Walther (2009), Samworth (2018), and Dümbgen (2024).

The idea of local likelihood appeared in Brillinger (1977) and was developed in Tibshirani (1984) and Tibshirani and Hastie (1987). Local likelihood estimation extends previous work on scatterplot smoothing by local fitting (Cleveland, 1979; Friedman and Stuetzle, 1981) and locally weighted regression (Cleveland and Devlin, 1988).

Fan et al. (1998) developed methodology for bandwidth selection and computation of confidence intervals in local likelihood estimation. Loader (1999a) gave a detailed account of local fitting methods with associated software. Fan et al. (1997) studied local likelihoods and local partial likelihood for survival analysis. Hjort and Jones (1996) and Loader (1996) discussed local likelihood for density estimation.

The sieve estimator obtained by truncating the series representation (6.28) is sometimes referred to as the series estimator; see for example Eubank (1999, Chapter 3). A classical reference on sieve estimation is Geman and Hwang (1982). Shen and Wong (1994) studied the convergence rate of sieve estimators, Shen (1997) compared the relative efficiency of sieve estimators with those of maximum likelihood and maximum penalized likelihood estimators, and Chen (2007) gave a detailed account of sieve estimation in semi-nonparametric models used in econometrics. Hansen (2014) discussed weighted averages of sieve regression estimators that have lower integrated mean-squared error than sieve estimators of a selected order.

Schoenberg (1946) formalized the mathematical concept of splines as piecewise polynomial functions. de Boor (1978) is a classical reference for spline interpolation using B-splines, detailing their properties and computation. Wahba (1978) was instrumental in bringing spline methods from numerical analysis into the statistical community, developing smoothing splines and generalized cross-validation for selection of the penalty parameter (Golub et al., 1979; Wahba, 1990). Silverman (1986) is a key reference for nonparametric density estimation, including the use of kernel methods and penalized likelihood smoothing with splines, while Green and Silverman (1993) detailed the roughness penalty approach for nonparametric regression and generalized linear models. Efficient P-splines, utilizing B-splines with a penalty on differences of coefficients, were introduced by Eilers and Marx (1996). Regression and smoothing splines were discussed in Hastie et al. (2009, Ch. 5) and James et al. (2013, Ch. 7).

Generalized additive models were introduced in Hastie and Tibshirani (1986) to incorporate multiple spline terms in generalized linear models. The backfitting algorithm was introduced by Friedman and Stuetzle (1981) in the context of projection pursuit regression. It was subsequently employed in the alternating conditional expectations algorithm (Breiman and Friedman, 1985) to identify optimal nonparametric transformations for both response and predictors. Hastie and Tibshirani (1986, 1990) later adapted the procedure to generalized additive models. Wood (2017) significantly advanced the theory and implementation of generalized additive models with splines, particularly through the development of stable and efficient fitting algorithms; see also Wood (2025). The mixed model approach to spline modelling is discussed in detail in Ruppert et al. (2003).

The attractive properties of empirical likelihood have generated a considerable literature as described in the monograph by Owen (2001) and in the review papers Lazar (2021) and Liu and Zhao (2023).

DiCiccio et al. (1991) showed that empirical likelihood is Bartlett-correctable, giving adjusted empirical likelihood confidence region with a coverage error of order $O(n^{-2})$. Chen (1993) discussed Bartlett correction of empirical likelihood for linear regression models. Chen and Cui (2006) proved that the Bartlett correction can be extended to the profile empirical likelihood to handle nuisance parameters.

Empirical likelihood for regression models in a variety of settings, including the presence of missing data and censored observations, is reviewed in Chen and Van Keilegom (2009). Newey and Smith (2004) discussed empirical likelihood for the generalized method of moments. Empirical likelihood is particularly attractive for the analysis of survival data. In fact, Owen's development of the empirical likelihood was motivated by earlier work of Thomas and Grunkemeier (1975) on nonparametric confidence intervals for survival probabilities under random censoring.

Empirical likelihood for high-dimensional models is reviewed in Liu and Zhao (2023). Hjort et al. (2018) introduced a hybrid likelihood function, a mixture of parametric likelihood and empirical likelihood, and study the properties of the resultant Wald and likelihood-ratio type statistics. Dæhlen and Hjort (2025) considered the robustness of this hybrid likelihood inference under model misspecification.

The book by Hjort et al. (2010) and the references in Section 6.8 provide an entry to the large literature on nonparametric Bayesian methods. Empirical Bayes approaches, where the parameters in the prior are estimated from the marginal distribution, were discussed in Petrone et al. (2014) and Zhang and Gao (2020) and references therein. Other choices of prior distribution for the sequence model of Example 6.24 and its linear regression extension include the spike-and-slab prior (e.g., Ročková and George, 2018) and the horseshoe prior (Carvalho et al., 2010; van der Pas et al., 2017).

The distinction between prediction intervals and tolerance intervals was clarified in Carroll and Ruppert (1991). In Fraser and Guttman (1956) prediction intervals were called expectation-tolerance intervals. Conformal prediction was reviewed in Angelopoulos and Bates (2023). The discussion here draws heavily on Lei and Wasserman (2014), Lei et al. (2018), and the online notes by Tibshirani (2023). Section 2 in Lei et al. (2018) reviews both split conformal inference and full conformal inference, and emphasizes the link to hypothesis testing of candidate values.

6.11 Further results and exercises

6.1. Verify that if there is no censoring, then the product-limit estimator is $\widehat{S}_n(y) = 1 - \widehat{F}_n(y)$. Derive Greenwood's formula for this case. *Davison (2003, Exercise 5.4.1)*

6.2. Replicate the simulation study described at the end of Example 6.2 and summarized in Figure 6.2.

6.3. Instead of central differences (6.13), some presentations of the nonparametric density estimator use forward differences

$$\widetilde{f}_h(y_0) = \frac{\widehat{F}_n(y_0 + h) - \widehat{F}_n(y_0)}{h}.$$

Verify that the convergence for the forward difference estimator $\widetilde{f}_h(y_0)$ is slower than that of the central difference estimator $\widehat{f}_h(y_0)$ in the sense that the fastest rate of convergence of the mean squared error of $\widetilde{f}_h(y_0)$ is $O(n^{-2/3})$, compared to $O(n^{-4/5})$ for $\widehat{f}_h(y_0)$.

6.4. The derivation of the approximately optimal bandwidth for nonparametric density estimation, as found in the rules of thumb proposed by Scott (1992) and Silverman (1986), relies on the identity

$$\|f''(y)\|_2^2 = \int_{-\infty}^{\infty} \{f''(y)\}^2 \, dy = \frac{3}{8\sigma^5\sqrt{\pi}},$$

where $f(y)$ is the density of a normal random variable with zero mean and variance σ^2. Verify this identity. Discuss the impact on the smoothing properties of the estimated density if normality is assumed when the true density $f(y)$ actually follows a t-distribution.

6.5. Verify the expression for AMISE$(\widehat{f}_h)$ given in (6.17) and compute σ_K^2 and $\|K(y)\|_2^2$ for the Epanechikov kernel $K(u) = 3/4(1 - u^2)$.

6.6. Verify the expression of the biased cross-validation objective function given in (6.19).

6.7. Show that the trigonometric polynomials (6.30) constitute a set of orthogonal basis functions. Explain why orthogonality of the basis functions is important for sieve estimation.

6.8. Simulate 100 observations from the model $Y = \Phi(x) + \sigma\epsilon$, where $\Phi(\cdot)$ is the cumulative distribution function of a standard normal variable, x is drawn from a uniform distribution in the interval $(-5, 5)$, $\sigma = 0.1$. and ϵ is drawn from a standard normal distribution. Using some available software for local fitting, such as the R package `locfit` (Loader, 2024), fit the simulated data and compare the results obtained by varying the kernel function, the polynomial degree and the bandwidth.

6.9. The extended log-likelihood function $\ell(f)$ in (6.25) enables maximum likelihood estimation for any non-negative integrable function f, not necessarily a density function. To show this, verify that $\ell(f)$ satisfies the inequality $\mathrm{E}_f\{\ell(f_1)\} \leq \mathrm{E}_f\{\ell(f)\}$ for any non-negative integrable function f_1, with equality if and only if $f_1 = f$. The proof of this result uses the inequality

$$\log f_1(y) \leq \log f(y) + \frac{1}{f(y)}\{f_1(y) - f(y)\}.$$

As a special case, check that maximizer of the extended log-likelihood function for $f(y; \mu, C) = C\exp\{-(y-\mu)^2/2\}$ gives $\widehat{C} = (2\pi)^{-1/2}$. *Loader (1996, Section 5.4.1)*

6.10. In the case that Y follows a normal distribution with mean $\mu(x)$ and variance σ^2, the normal posterior (6.42) for β is used for Bayesian approximate confidence intervals around the smoothing spline fitted line $\mu(x_i; \widehat{\beta}_\lambda)$. These intervals depend on the tuning parameter λ, which can be estimated by empirical Bayes. Verify that the empirical Bayes estimate of λ is equivalent to the restricted maximum likelihood estimate of Section 3.2.5.

6.11. A B-spline of degree k is defined recursively. For degree 0 (constant piecewise functions), the B-spline is the rectangular pulse,

$$\varphi_{r,0}(x) = 1(\tau_r \leq x < \tau_{r+1}), \quad r = 0, \ldots, m. \tag{6.65}$$

For degree $k > 0$, the B-spline is

$$\varphi_{r,k}(x) = \frac{x - \tau_r}{\tau_{r+k} - \tau_r}\varphi_{r,k-1}(x) + \frac{\tau_{r+k+1} - x}{\tau_{r+k+1} - \tau_{r+1}}\varphi_{r+1,k-1}(x), \quad r = 0, \ldots, m. \tag{6.66}$$

(a) Show that each $\varphi_{r,k}(x)$ is non-zero only over the interval $[\tau_r, \tau_{r+k+1})$. This local support means that changing a spline coefficient β_r affects the spline curve in a localized region.

(b) Verify that the integrals (6.35) that form the penalty matrix Ω used in smoothing splines have a closed-form expression when the basis functions (6.65)–(6.66) are used.

6.12. Verify that ties in the sample data $\{Y_1, \ldots, Y_n\}$ do not affect the theory of empirical likelihood inference. Suppose that there are $k \leq n$ distinct observations in the sample and introduce weights w_i associated with each observation such that $\pi_j = \Sigma_i 1(Y_j = Y_i)w_i$, for $j = 1, \ldots, k$. Empirical likelihood inference can then be formulated as described in Section 6.6 using the nonparametric likelihood $L(F) = \Pi_i w_i$. *Owen (2001, Section 2.3)*

6.13. Consider a sample $\{Y_1, \ldots, Y_n\}$ of independent variables from the Laplace distribution

$$f(y; \mu, \sigma) = \frac{1}{2\sigma}\exp\left(-\frac{|y-\mu|}{\sigma}\right), \quad y \in \mathbb{R},\ \mu \in \mathbb{R},\ \sigma \in \mathbb{R}_+.$$

(a) Show that the likelihood ratio statistic for μ based on the parametric model, after profiling over σ, is

$$w_{\mathrm{p}}(\mu) = -2n\log\left\{\frac{\sum_{i=1}^n |y_i - \mathrm{med}(y)|}{\sum_{i=1}^n |y_i - \mu|}\right\},$$

where $\mathrm{med}(y)$ is the median of the sample observations.

(b) Compare numerically empirical likelihood confidence intervals for the population mean with the parametric likelihood ratio confidence intervals calculated assuming (i) the correct Laplace distribution and (ii) an incorrect normal distribution for the observations. Evaluate the confidence intervals for $\mu = 0$ with varying values of σ.

(c) Repeat the exercise from (b), but consider inference for the population median. Discuss the differences between the results obtained with those obtained in (b) for the mean.

6.14. Example 6.18 discusses empirical likelihood inference for the pth quantile, θ_p, defined as $\mathrm{E}\{\Psi(\theta_p; Y)\} = 0$ with $\Psi(\theta_p; Y) = 1(Y \leq \theta_p) - p$. In the continuous case, this definition corresponds to the usual definition of the pth quantile,

$$\mathrm{pr}(Y \leq \theta) \geq p \quad \text{and} \quad \mathrm{pr}(Y \geq \theta) \geq 1 - p, \quad p \in (0, 1).$$

When the distribution of Y is discrete, the two definitions are not equivalent, and this can be problematic when there are ties in the observations. Explain why a simple remedy for empirical likelihood inference on θ_p is jittering of the observations, and suggest how it could be implemented. *Owen (2001, Section 3.7)*

6.15. Instead of computing $\widehat{\tau}$ as the root of (6.46) subject to the n inequality constraints (6.47), it is possible to compute $\widehat{\tau}$ by solving a simpler unconstrained minimization. Define the pseudo-logarithm function

$$\log_*(z) = \begin{cases} \log(z), & z > 1/n, \\ \log(1/n) - 1.5 + 2nz - (nz)^2/2, & z \leq 1/n. \end{cases}$$

Verify that $\widehat{\tau}$ is the minimizer of $\varphi(\tau) = -\Sigma_i \log_*\{1 + \tau^{\mathsf{T}} \Psi(\theta; y_i)\}$ over the unconstrained space $\mathbb{R}^d$. *Owen (2001, Section 3.14).*

6.16. Consider the prediction interval described above (6.63). Show that the fully conditional coverage

$$\mathrm{pr}\{Y_{n+1} \in C_{n,r}(X_{n+1}) \mid (x_1, y_1), \ldots, (x_n, y_n)\}$$

follows a Beta$(k_\alpha, n_2 + 1 - k_\alpha)$ distribution, in the event of no ties among the scores, where $k_\alpha = \lceil (1 - \alpha)(n_2 + 1) \rceil$, and n_2 is the size of the calibration set. What are the expected value and variance of this distribution?

Tibshirani (2023)

6.17. In the fully exchangeable case described at (6.57), show that the prediction set $C_{n,r}$ is equivalent to

$$\{y : p(y) \geq \lceil \alpha n/(n + 1) \rceil\},$$

where $p(y)$ is the p-value for a test of the hypothesis $H : Y_{n+1} = y$.

Tibshirani (2023)

7

Regularized Likelihood

7.1 Introduction

Inference based on the maximum likelihood estimator or the likelihood ratio statistic typically requires the sample size n to be larger than the number of variables p, and the asymptotic theory discussed in Chapters 1 and 2 usually considers p fixed as $n \to \infty$. In more complex or high-dimensional models, however, p can exceed n and may even grow with n at polynomial or exponential rates. Sometimes special techniques allow construction of marginal or conditional likelihoods, as in the Neyman–Scott problems discussed in Section 2.6. However, when the full likelihood function is used, the maximum likelihood estimator may encounter challenges such as non-identifiability, overfitting, and unstable estimates.

A widely used class of methods for addressing these issues consists of regularized likelihood approaches, which incorporate prior knowledge or preferences (such as model simplicity) into the estimation process. One common form of regularized likelihood is penalized likelihood, though the terms are often used interchangeably. Penalized likelihood methods typically add a penalty term to the log-likelihood function to regularize the resulting parameter estimates. While these techniques are effective in mitigating the challenges of high-dimensional settings, making valid inferences, such as constructing confidence intervals, performing hypothesis tests, or controlling false discovery rates, remains considerably more challenging and is an area of ongoing research. Penalization in nonparametric models was introduced in Section 6.5.

In this chapter, we introduce penalized likelihood and discuss relevant concepts. For simplicity in this chapter we use uppercase letters for both random variables and their realizations. The proofs of the theorems are provided in the appendix.

7.2 Regression with $p > n$

Suppose $\{Y_1, \ldots, Y_n\}$ is a sequence of independent random variables, and for each $i = 1, \ldots, n$, X_i is an associated p-dimensional covariate vector. Consider the linear regression model

$$Y_i = X_i^\top \beta + \epsilon_i, \tag{7.1}$$

where ϵ_i is the random error term assumed to be independent of X_i, with mean zero and variance σ^2. For simplicity, Y_i and X_i are centred by subtracting their sample means, so that an intercept is not needed in the model and does not appear in $\beta = (\beta_1, \ldots, \beta_p)^\top$.

In inference about the regression parameter β, σ^2 is often treated as a nuisance parameter. To illustrate the use of the likelihood method for estimating the parameters, we assume that ϵ_i follows a normal distribution $N(0, \sigma^2)$.

DOI: 10.1201/9780429090936-7

The likelihood function for β is

$$L(\beta) = \prod_{i=1}^{n} \frac{1}{\sqrt{2\pi}\sigma} \exp\left\{ -\frac{(Y_i - X_i^{\mathsf{T}}\beta)^2}{2\sigma^2} \right\},$$

and the maximum likelihood estimator $\widehat{\beta}$ is found by solving the optimization problem

$$\widehat{\beta} = \text{argmin}_{\beta} \left\{ \frac{1}{n} \sum_{i=1}^{n} (Y_i - X_i^{\mathsf{T}}\beta)^2 \right\} = \text{argmin}_{\beta} \left\{ \frac{1}{n} (Y - X\beta)^{\mathsf{T}} (Y - X\beta) \right\}, \tag{7.2}$$

where $X = [X_1 \; X_2 \; \dots \; X_n]^{\mathsf{T}}$ is an $n \times p$ design matrix, and $Y = (Y_1, \dots, Y_n)^{\mathsf{T}}$.

Differentiating the objective function in (7.2) with respect to β and setting it to zero yields

$$X^{\mathsf{T}}Y - X^{\mathsf{T}}X\widehat{\beta} = 0, \tag{7.3}$$

leading to the least-squares solution $\widehat{\beta} = (X^{\mathsf{T}}X)^{-1}X^{\mathsf{T}}Y$ if the matrix $X^{\mathsf{T}}X$ is invertible. However, if $p \geq n$, X will not be of full column rank, so (7.3) does not have a unique solution and admits infinitely many possible solutions, indicating that the model is not identifiable and the likelihood method will overfit the data by perfectly fitting all observations.

One possible approach to addressing this issue is to manipulate the parameter space by reducing its size. This can be achieved by imposing constraints on the parameters β. One may opt to reduce the number of covariates entering the model to make it smaller than the sample size or, more generally, to restrict the range of the parameters so that certain regions of the parameter space become inadmissible. This consideration is also driven by the presumption that, in applications with a large number of potential covariates, the majority do not impact the response; so the true vector β has many zeros, which is described as sparsity.

While there is no single approach to constraining the parameter space, a few methods have emerged as useful in applications. Here, we present two examples focusing on imposing constraints on the optimization problem (7.2).

Example 7.1 (Best subset selection) A natural approach is to impose the following constraint on the optimization problem (7.2) to limit the number of covariates entering the model, solving

$$\min_{\beta} \left\{ \frac{1}{n} (Y - X\beta)^{\mathsf{T}} (Y - X\beta) \right\}, \quad \text{subject to } \sum_{j=1}^{p} 1(\beta_j \neq 0) \leq s, \tag{7.4}$$

where $s < n$ is the sparsity, here a pre-specified positive integer, and $1(\cdot)$ is the indicator function.

Best subset selection essentially reduces the parameter space from $\mathbb{R}^p$ to $\mathbb{R}^s$, yielding a model in which the number of parameters to be estimated is smaller than the sample size. While this procedure is conceptually intuitive, it has a notable drawback due to the difficulty of solving a non-convex optimization problem with a discontinuous constraint. While in principle the minimizer can be obtained by fitting all possible $\binom{p}{s}$ models by least squares, this is computationally prohibitive when p is large. □

An alternative strategy is to impose a shrinkage constraint on the parameter space so that the parameter space is bounded, making it significantly smaller than $\mathbb{R}^p$, as illustrated in the following example.

Example 7.2 (Ridge regression and lasso) Ridge regression and least absolute shrinkage and selection operator (lasso) regression were introduced in Example 2.26. The estimators constrain the regression parameter so that large values in the parameter space $\mathbb{R}^p$ become inadmissible. Without additional information about the covariates, these methods treat all parameters equally and shrink their values by placing an upper bound on the total of their absolute values or squared values. For this reason, it is essential in applying the shrinkage that all covariates are measured on the same scale, and this is typically achieved by centring and scaling each column of the design matrix X to have sample mean zero and sample variance 1.

The lasso estimator solves the optimization problem

$$\min_{\beta}\left\{\frac{1}{n}(Y - X\beta)^{\mathsf{T}}(Y - X\beta)\right\}, \qquad \text{subject to } \sum_{j=1}^{p} |\beta_j| \leq c, \tag{7.5}$$

and ridge regression solves the optimization problem

$$\min_{\beta}\left\{\frac{1}{n}(Y - X\beta)^{\mathsf{T}}(Y - X\beta)\right\}, \qquad \text{subject to } \sum_{j=1}^{p} \beta_j^2 \leq c, \tag{7.6}$$

where c is a pre-specified positive value.

These two methods differ in how they constrain the parameter space. Let the constrained parameter spaces of the lasso and ridge regression methods be denoted respectively by

$$\Theta_{\text{lasso}}(c) = \left\{(\beta_1, \ldots, \beta_p)^{\mathsf{T}} : \sum_{j=1}^{p} |\beta_j| \leq c\right\}$$

and

$$\Theta_{\text{ridge}}(c) = \left\{(\beta_1, \ldots, \beta_p)^{\mathsf{T}} : \sum_{j=1}^{p} \beta_j^2 \leq c\right\}$$

for $c \geq 0$. When $p = 1$, $\Theta_{\text{lasso}}(c)$ and $\Theta_{\text{ridge}}(c)$ have the same shape. However, for $p \geq 2$, while both are convex, $\Theta_{\text{lasso}}(c)$ may be regarded as diamond-like and $\Theta_{\text{ridge}}(c)$ as sphere-like. More precisely, $\Theta_{\text{lasso}}(c)$ is a polytope whose boundary consists of flat faces of dimension $p - 1$ meeting at lower-dimensional edges and corners, whereas the boundary of $\Theta_{\text{ridge}}(c)$ is a single smooth surface. This difference leads to distinct behaviours in parameter shrinkage between the two methods, resulting in different estimators. In particular, the lasso solution has some parameters estimated to be exactly zero, so it can be used to select covariates. Ridge regression, by contrast, yields non-zero estimates of all p components of β. Centring the responses in (7.1) ensures that no shrinkage is applied to the intercept.

□

7.3 Penalized likelihood functions

The examples in Section 7.2 illustrate that the best subset selection, lasso, and ridge regression methods can be expressed as constrained minimization problems, where the objective function is derived from the negative log-likelihood function of the model parameter β, with a constraint imposed to reduce the parameter space, reflecting the user's preference for the form of the restriction. A reformulation using the Lagrangian converts these constrained minimization problems to unconstrained minimization problems for a new objective function; for details see Appendix 7.A.

Example 7.3 (Best subset selection, ridge regression, and lasso, cont'd) With the introduction of an additional parameter $\lambda \geq 0$, the constrained minimization problem in (7.4) can be transformed into an unconstrained minimization problem

$$\min_{\beta}\left\{\frac{1}{n}(Y - X\beta)^{\mathsf{T}}(Y - X\beta) + \lambda\sum_{j=1}^{p} 1(\beta_j \neq 0)\right\}, \tag{7.7}$$

and the constrained minimization problems in (7.5) and (7.6) can be implemented respectively by the unconstrained minimization problems

$$\min_{\beta}\left\{\frac{1}{n}(Y - X\beta)^{\mathsf{T}}(Y - X\beta) + \lambda\sum_{j=1}^{p} |\beta_j|\right\} \tag{7.8}$$

and

$$\min_{\beta}\left\{\frac{1}{n}(Y - X\beta)^{\mathsf{T}}(Y - X\beta) + \lambda\sum_{j=1}^{p} \beta_j^2\right\}. \tag{7.9}$$

Minimization problems (7.7), (7.8), and (7.9) share a common structure,

$$\min_{\beta}\left\{-\frac{1}{n}\log L(\beta) + \lambda\sum_{j=1}^{p} g(\beta_j)\right\}, \tag{7.10}$$

where $L(\beta)$ is the likelihood function for β, λ is a non-negative constant, and $g(\cdot)$ is a real-valued function. □

Example 7.4 (Orthonormal design) As noted in Exercise 7.3, there is a closed-form solution to (7.9) for ridge regression. However, the best subset and lasso estimators, derived from (7.7) and (7.8), generally do not have a closed-form expression, except in special cases, such as when the design matrix is orthonormal, i.e., $X^{\mathsf{T}}X = \mathcal{I}_{p\times p}$. In this case, the best subset and lasso estimators have the jth elements

$$\widehat{\beta}_{j,\text{best}} = \widehat{\beta}_j 1(|\widehat{\beta}_j| \geq |\widehat{\beta}_{(s)}|), \quad \text{and} \quad \widehat{\beta}_{j,\text{lasso}} = \text{sign}(\widehat{\beta}_j)(|\widehat{\beta}_j| - \lambda)_+,$$

for $j = 1, \ldots, p$, where $\widehat{\beta}_j = X_{\bullet j}^{\mathsf{T}}Y$ is the jth ordinary least squares estimate, $X_{\bullet j} = (X_{1j}, \ldots, X_{nj})^{\mathsf{T}}$ is the n-dimensional vector of the jth covariate for all subjects in the sample, $(u)_+ = \max(u, 0)$ is the positive part of u, and $|\widehat{\beta}_{(s)}|$ is the sth largest ordinary least squares estimate in absolute value, chosen to satisfy the sparsity constraint in (7.4) by retaining only the s largest estimates in absolute value. The jth element of the ridge estimator is given by $\widehat{\beta}_{j,\text{ridge}} = \widehat{\beta}_j/(1 + n\lambda)$. Figure 7.1 illustrates the three regularized estimators as functions of the ordinary least squares estimate $\widehat{\beta}_j$, taking $|\widehat{\beta}_{(s)}| = 1$ and $\lambda = 1$ for illustration.

□

More generally, extending (7.10) to a broader context and multiplying it by the sample size n, we may consider the minimization problem $\min_\theta\{-Q(\theta)\}$, or equivalently the maximization problem

$$\max_{\theta} Q(\theta) = \max_{\theta}\{\ell(\theta) - np_\lambda(\beta)\}, \tag{7.11}$$

where $\ell(\theta)$ is the log-likelihood function for the model parameter θ, and $\theta = (\vartheta^{\mathsf{T}}, \beta^{\mathsf{T}})^{\mathsf{T}}$ has an unpenalized parameter ϑ, and a penalized parameter β. Centring the responses to avoid penalizing the intercept only works in the linear model (7.1), so ϑ could for example be the intercept in a logistic regression. In some applications certain covariates, such as a

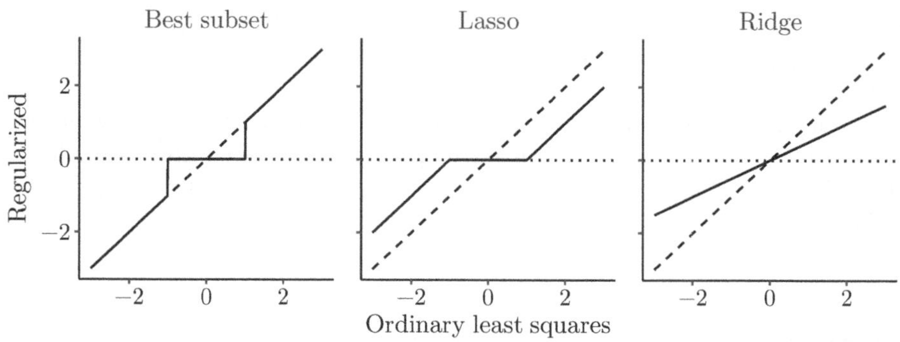

FIGURE 7.1
Best subset, lasso, and ridge estimators in the orthonormal case. The plots show how a generic ordinary least squares estimate $\widehat{\beta}_j$ (on the x-axis) is converted into a regularized estimate (on the y-axis). For the best subset selector, the threshold is the sth largest ordinary least squares coefficient $|\hat{\beta}_{(s)}|$, which is set to 1 for illustration. Similarly, the tuning parameter is set to $\lambda = 1$ for the lasso and ridge estimators. The dashed line represents the ordinary least squares benchmark.

treatment indicator, might also not be included in the penalization step. The function $p_\lambda(\beta)$ is called the penalty (or regularization) function, with $\lambda \geq 0$ as the tuning or regularization parameter.

The function $Q(\theta)$ is often called the penalized likelihood function, although, strictly speaking, it is the penalized log-likelihood function. Sometimes $-Q(\theta)$ is loosely referred to as the penalized likelihood function. Solving the minimization problem $\min_\theta\{-Q(\theta)\}$ or the maximization problem $\max_\theta Q(\theta)$ is known as the penalized likelihood or regularized likelihood method, resulting in a penalized likelihood estimator. Both forms are widely used in applications. Maximization is often preferred to emphasize the connection to the likelihood function, while minimization treats the negative log-likelihood function as a loss function.

Other versions of $Q(\theta)$ also appear in the literature. For example, sometimes $-Q(\theta)$ is defined as $\ell(\theta) - p_\lambda(\beta)$, without multiplying the penalty function $p_\lambda(\beta)$ by the sample size n; or the form $n^{-1}\ell(\theta) - p_\lambda(\beta)$ is proposed. With the same penalty function, different definitions of $Q(\theta)$ will result in the same class of penalized estimators, with appropriate adjustment to the tuning parameter. For example, in ridge regression (7.9) for the linear model, all parameters are penalized so $\theta = \beta$. Using $-Q(\theta) = n^{-1}(Y - X\beta)^\mathsf{T}(Y - X\beta) + \lambda_1 \sum_{j=1}^p \beta_j^2$ leads to the ridge regression estimator

$$\widehat{\beta}_{\text{ridge1}} = (X^\mathsf{T}X + n\lambda_1 \mathcal{I}_{p\times p})^{-1}X^\mathsf{T}Y.$$

Alternatively, defining $-Q(\theta) = (Y - X\beta)^\mathsf{T}(Y - X\beta) + \lambda_2 \sum_{j=1}^p \beta_j^2$ yields the ridge regression estimator

$$\widehat{\beta}_{\text{ridge2}} = (X^\mathsf{T}X + \lambda_2 \mathcal{I}_{p\times p})^{-1}X^\mathsf{T}Y, \tag{7.12}$$

While the estimators $\widehat{\beta}_{\text{ridge1}}$ and $\widehat{\beta}_{\text{ridge2}}$ appear different, they are equivalent. Setting $\lambda_1 = n^{-1}\lambda_2$ makes $\widehat{\beta}_{\text{ridge1}}$ identical to $\widehat{\beta}_{\text{ridge2}}$. Thus, varying λ over $[0, \infty)$ produces the same class of estimators from either form.

Penalized likelihood methods provide a viable solution for handling high-dimensional data. While rooted in the likelihood formation, they significantly extend the traditional likelihood inference framework. Typically, usual likelihood-based methods assume the dimension p of variables to be fixed and smaller than the sample size n. However, in high-dimensional settings, p may be not only larger than n but also allowed to increase with n in the conceptual framework. In these contexts, penalized likelihood methods become particularly relevant for selecting important variables and excluding unimportant ones, or regularizing parameter estimates.

When p is allowed to increase with n, it is often called a diverging dimension, with p expressed as p_n to emphasize its dependence on n. Two diverging dimensionality scenarios are often differentiated in the literature: (1) $p_n = O(n^\alpha)$ for $\alpha > 0$, and (2) $p_n = O\{\exp(n^\alpha)\}$ for $\alpha > 0$; (2) is sometimes referred to as the ultra-high dimensional case.

7.4 Asymptotic results

In this section, we examine the asymptotic properties of penalized likelihood estimators under regularity conditions. We consider a class of models for the conditional distribution of Y, given X, where, unlike those in Sections 7.2–7.3, X and Y represent the covariate and response variables, respectively. Let $f(y \mid x; \theta)$ denote the probability density or mass function for this conditional distribution. As in (7.11), we write the parameter θ as $\theta = (\vartheta^\top, \beta^\top)^\top$, where ϑ is the unpenalized parameter vector, and β is the p-dimensional penalized parameter vector with components β_j, $j = 1, \ldots, p$.

Given a sample of observations $\{\{x_i, y_i\} : i = 1, \ldots, n\}$ from $f(y \mid x; \theta)$, $Q(\theta)$ in (7.11) is

$$Q(\theta) = \ell(\theta) - n \sum_{j=1}^{p} p_\lambda(|\beta_j|), \tag{7.13}$$

where $\ell(\theta) = \Sigma_{i=1}^n \ell_i(\theta) = \Sigma_{i=1}^n \log f(y_i \mid x_i; \theta)$. The penalty function $\Sigma_{j=1}^p p_\lambda(|\beta_j|)$ with the tuning parameter λ, is used to regularize the regression coefficients of the covariates, leaving the parameter ϑ unpenalized.

Given a tuning parameter λ that may depend on the sample size n, now denoted by λ_n, maximizing $Q(\theta)$ with respect to θ yields an estimator $\widehat{\theta}_{\lambda_n}$ of θ. This estimator has desirable asymptotic properties if $p_{\lambda_n}(\cdot)$ and the tuning parameter λ_n are properly chosen and regularity conditions on the model hold.

Let $\theta_0 = (\vartheta_0^\top, \beta_0^\top)^\top$ denote the true parameter value of θ, where ϑ_0 and β_0 are the true values for ϑ and β, respectively. The jth component of β_0 is denoted by β_{0j}. Assume that p is fixed and that β_0 has only $s < p$ non-zero elements, reflecting that only the corresponding s covariates are associated with the response variable. This set of s columns of X is sometimes called the active set. We write $\beta_0 = (\beta_{0(1)}^\top, \beta_{0(2)}^\top)^\top$, where $\beta_{0(1)}$ are the s non-zero elements associated with the active set and $\beta_{0(2)}$ is a $(p-s)$-vector of zeroes. We also partition θ_0 as $\theta_0 = (\theta_{0(1)}^\top, \beta_{0(2)}^\top)^\top = (\vartheta_0^\top, \beta_{0(1)}^\top, \beta_{0(2)}^\top)^\top$; $\theta_{0(1)}$ contains all the non-zero elements, whether penalized or unpenalized. For arbitrary θ, we write $\theta = (\theta_{(1)}^\top, \beta_{(2)}^\top)^\top = (\vartheta^\top, \beta_{(1)}^\top, \beta_{(2)}^\top)^\top$. For the penalized estimator $\widehat{\theta}_{\lambda_n}$, we write $\widehat{\theta}_{\lambda_n} = (\widehat{\theta}_{(1)\lambda_n}^\top, \widehat{\beta}_{(2)\lambda_n}^\top)^\top = (\widehat{\vartheta}_{\lambda_n}^\top, \widehat{\beta}_{(1)\lambda_n}^\top, \widehat{\beta}_{(2)\lambda_n}^\top)^\top$.

Theorem 7.5 Assume that conditions (a)–(d) in Appendix 7.B hold. For a given sequence of tuning parameters λ_n, if

$$\max\left\{|p''_{\lambda_n}(|\beta_{0j}|)| : \beta_{0j} \neq 0\right\} \to 0 \text{ as } n \to \infty,$$

then there exists a local maximizer of $Q(\theta)$ in (7.13), denoted by $\widehat{\theta}_{\lambda_n}$, such that

$$\|\widehat{\theta}_{\lambda_n} - \theta_0\|_2 = O_p(n^{-1/2} + a_n),$$

where $a_n = \max\{p'_{\lambda_n}(|\beta_{0j}|) : \beta_{0j} \neq 0\}$, and $\|c\|_2 = (\Sigma_{j=1}^d c_j^2)^{1/2}$ is the L_2 norm of $c = (c_1, \ldots, c_d)^\top$. □

Unlike the maximum likelihood estimator, the convergence rate of the penalized likelihood estimator $\widehat{\theta}_{\lambda_n}$ is influenced not only by $\sqrt{n}$ but also by a_n. To ensure that the estimator $\widehat{\theta}_{\lambda_n}$ is $\sqrt{n}$–consistent, we require that $a_n = O(n^{-1/2})$, which in turn depends on the specific form of the penalty function, and can be influenced by both the tuning parameter and the size of β. For some penalty functions, a_n depends solely on the tuning parameter; for example, the lasso penalty has $a_n = \lambda_n$, so to achieve $\sqrt{n}$–consistency, it suffices that $\lambda_n = O(n^{-1/2})$. In contrast, for the ridge regression penalty, a_n depends on both λ_n and the magnitude of β, as $a_n = \max\{2\lambda_n|\beta_{0j}| : \beta_{0j} \neq 0\}$. Requiring $a_n = O(n^{-1/2})$ necessitates knowledge of the largest absolute value among the unknown true values of β, in combination with the requirement of λ_n.

To study the smoothness and curvature of the penalty function, we define the $\dim(\theta_{(1)}) \times 1$ vector

$$b = \left(0, p'_{\lambda_n}(|\beta_{01}|)\text{sgn}(\beta_{01}), \ldots, p'_{\lambda_n}(|\beta_{0s}|)\text{sgn}(\beta_{0s})\right)^\top$$

and the $\dim(\theta_{(1)}) \times \dim(\theta_{(1)})$ matrix

$$\Sigma = \text{diag}\left\{0, p''_{\lambda_n}(|\beta_{01}|), \ldots, p''_{\lambda_n}(|\beta_{0s}|)\right\},$$

where $\text{sgn}(u) = 1$ if $u > 0$, 0 if $u = 0$, and -1 if $u < 0$. Here and elsewhere, 0 is used to represent the number zero, a zero column or row vector, or a zero matrix; its meaning is clear from the context. Define

$$i_{11}(\theta_0) = \text{E}_{\theta_0}\left\{-\frac{\partial^2 \ell_i(\theta)}{\partial\theta_{(1)}\partial\theta_{(1)}^\top}\right\} \quad \text{and} \quad B = i_{11}(\theta_0) + \Sigma. \tag{7.14}$$

Theorem 7.6 Assume that conditions (a)–(e) in Appendix 7.B hold. Further, assume that

$$\liminf_{n\to\infty} \liminf_{v\to 0+} \frac{p'_{\lambda_n}(v)}{\lambda_n} > 0,$$

and that $\lambda_n \to 0$ and $n^{1/2}\lambda_n \to \infty$ as $n \to \infty$. Then a $\sqrt{n}$–consistent estimator $\widehat{\theta}_{\lambda_n} = (\widehat{\theta}^\top_{(1)\lambda_n}, \widehat{\beta}^\top_{(2)\lambda_n})^\top$ obtained by maximizing $Q(\theta)$ satisfies

(a) $\lim_{n\to\infty} \text{pr}(\widehat{\beta}_{(2)\lambda_n} = 0) = 1$,

(b) $\sqrt{n}(\widehat{\theta}_{(1)\lambda_n} - \theta_{0(1)} + B^{-1}b) \xrightarrow{d} N\{0, B^{-1}i_{11}(\theta_0)B^{-\top}\}$, as $n \to \infty$.

□

Part (a) states that with a properly selected λ_n and suitable penalty function, the parameter that maximizes the penalized likelihood function can correctly exclude unimportant covariates with probability approaching 1 as $n \to \infty$. However, this property and $\sqrt{n}$–consistency do not hold simultaneously for the estimators derived using an arbitrary penalty function. For example, for the lasso penalty, $a_n = \lambda_n$, so the requirement $a_n = O(n^{-1/2})$ for $\sqrt{n}$–consistency and the requirement that $n^{1/2}\lambda_n \to \infty$ as $n \to \infty$ for correctly identifying sparsity cannot both be met.

Part (b) shows that the asymptotic distribution of $\widehat{\theta}_{\lambda_n}$ does not follow the same form as the usual maximum likelihood estimator. This is reflected in two aspects: the asymptotic bias term $B^{-1}b$ and the sandwich form of the asymptotic covariance matrix. With certain penalty functions, $b = 0$; one example is the SCAD penalty function discussed in Section 7.5. In this case, the asymptotic normality of $n^{1/2}(\widehat{\theta}_{(1)\lambda_n} - \theta_{0(1)})$ can be established, but its asymptotic covariance matrix is not the inverse of the Fisher information matrix, as is the case for the maximum likelihood estimator.

Theorems 7.5 and 7.6 establish asymptotic properties of penalized likelihood estimators when the covariate dimension p is fixed. The conditions imposed on penalty functions provide sufficient conditions for these results. In the literature, various sets of sufficient conditions have been proposed to establish theoretical properties of penalized likelihood estimators in different contexts. Specific conditions depend heavily on whether p is fixed or diverging and on the particular statistical properties one aims to establish.

7.5 Specification of penalty functions

The asymptotic results in Section 7.4 show that establishing desirable properties for penalized likelihood estimators typically requires conditions on the penalty function. These conditions may contain simultaneous constraints on both the functional form of the penalty and the tuning parameter.

A simple way to specify a penalty function is to separate the involvement of the tuning parameter λ and the parameter $\beta = (\beta_1, \dots, \beta_p)^\top$ in $p_\lambda(\beta)$ in (7.11) or (7.13) using the multiplicative form,

$$p_\lambda(\beta) = \lambda \sum_{j=1}^{p} g(\beta_j), \tag{7.15}$$

with $g(u)$ taken as a non-negative real-valued function. For example, setting $g(u) = \min(1, |u|/\gamma)$, with $\gamma > 0$, gives the truncated L_1 penalty $p_\lambda(\beta) = \lambda \sum_{j=1}^{p} \min(1, |\beta_j|/\gamma)$. Letting $g(u) = |u|^q$ for a positive constant q yields the bridge penalty function, also known as the L_q penalty function, $p_\lambda(\beta) = \lambda \sum_{j=1}^{p} |\beta_j|^q$.

The bridge penalty accommodates several useful methods under the same umbrella. Taking $q = 0$ can yield well-known model selection criteria such as the Akaike information criterion (AIC) and the Bayesian information criterion (BIC). Subset selection emerges as $q \to 0$ when using the L_q penalty because

$$\lim_{q \to 0+} \sum_{j=1}^{p} |\beta_j|^q = \sum_{j=1}^{p} 1(\beta_j \neq 0).$$

Thus, model selection methods that penalize the number of non-zero parameters (such as AIC and BIC) can be viewed as limiting cases of bridge estimation as $q \to 0$. When $q \in (0, 1)$, the L_q penalty is a non-convex penalty. The lasso uses the smallest q that preserves convexity, which is attractive computationally. The bridge penalty with $q = 2$ reduces to ridge regression.

Specifying a penalty function in the multiplicative form (7.15) provides an intuitive way to interpret regularized problems by visualizing the constraints on the parameter space through the Lagrangian formulation, as noted in Example 7.3. However, this specification may not offer enough flexibility to ensure that the resulting estimators possess the desired properties for diverse settings. A more general form of penalty function is

$$p_\lambda(\beta) = \sum_{j=1}^{p} g_\lambda(|\beta_j|),$$

where, for a given $\lambda > 0$, $g_\lambda(u)$ is a non-decreasing function mapping $[0, +\infty)$ to $[0, +\infty)$ with $g(0) = 0$. The monotonicity of $g_\lambda(u)$ reflects that, for a given tuning parameter, a larger parameter value contributes a larger value in forming the regularized function (7.11). Requiring $g(0) = 0$ guarantees that a zero coefficient incurs no penalty, aligning with variable-exclusion (sparsity) behaviour.

In applications, additional requirements may be imposed on $g_\lambda(u)$ to ensure that the resulting estimators have desired statistical properties, such as those presented in Theorems 7.5 and 7.6 or their variants. Since the penalized likelihood method essentially involves a constrained optimization problem, solving it requires examining the Karush–Kuhn–Tucker (KKT) conditions (outlined in Appendix 7.A), which involve differential operations. It is often more instructive to impose requirements on its derivative rather than on the penalty function itself.

Penalty functions are also categorized based on their convexity or non-convexity. Convex functions such as the L_1 and L_2 norms are frequently used, while various non-convex penalties have also been proposed. One commonly used class of non-convex penalty functions are folded concave penalty functions. These are neither convex nor concave over the entire real line, but they are symmetric about 0 and concave on both the positive and negative real lines. This symmetry resembles folding the argument due to its form involving the absolute value.

Unlike L_1 and L_2 penalties, many concave penalties vary with λ in a non-multiplicative manner and may involve an additional tuning parameter a that controls the degree of concavity of the penalty, thereby influencing the bias of the estimates and the stability of the estimators.

Example 7.7 (Folded concave penalty functions) Popular folded concave penalty functions include the smoothly clipped absolute deviation (SCAD) penalty (Fan, 1997; Fan and Li, 2001) and the minimax concave penalty (MCP) (Zhang, 2010).

The SCAD penalty function $g_\lambda(u)$ is

$$g_\lambda(u) = \begin{cases} \lambda|u|, & |u| \le \lambda, \\ -\dfrac{|u|^2 - 2a\lambda|u| + \lambda^2}{2(a-1)}, & \lambda < |u| \le a\lambda, \\ \dfrac{(a+1)\lambda^2}{2}, & |u| > a\lambda, \end{cases}$$

with derivative

$$g'_\lambda(u) = \begin{cases} ll\lambda, & |u| \le \lambda, \\ \dfrac{a\lambda - |u|}{a-1}, & \lambda < |u| \le a\lambda, \\ 0, & |u| > a\lambda. \end{cases}$$

The SCAD penalty function is continuously differentiable on $(-\infty, 0) \cup (0, \infty)$, singular at $u = 0$, and its derivatives are zero outside the interval $[-a\lambda, a\lambda]$. This leads to estimates for which some coefficients are zero, some are shrunk towards zero, and large coefficients are left relatively unchanged. As a result, using the SCAD penalty can produce sparse solutions and shrink parameter estimates.

The SCAD penalty coincides with the lasso for $|u| \le \lambda$, and smoothly transitions to a quadratic function for $|u| \le a\lambda$. Beyond this threshold, the SCAD penalty remains constant for all $|u| > a\lambda$. It maintains the same penalization strength (and corresponding bias) as the lasso for small coefficients but gradually weakens the penalty as the absolute value of the coefficient increases.

The MCP penalty function is

$$g_\lambda(u) = \begin{cases} \lambda|u| - \dfrac{u^2}{2a}, & |u| \le a\lambda, \\ \dfrac{1}{2}a\lambda^2, & |u| > a\lambda, \end{cases}$$

with derivative

$$g'_\lambda(u) = \begin{cases} \left(\lambda - \dfrac{|u|}{a}\right)\operatorname{sign}(u), & |u| \le a\lambda, \\ 0, & |u| > a\lambda. \end{cases}$$

The MCP penalty has a derivative that decreases linearly from the lasso rate to zero as the absolute value of the coefficient increases, whereas the SCAD penalty maintains a constant derivative near zero and only begins to decrease after a threshold.

The additional tuning parameter a influences the concavity level of the penalty function. As $a \to \infty$, both the SCAD and MCP penalties converge to the L_1 penalty. Smaller values of a increase concavity, which tends to reduce bias but make the estimates become unstable. Here stability refers to the property that optimization problems with a single, well-defined minimum are stable, whereas those with multiple local minima tend to be unstable. Similar considerations apply to the statistical properties of the estimators. Estimators with higher variability are less stable, and for SCAD and MCP, lower values of a result in more variable but less biased estimates.

In principle, (λ, a) may be chosen by a two–dimensional grid search with cross-validation, but this can be costly. Motivated by Bayesian reasoning and simulations, Fan and Li (2001) suggested $a = 3.7$ for SCAD, which works well in many settings. Fixing a reduces tuning to selecting λ only, in line with (7.15).

□

Figure 7.2 compares several penalty functions, including lasso, elastic net $g_\lambda(u) = \lambda\{\alpha|u| + (1-\alpha)u^2\}$ with $\alpha = 0.7$, ridge regression, lasso, the hard thresholding penalty $g_\lambda(u) = \lambda^2 - (|u| - \lambda)^2 1(|u| < \lambda)$, MCP, and SCAD. All these penalties attain their minimum at $u = 0$ and are continuous at $u = 0$. However, except for ridge regression, none is differentiable at $u = 0$. Ridge regression, lasso, and elastic net are convex and unbounded, whereas the hard-thresholding penalty, MCP, and SCAD are non-convex and bounded.

7.6 Determination of tuning parameters

Given a statistical model and hence likelihood function, the choice of a penalty function is guided by the nature of the data, considerations of model complexity, desired estimator properties, and computational cost. A penalty typically depends on one or more tuning parameters (but not on the data).

The appropriate selection of tuning parameters depends on several factors, including research objectives (e.g., whether the focus is on prediction accuracy or on identifying important variables), the trade-off between the model fit and achieving sparsity, the balance between finite sample performance and asymptotic guarantees, and whether the parameter dimension is fixed (possibly larger than the sample size) or diverging with the sample size. Criteria for selecting tuning parameter values include minimizing prediction error, minimizing

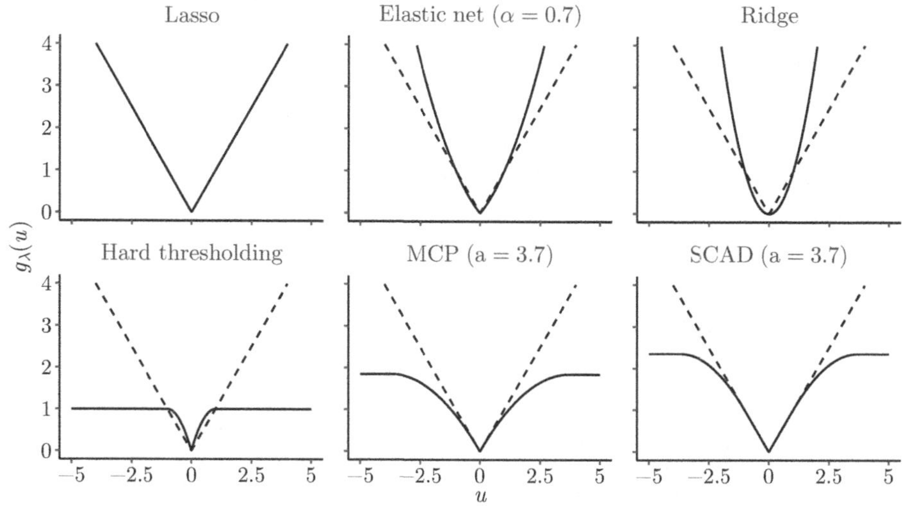

FIGURE 7.2
Commonly used penalty functions with a tuning parameter of $\lambda = 1$. The lasso penalty, shown as a dashed line, serves as a reference in all other penalty graphs.

estimation error, maximizing posterior model probability, among others. It is important to note that prediction error and estimation error are often different, as illustrated below.

Consider the random sample $\mathcal{S} = \{\{Y_i, X_i\} : i = 1, \ldots, n\}$, drawn from the joint distribution of $\{Y, X\}$ as described in Section 7.4, where Y is a continuous response variable and X is a vector of associated covariates. Assume that Y and X are related by the model

$$Y = \eta(X) + \epsilon, \tag{7.16}$$

where $\eta(\cdot)$ is an unknown function, and ϵ is independent of X, with mean zero and variance σ^2.

Suppose $\widehat{\eta}_{\mathcal{S}}(\cdot)$ is an estimator of $\eta(\cdot)$ based on fitting the model (7.16) to the sample $\mathcal{S}$. Using $\widehat{\eta}_{\mathcal{S}}(\cdot)$ as a prediction model, we aim to predict the response for a future covariate vector $\widetilde{X}$. The accuracy of the prediction performance is usually measured by a loss function that measures the discrepancy between the predicted response $\widehat{\eta}_{\mathcal{S}}(\widetilde{X})$ and the true response $\widetilde{Y}$ associated with $\widetilde{X}$. The loss function $L(u, v)$ can take various forms; common choices are squared error loss $(u - v)^2$, absolute error loss $|u - v|$, and the Huber loss. The choice depends on the characteristics of the data and relevant performance metrics.

The expected loss, or expected prediction error (EPE), is defined as

$$\text{EPE} = \text{E}[L\{\widetilde{Y}, \widehat{\eta}_{\mathcal{S}}(\widetilde{X})\}],$$

commonly referred to as the generalization error or test error. The expectation here is taken over the joint distribution of the estimator $\widehat{\eta}_{\mathcal{S}}(\cdot)$, which depends on the training sample $\mathcal{S}$, and the future test data point $(\widetilde{X}, \widetilde{Y})$.

To highlight the roles of the two sources of randomness, consider the conditional prediction error $\text{PE}_{\mathcal{S}|(\widetilde{X},\widetilde{Y})} = \text{E}[L\{\widetilde{Y}, \widehat{\eta}_{\mathcal{S}}(\widetilde{X})\} \mid (\widetilde{X}, \widetilde{Y})]$. The expected prediction error is then

$$\text{EPE} = \text{E}\{\text{PE}_{\mathcal{S}|(\widetilde{X},\widetilde{Y})}\},$$

where the outer expectation is the average over all possible future points, of the expected loss $\text{PE}_{\mathcal{S}|(\widetilde{X},\widetilde{Y})}$ induced by the randomness in the training sample.

Alternatively, conditional on $\widetilde{X} = x$, define $\mathrm{PE}_{(\mathcal{S},\widetilde{Y})|\widetilde{X}}(x) = \mathrm{E}[L\{\widetilde{Y}, \widehat{\eta}_{\mathcal{S}}(\widetilde{X})\} \mid \widetilde{X} = x]$. EPE is then given by

$$\mathrm{EPE} = \mathrm{E}\{\mathrm{PE}_{(\mathcal{S},\widetilde{Y})|\widetilde{X}}(\widetilde{X})\}.$$

This expression is convenient for calculation.

As an example, consider squared-error loss $L(u, v) = (u - v)^2$ and a fixed test point $\widetilde{X} = x$. Using (7.16) gives

$$\begin{aligned}
\mathrm{PE}_{(\mathcal{S},\widetilde{Y})|\widetilde{X}}(x) &= \mathrm{E}[\{\widetilde{Y} - \widehat{\eta}_{\mathcal{S}}(\widetilde{X})\}^2 \mid \widetilde{X} = x] \\
&= \mathrm{E}[\{\eta(\widetilde{X}) + \epsilon - \widehat{\eta}_{\mathcal{S}}(\widetilde{X})\}^2 \mid \widetilde{X} = x] \\
&= \mathrm{E}[\{\eta(\widetilde{X}) - \widehat{\eta}_{\mathcal{S}}(\widetilde{X})\}^2 \mid \widetilde{X} = x] + \mathrm{E}(\epsilon^2 \mid \widetilde{X} = x) \\
&= \mathrm{E}\left(\left[\eta(\widetilde{X}) - \mathrm{E}\left\{\widehat{\eta}_{\mathcal{S}}(\widetilde{X}) \mid \widetilde{X} = x\right\}\right]^2 \middle| \widetilde{X} = x\right) + \mathrm{var}\{\widehat{\eta}_{\mathcal{S}}(\widetilde{X}) \mid \widetilde{X} = x\} + \sigma^2 \\
&= \mathrm{bias}^2\{\widehat{\eta}_{\mathcal{S}}(\widetilde{X}) \mid \widetilde{X} = x\} + \mathrm{var}\{\widehat{\eta}_{\mathcal{S}}(\widetilde{X}) \mid \widetilde{X} = x\} + \sigma^2. \qquad (7.17)
\end{aligned}$$

Averaging over the distribution of $\widetilde{X}$ yields EPE. The expected prediction error includes an irreducible component, σ^2, reflecting inherent variability in response in (7.16), regardless of the method used to estimate $\eta_{\mathcal{S}}(\cdot)$ or the amount of data in the training sample $\mathcal{S}$. The first two terms in EPE derived from (7.17) reflect the fit to the underlying model and the effectiveness of the prediction model.

Finally, we comment that for prediction, the model needs to perform well on future unseen data. Ideally, distinct training and testing datasets are available, both originating from the same population. However, in many real-world scenarios, such separate datasets are unavailable. A widely used strategy is cross-validation, which splits the data, either once or repeatedly, into training and test subsets. While cross-validation is simple and broadly applicable, Bates et al. (2024) observed that it usually yields a higher mean squared error when estimating the prediction error of the final model compared to the average prediction error across unseen datasets, because cross-validation underestimates the variance by not accounting for the correlation induced by reusing observations for both training and testing. They proposed a nested cross-validation approach to obtain more accurate variance estimates. Additionally, cross-validation focuses on tuning parameters for optimal predictive performance, which may not correspond to identifying the true underlying model. Meinshausen and Bühlmann (2006) showed that tuning solely for prediction does not guarantee model–selection consistency, so balancing variable selection and prediction accuracy can be important in practice.

7.7 Penalized likelihood estimators and variants

Example 7.8 (Graphical Lasso) Suppose a d-dimensional random vector Y follows a multivariate normal distribution $N_d(0, \Sigma)$ with mean zero and a nonsingular covariance matrix Σ, where we assume the diagonal elements are equal to one. We define the concentration, or precision, matrix to be $\Omega = \Sigma^{-1}$. Let θ denote the vector containing unknown parameters in $\Omega = [\theta_{jk}]$, where θ_{jk} is the (j, k) element of Ω.

Given a random sample $\{Y_1, \ldots, Y_n\}$ of independent and identically distributed variables from this normal distribution, the log-likelihood function for θ, up to a constant, is

$$\ell(\theta) = n \log \det(\Omega) - \sum_{i=1}^{n} Y_i^{\mathsf{T}} \Omega Y_i,$$

and the maximum likelihood estimators are (Example 1.19)

$$\widehat{\Sigma} = \frac{1}{n} \sum_{i=1}^{n} Y_i Y_i^{\mathsf{T}}, \quad \widehat{\Omega} = \widehat{\Sigma}^{-1},$$

provided that $\widehat{\Sigma}$ is invertible. The estimator $\widehat{\Omega}$ is unstable when d is moderate or large, and the entries are typically non-zero due to sampling variability, even when the true values are zero. To identify these zero entries in Ω, we consider an alternative perspective, viewing the random vector $Y \in \mathbb{R}^d$ through the lens of an undirected graph.

Let $G = \{V, E\}$ represent an undirected graph, where V contains d vertices corresponding to the d components of Y, and the edge set $E = \{e_{jk} : 1 \leq j < k \leq d\}$ describes the conditional dependence relationships among the components. An edge e_{jk} between the jth and kth components of Y is absent if and only if Y_j and Y_k are conditionally independent, given the remaining components of Y, which is the same as the constraint on the concentration matrix that $\theta_{jk} = 0$. The initial goal of estimating parameters and identifying zeros in the concentration matrix Ω is now converted to the equivalent problem of parameter estimation and model selection for a Gaussian graphical model.

To account for the sparsity in the concentration matrix, we adapt the lasso penalty in (7.5) and consider the problem

$$\min_{\theta \in \Theta} \left\{ -\frac{1}{n} \ell(\theta) \right\}, \quad \text{subject to } \sum_{j \neq k} |\theta_{jk}| \leq c,$$

where Θ is the set of parameters corresponding to all positive definite matrices, and c is a non-negative tuning parameter. If $c = \infty$, the solution is identical to the maximum likelihood estimator $\widehat{\Omega}$; if $c = 0$, then the constraint forces Ω to be diagonal.

This constrained minimization problem can be expressed as the minimization of a single function using the Lagrangian form,

$$-\frac{1}{n} \ell(\theta) + \lambda \sum_{j \neq k} |\theta_{jk}|$$

with a non-negative tuning parameter λ. Noting that

$$\sum_{i=1}^{n} Y_i^{\mathsf{T}} \Omega Y_i = \operatorname{tr} \left(\Omega \sum_{i=1}^{n} Y_i Y_i^{\mathsf{T}} \right),$$

the minimization problem becomes

$$\widehat{\theta}_{\text{lasso}} = \operatorname{argmin}_{\theta \in \Theta} \left\{ -\log \det(\Omega) + \operatorname{tr}\left(\Omega \widehat{\Sigma}\right) + \lambda \sum_{j \neq k} |\theta_{jk}| \right\},$$

leading to the estimate $\widehat{\Omega}_{\text{lasso}}$, where $\widehat{\Omega}_{\text{lasso}}$ is determined by Ω with its parameters replaced by $\widehat{\theta}_{\text{lasso}}$. This is known as the graphical lasso.

□

Example 7.9 (Mixture regression model) Suppose that X is a vector of covariates that may be associated with random variable Y. Assume that, given $X = x$, the conditional probability density or mass function of Y is of the finite mixture form

$$\sum_{k=1}^{K} \pi_k f\{y; \eta_k(x), \phi_k\}, \tag{7.18}$$

where $f\{y; \eta_k(x), \phi_k\}$ is a probability density or mass function, with $\eta_k(x) = g(x^\top \beta_k)$, $k = 1, \ldots, K$. Here, $g(\cdot)$ is a given link function, K is a positive integer, $\pi = (\pi_1, \ldots, \pi_K)^\top$ with $0 < \pi_k < 1$ and $\Sigma_{k=1}^{K} \pi_k = 1$, ϕ_k is the dispersion parameter, and $\beta_k = (\beta_{k1}, \ldots, \beta_{kp})^\top$ is a p-dimensional regression parameter. We assume (7.18) is identifiable, meaning that no two different sets of parameter values lead to the same distribution. Writing $\beta = (\beta_1^\top, \ldots, \beta_K^\top)^\top$ and $\phi = (\phi_1, \ldots, \phi_K)^\top$, the vector of model parameters is $\theta = (\beta^\top, \phi^\top, \pi^\top)^\top$.

For a random sample $\{\{x_i, y_i\} : i = 1, \ldots, n\}$ from (7.18) the log-likelihood function of θ is

$$\ell(\theta) = \sum_{i=1}^{n} \log \left\{ \sum_{k=1}^{K} \pi_k f(y_i; \eta_k(x_i), \phi_k) \right\}.$$

To enhance predictability and achieve a parsimonious model, we aim to include only the important covariates in the model, so we consider the penalized log-likelihood function $Q(\theta) = \ell(\theta) - p(\theta; \lambda)$, where

$$p(\theta; \lambda) = \sum_{k=1}^{K} \pi_k \sum_{j=1}^{p} p_\lambda(|\beta_{kj}|),$$

and $p_\lambda(|\beta_{kj}|)$ is a non-negative and nondecreasing penalty function with tuning parameter λ.

In line with common practice, where the penalty is often related to the sample size, we choose the penalty applied to the regression coefficients within the kth component of the finite mixture regression model to be proportional to π_k. This aligns with the notion that the effective sample size from the kth subpopulation is proportional to π_k.

With a properly chosen penalty function $p_\lambda(|\beta_{kj}|)$, maximizing the penalized log-likelihood function $Q(\theta)$ produces the penalized likelihood estimator for θ. Its asymptotic properties can be derived using approaches similar to those applied in establishing the theorems in Section 7.4, assuming regularity conditions are met for both the penalty function and the model. Khalili and Chen (2007) examine different penalties, including the lasso and SCAD penalties, and establish theoretical results for the corresponding penalized likelihood estimators.

□

Penalization can be adapted to other versions of likelihood functions. We illustrate here penalized profile likelihood for a linear mixed effects model.

Example 7.10 (Penalized profile likelihood) Suppose $Y_i = (Y_{i1}, \ldots, Y_{im})^\top$ is an m-dimensional random vector. The random effects model is given by

$$Y_{ij} = \mu + b_i + \epsilon_{ij}, \tag{7.19}$$

where μ is the baseline mean, b_i is the subject-level random effect, and ϵ_{ij} is the within-subject variation. We assume that $b_i \sim N(0, \sigma^2)$ and $\epsilon_{ij} \sim N(0, \sigma_0^2)$, with non-negative parameters σ^2 and σ_0^2, and that b_i and ϵ_{ij} are independent.

Our goal is to estimate σ^2, treating μ as a nuisance parameter. For simplicity, assume that σ_0^2 is known. For a random sample $\{y_1, \ldots, y_n\}$ with $y_i = (y_{i1}, \ldots, y_{im})^\top$, let $\bar{y}_{i\bullet} = m^{-1}\sum_{j=1}^m y_{ij}$ and $\bar{y}_{\bullet\bullet} = n^{-1}\sum_{i=1}^n \bar{y}_{i\bullet}$. Define the total sum of squares (SST) and the sum of squares between groups (SSB)

$$\text{SST} = \sum_{i=1}^n \sum_{j=1}^m (y_{ij} - \bar{y}_{\bullet\bullet})^2, \quad \text{SSB} = m\sum_{i=1}^n (\bar{y}_{i\bullet} - \bar{y}_{\bullet\bullet})^2.$$

The profile log-likelihood function for σ, up to an additive constant, is (see Exercise 7.8)

$$\ell_{\text{p}}(\sigma) = -\frac{n}{2}\log(\sigma_0^2 + m\sigma^2) - \frac{1}{2}\left(\frac{\text{SST}}{\sigma_0^2} - \frac{m\sigma^2\text{SSB}}{\sigma_0^2(\sigma_0^2 + m\sigma^2)}\right). \tag{7.20}$$

We then consider the penalized profile log-likelihood function

$$\ell_{\text{p}}(\sigma) + p_\lambda(\sigma), \tag{7.21}$$

where the penalty function $p_\lambda(\sigma)$ is the logarithm of the gamma density function with scale parameter $1/\lambda$ and shape parameter α

$$p_\lambda(\sigma) = \log\left\{\frac{\lambda^\alpha}{\Gamma(\alpha)}\sigma^{\alpha-1}\exp(-\lambda\sigma)\right\},$$

with $\alpha > 0$ and $\lambda > 0$.

To compare the profile likelihood and penalized profile likelihood estimators, let $\widehat{\sigma}$ be the maximum likelihood estimator of σ. Taylor series expansion of $\ell_{\text{p}}(\sigma)$ around $\widehat{\sigma}$ gives

$$\ell_{\text{p}}(\sigma) \doteq c - \frac{(\sigma - \widehat{\sigma})^2}{2\text{var}(\widehat{\sigma})}, \tag{7.22}$$

where $c = \ell_{\text{p}}(\widehat{\sigma})$ is a constant free of σ, and we have used the approximation

$$\text{var}(\widehat{\sigma}) \doteq \left\{-\left.\frac{\partial^2 \ell_{\text{p}}(\sigma)}{\partial\sigma^2}\right|_{\sigma=\widehat{\sigma}}\right\}^{-1}. \tag{7.23}$$

Using (7.22), the maximum of the approximate penalized profile log-likelihood function (7.21) is

$$\widehat{\sigma}_{\text{MPL}} = -\frac{\lambda\text{var}(\widehat{\sigma})}{2} + \frac{\widehat{\sigma}}{2} + \frac{1}{2}\left[\{\lambda\text{var}(\widehat{\sigma}) - \widehat{\sigma}\}^2 + 4(\alpha-1)\text{var}(\widehat{\sigma})\right]^{1/2}, \tag{7.24}$$

where α and λ must satisfy the constraint

$$\alpha \geq 1 - b(\lambda) \ \text{ with } \ b(\lambda) = \frac{\{\lambda\text{var}(\widehat{\sigma}) - \widehat{\sigma}\}^2}{4\text{var}(\widehat{\sigma})}. \tag{7.25}$$

Equation (7.24) provides an approximate relationship between the maximum likelihood and maximum penalized profile likelihood estimators, highlighting the dependence on the parameters λ and α, as well as $\text{var}(\widehat{\sigma})$. When $\lambda \to 0$,

$$\widehat{\sigma}_{\text{MPL}} \doteq \frac{\widehat{\sigma} + \sqrt{\widehat{\sigma}^2 + 4(\alpha-1)\text{var}(\widehat{\sigma})}}{2},$$

provided $\alpha \geq 1 - b(0)$. In particular, if $\alpha \geq 1$, then $\widehat{\sigma}_{\text{MPL}} \geq \widehat{\sigma}$; if $1 - b(0) \leq \alpha < 1$, then $\widehat{\sigma}_{\text{MPL}} < \widehat{\sigma}$. Thus, when $\lambda \to 0$, the parameter α controls whether the penalized estimator $\widehat{\sigma}_{\text{MPL}}$ inflates or shrinks the maximum likelihood estimator $\widehat{\sigma}$.

Given a fixed value of α and the data (where $\text{var}(\widehat{\sigma})$ is fixed), the magnitude of λ governs the relationship between the maximum likelihood and maximum penalized profile likelihood estimators. As λ approaches infinity, $\widehat{\sigma}_{\text{MPL}} \to 0$ in probability, regardless of the value of the maximum likelihood estimator (see Problem 7.8).

□

Objective functions other than the profile log-likelihood function can also be regularized using (7.11) or $\min_\theta\{-Q(\theta)\}$. For example, for a semi-parametric model such as the Cox proportional hazards model (3.63) considered in Section 3.6.3, $-\ell(\theta)$ would be based on the negative log partial likelihood function. Another example is to specify $-\ell(\theta)$ using a linear combination of convex loss functions, $\sum_{j=1}^{K} w_j\phi_j(\theta)$, where K is a positive integer, the $\phi_j(\cdot)$ are convex loss functions, and w_j are positive weights for $j = 1, \ldots, K$. This formulation provides flexibility in specifying the objective function.

Further extensions can also be pursued. For example, $\ell(\theta)$ in (7.11) can be replaced by an unbiased estimating function, transforming the maximization problem in (7.11) into that of solving a penalized estimating equation.

7.8 Bibliographic notes

Penalized likelihood, which involves adding a penalty function to the log-likelihood function derived from a model, has origins dating back to the early 20th century. Although not initially associated with likelihood functions, Tikhonov (1943) introduced the concept of regularization to address ill-posed problems in integral equations, which laid the foundation for modern regularization techniques, including those applied to likelihood functions (Bickel and Li, 2006).

The asymptotic results in Section 7.4 build upon the work of Fan and Li (2001), with extended proofs provided in Appendix 7.C. Section 7.5 discusses various penalty functions explored by authors such as Hoerl and Kennard (1970), Frank and Friedman (1993), Donoho and Johnstone (1994), Tibshirani (1996), Fan (1997), Fan and Li (2001), Johnson et al. (2008), Zhang (2010), Fan and Lv (2010), and Shen et al. (2012). Examples in Section 7.7 are based on Yuan and Lin (2007), Friedman et al. (2008), and Khalili and Chen (2007). Example 7.10 adapts one presented by Chung et al. (2013).

Criteria for tuning parameter selection vary depending on whether the goal is to minimize prediction error, minimize estimation error, or maximize posterior model probability. Common approaches include cross-validation and classical criteria such as Mallows' C_p (Mallows, 1973), Akaike's information criterion (AIC) (Akaike, 1973), the Bayesian information criterion (BIC) (Schwarz, 1978), and final prediction error (Akaike, 1969, 1970, 1971). Generalized cross-validation is discussed in Craven and Wahba (1979) and Tibshirani (1996). An extended BIC is introduced in Chen and Chen (2008); see Zhang et al. (2010) and Fan and Tang (2013) for discussions on the generalized information criterion. Comprehensive reviews of these methods can be found in Arlot and Celisse (2010) and Wu and Wang (2020).

The general shrinkage framework summarized in (7.11) includes penalized likelihood methods, regularized methods, and shrinkage methods. Hoerl and Kennard (1970) introduced ridge regression, which addresses multicollinearity in linear regression models by adding an L_2 penalty term. Donoho and Johnstone (1994) and Tibshirani (1996) replaced the L_2 penalty by an L_1 penalty. Fan and Li (2001) introduced the smoothly clipped absolute deviation method. Other regularization and variable selection methods include the group lasso (Yuan and Lin, 2006), adaptive lasso (Zou, 2006), elastic net (Zou and Hastie, 2005), minimax concave penalty (Zhang, 2010), and the debiased lasso or de-sparsifying lasso estimator (Zhang and Zhang, 2014), among others. For comprehensive reviews, see Fan and Lv (2010), Bickel and Li (2006), Heinze et al. (2018), and Desboulets (2018). Several monographs on high-dimensional inference provide further insights, including those by Hastie et al. (2009), Bühlmann and van de Geer (2011), James et al. (2013), Hastie et al. (2015), and Wainwright (2019).

Asymptotic properties of various penalized estimators have been obtained under various regularity conditions. For in-depth discussions, see Bühlmann and van de Geer (2011). It is difficult to list all relevant work here, but some notable examples include Knight and Fu (2000), Efron et al. (2004), Fan and Peng (2004), Lockhart et al. (2014), Meinshausen and Bühlmann (2006), Zhao and Yu (2006), Huang et al. (2008), Liang and Li (2009), Bradic et al. (2011), Chatterjee and Lahiri (2011), Lu et al. (2012), Stefanski et al. (2014) and the references therein.

As noted at the end of Section 7.7, penalized likelihood can be extended to give different regularization methods, and there is a large literature on this. For example, Wood (2011) proposed an effective method for estimating smoothing parameters in penalized generalized linear models. Wood et al. (2016) developed a framework for estimating smoothing parameters in models with likelihoods based on unknown smooth functions of covariates. Tang and Leng (2010) discussed penalized empirical likelihood methods for parameter estimation and variable selection with diverging numbers of parameters. Fu (2003) introduced penalized estimating equations to address collinearity in longitudinal studies. Bradic et al. (2011) explored penalized composite quasi-likelihood for ultrahigh dimensional variable selection, and Johnson et al. (2008) proposed a variable selection method for semi-parametric regression models by penalizing the corresponding estimating functions. Candès and Tao (2007) initiated the Dantzig selector by solving an L_1 regularization problem for settings where the dimension of covariates exceeds the sample size, and Bickel et al. (2009) showed that the resulting estimator exhibits behaviour similar to the lasso estimator under a sparsity assumption. For more details, see Hastie et al. (2015) and Fan et al. (2020).

Classical likelihood inference proceeds under a fully specified model selected independently of the data yet calibrated to its nature (e.g., a logistic model for binary responses). By contrast, post-selection inference addresses scenarios in which the model or variable subset is chosen data-dependently (e.g., via penalized likelihood in high-dimensional problems), which introduces additional variability. Unlike traditional procedures that ignore selection uncertainty, post-selection methods adjust inference to account for it. We do not cover this topic in the present book. It is an active and evolving research area motivated by the growing use of data-driven model selection; see, for example, Berk et al. (2013), Chernozhukov et al. (2015), Lee et al. (2016), Tibshirani et al. (2016), Kuchibhotla et al. (2020), and García Rasines and Young (2023).

7.9 Further results and exercises

7.1. Suppose $f(x) = x_1^2 + x_2^2$ with $x = (x_1, x_2)^\top \in \mathbb{R}^2$. Let

$$h_1(x) = \frac{x_1}{1 + x_2^2}, \quad h_2(x) = (x_1 + x_2)^2.$$

Find the solution to the constrained optimization problem

$$\min_{x \in \mathbb{R}^2} f(x), \quad \text{subject to } h_1(x) \le 0,\ h_2(x) = 0.$$

(Boyd and Vandenberghe, 2004, Section 4.2)

7.2. Consider the linear regression model (7.1). The maximum likelihood estimator is $\widehat{\beta} = (X^\top X)^{-1}X^\top Y$, under the assumption that $\epsilon \sim N(0, \sigma^2 \mathcal{I}_{n\times n})$

(a) The Euclidean distance between $\widehat{\beta}$ and β is $D = (\widehat{\beta} - \beta)^\top(\widehat{\beta} - \beta)$. Show that

$$\mathrm{E}(D) = \sigma^2 \mathrm{tr}\{(X^\top X)^{-1}\}, \quad \mathrm{var}(D) = 2\sigma^4 \mathrm{tr}\{(X^\top X)^{-2}\}.$$

(b) Let $\xi_1, \ldots, \xi_p$ be the eigenvalues of $X^\top X$. Then

$$\mathrm{E}(D) = \sigma^2 \sum_{j=1}^{p} \frac{1}{\xi_j}, \quad \mathrm{var}(D) = 2\sigma^4 \sum_{j=1}^{p} \frac{1}{\xi_j^2}.$$

(Hoerl and Kennard, 1970)

7.3. Consider the linear model (7.1).

(a) Show that the minimization problem (7.2) is equivalent to $\min_\beta (\beta - \widehat{\beta}\,)^\top(X^\top X)(\beta - \widehat{\beta}\,)$, where $\widehat{\beta}$ is given by (7.3).

(b) Show that the ridge regression estimator (7.12) is

$$\widehat{\beta}_{\text{ridge}} = (X^\top X + \lambda \mathcal{I}_{p\times p})^{-1} X^\top Y,$$

or equivalently,

$$\widehat{\beta}_{\text{ridge}} = X^\top (X X^\top + \lambda \mathcal{I}_{n\times n})^{-1} Y, \tag{7.26}$$

where $\lambda \geq 0$, denoting λ_2 in (7.12).

7.4. For the model and estimator considered in Problem 7.3, let $D_{\text{ridge}}(\lambda) = (\widehat{\beta}_{\text{ridge}} - \beta)^\top(\widehat{\beta}_{\text{ridge}} - \beta)$ be the distance from $\widehat{\beta}_{\text{ridge}}$ to β, where the dependence on λ is suppressed in the notation $\widehat{\beta}_{\text{ridge}}$ but explicitly expressed in $D_{\text{ridge}}(\lambda)$. Let $\xi_1, \ldots, \xi_p$ be the eigenvalues of $X^\top X$.

(a) Show that

$$\mathrm{E}\{D_{\text{ridge}}(\lambda)\} = \sigma^2 \sum_{j=1}^{p} \frac{\xi_j}{(\xi_j + \lambda)^2} + \lambda^2 \beta^\top (X^\top X + \lambda \mathcal{I}_{p\times p})^{-2}\beta.$$

(b) Show that there always exists $\lambda \geq 0$ such that

$$\mathrm{E}\{D_{\text{ridge}}(\lambda)\} \leq \mathrm{E}\{D_{\text{ridge}}(0)\}.$$

(Hoerl and Kennard, 1970)

7.5. We continue with the model (7.1), with $\widehat{\beta}$ as in (7.3), and $\widehat{\beta}_{\text{ridge}}$ defined in (7.26).

(a) Is $\mathrm{var}(\widehat{\beta}) - \mathrm{var}(\widehat{\beta}_{\text{ridge}})$ non-negative definite?

(b) For what values of λ is

$$\mathrm{E}\{(\widehat{\beta} - \beta)(\widehat{\beta} - \beta)^\top\} - \mathrm{E}\{(\widehat{\beta}_{\text{ridge}} - \beta)(\widehat{\beta}_{\text{ridge}} - \beta)^\top\}$$

positive definite?

7.6. In Exercise 7.3, when $p \gg n$, the second expression of the ridge regression estimator in (7.26) provides a computationally efficient formulation. Consider the singular-value decomposition of $X = UDV^\top$, where U is an $n \times n$ orthogonal matrix, V is a $p \times n$ matrix with orthonormal columns, and D is an $n \times n$ diagonal matrix with elements $d_1 \geq d_2 \geq \ldots \geq d_n \geq 0$. Let $R = UD$.

(a) Show that the ridge regression estimator (7.26) is

$$\widehat{\beta}_{\text{ridge}} = V(R^\top R + \lambda \mathcal{I}_{n\times n})^{-1} R^\top Y.$$

This reformulation reduces the computational cost from $O(p^3)$ to $O(pn^2)$.

(b) To reflect the use of a real-valued function of $g(X_i)$ to predict Y_i, let $L\{y, g(x)\}$ be a loss function, which could be squared error, negative log-likelihood, or negative partial log-likelihood, as mentioned in Section 7.7.

Let r_i be the ith row of R for $i = 1, \ldots, n$. Define

$$\widehat{\beta} = \operatorname{argmin}_{\beta \in \mathbb{R}^p} \left\{ \sum_{i=1}^{n} L(y_i, X_i^\top \beta) + \lambda \beta^\top \beta \right\}$$

and

$$\widehat{\theta} = \operatorname{argmin}_{\theta \in \mathbb{R}^n} \left\{ \sum_{i=1}^{n} L(y_i, r_i^\top \theta) + \lambda \theta^\top \theta \right\}.$$

Show that $\widehat{\beta} = V\widehat{\theta}$. *(Hastie et al., 2009, Section 3.4)*

7.7. Suppose that $Y_1, \ldots, Y_n$ are independent and identically distributed, each following the normal distribution $N(\theta, 1)$. Let $\overline{Y} = n^{-1}\Sigma_{i=1}^n Y_i$. Verify the form of the penalized likelihood estimator $\widehat{\theta}$ from (7.11) with the log-likelihood function

$$\ell(\theta) = -\sum_{i=1}^{n} (Y_i - \theta)^2,$$

and the penalty functions $p_\lambda(v)$ specified as follows, where λ is a tuning parameter.

(a) If $p_\lambda(u) = 2\lambda|u|$, then $\widehat{\theta} = \text{sign}(\overline{Y})(|\overline{Y}| - \lambda)_+$, where $u_+ = \max(u, 0)$ for a given u; and $\text{sign}(u)$ is the sign function, taking values -1, 0, and 1 if $u < 0, u = 0$, and $u > 0$, respectively.

(b) If $p_\lambda(u) = \lambda^2 - (|u| - \lambda)^2 1(|u| < \lambda)$, then $\widehat{\theta} = \overline{Y} 1(|\overline{Y}| > \lambda)$.

(c) If $p_\lambda(u)$ is the SCAD penalty function, find $\widehat{\theta}$. *(Pötscher and Leeb, 2009)*

7.8. Consider the development in Example 7.10.

(a) Show that the profile likelihood function for σ is given by (7.20) when σ_0 is known.

(b) Discuss the conditions for the approximation (7.23) to be reasonable, and give specific examples.

(c) Show that $\widehat{\sigma}_{\text{MPL}} \to 0$ in probability as $\lambda \to \infty$.

(d) If σ_0 is unknown, can you construct the profile likelihood function for σ and repeat the development of Section 7.7?

(e) In model (7.19) assume the parameters μ, σ_0 and σ are all unknown. Using the same ideas as in Section 7.7, compare the maximum likelihood estimators and the maximum penalized likelihood estimators for μ, σ_0 and σ by considering suitable penalty functions for μ, σ_0 and σ. *(Chung et al., 2013)*

Appendix 7.A Optimization and the Lagrangian method

Let d and m be given positive integers, and let $\mathcal{X}$ be a subset of $\mathbb{R}^d$. Consider real-valued functions $f : \mathcal{X} \to \mathbb{R}$ and $g_j : \mathcal{X} \to \mathbb{R}$ for $j = 1, \ldots, m$. A constrained optimization problem is formulated as

$$\min_{x \in \mathcal{X}} f(x) \text{ subject to } g_j(x) \le 0,\ j = 1, \ldots, m. \tag{7.27}$$

Definition 7.11 (Lagrangian) The Lagrange function, or Lagrangian, associated with (7.27) is the function

$$\mathcal{L}(x, \alpha) = f(x) + \sum_{j=1}^{m} \alpha_j g_j(x), \quad x \in \mathcal{X},\ \alpha_j \ge 0,\ j = 1, \ldots, m.$$

Here, $\alpha = (\alpha_1, \ldots, \alpha_m)^\mathsf{T}$. The variables α_j are called the Lagrange or dual variables.

Let $g(x) = (g_1(x), \ldots, g_m(x))^\mathsf{T}$. Then the Lagrangian can also be expressed as

$$\mathcal{L}(x, \alpha) = f(x) + \alpha^\mathsf{T} g(x).$$

Its derivatives are

$$\nabla_x \mathcal{L}(x, \alpha) = \nabla_x f(x) + \alpha^\mathsf{T} \nabla_x g(x), \quad \nabla_\alpha \mathcal{L}(x, \alpha) = g(x),$$

where $\nabla_z h(z)$ represents the derivative of the function $h(z)$ with respect to z. □

Theorem 7.12 (Karush–Kuhn–Tucker) Assume that $f : \mathcal{X} \to \mathbb{R}$ and $g_j : \mathcal{X} \to \mathbb{R}$ with $j = 1, \ldots, m$ are convex and differentiable, and that the constraints in (7.27) are satisfied. Then, x^* is a solution to (7.27) if and only if there exists $\alpha^* \succeq 0$ such that

$$\nabla_x \mathcal{L}(x, \alpha) = 0, \tag{7.28}$$

$$\nabla_\alpha \mathcal{L}(x, \alpha) \preceq 0, \tag{7.29}$$

$$\alpha^{*\mathrm{T}} g(x^*) = 0. \tag{7.30}$$

□

Conditions (7.28)–(7.30) are called the Karush–Kuhn–Tucker conditions. Conditions (7.29)–(7.30) are equivalent to

$$g(x^*) \preceq 0 \quad \text{and} \quad \alpha_j^* g_j(x^*) = 0, \quad j = 1, \ldots, m,$$

where $\alpha^* = (\alpha_1^*, \ldots, \alpha_m^*)^\mathsf{T}$. For details, see Mohri et al. (2018, Appendix B3).

Appendix 7.B Regularity conditions

This appendix presents regularity conditions for the model, penalty function, and data that are used to establish Theorems 7.5 and 7.6. The model assumptions in (b)–(d) are similar to the usual regularity conditions for likelihood inference, as in Sections 1.7.1 and 1.7.2. Condition (e) combines requirements for the model and the penalty function.

(a) $\{(Y_i, X_i) : i = 1, \ldots, n\}$ are independent random variables, each following the same distribution as (Y, X), with joint probability density or mass function $f(y, x; \theta)$.

(b) The support of $f(y, x; \theta)$ does not depend on θ, and $f(y, x; \theta_1) = f(y, x; \theta_2)$ implies $\theta_1 = \theta_2$.

(c) The first and second derivatives of $f(y, x; \theta)$ satisfy the first and second Bartlett identities

$$\mathrm{E}\left\{\frac{\partial \log f(Y, X; \theta)}{\partial \theta}\right\} = 0,$$

$$\mathrm{E}\left[\left\{\frac{\partial \log f(Y, X; \theta)}{\partial \theta}\right\}\left\{\frac{\partial \log f(Y, X; \theta)}{\partial \theta^{\mathsf{T}}}\right\}\right] = \mathrm{E}\left\{-\frac{\partial^2 \log f(Y, X; \theta)}{\partial \theta \partial \theta^{\mathsf{T}}}\right\},$$

and this matrix is positive definite at the true parameter value θ_0.

(d) There exists an open subset Ω of the parameter space containing θ_0 such that $f(y, x; \theta)$ has third-order partial derivatives

$$\frac{\partial^3 f(y, x; \theta)}{\partial \theta_j \partial \theta_k \partial \theta_l} \text{ for all } \theta \in \Omega \text{ and } j, k, l = 1, \ldots, \dim(\theta).$$

Furthermore, for $j, k, l = 1, \ldots, \dim(\theta)$, there exist functions $M_{jkl}(\cdot)$, satisfying $\mathrm{E}\{M_{jkl}(Y, X)\} < \infty$, such that

$$\left|\frac{\partial^3 f(y, x; \theta)}{\partial \theta_j \partial \theta_k \partial \theta_l}\right| \leq M_{jkl}(y, x), \quad \theta \in \Omega.$$

(e) The matrix B defined at (7.14) is invertible.

These conditions are imposed on the joint distribution of the response Y and covariate vector X, although regression parameters are usually viewed as pertaining to the conditional distribution of Y given X. See, for example, the discussion in Section 1.1 preceding Example 1.13. In the settings considered here, where interest focuses on both model selection and inference, asymptotic theory is developed under conditions on the joint distribution, as in Fan and Li (2001), among many others.

For ease of exposition, in the following proofs, let the nonzero subvector $\beta_{0(1)}$ consist of the first s components of β_0, and let the zero subvector $\beta_{0(2)}$ consist of the remaining $(p - s)$ components of β_0.

Appendix 7.C Proof of theorems

Proof of Theorem 7.5

Using conditions (a)–(d) in Appendix 7.B and following the development in Sections 1.7.1 and 1.7.2, we have

$$\frac{1}{\sqrt{n}} \frac{\partial \ell(\theta_0)}{\partial \theta} = O_p(1) \tag{7.31}$$

and

$$\frac{1}{n} \frac{\partial^2 \ell(\theta)}{\partial \theta \partial \theta^{\mathsf{T}}}\bigg|_{\theta = \theta_0} = -i(\theta_0) + o_p(1), \tag{7.32}$$

where $i(\theta)$ is the expected Fisher information in a single observation, using the second Bartlett identity in condition (c) of Appendix 7.B. We use $O_p(1)$ and $o_p(1)$ to denote probability orders for vectors or matrices, as needed.

Let $\alpha_n = n^{-1/2} + a_n$, and let u be a vector of constants of the same dimension as θ. We want to show that for any given $0 < \epsilon < 1$, there exists a constant $C_\epsilon > 0$ such that for sufficiently large n

$$\text{pr}\left\{\sup_{\|u\|_2 = C_\epsilon} Q(\theta_0 + \alpha_n u) < Q(\theta_0)\right\} \geq 1 - \epsilon. \tag{7.33}$$

This implies that with probability at least $1-\epsilon$, there exists a local maximizer $\widehat{\theta}$ of $Q(\theta)$ in $\{\theta_0 + \alpha_n u : \|u\|_2 \leq C_\epsilon\}$ such that

$$\|\widehat{\theta} - \theta_0\|_2 = O_p(\alpha_n).$$

To show (7.33), for any $C > 0$, we examine the difference

$$D_n(u) = Q(\theta_0 + \alpha_n u) - Q(\theta_0)$$

for any u with $\|u\|_2 = C$.

For $j = 1, \ldots, p$, let u_j denote the jth element of u, corresponding to the jth entry β_{0j} of β_0. By the definitions of $\beta_{0(1)}$ and $\beta_{0(2)}$, the condition $p_{\lambda_n}(0) = 0$, and the nonnegativity of $p_{\lambda_n}(\cdot)$, we have

$$\sum_{j=s+1}^{p} \{p_{\lambda_n}(|\beta_{0j} + \alpha_n u_j|) - p_{\lambda_n}(|\beta_{0j}|)\} \geq 0.$$

The definition of $Q(\theta)$ gives

$$\begin{aligned}
D_n(u) &= \ell(\theta_0 + \alpha_n u) - \ell(\theta_0) - n\sum_{j=1}^{p}\{p_{\lambda_n}(|\beta_{0j} + \alpha_n u_j|) - p_{\lambda_n}(|\beta_{0j}|)\} \\
&= \ell(\theta_0 + \alpha_n u) - \ell(\theta_0) \\
&\quad - n\sum_{j=1}^{s}\{p_{\lambda_n}(|\beta_{0j} + \alpha_n u_j|) - p_{\lambda_n}(|\beta_{0j}|)\} - n\sum_{j=s+1}^{p}\{p_{\lambda_n}(|\beta_{0j} + \alpha_n u_j|) - p_{\lambda_n}(|\beta_{0j}|)\} \\
&\leq \ell(\theta_0 + \alpha_n u) - \ell(\theta_0) - n\sum_{j=1}^{s}\{p_{\lambda_n}(|\beta_{0j} + \alpha_n u_j|) - p_{\lambda_n}(|\beta_{0j}|)\}.
\end{aligned} \tag{7.34}$$

By (7.32), Taylor series expansion of the right–hand side of (7.34) gives

$$\begin{aligned}
D_n(u) \leq &\left\{\left.\frac{\partial \ell(\theta)}{\partial \theta^{\mathsf{T}}}\right|_{\theta=\theta_0}\right\}\alpha_n u - \frac{1}{2}u^{\mathsf{T}} i(\theta_0) u n \alpha_n^2\{1 + o_p(1)\} \\
&- n\sum_{j=1}^{s}\left[\alpha_n p'_{\lambda_n}(|\beta_{0j}|)\text{sgn}(\beta_{0j})u_j + \frac{1}{2}\alpha_n^2 p''_{\lambda_n}(|\beta_{0j}|)u_j^2\{1 + o(1)\}\right],
\end{aligned}$$

which is denoted by $A_1 - A_2 - A_3$.

We now examine bounds for these three terms for any $\epsilon > 0$ and u satisfying $||u||_2 = C$, where $C > 0$. First, by (7.31), there exist $K_1 > 0$ and $N_1 \in \mathbb{N}$ such that

$$\text{P}_{\text{r}}\left(\left\|\left.\frac{\partial \ell(\theta)}{\partial \theta}\right|_{\theta=\theta_0}\right\|_2 \leq K_1 n^{1/2}\right) \geq 1 - \epsilon/3 \quad \text{for all } n \geq N_1.$$

Therefore, with probability at least $1 - \epsilon/3$,

$$|A_1| = \left|\left\{\left.\frac{\partial \ell(\theta)}{\partial \theta^{\mathsf{T}}}\right|_{\theta=\theta_0}\right\}\alpha_n u\right| \leq \alpha_n \left\|\left.\frac{\partial \ell(\theta)}{\partial \theta}\right|_{\theta=\theta_0}\right\|_2 \|u\|_2 \leq K_1 n^{1/2}\alpha_n \|u\|_2 \tag{7.35}$$

for all $n \geq N_1$.

Second, since $i(\theta_0)$ is positive definite by condition (c) in Appendix 7.B, there exists $K_2 > 0$ such that

$$v^\top i(\theta_0) v \geq K_2 \|v\|_2^2$$

for all v. Therefore, for $\|u\|_2 = C$, there exists $N_2 \in \mathbb{N}$ such that for $n \geq N_2$,

$$A_2 = \frac{1}{2} u^\top i(\theta_0) u n \alpha_n^2 \{1 + o_p(1)\} \geq \frac{K_2}{4} n \alpha_n^2 C^2 \tag{7.36}$$

with probability at least $1 - \epsilon/3$.

Third, by the definition of a_n, $|p'_{\lambda_n}(|\beta_{0j}|)| \leq a_n, j \leq s$. Moreover, for $\|u\|_2 = C$, we have

$$\sum_{j=1}^{s} |u_j| \leq \sqrt{s}\, \|u\|_2 = \sqrt{s} C, \qquad \sum_{j=1}^{s} u_j^2 \leq \|u\|_2^2 = C^2.$$

Hence,

$$\begin{aligned} |A_3| &\leq n \alpha_n \sum_{j=1}^{s} |p'_{\lambda_n}(|\beta_{0j}|)|\, |u_j| + \frac{n}{2} \alpha_n^2 \sum_{j=1}^{s} |p''_{\lambda_n}(|\beta_{0j}|)|\, u_j^2 \{1 + o(1)\} \\ &\leq n \alpha_n a_n \sqrt{s} C + \frac{n}{2} \alpha_n^2 M_n C^2 \{1 + o(1)\}, \end{aligned}$$

where $M_n = \max\{|p''_{\lambda_n}(|\beta_{0j}|)| : \beta_{0j} \neq 0\}$.

Comparing each term in the upper bound of $|A_3|$ to A_2 and using (7.36), we have

$$\frac{n \alpha_n a_n \sqrt{s} C}{A_2} \leq \frac{4\sqrt{s}}{K_2} \cdot \frac{a_n}{\alpha_n C} \leq \frac{4\sqrt{s}}{K_2 C},$$

since $a_n \leq \alpha_n$. Let $C_1 = 16\sqrt{s}/K_2$. Then choosing $C \geq C_1$ gives

$$\frac{n \alpha_n a_n \sqrt{s} C}{A_2} \leq \frac{1}{4}. \tag{7.37}$$

Similarly, for C in (7.37), using (7.36) gives

$$\frac{(n/2) \alpha_n^2 M_n C^2}{A_2} \leq \frac{2 M_n}{K_2}.$$

By the assumption that $M_n \to 0$ as $n \to \infty$, there exists $N_3 \in \mathbb{N}$ such that for all $n \geq N_3$,

$$\frac{2 M_n}{K_2} \leq \frac{1}{4}.$$

Therefore, for $n \geq N_3$ and $C \geq C_1$, we have

$$\frac{|A_3|}{A_2} \leq \frac{1}{4} + \frac{1}{4}, \quad \text{so} \quad |A_3| \leq \frac{1}{2} A_2. \tag{7.38}$$

Let $C_2 = 16 K_1 / K_2$. Define $C_\epsilon = \max(C_1, C_2)$ and $N = \max(N_1, N_2, N_3)$. Now consider the event

$$\mathcal{E}_n = \left\{ |A_1| \leq K_1 n^{1/2} \alpha_n C_\epsilon \right\} \cap \left\{ A_2 \geq \frac{K_2}{4} n \alpha_n^2 C^2 \right\} \cap \left\{ |A_3| \leq \frac{1}{2} A_2 \right\}.$$

Then by (7.35), (7.36), and (7.38), we have that for u satisfying $\|u\|_2 = C_\epsilon$,

$$\mathrm{pr}(\mathcal{E}_n) \geq 1 - \epsilon \quad \text{for all } n \geq N.$$

Hence, on $\mathcal{E}_n$, for any u with $\|u\|_2 = C_\epsilon$ and for all $n \geq N$,

$$\begin{aligned} D_n(u) &\leq A_1 - A_2 - A_3 \leq |A_1| - A_2 + |A_3| \\ &\leq K_1 n^{1/2} \alpha_n C_\epsilon - \frac{K_2}{8} n \alpha_n^2 C_\epsilon^2 \\ &\leq \frac{K_2}{16} n \alpha_n^2 C_\epsilon^2 - \frac{K_2}{8} n \alpha_n^2 C_\epsilon^2 \\ &< 0, \end{aligned}$$

since $\alpha_n \geq n^{-1/2}$ and $C_\epsilon \geq C_2$. Therefore, for sufficiently large n,

$$\mathrm{pr}\left\{ \sup_{\|u\|_2 = C_\epsilon} Q(\theta_0 + \alpha_n u) < Q(\theta_0) \right\} \geq \mathrm{pr}(\mathcal{E}_n) \geq 1 - \epsilon,$$

which establishes (7.33), thus, completing the proof.

□

Proof of Theorem 7.6

(a) To establish Theorem 7.6 (a), it suffices to show that with probability tending to 1 as $n \to \infty$, for any $\theta_{(1)}$ satisfying $||\theta_{(1)} - \theta_{0(1)}||_2 = O(n^{-1/2})$ and for any positive constant C,

$$Q\left\{ \begin{pmatrix} \theta_{(1)} \\ 0 \end{pmatrix} \right\} = \max_{||\beta_{(2)}||_2 \leq \epsilon_n} Q\left\{ \begin{pmatrix} \theta_{(1)} \\ \beta_{(2)} \end{pmatrix} \right\},$$

where $\epsilon_n = Cn^{-1/2}$. It then suffices to show that for $j = s+1, \ldots, p$,

$$\begin{aligned} &\frac{\partial Q(\theta)}{\partial \beta_j} < 0 \quad \text{for } 0 < \beta_j < \epsilon_n, \\ &\frac{\partial Q(\theta)}{\partial \beta_j} > 0 \quad \text{for } -\epsilon_n < \beta_j < 0. \end{aligned} \tag{7.39}$$

To show (7.39), for each $j = s+1, \ldots, p$, we apply Taylor series expansion to $\partial Q(\theta)/\partial \beta_j$ around θ_0. Specifically, for $k = 1, \ldots, \dim(\theta)$, let θ_k and θ_{0k} be the kth element of θ and θ_0, respectively. Then

$$\begin{aligned} \frac{\partial Q(\theta)}{\partial \beta_j} &= \frac{\partial \ell(\theta)}{\partial \beta_j} - np'_{\lambda_n}(|\beta_j|)\mathrm{sgn}(\beta_j) \\ &= \left.\frac{\partial \ell(\theta)}{\partial \beta_j}\right|_{\theta=\theta_0} + \sum_{k=1}^{\dim(\theta)} \left\{ \left.\frac{\partial^2 \ell(\theta)}{\partial \beta_j \partial \theta_k}\right|_{\theta=\theta_0} \right\} (\theta_k - \theta_{0k}) \\ &\quad + \frac{1}{2} \sum_{k=1}^{\dim(\theta)} \sum_{l=1}^{\dim(\theta)} \left\{ \left.\frac{\partial^3 \ell(\theta)}{\partial \beta_j \partial \theta_k \partial \theta_l}\right|_{\theta=\theta^*} \right\} (\theta_k - \theta_{0k})(\theta_l - \theta_{0l}) - np'_{\lambda_n}(|\beta_j|)\mathrm{sgn}(\beta_j), \end{aligned}$$

where θ^* lies on the straight line from θ to θ_0.

By the properties of the likelihood function, for $k, l = 1, \ldots, \dim(\theta)$,

$$n^{-1} \frac{\partial \ell(\theta)}{\partial \theta_k} = O_p(n^{-1/2})$$

and

$$n^{-1} \left.\frac{\partial^2 \ell(\theta)}{\partial \theta_k \partial \theta_l}\right|_{\theta=\theta_0} = \mathrm{E}\left\{ \left.\frac{\partial^2 \ell(\theta)}{\partial \theta_k \partial \theta_l}\right|_{\theta=\theta_0} \right\} + o_p(1).$$

By the assumption that $\theta - \theta_0 = O(n^{-1/2})$ and conditions (c) and (d) in Appendix 7.B,

$$\begin{aligned}\frac{\partial Q(\theta)}{\partial \beta_j} &= O_p(n^{1/2}) - np'_{\lambda_n}(|\beta_j|)\text{sgn}(\beta_j) \\ &= n\lambda_n\{O_p(n^{-1/2}/\lambda_n) - \lambda_n^{-1}p'_{\lambda_n}(|\beta_j|)\text{sgn}(\beta_j)\}.\end{aligned}$$

By the conditions that $\lambda_n \geq 0$, and

$$\liminf_{n\to\infty}\liminf_{v\to 0+}\frac{p'_{\lambda_n}(v)}{\lambda_n} > 0 \quad \text{and } n^{-1/2}/\lambda_n \to 0, \quad \text{as } n \to \infty,$$

the sign of the derivative $\partial Q(\theta)/\partial\beta_j$ is determined by the negative sign of β_j. Hence, (7.39) follows.

(b) With the result in Theorem 7.5(a), we consider $Q(\theta)$ in (7.13) as a function of $\theta_{(1)}$ by restricting $\beta_{(2)}$ to be 0. To emphasize this, we let $\eta = (\theta_{(1)}^\mathsf{T},\ 0)^\mathsf{T}$ denote the parameter θ with $\beta_{(2)}$ constrained to be 0, and write $\eta_0 = (\theta_{0(1)}^\mathsf{T},\ 0)^\mathsf{T}$ accordingly. Then (7.13) becomes

$$Q(\eta) = \ell(\eta) - n\sum_{j=1}^{s} p_{\lambda_n}(|\beta_j|).$$

It can be shown that given λ_n, there exists a $\sqrt{n}$–consistent local maximizer of $Q(\eta)$, $\widehat{\eta}_{\lambda_n} = (\widehat{\theta}_{(1)\lambda_n}^\mathsf{T},\ 0)^\mathsf{T}$ that satisfies

$$\left.\frac{\partial Q(\eta)}{\partial \theta_{(1)}}\right|_{\eta=\widehat{\eta}_{\lambda_n}} = 0,$$

i.e.,

$$\left.\frac{\partial \ell(\eta)}{\partial \theta_{(1)}}\right|_{\eta=\widehat{\eta}_{\lambda_n}} - n\sum_{j=1}^{s}\left\{\left.\frac{\partial p_{\lambda_n}(|\beta_j|)}{\partial \theta_{(1)}}\right|_{\eta=\widehat{\eta}_{\lambda_n}}\right\} = 0.$$

Applying Taylor series expansion to the left-hand side of these equations around $\theta_{(1)} = \theta_{0(1)}$, we have

$$\begin{aligned}&\left.\frac{\partial \ell(\eta)}{\partial \theta_{(1)}}\right|_{\eta=\eta_0} + \left\{\left.\frac{\partial^2 \ell(\eta)}{\partial \theta_{(1)}\partial \theta_{(1)}^\mathsf{T}}\right|_{\eta=\eta^{(0)}} + o_p(1)\right\}(\widehat{\theta}_{(1)} - \theta_{0(1)}) - n\sum_{j=1}^{s}\left\{\left.\frac{\partial p_{\lambda_n}(|\beta_j|)}{\partial \theta_{(1)}}\right|_{\eta=\eta_0}\right\} \\ &- \left[n\sum_{j=1}^{s}\left\{\left.\frac{\partial^2 p_{\lambda_n}(|\beta_j|)}{\partial \theta_{(1)}\partial \theta_{(1)}^\mathsf{T}}\right|_{\eta=\eta_0}\right\} + o(1)\right](\widehat{\theta}_{(1)} - \theta_{0(1)}) = 0.\end{aligned}$$

Therefore,

$$\begin{aligned}&\left[-\frac{1}{n}\left.\frac{\partial^2 \ell(\eta)}{\partial \theta_{(1)}\partial \theta_{(1)}^\mathsf{T}}\right|_{\eta=\eta_0} + \sum_{j=1}^{s}\left\{\left.\frac{\partial^2 p_{\lambda_n}(|\beta_j|)}{\partial \theta_{(1)}\partial \theta_{(1)}^\mathsf{T}}\right|_{\eta=\eta_0}\right\}\right](\widehat{\theta}_{(1)} - \theta_{0(1)}) \\ &+ \sum_{j=1}^{s}\left\{\left.\frac{\partial p_{\lambda_n}(|\beta_j|)}{\partial \theta_{(1)}}\right|_{\eta=\eta_0}\right\} + o_p(1) \\ &= \frac{1}{n}\left.\frac{\partial \ell(\eta)}{\partial \theta_{(1)}}\right|_{\eta=\eta_0},\end{aligned} \tag{7.40}$$

From the definition of $i_{11}(\theta_0)$, b, and Σ, we have that

$$\mathrm{E}\left\{-\frac{\partial^2 \ell_i(\eta)}{\partial\theta_{(1)}\partial\theta_{(1)}^{\mathsf{T}}}\right\}\Bigg|_{\eta=\eta_0} = i_{11}(\theta_0),$$

$$\sum_{j=1}^{s}\frac{\partial p_{\lambda_n}(|\beta_j|)}{\partial\theta_{(1)}}\Bigg|_{\eta=\eta_0} = b, \quad \text{and} \quad \sum_{j=1}^{s}\frac{\partial^2 p_{\lambda_n}(|\beta_j|)}{\partial\theta_{(1)}\partial\theta_{(1)}^{\mathsf{T}}}\Bigg|_{\eta=\eta_0} = \Sigma.$$

Using the fact that

$$-\frac{1}{n}\frac{\partial^2 \ell(\eta)}{\partial\theta_{(1)}\partial\theta_{(1)}^{\mathsf{T}}}\Bigg|_{\eta=\eta_0} = \mathrm{E}\left[\left\{-\frac{\partial^2 \ell_i(\eta)}{\partial\theta_{(1)}\partial\theta_{(1)}^{\mathsf{T}}}\Bigg|_{\eta=\eta_0}\right\}\right] + o_p(1),$$

we write (7.40) as

$$\{i_{11}(\eta_0) + \Sigma\}\,(\widehat{\theta}_{(1)} - \theta_{0(1)}) + b + o_p(1) = \frac{1}{n}\sum_{i=1}^{n}\frac{\partial \ell_i(\eta)}{\partial\theta_{(1)}}\Bigg|_{\eta=\eta_0},$$

which by condition (e) in Appendix 7.B,

$$\sqrt{n}B\left\{(\widehat{\theta}_{(1)} - \theta_{0(1)}) + B^{-1}b\right\} + o_p(1) = \frac{1}{\sqrt{n}}\sum_{i=1}^{n}\frac{\partial \ell_i(\eta)}{\partial\theta_{(1)}}\Bigg|_{\eta=\eta_0}. \tag{7.41}$$

By the definition of η_0, for $i = 1, \ldots, n$, $\partial\ell_i(\eta)/\partial\theta_{(1)}|_{\eta=\eta_0}$ are independent and identically distributed with mean zero and covariance matrix $i_{11}(\theta_0)$. Applying the central limit theorem to the right-hand side of (7.41) gives

$$\frac{1}{\sqrt{n}}\sum_{i=1}^{n}\frac{\partial \ell_i(\eta)}{\partial\theta_{(1)}}\Bigg|_{\eta=\eta_0} \xrightarrow{d} N\{0, i_{11}(\theta_0)\} \quad \text{as } n \to \infty.$$

Left-multiplying (7.41) by B^{-1} and applying Slutsky's theorem then yields, as $n \to \infty$,

$$\sqrt{n}\left\{(\widehat{\theta}_{(1)\lambda_n} - \theta_{0(1)}) + B^{-1}b\right\} \xrightarrow{d} N\{0, B^{-1}i_{11}(\theta_0)B^{-\mathsf{T}}\},$$

which completes the proof.

□

Bibliography

Aalen, O. O. (1978). Non-parametric inference for a family of counting processes. *Annals of Statistics 6*, 701–726.

Aalen, O. O., Ø. Borgan, and H. K. Gessing (2008). *Survival and Event History Analysis: a Process Point of View.* Heidelberg: Springer.

Abadie, A., S. Athey, G. W. Imbens, and J. M. Wooldridge (2023). When should you adjust standard errors for clustering? *Quarterly Journal of Economics 138*, 1–35.

Agresti, A. (2002). *Categorical Data Analysis* (2nd ed.). Hoboken, NJ: Wiley.

Aitkin, M. (1996). A general maximum likelihood analysis of overdispersion in generalized linear models. *Statistics and Computing 6*, 251–262.

Aitkin, M. and D. B. Rubin (1985). Estimation and hypothesis testing in finite mixture models. *Journal of the Royal Statistical Society B 47*, 67–75.

Akaike, H. (1969). Fitting autoregressive models for prediction. *Annals of the Institute of Statistical Mathematics 21*, 243–247.

Akaike, H. (1970). Statistical predictor identification. *Annals of the Institute of Statistical Mathematics 22*, 203–217.

Akaike, H. (1971). Autoregressive model fitting for control. *Annals of the Institute of Statistical Mathematics 23*, 163–180.

Akaike, H. (1973). Information theory and an extension of the maximum likelihood principle. In B. N. Petrov and F. Csaki (Eds.), *Proceedings of the 2nd International Symposium on Information Theory*, Tsahkadsor, Armenia, pp. 267–281. Akadémiai Kiadó.

Albert, A. and J. A. Anderson (1984). On the existence of maximum likelihood estimates in logistic regression models. *Biometrika 71*, 1–10.

Alfonzetti, G., R. Bellio, Y. Chen, and I. Moustaki (2025). When composite likelihood meets stochastic approximation. *Journal of the American Statistical Association 120*, 1906–1918.

Andersen, P. K. and R. D. Gill (1988). Cox's regression model for counting processes: a large-sample study. *Annals of Statistics 10*, 1100–1120.

Andrei, A. and C. Kendziorski (2009). An efficient method for identifying statistical interactors in gene association networks. *Biostatistics 10*, 706–718.

Andrews, D. W. K. (1991). Heteroskedasticity and autocorrelation consistent covariance matrix estimation. *Econometrica 59*, 817–858.

Angelopoulos, A. N. and S. Bates (2023). Conformal prediction: A gentle introduction. *Foundations and Trends in Machine Learning 16*, 494–591.

Arellano, M. (1987). Computing robust standard errors for within-groups estimators. *Oxford Bulletin of Economics & Statistics 49*, 431–434.

Arlot, S. and A. Celisse (2010). A survey of cross-validation procedures for model selection. *Statistics Surveys 4*, 40–79.

Azzalini, A. (1985). A class of distributions which includes the normal ones. *Scandinavian Journal of Statistics 12*, 171–178.

Azzalini, A. (1996). *Statistical Inference Based on the Likelihood.* London: Chapman and Hall/CRC.

Azzalini, A. (2005). The skew-normal distribution and related multivariate families. *Scandinavian Journal of Statistics 32*, 159–188.

Baddeley, A., E. Rubak, and R. Turner (2015). *Spatial Point Patterns: Methodology and Applications with R.* Boca Raton, FL: Chapman & Hall/CRC.

Balabdaoui, F., K. Rufibach, and J. A. Wellner (2009). Limit distribution theory for maximum likelihood estimation of a log-concave density. *Annals of Statistics 37*, 1299–1331.

Bang, H. and J. M. Robins (2005). Doubly robust estimation in missing data and causal inference models. *Biometrics 61*, 962–972.

Barber, R. F. and M. Drton (2015). High-dimensional Ising model selection with Bayesian information criteria. *Electronic Journal of Statistics 9*, 567–607.

Barlow, A. M., C. Sherlock, and J. A. Tawn (2020). Inference for extreme values under threshold-based stopping rules. *Journal of the Royal Statistical Society C 69*, 765–789.

Barndorff-Nielsen, O. E. (1980). Conditionality resolutions. *Biometrika 67*, 203–310.

Barndorff-Nielsen, O. E. (1983). On a formula for the distribution of the maximum likelihood estimator. *Biometrika 70*, 343–365.

Barndorff-Nielsen, O. E. (1987). Contribution to discussion of "Parameter orthogonality and approximate conditional inference" by D. R. Cox and N. Reid. *Journal of the Royal Statistical Society B 49*, 18–20.

Barndorff-Nielsen, O. E. and D. R. Cox (1989). *Asymptotic Techniques for Use in Statistics.* London: Chapman and Hall/CRC.

Barndorff-Nielsen, O. E. and D. R. Cox (1994). *Inference and Asymptotics.* London: Chapman and Hall/CRC.

Barndorff-Nielsen, O. E. and P. McCullagh (1993). A note on the relation between modified profile likelihood and the Cox–Reid adjusted profile likelihood. *Biometrika 80*, 321–328.

Bartlett, M. S. (1936a). The information available in small samples. *Proceedings of the Cambridge Philosophical Society 32*, 560–566.

Bartlett, M. S. (1936b). Statistical information and properties of sufficiency. *Proceedings of the Royal Society of London A 154*, 124–137.

Bartlett, M. S. (1937). Properties of sufficiency and statistical tests. *Proceedings of the Royal Society of London A 160*, 268–282.

Bates, D., M. Mächler, B. Bolker, and S. Walker (2015). Fitting linear mixed-effects models using lme4. *Journal of Statistical Software 67*, 1–48.

Bates, S., T. J. Hastie, and R. Tibshirani (2024). Cross-validation: What does it estimate and how well does it do it? *Journal of the American Statistical Association 119*, 1434–1445.

Battey, H. S., D. R. Cox, and S. H. Lee (2024). On partial likelihood and the construction of factorisable transformations. *Information Geometry 7*, 9–28.

Battey, H. S. and P. McCullagh (2024). An anomaly arising in the analysis of processes with more than one source of variability. *Biometrika 111*, 677–689.

Battey, H. S. and N. Reid (2024). On the role of parameterization in models with a misspecified nuisance component. *Proceedings of the National Academy of Sciences 121*, e2402736121.

Beaton, A. E. and J. W. Tukey (1974). The fitting of power series, meaning polynomials, illustrated on band-spectroscopic data. *Technometrics 16*, 147–185.

Bellio, R., S. Ghosh, A. B. Owen, and C. Varin (2025). Consistent and scalable composite likelihood estimation of probit models with crossed random effects. *Biometrika 112*, asaf037.

Bellio, R. and D. Pierce (2020). *likelihoodAsy: Functions for Likelihood Asymptotics.* R package version 0.51.

Beran, J. (1994). *Statistics for Long-Memory Processes.* New York: Chapman & Hall/CRC.

Berger, J. O. and J. M. Bernardo (1992). On the development of reference priors (with discussion). In J. M. Bernardo, J. O. Berger, A. P. Dawid, and A. F. M. Smith (Eds.), *Bayesian Statistics 4*, pp. 35–60. Oxford: Oxford University Press.

Berger, J. O., B. Liseo, and R. L. Wolpert (1999). Integrated likelihood methods for eliminating nuisance parameters. *Statistical Science 14*, 1–29.

Berk, R., L. Brown, A. Buja, K. Zhang, and L. H. Zhao (2013). Valid post-selection inference. *Annals of Statistics 41*, 802–837.

Berk, R. H. (1966). Limiting behavior of posterior distributions when the model is incorrect. *Annals of Mathematical Statistics 37*, 51–58.

Bernton, E., P. E. Jacob, M. Gerber, and C. P. Robert (2019). Approximate Bayesian computation with the Wasserstein distance. *Journal of the Royal Statistical Society B 81*, 235–269.

Besag, J. (1974). Spatial interaction and the statistical analysis of lattice systems (with discussion). *Journal of the Royal Statistical Society B 36*, 192–236.

Besag, J. (1986). On the statistical analysis of dirty pictures (with discussion). *Journal of the Royal Statistical Society B 48*, 259–279.

Bevilacqua, M. and C. Gaetan (2015). Comparing composite likelihood methods based on pairs for spatial Gaussian random fields. *Statistics and Computing 25*, 877–892.

Bhapkar, V. P. (1972). On a measure of efficiency of an estimating equation. *Sankhyā A 34*, 467–472.

Bickel, P. J. and J. K. Ghosh (1990). A decomposition for the likelihood ratio statistic and the Bartlett correction—a Bayesian argument. *Annals of Statistics 18*, 1070–1090.

Bickel, P. J. and B. Li (2006). Regularization in statistics (with discussion). *Test 15*, 271–344.

Bickel, P. J., Y. Ritov, and A. B. Tsybakov (2009). Simultaneous analysis of Lasso and Dantzig selector. *Annals of Statistics 37*, 1705–1732.

Bjørnstadt, J. (1990). Predictive likelihood: a review. *Statistical Science 5*, 242–254.

Blier-Wong, C. and R. Wang (2026). Improved thresholds for e-values. *Annals of Statistics 54*. to appear.

Bollerslev, T. (1986). Generalized autoregressive conditional heteroskedasticity. *Journal of Econometrics 31*, 307–327.

Bollerslev, T., R. F. Engle, and D. B. Nelson (1994). ARCH models. *Handbook of Econometrics 4*, 2959–3038.

Bollerslev, T. and J. M. Wooldridge (1992). Quasi-maximum likelihood estimation and inference in dynamic models with time-varying covariances. *Econometric Reviews 11*, 143–172.

Booth, J. G. and J. P. Hobert (1999). Maximizing generalized linear mixed model likelihoods with an automated Monte Carlo EM algorithm. *Journal of the Royal Statistical Society B 61*, 265–285.

Borgan, Ø. (2014). Nelson–Aalen estimator. In *Wiley StatsRef: Statistics Reference Online*. New York: Wiley.

Bottai, M. (2003). Confidence regions when the Fisher information is zero. *Biometrika 90*, 73–84.

Box, G. E. P. and G. M. Jenkins (1970). *Time Series Analysis: Forecasting and Control.* Oakland: Holden-Day.

Boyd, S. and L. Vandenberghe (2004). *Convex Optimization.* Cambridge: Cambridge University Press.

Bradic, J., J. Fan, and W. Wang (2011). Penalized composite quasi-likelihood for ultrahigh dimensional variable selection. *Journal of the Royal Statistical Society B 73*, 325–349.

Bradley, R. A. and M. E. Terry (1952). Rank analysis of incomplete block designs: I. the method of paired comparisons. *Biometrika 39*, 324–345.

Brazzale, A. R. (2005). hoa: An R package bundle for higher order likelihood inference. *Rnews 5/1 May 2005*, 20–27.

Brazzale, A. R., A. C. Davison, and N. Reid (2007). *Applied Asymptotics: Case Studies in Small-Sample Statistics.* Cambridge: Cambridge University Press.

Brazzale, A. R. and V. Mameli (2024). Likelihood asymptotics in nonregular settings: A review with emphasis on the likelihood ratio. *Statistical Science 39*, 322–345.

Breiman, L. and J. H. Friedman (1985). Estimating optimal transformations for multiple regression and correlation. *Journal of the American statistical Association 80*, 580–598.

Brent, R. P. (1973). *Algorithms for Minimization Without Derivatives.* Englewood Cliffs NJ: Prentice-Hall.

Breslow, N. E. and D. G. Clayton (1993). Approximate inference in generalized linear mixed models. *Journal of the American Statistical Association 88*, 9–25.

Breslow, N. E. and X. Lin (1995). Bias correction in generalised linear mixed models with a single component of dispersion. *Biometrika 82*, 81–91.

Brillinger, D. R. (1977). Contribution to discussion of "Consistent nonparametric regression" by C. J. Stone. *Annals of Statistics 5*, 622–623.

Brockwell, P. J. and R. A. Davis (1987). *Time Series: Theory and Methods.* New York: Springer.

Brockwell, P. J. and R. A. Davis (2002). *Introduction to Time Series and Forecasting* (2nd ed.). New York: Springer.

Brown, L. D. (1986). *Fundamentals of Statistical Exponential Families with Applications to Statistical Decision Theory.* Hayward, CA: Institute of Mathematical Statistics.

Bücher, A., J. Segers, and T. Staud (2025). Consistency of M-estimators for non-identically distributed data: the case of fixed-design distributional regression. https://arxiv.org/pdf/2511.11067.

Bühlmann, P. and S. van de Geer (2011). *Statistics for High-dimensional Data: Methods, Theory and Applications.* Heidelberg: Springer.

Burnham, K. P. and D. R. Anderson (2002). *Model Selection and Multimodel Inference: a Practical Information-theoretic Approach* (2nd ed.). New York: Springer.

Cameron, A. C. and P. K. Trivedi (2005). *Microeconometrics: Methods and Applications.* Cambridge: Cambridge University Press.

Candès, E. J. and P. Sur (2020). The phase transition for the existence of the maximum likelihood estimate in high-dimensional logistic regression. *Annals of Statistics 48*, 27–42.

Candès, E. J. and T. Tao (2007). The Dantzig selector: Statistical estimation when p is much larger than n. *Annals of Statistics 35*, 2313–2351.

Candès, E. J., Y. Fan, L. Janson, and J. Lv (2018). Panning for gold: 'Model-X' knockoffs for high dimensional controlled variable selection. *Journal of the Royal Statistical Society B 80*, 551–577.

Carpenter, B., A. Gelman, M. D. Hoffman, D. Lee, B. Goodrich, M. Betancourt, M. Brubaker, J. Guo, P. Li, and A. Riddell (2017). Stan: A probabilistic programming language. *Journal of Statistical Software 76*, 1–32.

Carroll, R. J. and D. Ruppert (1991). Prediction and tolerance intervals with transformation and/or weighting. *Technometrics 33*, 197–210.

Carvalho, C. M., N. G. Polson, and J. G. Scott (2010). The horseshoe estimator for sparse signals. *Biometrika 97*, 465–480.

Castillo, I., J. Schmidt-Hieber, and A. van der Vaart (2015). Bayesian linear regression with sparse priors. *Annals of Statistics 43*, 1986–2018.

Castillo, I. and A. van der Vaart (2012). Needles and straw in a haystack: Posterior concentration for possibly sparse sequences. *Annals of Statistics 40*, 2069–2101.

Chamberlain, G. (1987). Asymptotic efficiency in estimation with conditional moment restrictions. *Journal of Econometrics 34*, 305–334.

Chandler, R. E. and S. Bate (2007). Inference for clustered data using the independence log-likelihood. *Biometrika 94*, 167–183.

Chatfield, C. (1996). *The Analysis of Time Series: An Introduction.* London: Chapman & Hall.

Chatterjee, A. and S. N. Lahiri (2011). Bootstrapping Lasso estimators. *Journal of the American Statistical Association 106*, 608–625.

Chen, B., G. Y. Yi, and R. J. Cook (2010). Weighted generalized estimating functions for longitudinal response and covariate data that are missing at random. *Journal of the American Statistical Association 105*, 336–353.

Chen, H. and J. Chen (2003). Tests for homogeneity in normal mixtures in the presence of a structural parameter. *Statistics and Probability Letters 52*, 125–133.

Chen, J. and Z. Chen (2008). Extended Bayesian information criteria for model selection with large model spaces. *Biometrika 95*, 759–771.

Chen, S. X. (1993). On the accuracy of empirical likelihood confidence regions for linear regression model. *Annals of the Institute of Statistical Mathematics 45*, 621–637.

Chen, S. X. and H. Cui (2006). On Bartlett correction of empirical likelihood in the presence of nuisance parameters. *Biometrika 93*, 215–220.

Chen, S. X. and I. Van Keilegom (2009). A review on empirical likelihood methods for regression. *Test 18*, 415–447.

Chen, W., M. G. Genton, and Y. Sun (2021). Space-time covariance structures and models. *Annual Review of Statistics and Its Application 8*, 191–215.

Chen, X. (2007). Large sample sieve estimation of semi-nonparametric models. In J. J. Heckman and E. E. Leamer (Eds.), *Handbook of Econometrics*, Volume 6, pp. 5549–5632. Amsterdam: Elsevier.

Chernoff, H. (1954). On the distribution of the likelihood ratio. *Annals of Mathematical Statistics 25*, 573–578.

Chernozhukov, V., D. Chetverikov, M. Demirer, E. Duflo, C. Hansen, W. Newey, and J. Robins (2018). Double/debiased machine learning for treatment and structural parameters. *Econometrics Journal 21*, C1–C68.

Chernozhukov, V., C. Hansen, and M. Spindler (2015). Valid post-selection and post-regularization inference: An elementary, general approach. *Annual Review of Economics 7*, 649–688.

Chung, Y., S. Rabe-Hesketh, V. Dorie, A. Gelman, and J. Liu (2013). A nondegenerate penalized likelihood estimator for variance parameters in multilevel models. *Psychometrika 78*, 685–709.

Claeskens, G. and N. L. Hjort (2003). The focused information criterion. *Journal of the American Statistical Association 98*, 900–916.

Claeskens, G. and N. L. Hjort (2008). *Model Selection and Model Averaging.* Cambridge: Cambridge University Press.

Clayton, D. G. (1996). Generalized linear mixed models. In W. R. Gilks, S. Richardson, and D. J. Spiegelhalter (Eds.), *Markov Chain Monte Carlo in Practice*, pp. 275–301. London: Chapman and Hall.

Cleveland, W. S. (1979). Robust locally weighted regression and smoothing scatterplots. *Journal of the American Statistical Association 74*, 829–836.

Cleveland, W. S. and S. J. Devlin (1988). Locally weighted regression: an approach to regression analysis by local fitting. *Journal of the American Atatistical Association 83*, 596–610.

Clifford, P. (1990). Markov random fields in statistics. In G. R. Grimmett and D. J. A. Welsh (Eds.), *Disorder in Physical Systems: A Volume in Honour of John M. Hammersley*, pp. 19–32. Oxford: Clarendon Press.

Cook, R. D., C.-L. Tsai, and B. C. Wei (1986). Bias in nonlinear regression. *Biometrika 73*, 615–623.

Cook, R. J. and J. F. Lawless (2018). *Multistate Models for the Analysis of Life History Data.* London: Chapman and Hall/CRC.

Cook, R. J. and J. F. Lawless (2022). Life history analysis with multistate models: a review and some current issues. *Canadian Journal of Statistics 50*, 1270–1298.

Copas, J. (1999). What works?: selectivity models and meta-analysis. *Journal of the Royal Statistical Society A 162*, 95–109.

Copas, J. B. and H. G. Li (1997). Inference for non-random samples (with discussion). *Journal of the Royal Statistical Society B 59*, 55–95.

Corbeil, R. R. and S. R. Searle (1976). Restricted maximum likelihood (REML) estimation of variance components in the mixed model. *Technometrics 18*, 31–38.

Cordeiro, G. M. and P. McCullagh (1991). Bias correction in generalized linear models. *Journal of the Royal Statistical Society B 53*, 629–643.

Cox, D. R. (1955). Some statistical methods connected with series of events (with discussion). *Journal of the Royal Statistical Society B 17*, 129–157.

Cox, D. R. (1958). Some problems connected with statistical inference. *Annals of Mathematical Statistics 29*, 357–372.

Cox, D. R. (1961). Tests of separate families of hypotheses. *Proceedings of the Fourth Berkeley Symposium on Mathematical Statistics and Probability 1*, 105–123.

Cox, D. R. (1962). Further results on tests of separate families of hypotheses. *Journal of the Royal Statistical Society B 24*, 406–424.

Cox, D. R. (1972). Regression models and life-tables (with discussion). *Journal of the Royal Statistical Society B 34*, 187–220.

Cox, D. R. (1975). Partial likelihood. *Biometrika 62*, 269–276.

Cox, D. R. (1981). Statistical analysis of time series: Some recent developments (with discussion). *Scandinavian Journal of Statistics 8*, 93–115.

Cox, D. R. (1990). Role of models in statistical analysis. *Statistical Science 5*, 169–174.

Cox, D. R. (2006). *Principles of Statistical Inference.* Cambridge: Cambridge University Press.

Cox, D. R. (2013). A return to an old paper: "Tests of separate families of hypotheses". *Journal of the Royal Statistical Society B 75*, 207–215.

Cox, D. R. and C. A. Donnelly (2011). *Principles of Applied Statistics.* Cambridge: Cambridge University Press.

Cox, D. R. and D. V. Hinkley (1974). *Theoretical Statistics.* London: Chapman and Hall.

Cox, D. R. and V. Isham (1980). *Point Processes.* London: Chapman & Hall/CRC.

Cox, D. R. and P. A. W. Lewis (1966). *The Statistical Analysis of Series of Events.* New York: Springer.

Cox, D. R. and N. Reid (1987). Parameter orthogonality and approximate conditional inference (with discussion). *Journal of the Royal Statistical Society B 49*, 1–39.

Cox, D. R. and N. Reid (1992). A note on the difference between profile and modified profile likelihood. *Biometrika 79*, 408–411.

Cox, D. R. and N. Reid (2004). A note on pseudolikelihood constructed from marginal densities. *Biometrika 91*, 729–737.

Cox, D. R. and E. J. Snell (1968). A general definition of residuals (with discussion). *Journal of the Royal Statistical Society B 30*, 248–265.

Cox, D. R. and N. Wermuth (2002). On some models for multivariate binary variables parallel in complexity with the multivariate Gaussian distribution. *Biometrika 89*, 462–469.

Crainiceanu, C. M. and D. Ruppert (2004). Likelihood ratio tests in linear mixed models with one variance component. *Journal of the Royal Statistical Society B 66*, 165–185.

Craven, P. and G. Wahba (1979). Smoothing noisy data with spline functions: Estimating the correct degree of smoothing by the method of generalized cross-validation. *Numerische Mathematik 31*, 377–403.

Cressie, N. A. C. (1990). The origins of kriging. *Mathematical Geology 22*, 239–252.

Cressie, N. A. C. (1993). *Statistics for Spatial Data.* New York: Wiley.

Cressie, N. A. C. and C. K. Wikle (2011). *Statistics for Spatio-temporal Data.* New York: Wiley.

Dæhlen, I. and N. L. Hjort (2025). Model robust hybrid likelihood. *Journal of Statistical Planning and Inference 241*, 106327.

Dæhlen, I., N. L. Hjort, and I. Hobæk Haff (2024). Accurate bias estimation with applications to focused model selection. *Scandinavian Journal of Statistics 51*, 724–759.

Daniels, M. and Y.-D. Zhao (2003). Modelling the random effects covariance matrix in longitudinal data. *Statistics in Medicine 22*, 1631–1647.

DasGupta, A. (2008). *Asymptotic Theory of Statistics and Probability.* New York: Springer.

Davies, R. B. (1977). Hypothesis testing when a nuisance parameter is present only under the alternative. *Biometrika 64*, 247–254.

Davies, R. B. (1987). Hypothesis testing when a nuisance parameter is present only under the alternative. *Biometrika 74*, 33–43.

Davies, R. B. (2002). Hypothesis testing when a nuisance parameter is present only under the alternative: Linear model case. *Biometrika 89*, 484–489.

Davis, R. A., W. T. M. Dunsmuir, and S. B. Streett (2003). Observation-driven models for Poisson counts. *Biometrika 90*, 777–790.

Davis, R. A., K. Fokianos, S. H. Holan, H. Joe, J. Livsey, R. Lund, V. Pipiras, and N. Ravishanker (2021). Count time series: A methodological review. *Journal of the American Statistical Association 116*, 1533–1547.

Davis, R. A. and C. Y. Yau (2011). Comments on pairwise likelihood in time series. *Statistica Sinica 21*, 255–277.

Davison, A. C. (1988). Approximate conditional inference in generalized linear models. *Journal of the Royal Statistical Society B 50*, 445–461.

Davison, A. C. (2003). *Statistical Models.* Cambridge: Cambridge University Press.

Davison, A. C. and D. V. Hinkley (1997). *Bootstrap Methods and Their Application.* Cambridge: Cambridge University Press.

Davison, A. C., D. V. Hinkley, and B. J. Worton (1992). Bootstrap likelihoods. *Biometrika 79*, 113–130.

Davison, A. C., D. V. Hinkley, and B. J. Worton (1995). Accurate and efficient construction of bootstrap likelihoods. *Statistics and Computing 5*, 257–264.

Davison, A. C. and R. Huser (2015). Statistics of extremes. *Annual Review of Statistics and Its Application 2*, 203–235.

Davison, A. C., S. A. Padoan, and M. Ribatet (2012). Statistical modeling of spatial extremes (with discussion). *Statistical Science 27*, 161–186.

Dawid, A. P., M. Stone, and J. V. Zidek (1973). Marginalization paradoxes in Bayesian and structural inference (with discussion). *Journal of the Royal Statistical Society B 35*, 189–233.

de Boor, C. (1978). *A Practical Guide to Splines.* New York: Springer-Verlag.

Dempster, A. P., N. M. Laird, and D. B. Rubin (1977). Maximum likelihood from incomplete data via the EM algorithm (with discussion). *Journal of the Royal Statistical Society B 39*, 1–22.

DerSimonian, R. and N. Laird (1986). Meta-analysis in clinical trials. *Controlled Clinical Trials 7*, 177–188.

Desboulets, L. D. D. (2018). A review on variable selection in regression analysis. *Econometrics 6*, 45.

Desmond, A. F. (1997). Optimal estimating functions, quasi-likelihood and statistical modelling. *Journal of Statistical Planning and Inference 60*, 77–123.

DiCaterina, C., N. Reid, and N. Sartori (2025). Directional tests in Gaussian graphical models. *Statistica Sinica 35*, 361–387.

DiCiccio, T., P. Hall, and J. Romano (1991). Empirical likelihood is Bartlett-correctable. *Annals of Statistics 19*, 1053–1061.

Diggle, P. J. (2013). *Statistical Analysis of Spatial and Spatio-temporal Point Patterns.* Boca Raton, FL: Chapman & Hall/CRC.

Diggle, P. J. and E. Giorgi (2016). Model-based geostatistics for prevalence mapping in low-resource settings. *Journal of the American Statistical Association 111*, 1096–1120.

Diggle, P. J. and R. J. Gratton (1984). Monte Carlo methods of inference for implicit statistical models (with discussion). *Journal of the Royal Statistical Society B 46*, 193–212.

Diggle, P. J., P. Heagerty, K. Y. Liang, and S. L. Zeger (2013). *Longitudinal Data Analysis* (2nd ed.). Oxford: Oxford University Press.

Diggle, P. J., P. Moraga, B. Rowlingson, and B. M. Taylor (2013). Spatial and spatio-temporal log-gaussian Cox processes: Extending the geostatistical paradigm. *Statistical Science 28*, 542–563.

Diggle, P. J. and P. J. Ribeiro, Jr. (2007). *Model-based Geostatistics.* New York: Springer.

Diggle, P. J., J. A. Tawn, and R. A. Moyeed (1998). Model-based geostatistics (with discussion). *Journal of the Royal Statistical Society C 47*, 299–350.

Donoho, D. L. and I. M. Johnstone (1994). Ideal spatial adaptation by wavelet shrinkage. *Biometrika 81*, 425–455.

Drovandi, C. C. (2018). ABC and indirect inference. In S. A. Sisson, Y. Fan, and M. Beaumont (Eds.), *Handbook of Approximate Bayesian Computation*, pp. 179–209. Boca Raton: Chapman and Hall/CRC.

Dümbgen, L. (2024). Shape-constrained statistical inference. *Annual Review of Statistics and Its Application 11*, 373–391.

Dümbgen, L. and K. Rufibach (2009). Maximum likelihood estimation of a log-concave density and its distribution function: Basic properties and uniform consistency. *Bernoulli 15*, 40–68.

Dümbgen, L. and K. Rufibach (2011). logcondens: Computations related to univariate log-concave density estimation. *Journal of Statistical Software 39*, 1–28.

Dunsmuir, W. T. M. and D. J. Scott (2015). The glarma package for observation-driven time series regression of counts. *Journal of Statistical Software 67*, 1–36.

Durbin, J. (1960). Estimation of parameters in time-series regression models. *Journal of the Royal Statistical Society B 22*, 139–153.

Durbin, J. and S. J. Koopman (2012). *Time Series Analysis by State Space Methods.* Oxford: Oxford University Press.

Edwards, A. W. F. (1992). *Likelihood.* Cambridge: Cambridge University Press.

Efron, B. (1967). The two-sample problem with censored data. *Proceedings of the Fourth Berkeley Symposium on Mathematical Statistics and Probability 4*, 831–853.

Efron, B. (1977). The efficiency of Cox's likelihood function for censored data. *Journal of the American Statistical Association 73*, 557–565.

Efron, B. (1982). *The Jackknife, the Bootstrap, and Other Resampling Plans.* Society for Industrial and Applied Mathematics.

Efron, B. (1993). Bayes and likelihood calculations from confidence intervals. *Biometrika 80*, 3–26.

Efron, B. (2014). Two modeling strategies for empirical Bayes estimation. *Statistical Science 29*, 285.

Efron, B. and G. Gong (1983). A leisurely look at the bootstrap, the jackknife, and cross-validation. *American Statistician 37*, 36–48.

Efron, B., T. J. Hastie, I. M. Johnstone, and R. Tibshirani (2004). Least angle regression. *Annals of Statistics 32*, 407–451.

Efron, B. and D. V. Hinkley (1978). Assessing the accuracy of the maximum likelihood estimator: Observed versus expected Fisher information. *Biometrika 65*, 457–482.

Efron, B. and C. Morris (1972). Empirical Bayes on vector observations: An extension of Stein's method. *Biometrika 59*, 335–347.

Efron, B. and R. Tibshirani (1994). *An Introduction to the Bootstrap.* Boca Raton, FL: Chapman & Hall/CRC.

Efron, B., R. Tibshirani, J. D. Storey, and V. Tusher (2001). Empirical Bayes analysis of a microarray experiment. *Journal of the American Statistical Association 96*, 1151–1160.

Eilers, P. H. and B. D. Marx (1996). Flexible smoothing with B-splines and penalties. *Statistical Science 11*, 89–121.

Ekvall, K. O. and M. Bottai (2022). Confidence regions near singular information and boundary points with applications to mixed models. *Annals of Statistics 50*, 1806–1832.

Elliott, P., J. Wakefield, N. Best, and D. Briggs (2001). *Spatial Epidemiology: Methods and Applications.* Oxford: Oxford University Press.

Engle, R. F. (1982). Autoregressive conditional heteroscedasticity with estimates of the variance of United Kingdom inflation. *Econometrica 50*, 987–1007.

Eubank, R. L. (1999). *Nonparametric Regression and Spline Smoothing* (2nd ed.). Boca Raton, FL: Chapman & Hall/CRC Press.

Fan, J. (1997). Comments on "Wavelets in statistics: A Review" by A. Antoniadis. *Journal of the Italian Statistical Association 6*, 131–138.

Fan, J., M. Farmen, and I. Gijbels (1998). Local maximum likelihood estimation and inference. *Journal of the Royal Statistical Society B 60*, 591–608.

Fan, J., I. Gijbels, and M. King (1997). Local likelihood and local partial likelihood in hazard regression. *Annals of Statistics 25*, 1661–1690.

Fan, J. and R. Li (2001). Variable selection via nonconcave penalized likelihood and its oracle properties. *Journal of the American Statistical Association 96*, 1348–1360.

Fan, J., R. Li, C.-H. Zhang, and H. Zou (2020). *Statistical Foundations of Data Science*. Boca Raton, FL: Chapman and Hall/CRC.

Fan, J. and J. Lv (2010). A selective overview of variable selection in high dimensional feature space. *Statistica Sinica 20*, 101–148.

Fan, J. and H. Peng (2004). Nonconcave penalized likelihood with a diverging number of parameters. *Annals of Statistics 32*, 928–961.

Fan, Y., E. Demirkaya, and J. Lv (2019). Nonuniformity of P-values can occur early in diverging dimensions. *Journal of Machine Learning Research 20*, 1–33.

Fan, Y. and C. Y. Tang (2013). Tuning parameter selection in high dimensional penalized likelihood. *Journal of the Royal Statistical Society B 75*, 531–552.

Far, S. S., M. Papathomas, and R. King (2021). Parameter redundancy and the existence of maximum likelihood estimates in log-linear models. *Statistica Sinica 31*, 1125–1143.

Fearnhead, P. and D. Prangle (2012). Approximate likelihood methods for estimating local recombination rates. *Journal of the Royal Statistical Society B 64*, 657–680.

Feigin, P. D. and B. Reiser (1979). On asymptotic ancillarity and inference for Yule and regular nonergodic processes. *Biometrika 66*, 279–283.

Ferguson, T. S. (1974). Prior distributions on spaces of probability measures. *Annals of Statistics 2*, 615–629.

Fienberg, S. E. and A. Rinaldo (2012). Maximum likelihood estimation in log-linear models. *Annals of Statistics 40*, 996–1023.

Firth, D. (1987). On the efficiency of quasi-likelihood estimation. *Biometrika 74*, 233–245.

Firth, D. (1993a). Bias reduction of maximum likelihood estimates. *Biometrika 80*, 27–38.

Firth, D. (1993b). Recent developments in quasi-likelihood methods. *Bulletin of the International Statistical Institute 55*, 341–358.

Firth, D., J. Glosup, and D. V. Hinkley (1991). Model checking with nonparametric curves. *Biometrika 78*, 245–252.

Fisher, R. A. (1956). *Statistical Methods and Scientific Inference*. Edinburgh: Oliver and Boyd.

Fitzmaurice, G. M., N. M. Laird, and J. H. Ware (2011). *Applied Longitudinal Analysis* (2nd ed.). Hoboken, NJ: Wiley.

Fong, Y., H. Rue, and J. Wakefield (2010). Bayesian inference for generalized linear mixed models. *Biostatistics 11*, 397–412.

Frank, I. E. and J. H. Friedman (1993). A statistical view of some chemometrics regression tools. *Technometrics 35*, 109–135.

Fraser, D. A. S. (1965). On information in statistics. *Annals of Mathematical Statistics 36*, 890–896.

Fraser, D. A. S. (1979). *Inference and Linear Models.* Toronto: McGraw-Hill.

Fraser, D. A. S. (1990). Tail probabilities from observed likelihoods. *Biometrika 77*, 65–76.

Fraser, D. A. S. (2003). Likelihood for component parameters. *Biometrika 90*, 327–339.

Fraser, D. A. S. and I. Guttman (1956). Tolerance regions. *Annals of Mathematical Statistics 27*, 162–179.

Fraser, D. A. S. and A. Naderi (2007). Minimal sufficient statistics emerge from the observed likelihood function. *International Journal of Statistical Science 6*, 55–61.

Friedman, J. H., T. J. Hastie, and R. Tibshirani (2008). Sparse inverse covariance estimation with the graphical lasso. *Biostatistics 9*, 432–441.

Friedman, J. H. and W. Stuetzle (1981). Projection pursuit regression. *Journal of the American Statistical Association 76*, 817–823.

Fu, W. J. (2003). Penalized estimating equations. *Biometrics 59*, 126–132.

Fu, Y., J. Chen, and P. Li (2008). Modified likelihood ratio tests for homogeneity in a mixture of von Mises distributions. *Journal of Statistical Planning and Inference 138*, 667–681.

Fuller, W. A. (1976). *Introduction to Statistical Time Series.* New York: Wiley.

Gaetan, C. and X. Guyon (2010). *Spatial Statistics and Modeling.* New York: Springer.

Gamerman, D. (1997). Sampling from the posterior distribution in generalized linear mixed models. *Statistics and Computing 7*, 57–68.

Gao, X. and P. X.-K. Song (2010). Composite likelihood Bayesian information criteria for model selection in high-dimensional data. *Journal of the American Statistical Association 105*, 1531–1540.

García Rasines, D. and G. A. Young (2023). Splitting strategies for post-selection inference. *Biometrika 110*, 597–614.

Geman, S. and D. Geman (1984). Stochastic relaxation, Gibbs distributions, and the Bayesian restoration of images. *IEEE Transactions on Pattern Analysis and Machine Intelligence PAMI-6*, 721–741.

Geman, S. and C.-R. Hwang (1982). Nonparametric maximum likelihood estimation by the method of sieves. *Annals of Statistics 10*, 401–414.

Geyer, C. J. (1994). On the convergence of Monte Carlo maximum likelihood calculations. *Journal of the Royal Statistical Society B 56*, 261–274.

Geyer, C. J. and E. A. Thompson (1992). Constrained Monte Carlo maximum likelihood for dependent data (with discussion). *Journal of the Royal Statistical Society B 54*, 657–683.

Gill, R. D. and S. Johansen (1990). Product integrals and counting processes. *Annals of Statistics 18*, 1501–1555.

Giorgi, E. and P. J. Diggle (2017). PrevMap: an R package for prevalence mapping. *Journal of Statistical Software 78*, 1–29.

Godambe, V. P. (1960). An optimum property of regular maximum likelihood estimation. *Annals of Mathematical Statistics 31*, 1208–1211.

Godambe, V. P. (1976). Conditional likelihood and unconditional optimum estimating equations. *Biometrika 63*, 277–284.

Godambe, V. P. (1985). The foundations of finite sample estimation in stochastic processes. *Biometrika 72*, 419–428.

Godambe, V. P. (1991). *Estimating Functions.* Oxford: Oxford University Press.

Godambe, V. P. and C. C. Heyde (1987). Quasi-likelihood and optimal estimation. *International Statistical Review 55*, 231–244.

Godambe, V. P. and M. E. Thompson (1974). Estimating equations in the presence of a nuisance parameter. *Annals of Statistics 2*, 568–571.

Godambe, V. P. and M. E. Thompson (1984). Robust estimation through estimating equations. *Biometrika 71*, 115–125.

Godambe, V. P. and M. E. Thompson (1989). An extension of quasi-likelihood estimation. *Journal of Statistical Planning and Inference 22*, 137–152.

Golub, G. H., M. Heath, and G. Wahba (1979). Generalised cross-validation as a method for choosing a good ridge parameter. *Technometrics 21*, 215–223.

Gong, G. and F. J. Samaniego (1981). Pseudo maximum likelihood estimation: theory and applications. *Annals of Statistics 9*, 861–869.

Good, I. J. and G. H. Toulmin (1956). The number of new species, and the increase in population coverage, when a sample is increased. *Biometrika 43*, 45–63.

Gouriéroux, C. (2012). *ARCH Models and Financial Applications.* New York: Springer.

Gouriéroux, C. and A. Monfort (1996). *Simulation-based Econometric Methods.* Oxford: Oxford University Press.

Gouriéroux, C., A. Monfort, and E. Renault (1993). Indirect inference. *Journal of Applied Econometrics 8*, S85–S118.

Gouriéroux, C., A. Monfort, and A. Trognon (1984). Pseudo maximum likelihood methods: Theory. *Econometrica 52*, 681–700.

Green, P. J. (1987). Penalized likelihood for general semi-parametric regression models. *International Statistical Review 55*, 245–259.

Green, P. J. and B. W. Silverman (1993). *Nonparametric Regression and Generalized Linear Models: A Roughness Penalty Approach.* London: Chapman and Hall/CRC.

Grenander, U. (1956). On the theory of mortality measurements. *Skandinavisk Aktuarietidskrift 39*, 1–55.

Grenander, U. (1981). *Abstract Inference.* New York: Wiley.

Guolo, A. (2012). Higher-order likelihood inference in meta-analysis and meta-regression. *Statistics in Medicine 31*, 313–327.

Guolo, A. and C. Varin (2017). Random-effects meta-analysis: the number of studies matters. *Statistical Methods in Medical Research 26*, 1500–1518.

Hall, A. R. (2005). *Generalized Method of Moments.* Oxford: Oxford University Press.

Hall, P. and J. S. Marron (1987). Estimation of integrated squared density derivatives. *Statistics and Probability Letters 6*, 109–115.

Hamilton, J. D. (1994). *Time Series Analysis.* Princeton: Princeton University Press.

Hampel, F. R. (1974). The influence curve and its role in robust estimation. *Journal of the American Statistical Association 69*, 383–393.

Hanfelt, J. J. and K. Y. Liang (1995). Approximate likelihood ratios for general estimating functions. *Biometrika 82*, 461–477.

Hansen, B. E. (2014). Nonparametric sieve regression: Least squares, averaging least squares, and cross-validation. In J. S. Racine, L. Su, and A. Ullah (Eds.), *The Oxford Handbook of Applied Nonparametric and Semiparametric Econometrics and Statistics*, pp. 214–248. Oxford: Oxford Academics.

Hansen, L. P. (1982). Large sample properties of generalized method of moments estimators. *Econometrica 50*, 1029–1054.

Hansen, L. P. (1985). A method for calculating bounds on asymptotic covariance matrices of generalized method of moments estimators. *Journal of Econometrics 30*, 203–238.

Hansen, L. P. (2001). Method of moments. In N. J. Smelser and P. B. Baltes (Eds.), *International Encyclopedia of the Social and Behavioral Sciences*, pp. 9743–9751. Oxford: Pergamon.

Hardin, J. W. and J. M. Hilbe (2013). *Generalized Estimating Equations* (2nd ed.). Boca Raton, FL: Chapman & Hall/CRC.

Harville, D. A. (1974). Bayesian inference for variance components using only error contrasts. *Biometrika 61*, 383–385.

Harville, D. A. (1977). Maximum likelihood approaches to variance component estimation and to related problems. *Journal of the American Statistical Association 72*, 320–338.

Hastie, T. J. and R. Tibshirani (1986). Generalized additive models. *Statistical Science 1*, 297–310.

Hastie, T. J. and R. Tibshirani (1990). *Generalized Additive Models.* Boca Raton, FL: Chapman & Hall.

Hastie, T. J., R. Tibshirani, and J. H. Friedman (2009). *The Elements of Statistical Learning: Data Mining, Inference, and Prediction* (2nd ed.). New York: Springer.

Hastie, T. J., R. Tibshirani, and M. Wainwright (2015). *Statistical Learning with Sparsity: The Lasso and Generalizations.* Boca Raton, FL: Chapman & Hall/CRC.

Hauck, W. W. and A. Donner (1977). Wald's test as applied to hypothesis testing in logit analysis. *Journal of the American Statistical Association 72*, 851–853.

Hauck, W. W. and A. Donner (1980). Corrigenda: Wald's test as applied to hypothesis testing in logit analysis. *Journal of the American Statistical Association 75*, 482.

Hausman, J. A. (1978). Specification tests in econometrics. *Econometrica 46*, 1251–1271.

Heagerty, P. J. and B. F. Kurland (2001). Misspecified maximum likelihood estimates and generalised linear mixed models. *Biometrika 88*, 973–985.

Heagerty, P. J. and S. R. Lele (1998). A composite likelihood approach to binary spatial data. *Journal of the American Statistical Association 93*, 1099–1111.

Heggland, K. and A. Frigessi (2004). Estimating functions in indirect inference. *Journal of the Royal Statistical Society B 66*, 447–462.

Heidenreich, N.-B., A. Schindler, and S. Sperlich (2013). Bandwidth selection for kernel density estimation: a review of fully automatic selectors. *AStA Advances in Statistical Analysis 97*, 403–433.

Heinze, G. and M. Schemper (2002). A solution to the problem of separation in logistic regression. *Statistics in Medicine 21*, 2409–2419.

Heinze, G., C. Wallisch, and D. Dunkler (2018). Variable selection – a review and recommendations for the practicing statistician. *Biometrical Journal 60*, 431–449.

Henmi, M. and S. Eguchi (2004). A paradox concerning nuisance parameters and projected estimating functions. *Biometrika 91*, 929–941.

Heyde, C. C. (1997). *Quasi-Likelihood and Its Application: A General Approach to Optimal Parameter Estimation.* New York: Springer.

Heyde, C. C. and R. Morton (1998). Multiple roots in general estimating equations. *Biometrika 85*, 954–959.

Higgins, J. P., S. G. Thompson, and D. J. Spiegelhalter (2009). A re-evaluation of random-effects meta-analysis. *Journal of the Royal Statistical Society A 172*, 137–159.

Hjort, N. L. (1992). On inference in parametric survival data models. *International Statistical Review 60*, 355–387.

Hjort, N. L. (1994). The exact amount of t-ness that the normal model can tolerate. *Journal of the American Statistical Association 89*, 665–675.

Hjort, N. L. and G. Claeskens (2003). Frequentist model average estimators (with discussion). *Journal of the American Statistical Association 98*, 879–899.

Hjort, N. L., C. Holmes, P. Müller, and S. G. Walker (2010). *Bayesian Nonparametrics.* Cambridge: Cambridge University Press.

Hjort, N. L. and M. C. Jones (1996). Locally parametric nonparametric density estimation. *Annals of Statistics 24*, 1619–1647.

Hjort, N. L., I. W. McKeague, and I. V. Keilegom (2018). Hybrid combinations of parametric and empirical likelihoods. *Statistica Sinica 28*, 2389–2407.

Hjort, N. L. and H. Omre (1994). Topics in spatial statistics (with discussion). *Scandinavian Journal of Statistics 21*, 289–357.

Hjort, N. L. and C. Varin (2008). ML, PL, QL in Markov chain models. *Scandinavian Journal of Statistics 35*, 64–82.

Hoerl, A. E. and R. W. Kennard (1970). Ridge regression: Biased estimation for nonorthogonal problems. *Technometrics 12*, 55–67.

Huang, J., S. Ma, and C.-H. Zhang (2008). Adaptive Lasso for sparse high-dimensional regression models. *Statistica Sinica 18*, 1603–1618.

Huber, P. J. (1964). Robust estimation of a location parameter. *Annals of Mathematical Statistics 35*, 73–101.

Huber, P. J. (1967). The behavior of maximum likelihood estimates under nonstandard conditions. *Proceedings of the Fifth Berkeley Symposium on Mathematical Statistics and Probability 1*, 221–233.

Huber, P. J. (1981). *Robust Statistics.* New York: Wiley.

Huber, P. J. and E. M. Ronchetti (2009). *Robust Statistics* (2nd ed.). New York: Wiley.

Hui, F. K., S. Müller, and A. H. Welsh (2021). Random effects misspecification can have severe consequences for random effects inference in linear mixed models. *International Statistical Review 89*, 186–206.

Hunter, D. R. and K. Lange (2004). A tutorial on MM algorithms. *American Statistician 58*, 30–37.

Hurvich, C. M., J. S. Simonoff, and C.-L. Tsai (1998). Smoothing parameter selection in nonparametric regression using an improved Akaike information criterion. *Journal of the Royal Statistical Society B 60*, 271–293.

Huser, R., C. Dombry, M. Ribatet, and M. G. Genton (2019). Full likelihood inference for max-stable data. *Stat 8*, e218.

Huser, R. and J. L. Wadsworth (2022). Advances in statistical modeling of spatial extremes. *Wiley Interdisciplinary Reviews: Computational Statistics 14*, e1537.

Huzurbazar, V. S. (1950). Probability distributions and orthogonal parameters. *Mathematical Proceedings of the Cambridge Philosophical Society 46*, 281–284.

Ibragimov, I. A. and R. Z. Has'minski (1981). *Statistical Estimation: Asymptotic Theory.* New York: Springer-Verlag.

Jacod, J. and M. Sørensen (2018). A review of asymptotic theory of estimating functions. *Statistical Inference for Stochastic Processes 21*, 415–434.

James, G., D. Witten, T. J. Hastie, and R. Tibshirani (2013). *An Introduction to Statistical Learning: with Applications in R* (2nd ed.). New York: Springer.

James, W. and C. Stein (1961). Estimation with quadratic loss. *Proceedings of the Fourth Berkeley Symposium on Mathematical Statistics and Probability 1*, 361–379.

Jeffreys, H. (1961). *Theory of Probability* (2nd ed.). Oxford: Oxford University Press.

Jensen, J. L. (1987). On asymptotic expansions in non-ergodic models. *Scandinavian Journal of Statistics 14*, 305–318.

Jesus, J. and R. E. Chandler (2011). Estimating functions and the generalized method of moments. *Interface Focus 1*, 871–885.

Jewell, S., P. Fearnhead, and D. Witten (2022). Testing for a change in mean after changepoint detection. *Journal of the Royal Statistical Society B 84*, 1082–1104.

Jiang, J. (2013). The subset argument and consistency in GLMM: Answer to an open problem. *Annals of Statistics 41*, 177–195.

Jiang, J. (2025). Asymptotic distribution of maximum likelihood estimator in generalized linear mixed models with crossed random effects. *Annals of Statistics 53*, 1298–1318.

Jiang, J. and T. Nguyen (2021). *Linear and Generalized Linear Mixed Models and Their Applications* (2nd ed.). New York: Springer.

Jiang, W. and B. Turnbull (2004). The indirect method: Inference based on intermediate statistics—a synthesis and examples. *Statistical Science 19*, 239–263.

Jin, S. and Y. Lee (2021). A review of *h*-likelihood and hierarchical generalized linear model. *Wiley Interdisciplinary Reviews: Computational Statistics 13*, e1527.

Johansen, S. (1983). An extension of Cox's regression model. *International Statistical Review 51*, 165–2174.

Johnson, B. A., D. Y. Lin, and D. Zeng (2008). Penalized estimating functions and variable selection in semiparametric regression models. *Journal of the American Statistical Association 103*, 672–680.

Johnson, R. A. (1978). An asymptotic expansion for posterior distributions. *Annals of Mathematical Statistics 38*, 1899–1907.

Jones, M. C., J. S. Marron, and S. J. Sheather (1996). A brief survey of bandwidth selection for density estimation. *Journal of the American Statistical Association 91*, 401–407.

Jordan, M. I., Z. Ghahramani, T. S. Jaakkola, and L. K. Saul (1999). An introduction to variational methods for graphical models. *Machine Learning 37*, 183–233.

Jørgensen, B. and S. J. Knudsen (2004). Parameter orthogonality and bias adjustment for estimating functions. *Scandinavian Journal of Statistics 31*, 93–114.

Kalbfleisch, J. D. and R. L. Prentice (1972). Discussion of "Regression models and life tables" by D. R. Cox. *Journal of the Royal Statistical Society B 34*, 215–216.

Kalbfleisch, J. D. and R. L. Prentice (2002). *The Statistical Analysis of Failure Time Data* (2nd ed.). Hoboken, NJ: Wiley.

Kalbfleisch, J. D. and D. E. Schnaubel (2023). Fifty years of the Cox model. *Annual Review of Statistics and Its Application 10*, 1–23.

Kale, B. K. (1987). Essential uniqueness of optimal estimating functions. *Journal of Statistical Planning and Inference 17*, 405–407.

Kaplan, E. L. and P. Meier (1958). Nonparametric estimation from incomplete observations. *Journal of the American Statistical Association 53*, 457–481.

Karr, A. (1991). *Point Processes and Their Statistical Inference* (2nd ed.). New York: Marcel Dekker.

Kass, R. E., L. Tierney, and J. B. Kadane (1987). The validity of posterior expansion based on Laplace's method. In S. Geisser, J. Hodges, S. Press, and A. Zellner (Eds.), *Bayesian and Likelihood Methods in Statistics and Econometrics*, pp. 473–488. Amsterdam: Elsevier.

Kauermann, G. and R. J. Carroll (2001). A note on the efficiency of sandwich covariance matrix estimation. *Journal of the American Statistical Association 96*, 1387–1396.

Kenne Pagui, E. C., A. Salvan, and N. Sartori (2017). Median bias reduction of maximum likelihood estimates. *Biometrika 104*, 923–938.

Kent, J. T. (1982). Robust properties of likelihood ratio test. *Biometrika 69*, 19–27.

Kent, J. T. (1986). The underlying structure of nonnested hypothesis tests. *Biometrika 73*, 333–343.

Kenward, M. G. and G. Molenberghs (1998). Likelihood based frequentist inference when data are missing at random. *Statistical Science 13*, 236–247.

Keogh, R. H. (2024). Introduction to the Special Issue based on contributions from "A celebration of 50 years of the Cox model in memory of Sir David Cox", 10 November 2022, London School of Hygiene and Tropical Medicine. *Journal of the Royal Statistical Society A 187*, 563–566.

Khalili, A. and J. Chen (2007). Variable selection in finite mixture of regression models. *Journal of the American Statistical Association 102*, 1025–1038.

Khare, K., S.-Y. Oh, and B. Rajaratnam (2015). A convex pseudolikelihood framework for high dimensional partial correlation estimation with convergence guarantees. *Journal of the Royal Statistical Society B 77*, 803–825.

Kiefer, J. and J. Wolfowitz (1956). Consistency of the maximum likelihood estimator in the presence of infinitely many incidental parameters. *Annals of Mathematical Statistics 27*, 887–906.

Kim, S., N. Shephard, and S. Chib (1998). Stochastic volatility: likelihood inference and comparison with arch models. *Review of Economic Studies 65*, 361–393.

Kimball, B. F. (1946). Sufficient statistical estimation functions for the parameters of the distribution of maximum values. *Annals of Mathematical Statistics 17*, 299–309.

Kimeldorf, G. S. and G. Wahba (1970). A correspondence between Bayesian estimation on stochastic processes and smoothing by splines. *Annals of Mathematical Statistics 41*, 495–502.

Kleijn, B. J. K. and A. W. van der Vaart (2012). The Bernstein–Von-Mises theorem under misspecification. *Electronic Journal of Statistics 6*, 354–381.

Klimko, L. A. and P. I. Nelson (1978). On conditional least squares estimation for stochastic processes. *Annals of Statistics 6*, 629–642.

Knight, K. (2000). *Mathematical Statistics.* Boca Raton, FL: Chapman and Hall/CRC.

Knight, K. and W. Fu (2000). Asymptotics for Lasso-type estimators. *Annals of Statistics 28*, 1356–1378.

Köhler, M., A. Schindler, and S. Sperlich (2014). A review and comparison of bandwidth selection methods for kernel regression. *International Statistical Review 82*, 243–274.

Kosmidis, I. (2014). Bias in parametric estimation: reduction and useful side-effects. *Wiley Interdisciplinary Reviews: Computational Statistics 6*, 185–196.

Kosmidis, I. and D. Firth (2009). Bias reduction in exponential family nonlinear models. *Biometrika 96*, 793–804.

Kosmidis, I. and D. Firth (2021). Jeffreys-prior penalty, finiteness and shrinkage in binomial-response generalized linear models. *Biometrika 108*, 71–82.

Kosmidis, I., A. Guolo, and C. Varin (2017). Improving the accuracy of likelihood-based inference in meta-analysis and meta-regression. *Biometrika 104*, 489–496.

Kosmidis, I., E. C. Kenne Pagui, and N. Sartori (2020). Mean and median bias reduction in generalized linear models. *Statistics and Computing 30*, 43–59.

Kosmidis, I. and N. Lunardon (2024). Empirical bias-reducing adjustments to estimating functions. *Journal of the Royal Statistical Society B 86*, 62–89.

Kosmidis, I. and P. Zietkiewicz (2026). Jeffreys' prior penalty for high-dimensional logistic regression: A conjecture about aggregate bias. *Statistical Science 41*. to appear.

Kuchibhotla, A. K., L. D. Brown, A. Buja, J. Cai, E. I. George, and L. H. Zhao (2020). Valid post-selection inference in model-free linear regression. *Annals of Statistics 48*, 2953–2981.

Kuchibhotla, A. K., J. E. Kolassa, and T. A. Kuffner (2022). Post-selection inference. *Annual Review of Statistics and Its Application 9*, 505–527.

Kuersteiner, G. (2002). Efficient IV estimation for autoregressive models with conditional heteroskedasticity. *Econometric Theory 18*, 547–583.

Kuk, A. Y. (1995). Asymptotically unbiased estimation in generalized linear models with random effects. *Journal of the Royal Statistical Society B 57*, 395–407.

Kwan, T.-K., F. Chen, and W. Dunsmuir (2024). Likelihood inference of the non-stationary Hawkes process with non-exponential kernel. https://arxiv.org/pdf/2408.09710.

Lai, T. L. and D. Small (2007). Marginal regression analysis of longitudinal data with time-dependent covariates: A generalized method-of-moments approach. *Journal of the Royal Statistical Society B 69*, 79–99.

Laird, N. (1978). Nonparametric maximum likelihood estimation of a mixing distribution. *Journal of the American Statistical Association 73*, 805–811.

Larribe, F. and P. Fearnhead (2011). On composite likelihoods in statistical genetics. *Statistica Sinica 21*, 43–69.

Lawless, J. F. (2003). *Statistical Models and Methods for Lifetime Data* (2nd ed.). New York: Wiley.

Lawless, J. F. and M. Fredette (2005). Frequentist prediction intervals and predictive distributions. *Biometrika 92*, 529–542.

Lawley, D. N. (1956). A general method for approximating to the distribution of the likelihood ratio criteria. *Biometrika 43*, 295–303.

Lawson, A. B. (2018). *Bayesian Disease Mapping: Hierarchical Modeling in Spatial Epidemiology*. Boca Raton, FL: Chapman and Hall/CRC.

Lazar, N. A. (2021). A review of empirical likelihood. *Annual Review of Statistics and Its Application 8*, 329–344.

Le Cam, L. (1960). Locally asymptotically normal families of distributions. *University of California—Publications in Statistics 3*, 37–98.

Lee, J. D. and T. J. Hastie (2015). Learning the structure of mixed graphical models. *Journal of Computational and Graphical Statistics 24*, 230–253.

Lee, J. D., D. L. Sun, Y. Sun, and J. E. Taylor (2016). Exact post-selection inference, with application to the lasso. *Annals of Statistics 44*, 907–927.

Lee, Y. and J. A. Nelder (1996). Hierarchical generalized linear models (with discussion). *Journal of the Royal Statistical Society B 58*, 619–678.

Lee, Y. and J. A. Nelder (2009). Likelihood inference for models with unobservables: another view (with discussion). *Statistical Science 24*, 255–269.

Lehmann, E. L. (1990). Model specification: the views of Fisher and Neyman, and later developments. *Statistical Science 5*, 160–168.

Lehmann, E. L. and G. Casella (1998). *Theory of Point Estimation* (2nd ed.). New York: Springer.

Lehmann, E. L. and J. P. Romano (2005). *Testing Statistical Hypotheses.* New York: Springer.

Lei, J. and L. Wasserman (2014). Distribution-free prediction bands for nonparametric regression. *Journal of the Royal Statistical Society B 76*, 71–96.

Lei, J., L. Wasserman, R. J. Tibshirani, M. G'Sell, and A. Rinaldo (2018). Distribution-free prediction inference for regression. *Journal of the American Statistical Association 113*, 1094–1111.

Lejay, A. and S. Mazzonetto (2024). Maximum likelihood estimator for skew Brownian motion: The convergence rate. *Scandinavian Journal of Statistics 51*, 612–642.

Lele, S. R. (2023). Profile likelihood for hierarchical models using data doubling. *Entropy 25*, 1262.

Liang, H. and R. Li (2009). Variable selection for partially linear models with measurement errors. *Journal of the American Statistical Association 104*, 234–248.

Liang, K.-Y. and S. L. Zeger (1986). Longitudinal data analysis using generalized linear models. *Biometrika 73*, 13–22.

Liang, K.-Y., S. L. Zeger, and B. Qaqish (1992). Multivariate regression analyses for categorical data (with discussion). *Journal of the Royal Statistical Society B 54*, 3–40.

Lin, D. Y. and L.-J. Wei (1989). The robust inference for the Cox proportional hazards model. *Journal of the American Statistical Association 84*, 1074–1078.

Lindgren, F., H. Rue, and J. Lindström (2011). An explicit link between Gaussian fields and Gaussian Markov random fields: the stochastic partial differential equation approach. *Journal of the Royal Statistical Society B 73*, 423–498.

Lindsay, B. G. (1988). Composite likelihood methods. *Contemporary Mathematics 80*, 221–239.

Lindsay, B. G. (1995). *Mixture Models: Theory, Geometry and Applications.* Hayward CA: Institute of Mathematical Statistics.

Lindsay, B. G., G. Y. Yi, and J. Sun (2011). Issues and strategies in the selection of composite likelihoods. *Statistica Sinica 21*, 71–105.

Linhart, H. and W. Zucchini (1986). *Model Selection.* New York: Wiley.

Little, R. J. (2021). Missing data assumptions. *Annual Review of Statistics and Its Application 8*, 89–107.

Little, R. J. and D. B. Rubin (1987). *Statistical Analysis with Missing Data.* New York: Wiley.

Liu, P. and Y. Zhao (2023). A review of recent advances in empirical likelihood. *Wiley Interdisciplinary Reviews: Computational Statistics 15*, e1599.

Liu, Q. and D. A. Pierce (1994). A note on Gauss—Hermite quadrature. *Biometrika 81*, 624–629.

Loader, C. (1999a). *Local Regression and Likelihood.* New York: Springer.

Loader, C. (2024). *locfit: Local Regression, Likelihood and Density Estimation.* The Comprehensive R Archive Network (CRAN). R package version 1.5-9.12.

Loader, C. R. (1996). Local likelihood density estimation. *Annals of Statistics 24*, 1602–1618.

Loader, C. R. (1999b). Bandwidth selection: classical or plug-in? *Annals of Statistics 27*, 415–438.

Lockhart, R., J. Taylor, R. J. Tibshirani, and R. Tibshirani (2014). A significance test for the lasso. *Annals of Statistics 42*, 413–468.

Loh, P. and M. Wainwright (2015). Regularized M-estimators with nonconvexity. *Journal of Machine Learning Research 16*, 559–616.

Louis, T. A. (1982). Finding the observed information matrix when using the EM algorithm. *Journal of the Royal Statistical Society B 44*, 226–233.

Lu, W., Y. Goldberg, and J. P. Fine (2012). On the robustness of the adaptive lasso to model misspecification. *Biometrika 99*, 717–731.

Lumley, T. and P. J. Heagerty (1999). Weighted empirical adaptive variance estimators for correlated data regression. *Journal of the Royal Statistical Society B 61*, 459–477.

Lunn, D. J., A. Thomas, N. Best, and D. Spiegelhalter (2000). WinBUGS—a Bayesian modelling framework: concepts, structure, and extensibility. *Statistics and Computing 10*, 325–337.

MacKinnon, J. G. and H. White (1985). Some heteroskedasticity-consistent covariance matrix estimators with improved finite sample properties. *Journal of Econometrics 29*, 305–325.

Maestrini, L., F. K. Hui, and A. H. Welsh (2026). Restricted maximum likelihood estimation in generalized linear mixed models. *Statistical Science 40*, to appear.

Mallows, C. L. (1973). Some comments on C_p. *Technometrics 15*, 661–675.

Mardia, K. V., J. T. Kent, G. Hughes, and C. C. Taylor (2009). Maximum likelihood estimation using composite likelihoods for closed exponential families. *Biometrika 96*, 975–982.

Mardia, K. V. and A. J. Watkins (1989). On multimodality of the likelihood in the spatial linear model. *Biometrika 76*, 289–295.

Marin, J.-M., P. Pudlo, C. P. Robert, and R. J. Ryder (2012). Approximate Bayesian computational methods. *Statistical Computing 22*, 1167–1180.

Maronna, R. A., R. D. Martin, V. J. Yohai, and M. Salibián-Barrera (2019). *Robust Statistics: Theory and Methods (with R).* New York: Wiley.

Marron, J. S. and M. P. Wand (1992). Exact mean integrated squared error. *Annals of Statistics 20*, 712–736.

Matheron, G. (1963). Principle of geostatistics. *Economic Geology 58*, 1246–1266.

McCullagh, P. (1983). Quasi-likelihood functions. *Annals of Statistics 11*, 59–67.

McCullagh, P. (1986). Quasi-likelihood. *In Encyclopedia of Statistical Sciences, eds. S. Kotz and N. L. Johnson. Wiley, New York 7*, 464–467.

McCullagh, P. (1987). *Tensor Methods in Statistics.* London: Chapman and Hall.

McCullagh, P. (1991). Quasi-likelihood and estimating functions. In *Statistical Theory and Modelling, eds. D. V. Hinkley, N. Reid and E. J. Snell*, pp. 265–286. London: Chapman and Hall.

McCullagh, P. (2002). What is a statistical model? *Annals of Statistics 30*, 1225–1310.

McCullagh, P. (2022). *Ten Projects in Applied Statistics.* New York: Springer.

McCullagh, P. and J. A. Nelder (1989). *Generalized Linear Models* (2nd ed.). London: Chapman and Hall/CRC.

McCullagh, P. and R. Tibshirani (1990). A simple method for the adjustment of profile likelihoods. *Journal of the Royal Statistical Society B 52*, 325–344.

McCullagh, P. and M. F. Tresoldi (2021). A likelihood analysis of quantile-matching transformations. *Biometrika 108*, 247–251.

McCulloch, C. E. (1997). Maximum likelihood algorithms for generalized linear mixed models. *Journal of the American Statistical Association 92*, 162–170.

McCulloch, C. E. and J. M. Neuhaus (2011). Prediction of random effects in linear and generalized linear models under model misspecification. *Biometrics 67*, 270–279.

McCulloch, C. E. and S. R. Searle (2004). *Generalized, Linear, and Mixed Models.* New York: Wiley.

McFadden, D. (1989). A method of simulated moments for estimation of discrete response models without numerical integration. *Econometrica 57*, 995–1026.

McGoff, K., S. Mukherjee, A. Nobel, and N. Pillai (2015). Consistency of maximum likelihood estimation for some dynamical systems. *Annals of Statistics 43*, 1–29.

McKenzie, E. (1986). Autoregressive moving-average processes with negative-binomial and geometric marginal distributions. *Advances in Applied Probability 18*, 679–705.

McLachlan, G. J. and T. Krishnan (2008). *The EM Algorithm and Extensions* (2nd ed.). New York: Wiley.

McLachlan, G. J., S. X. Lee, and S. I. Rathnayake (2019). Finite mixture models. *Annual Review of Statistics and Its Application 6*, 355–378.

Meinshausen, N. and P. Bühlmann (2006). High-dimensional graphs and variable selection with the Lasso. *Annals of Statistics 34*, 1436–1462.

Meng, X. L. and D. A. Van Dyk (1998). Fast EM-type implementations for mixed effects models. *Journal of the Royal Statistical Society B 60*, 559–578.

Miller, J. W. (2021). Asymptotic normality, concentration, and coverage of generalized posteriors. *Journal of Machine learning Research 22*, 1–53.

Mohri, M., A. Rostamizadeh, and A. Talwalkar (2018). *Foundations of Machine Learning* (2nd ed.). Cambridge, MA: MIT Press.

Molenberghs, G. and G. Verbeke (2005). *Models for Discrete Longitudinal Data.* New York: Springer.

Møller, J., A. R. Syversveen, and R. P. Waagepetersen (1998). Log–Gaussian Cox processes. *Scandinavian Journal of Statistics 25*, 451–482.

Møller, J. and R. P. Waagepetersen (2003). *Statistical Inference and Simulation for Spatial Point Processes.* Boca Raton, FL: Chapman & Hall/CRC.

Mukherjee, S., J. Son, and B. B. Bhattacharya (2025). Phase transitions of the maximum likelihood estimators in the p-spin Curie-Weiss model. *Bernoulli 31*, 1502–1526.

Mukhopadhyay, P. (2004). *An Introduction to Estimating Functions.* Oxford: Alpha Science International.

Murphy, S. A. and A. W. van der Vaart (2000). On profile likelihood. *Journal of the American Statistical Association 95*, 449–465.

Nelder, J. A. and D. Pregibon (1987). An extended quasi-likelihood function. *Biometrika 74*, 221–232.

Nelsen, R. B. (2007). *An Introduction to Copulas* (2nd ed.). New York: Springer.

Newey, W. K. (1997). Convergence rates and asymptotic normality for series estimators. *Journal of Econometrics 79*, 147–168.

Newey, W. K. and D. McFadden (1994). Large sample estimation and hypothesis testing. In *Handbook of Econometrics*, Volume 4, pp. 2111–2245. Amsterdam: Elsevier.

Newey, W. K. and R. J. Smith (2004). Higher order properties of GMM and generalized empirical likelihood estimators. *Econometrica 72*, 219–255.

Newey, W. K. and K. D. West (1987). A simple, positive-definite, heteroskedasticity and autocorrelation consistent covariance matrix. *Econometrica 55*, 703–708.

Newey, W. K. and K. D. West (1994). Automatic lag selection in covariance matrix estimation. *The Review of Economic Studies 61*, 631–653.

Newton, M. A. and C. J. Geyer (1994). Bootstrap recycling: A Monte Carlo alternative to the nested bootstrap. *Journal of the American Statistical Association 89*, 905–912.

Neyman, J. and E. L. Scott (1948). Consistent estimates based on partially consistent observations. *Econometrica 16*, 1–32.

Ning, Y. and H. Liu (2017). A general theory of hypothesis tests and confidence regions for sparse high dimensional models. *Annals of Statistics 45*, 158–195.

Ning, Y., G. Y. Yi, and N. Reid (2017). A class of weighted estimating equations for semiparametric transformation models with missing covariates. *Scandinavian Journal of Statistics 45*, 87–109.

Oakes, D. (1977). The asymptotic information in censored survival data. *Biometrika 64*, 487–493.

Ormerod, J. T. and M. P. Wand (2012). Gaussian variational approximate inference for generalized linear mixed models. *Journal of Computational and Graphical Statistics 21*, 2–17.

Owen, A. B. (1988). Empirical likelihood ratio confidence intervals for a single functional. *Biometrika 75*, 237–249.

Owen, A. B. (1990). Empirical likelihood ratio confidence regions. *Annals of Statistics 18*, 90–120.

Owen, A. B. (2001). *Empirical Likelihood.* Boca Raton, FL: Chapman and Hall/CRC.

Owen, A. B. (2013). Self-concordance for empirical likelihood. *Canadian Journal of Statistics 41*, 387–397.

Pace, L. and A. Salvan (1997). *Principles of Statistical Inference from a neo-Fisherian Perspective.* Singapore: World Scientific.

Pace, L., A. Salvan, and N. Sartori (2011). Adjusting composite likelihood ratio statistics. *Statistica Sinica 21*, 129–148.

Pan, J. and G. Mackenzie (2003). On modelling mean-covariance structures in longitudinal studies. *Biometrika 90*, 239–244.

Papageorgiou, G., K. Mauff, A. Tomer, and D. Rizopoulos (2019). An overview of joint modeling of time-to-event and longitudinal outcomes. *Annual Review of Statistics and its Application 6*, 223–240.

Papaspiliopoulos, O., T. Stumpf-Fétizon, and G. Zanella (2023). Scalable Bayesian computation for crossed and nested hierarchical models. *Electronic Journal of Statistics 17*, 3575–3612.

Parzen, E. (1962). On estimation of a probability density function and mode. *Annals of Mathematical Statistics 33*, 1065–1076.

Patterson, H. D. and R. Thompson (1971). Recovery of inter-block information when block sizes are unequal. *Biometrika 58*, 545–554.

Pauli, F., W. Racugno, and L. Ventura (2011). Bayesian composite marginal likelihoods. *Statistica Sinica 21*, 149–164.

Pawitan, Y. (2026). *In All Likelihood* (2nd ed.). Oxford: Oxford University Press.

Pearson, K. (1936). Method of moments and method of maximum likelihood. *Biometrika 28*, 34–47.

Pepe, M. S. (1992). Inference using surrogate outcome data and a validation sample. *Biometrika 79*, 355–365.

Pepe, M. S. and G. L. Anderson (1994). A cautionary note on inference for marginal regression models with longitudinal data and general correlated response data. *Communications in Statistics—Simulation and Computation 23*, 939–951.

Pepe, M. S., M. Reilly, and T. R. Fleming (1994). Auxiliary outcome data and the mean score method. *Journal of Statistical Planning and Inference 42*, 137–160.

Petrone, S., J. Rousseau, and C. Scricciolo (2014). Bayes and empirical Bayes: do they merge? *Biometrika 101*, 285–302.

Pierce, D. A. (1982). The asymptotic effect of substituting estimators for parameters in certain types of statistics. *Annals of Statistics 10*, 475–478.

Pierce, D. A. and R. Bellio (2017). Modern likelihood-frequentist inference. *International Statistical Review 85*, 519–541.

Pinheiro, J. C. and D. M. Bates (1995). Approximations to the log-likelihood function in the nonlinear mixed-effects model. *Journal of Computational and Graphical Statistics 4*, 12–35.

Pitt, M. K. and N. Shephard (1999). Filtering via simulation: Auxiliary particle filters. *Journal of the American Statistical Association 94*, 590–599.

Plummer, M. (2003). JAGS: A program for analysis of Bayesian graphical models using Gibbs sampling. In K. Hornik and A. Zeileis (Eds.), *Proceedings of the 3rd International Workshop on Distributed Statistical Computing*, pp. 1–10. Vienna: Vienna University of Economics and Business.

Plummer, M. (2023). Bayesian computational tools. *Annual Review of Statistics and Its Application 10*, 401–425.

Porcu, E., R. Furrer, and D. Nychka (2021). 30 years of space–time covariance functions. *Wiley Interdisciplinary Reviews: Computational Statistics 13*, e1512.

Portnoy, S. (1988). Asymptotic behavior of likelihood methods for exponential families when the number of parameters tends to infinity. *Annals of Statistics 16*, 356–366.

Pötscher, B. M. and H. Leeb (2009). On the distribution of penalized maximum likelihood estimators: The LASSO, SCAD, and thresholding. *Journal of Multivariate Analysis 100*, 2065–2082.

Pourahmadi, M. (1999). Joint mean-covariance models with applications to longitudinal data: Unconstrained parameterisation. *Biometrika 86*, 677–690.

Pregibon, D. (1984). Review of *Generalized Linear Models*, by P. McCullagh and J. A. Nelder. *Annals of Statistics 12*, 1589–1596.

Prentice, R. L. (1988). Correlated binary regression with covariates specific to each binary observation. *Biometrics 44*, 1033–1048.

Qin, J. and J. F. Lawless (1994). Empirical likelihood and general estimating equations. *Annals of Statistics 22*, 300–325.

Qu, A., B. G. Lindsay, and B. Li (2000). Improving generalised estimating equations using quadratic inference functions. *Biometrika 87*, 823–836.

Quenouille, M. H. (1956). Notes on bias in estimation. *Biometrika 43*, 353–360.

R Core Team (2025). *R: A Language and Environment for Statistical Computing.* Vienna, Austria: R Foundation for Statistical Computing.

Rao, C. R. (1945). Information and the accuracy attainable in the estimation of statistical parameters. *Bulletin of the Calcutta Mathematical Society 37*, 81–89.

Ravikumar, P., M. J. Wainwright, and J. Lafferty (2010). High-dimensional Ising model selection using ℓ_1-regularized regression. *Annals of Statistics 38*, 1287–1319.

Reid, N. (1988). Saddlepoint methods and statistical inference. *Statistical Science 3*, 213–227.

Reid, N. (1995). The roles of conditioning in inference (with discussion). *Statistical Science 10*, 138–157.

Reid, N. and H. S. Battey (2026). O. E. Barndorff-Nielsen's approximate conditional inference. *Bernoulli 32*, 81–95.

Reid, N. and Y. Sun (2010). Assessing sensitivity to priors using higher order approximations. *Communications in Statistics—Theory and Methods 39*, 1373–1386.

Ribatet, M., D. Cooley, and A. C. Davison (2012). Bayesian inference from composite likelihoods, with an application to spatial extremes. *Statistica Sinica 22*, 813–845.

Ripley, B. D. (1981). *Spatial Statistics.* London: Wiley.

Rizopoulos, D. J. M. (2012). *Joint Models for Longitudinal and Time-to-Event Data With Applications in R.* Boca Raton, FL: Chapman and Hall/CRC.

Robbins, H. (1956). An empirical Bayes approach to statistics. *Proceedings of the 3rd Berkeley Symposium on Mathematical Statistics and Probability 1*, 157–163.

Robert, C. P. (2014). Bayesian computational tools. *Annual Review of Statistics and Its Application 1*, 153–177.

Robins, J. M., A. Rotnitzky, and L. P. Zhao (1994). Estimation of regression coefficients when some regressors are not always observed. *Journal of the American Statistical Association 89*, 846–866.

Robins, J. M., A. Rotnitzky, and L. P. Zhao (1995). Analysis of semiparametric regression models for repeated outcomes in the presence of missing data. *Journal of the American Statistical Association 90*, 106–121.

Robinson, P. M. (1977). The estimation of a nonlinear moving average model. *Stochastic Processes and their Applications 5*, 81–90.

Rosenblatt, M. (1956). Remarks on some nonparametric estimates of a density function. *Annals of Mathematical Statistics 27*, 832–837.

Rothstein, H. R., A. J. Sutton, and M. Borenstein (2006). *Publication Bias in Meta-Analysis: Prevention, Assessment and Adjustments.* New York: Wiley.

Rotnitzky, A., D. R. Cox, M. Bottai, and J. Robins (2000). Likelihood-based inference with singular information matrix. *Bernoulli 6*, 243–284.

Rotnitzky, A. and N. P. Jewell (1990). Hypothesis testing of regression parameters in semiparametric generalized linear models for cluster correlated data. *Biometrika 77*(3), 485–497.

Rousseau, J. (2016). On the frequentist properties of Bayesian nonparametric methods. *Annual Review of Statistics and Its Application 3*, 211–231.

Ročková, V. and E. I. George (2018). The spike-and-slab LASSO. *Journal of the American Statistical Association 113*, 431–444.

Royall, R. J. (1997). *Statistical Evidence: A Likelihood Paradigm.* Boca Raton, FL: Chapman and Hall/CRC.

Rubin, D. B. (1976). Inference and missing data. *Biometrika 63*, 581–592.

Rue, H. and L. Held (2005). *Gaussian Markov Random Fields: Theory and Applications.* London: Chapman and Hall/CRC.

Rue, H., S. Martino, and N. Chopin (2009). Approximate Bayesian inference for latent Gaussian models by using integrated nested Laplace approximation (with discussion). *Journal of the Royal Statistical Society B 71*, 319–392.

Ruppert, D., M. P. Wand, and R. J. Carroll (2003). *Semiparametric Regression.* Cambridge: Cambridge University Press.

Salerno, S. and Y. Li (2023). High-dimensional survival analysis: methods and applications. *Annual Review of Statistics and Its Application 10*, 25–49.

Samworth, R. J. (2018). Recent progress in log-concave density estimation. *Statistical Science 33*, 493–509.

Sant, J., P. A. Jenkins, J. Koskela, and D. Spanò (2022). Convergence of likelihood ratios and estimators for selection in non-neutral Wright–Fisher diffusions. *Scandinavian Journal of Statistics 49*, 1728–1760.

Sartori, N. (2003). Modified profile likelihoods in models with stratum nuisance parameters. *Biometrika 90*, 533–549.

Sartori, N. and T. A. Severini (2004). Conditional likelihood inference in generalized linear mixed models. *Statistica Sinica 14*, 349–360.

Schabenberger, O. and C. A. Gotway (2005). *Statistical Methods for Spatial Data Analysis.* Boca Raton, FL: Chapman and Hall/CRC.

Schall, R. (1991). Estimation in generalized linear models with random effects. *Biometrika 78*, 719–727.

Schervish, M. J. (1995). *Theory of Statistics.* New York: Springer.

Schervish, M. J., J. B. Kadane, T. Seidenfeld, and R. B. Stern (2025). Information lost by conditioning on the random margins in a two-by-two table. *Biometrika 112*, asaf064.

Schoenberg, I. J. (1946). Contributions to the problem of approximation of equidistant data by analytic functions. *Quarterly of Applied Mathematics 4*, 45–99, 112–141.

Scholz, F. W. (2006). Maximum likelihood estimation. In S. Kotz, C. B. Read, N. Balakrishnan, B. Vidakovic, and N. L. Johnson (Eds.), *Encyclopedia of Statistical Science*, Volume 7. New York: Wiley.

Schwartz, L. (1965). On Bayes' procedures. *Zeitschrift für Wahrscheinlichkeitstheorie und verwandte Gebiete 4*, 10–26.

Schwarz, G. (1978). Estimating the dimension of a model. *Annals of Statistics 6*, 461–464.

Schweder, T. and N. L. Hjort (2016). *Confidence, Likelihood, Probability.* Cambridge: Cambridge University Press.

Scott, D. W. (1992). *Multivariate Density Estimation: Theory, Practice, and Visualization.* New York: Wiley.

Scott, D. W. and G. R. Terrell (1987). Biased and unbiased cross-validation in density estimation. *Journal of the American Statistical Association 82*, 1131–1146.

Self, S. and K.-Y. Liang (1987). Asymptotic properties of maximum likelihood estimators and likelihood ratio tests under nonstandard conditions. *Journal of the American Statistical Association 82*, 605–610.

Serfling, R. J. (1980). *Approximation Theorems of Mathematical Statistics.* New York: Wiley.

Severini, T. A. (2000). *Likelihood Methods in Statistics.* Oxford: Oxford University Press.

Shah, J., M. Cardone, C. Rush, and A. Dytso (2025). Generalized linear models with 1-bit measurements: Asymptotics of the maximum likelihood estimator. In *ICASSP 2025 - 2025 IEEE International Conference on Acoustics, Speech and Signal Processing (ICASSP)*, pp. 1–5.

Sheather, S. J. and M. C. Jones (1991). A reliable data-based bandwidth selection method for kernel density estimation. *Journal of the Royal Statistical Society B 53*, 683–690.

Shen, C., D. Li, and H. Tong (2025). A note on parameter orthogonality for multi-parameter distributions. https://arxiv.org/pdf/2501.08093v3.

Shen, J., R. Y. Liu, and M. Xie (2018). Prediction with confidence—A general framework for predictive inference. *Journal of Statistical Planning and Inference 195*, 126–140.

Shen, X. (1997). On methods of sieves and penalization. *Annals of Statistics 25*, 2555–2591.

Shen, X., W. Pan, and Y. Zhu (2012). Likelihood-based selection and sharp parameter estimation. *Journal of the American Statistical Association 107*, 223–232.

Shen, X. and W. H. Wong (1994). Convergence rate of sieve estimates. *Annals of Statistics 22*, 580–615.

Shestopaloff, A. Y. and R. M. Neal (2014). On Bayesian inference for the M/G/1 queue with efficient MCMC sampling. https://arxiv.org/abs/1401.5548.

Shi, H. and M. Drton (2025). On universal inference in Gaussian mixture models. https://arxiv.org/pdf/2407.19361.

Shumway, R. H. and D. S. Stoffer (2016). *Time Series Analysis and Its Applications* (3rd ed.). New York: Springer.

Shun, Z. and P. McCullagh (1995). Laplace approximation of high dimensional integrals. *Journal of the Royal Statistical Society B 57*, 749–760.

Silverman, B. W. (1985). Some aspects of the spline smoothing approach to non-parametric regression curve fitting (with discussion). *Journal of the Royal Statistical Society 47*, 1–21.

Silverman, B. W. (1986). *Density Estimation for Statistics and Data Analysis.* London: Chapman and Hall.

Skalski, T. and T. Stroiński (2025). Level sets and maximum likelihood estimation for the Ising model. https://arxiv.org/pdf/2511.20925.

Skovgaard, I. M. (1996). An explicit large-deviation approximation to one-parameter tests. *Bernoulli 2*, 145–165.

Skovgaard, I. M. (2001). Likelihood asymptotics. *Scandinavian Journal of Statistics 28*, 3–32.

Slutsky, E. (1927). The summation of random causes as the source of cyclic processes (in Russian). *Problems of Economic Conditions 3*, 34–64.

Small, C. G., J. Wang, and Z. Yang (2000). Eliminating multiple root problems in estimation. *Statistical Science 15*, 313–341.

Smith, Jr, A. A. (1993). Estimating nonlinear time-series models using simulated vector autoregressions. *Journal of Applied Econometrics 8*, S63–S84.

Smith, R. L. (1985). Maximum likelihood estimation in a class of nonregular cases. *Biometrika 72*, 67–90.

Smith, R. L. (1987). Contribution to the discussion of "Parameter orthogonality and approximate conditional inference" by D. R. Cox and N. Reid. *Journal of the Royal Statistical Society B 49*, 21–22.

Spokoiny, V. and M. Panov (2025). Accuracy of Gaussian approximation for high-dimensional posterior distributions. *Bernoulli 31*, 843–867.

Staicu, A.-M. and N. Reid (2008). On probability matching priors. *Canadian Journal of Statistics 36*, 613–622.

Stefanski, L. A. and D. D. Boos (2002). The calculus of M-estimation. *American Statistician 56*, 29–38.

Stefanski, L. A., Y. Wu, and K. White (2014). Variable selection in nonparametric classification via measurement error model selection likelihoods. *Journal of the American Statistical Association 109*, 574–589.

Stein, C. (1959). An example of wide discrepancy between fiducial and confidence intervals. *Annals of Mathematical Statistics 30*, 877–880.

Stein, M. L. (1999). *Interpolation of Spatial Data: Some Theory for Kriging.* New York: Springer.

Stoltenberg, E. A. and N. L. Hjort (2021). Models and inference for on–off data via clipped Ornstein–Uhlenbeck processes. *Scandinavian Journal of Statistics 48*, 908–929.

Stone, M. (1977). An asymptotic equivalence of choice of model by cross-validation and Akaike's criterion. *Journal of the Royal Statistical Society B 39*, 44–47.

Stram, D. O. and J. W. Lee (1994). Variance components testing in the longitudinal mixed effects model. *Biometrics 50*, 1171–1177.

Street, J. O., R. J. Carroll, and D. Ruppert (1988). A note on computing robust regression estimates via iteratively reweighted least squares. *American Statistician 42*, 152–154.

Strieder, D. and M. Drton (2022). On the choice of the splitting ratio for the split likelihood ratio test. *Electronic Journal of Statistics 16*, 6631–6650.

Sun, J. and C. R. Loader (1994). Simultaneous confidence bands for linear regression and smoothing. *Annals of Statistics 22*, 1328–1345.

Sur, P. and E. J. Candès (2019). A modern maximum likelihood theory for high-dimensional regression. *Proceedings of the National Academy of Sciences 116*, 14516–14525.

Sur, P., Y. Chen, and E. J. Candès (2019). The likelihood ratio test in high-dimensional logistic regression is asymptotically a rescaled Chi-square. *Probability Theory and Related Fields 175*, 487–558.

Sweeting, T. J. (1987). Contribution to the discussion of "Parameter orthogonality and approximate conditional inference" by D. R. Cox and N. Reid. *Journal of the Royal Statistical Society B 49*, 20–21.

Takeishi, S. (2023). Asymptotic properties of the smoothed maximum partial likelihood estimator in the change-plane Cox model. *Scandinavian Journal of Statistics 50*, 1503–1531.

Takeuchi, K. (1976). Distribution of information statistics and criteria for adequacy of models (in Japanese). *Mathematical Sciences 153*, 12–18.

Tang, C. Y. and C. Leng (2010). Penalized high-dimensional empirical likelihood. *Biometrika 97*, 905–919.

Tang, Y. and N. Reid (2025). Laplace and saddlepoint approximations in high dimensions. *Bernoulli 31*, 1759–1788.

Tavaré, S., D. J. Balding, and P. Donnelly (1997). Inferring coalescence times from DNA sequence data. *Genetics 145*, 505–518.

Taylor, S. J. (1994). Modeling stochastic volatility: A review and comparative study. *Mathematical Finance 4*, 183–204.

Thomas, D. R. and G. L. Grunkemeier (1975). Confidence interval estimation of survival probabilities for censored data. *Journal of the American Statistical Association 70*, 865–871.

Tibshirani, R. (1984). *Local Likelihood Estimation.* Phd thesis, Department of Statistics, Stanford University.

Tibshirani, R. (1996). Regression shrinkage and selection via the Lasso. *Journal of the Royal Statistical Society B 58*, 267–288.

Tibshirani, R. and T. J. Hastie (1987). Local likelihood estimation. *Journal of the American Statistical Association 82*, 559–568.

Tibshirani, R. J. (2023). Conformal prediction: advanced topics in statistical learning, Spring 2023. `https://www.stat.berkeley.edu/~ryantibs/statlearn-s23/lectures/conformal.pdf`.

Tibshirani, R. J., J. Taylor, R. Lockhart, and R. Tibshirani (2016). Exact post-selection inference for sequential regression procedures. *Journal of the American Statistical Association 111*, 600–620.

Tierney, L. and J. B. Kadane (1986). Accurate approximations for posterior moments and marginal densities. *Journal of the American Statistical Association 81*, 82–86.

Tikhonov, A. N. (1943). On the stability of inverse problems. *Comptes Rendus (Doklady) de l'Academie des Sciences de l'URSS 39*, 195–198.

Tong, H. (1988). A note on local parameter orthogonality and Levinson–Durbin algorithm. *Biometrika 75*, 788–789.

Tong, H. (1990). *Non-linear Time Series.* Oxford: Oxford University Press.

Tsao, M. (2024). Sparse maximum likelihood estimation for regression models. https://arxiv.org/abs/2403.09081.

Tsay, R. S. (2010). *Analysis of Financial Time Series* (3rd ed.). Hoboken, NJ: Wiley.

Tse, T. and A. C. Davison (2022). A note on universal inference. *Stat 11*, e501.

Tsiatis, A. A. (2006). *Semiparametric Theory and Missing Data.* New York: Springer.

Tsiatis, A. A. and M. Davidian (2004). Joint modeling of longitudinal and time-to-event data: an overview. *Statistica Sinica 14*, 809–834.

van der Pas, S., B. Szabó, and A. W. van der Vaart (2017). Uncertainty quantification for the horseshoe (with discussion). *Bayesian Analysis 12*, 1221–1274.

van der Vaart, A. W. (1998). *Asymptotic Statistics.* Cambridge: Cambridge University Press.

Varin, C. (2008). On composite marginal likelihoods. *AStA Advances in Statistical Analysis 92*, 1–28.

Varin, C., N. Reid, and D. Firth (2011). An overview of composite likelihood methods. *Statistica Sinica 21*, 5–42.

Vehtari, A. and J. Ojanen (2012). A survey of Bayesian predictive methods for model assessment, selection and comparison. *Statistics Surveys 6*, 142–228.

Venables, W. N. and B. D. Ripley (2002). *Modern Applied Statistics with S* (4th ed.). New York: Springer.

Verbeke, G. and E. Lesaffre (1997). The effect of misspecifying the random-effects distribution in linear mixed models for longitudinal data. *Computational Statistics and Data Analysis 23*, 541–556.

Verbeke, G. and G. Molenberghs (2000). *Linear Mixed Models for Longitudinal Data.* New York: Springer.

Viraswami, K. and N. Reid (1996). Higher-order asymptotics under model misspecification. *Canadian Journal of Statistics 24*, 263–278.

Viraswami, K. and N. Reid (1998). A note on the likelihood-ratio statistic under model misspecification. *Canadian Journal of Statistics 26*, 161–168.

Vovk, V. and R. Wang (2023). Confidence and discoveries with E-values. *Statistical Science 38*, 329–354.

Vuong, Q. H. (1989). Likelihood ratio tests for model selection and non-nested hypotheses. *Econometrica 57*, 307–333.

Wahba, G. (1978). Improper priors, spline smoothing and the problem of guarding against model errors in regression. *Journal of the Royal Statistical Society B 40*, 364–372.

Wahba, G. (1983). Bayesian "confidence intervals" for the cross-validated smoothing spline. *Journal of the Royal Statistical Society B 45*, 133–150.

Wahba, G. (1990). *Spline Models for Observational Data.* Philadelphia: SIAM.

Wainwright, M. (2019). *High-Dimensional Statistics: A Non-Asymptotic Viewpoint.* Cambridge: Cambridge University Press.

Wald, A. (1949). Note on the consistency of the maximum likelihood estimate. *Annals of Mathematical Statistics 20*, 595–601.

Walker, A. M. (1969). On the asymptotic behaviour of posterior distributions. *Journal of the Royal Statistical Society B 31*, 80–88.

Waller, L. A. and C. A. Gotway (2004). *Applied Spatial Statistics for Public Health Data.* New York: Wiley.

Walther, G. (2009). Inference and modeling with log-concave distributions. *Statistical Science 24*, 319–327.

Wand, M. P. and M. C. Jones (1994). *Kernel Smoothing.* London: Chapman & Hall/CRC.

Wang, C. Y., Y. Huang, E. C. Chao, and M. K. Jeffcoat (2008). Expected estimating equations for missing data, measurement error, and misclassification, with application to longitudinal nonignorable missing data. *Biometrics 64*, 85–95.

Wang, C. Y. and M. S. Pepe (2000). Expected estimating equations to accommodate covariate measurement error. *Journal of the Royal Statistical Society B 62*, 509–524.

Wang, C. Y. and X. Song (2013). Expected estimating equations via EM for proportional hazards regression with covariate misclassification. *Biostatistics 14*, 351–365.

Wang, N., J. Rauh, and H. Massam (2019). Approximating faces of marginal polytopes in discrete hierarchical models. *Annals of Statistics 47*, 1203–1233.

Wang, Q.-H. and B.-Y. Jing (2003). Empirical likelihood for partial linear models. *Annals of the Institute of Statistical Mathematics 55*, 585–595.

Wang, Y. (2019). Computation of the nonparametric maximum likelihood estimate of a univariate log-concave density. *Wiley Interdisciplinary Reviews: Computational Statistics 11*, e1452.

Wasserman, L. (2006). *All of Nonparametric Statistics.* New York: Springer.

Wasserman, L., A. Ramdas, and S. Balakrishnan (2020). Universal inference. *Proceedings of the National Academy of Sciences 117*, 16880–16890.

Wedderburn, R. W. M. (1974). Quasi-likelihood functions, generalized linear models, and the Gauss-Newton method. *Biometrika 61*, 439–447.

West, M., P. J. Harrison, and H. S. Migon (1985). Dynamic generalized linear models and Bayesian forecasting. *Journal of the American Statistical Association 80*, 73–83.

White, H. (1982). Maximum likelihood estimation of misspecified models. *Econometrica 50*, 1–25.

White, H. (1994). *Estimation, Inference and Specification Analysis.* Cambridge: Cambridge University Press.

Williams, E. J. (1959). *Regression Analysis.* New York: Wiley.

Wold, H. (1938). *A Study in the Analysis of Stationary Time Series.* Stockholm: Almquist & Wicksell.

Wolfinger, R. and M. O'Connell (1993). Generalized linear mixed models a pseudo-likelihood approach. *Journal of Statistical Computation and Simulation 48*, 233–243.

Wood, S. N. (2010). Statistical inference for noisy nonlinear ecological dynamic systems. *Nature 466*, 1102–1104.

Wood, S. N. (2011). Fast stable restricted maximum likelihood and marginal likelihood estimation of semiparametric generalized linear models. *Journal of the Royal Statistical Society B 73*, 3–36.

Wood, S. N. (2017). *Generalized Additive Models: an Introduction with R* (2nd ed.). Boca Raton, FL: Chapman and Hall.

Wood, S. N. (2025). Generalized additive models. *Annual Review of Statistics and Its Application 12*, 497–526.

Wood, S. N., N. Pya, and B. Säfken (2016). Smoothing parameter and model selection for general smooth models. *Journal of the American Statistical Association 111*, 1548–1575.

Wu, C. F. J. (1983). On the convergence properties of the EM algorithm. *Annals of Statistics 11*, 95–103.

Wu, Y. and L. Wang (2020). A survey of tuning parameter selection for high-dimensional regression. *Annual Review of Statistics and Its Application 7*, 209–226.

Xie, M. G. and K. Singh (2013). Confidence distribution, the frequentist distribution estimator of a parameter: A review (with discussion). *International Statistical Review 81*, 3–39.

Xu, X. and N. Reid (2011). On the robustness of maximum composite likelihood estimate. *Journal of Statistical Planning and Inference 141*, 3047–3054.

Xu, X., N. Reid, and L. Xu (2024). Note on information bias and efficiency of composite likelihood. *Statistica Sinica 34*, 523–526.

Xue, L., H. Zou, and T. Cai (2012). Nonconcave penalized composite conditional likelihood of sparse Ising models. *Annals of Statistics 40*, 1403–1429.

Yan, Y. and G. Y. Yi (2016). A class of functional methods for error-contaminated survival data under additive hazards models with replicate measurements. *Journal of the American Statistical Association 111*, 684–695.

Yanagimoto, T. and E. Yamamoto (1991). The role of unbiasedness in estimating equations. In V. P. Godambe (Ed.), *Estimating Functions*, pp. 89–102. Oxford: Oxford University Press.

Yee, T. W. (2022). On the Hauck-Donner effect in Wald's test: detection, tipping points, and parameter space characterization. *Journal of the American Statistical Association 117*, 1763–1774.

Yi, G. Y. (2017a). Composite likelihood/pseudolikelihood. In *Wiley StatsRef: Statistics Reference Online*, pp. 1–14. New York: Wiley.

Yi, G. Y. (2017b). *Statistical Analysis with Measurement Error or Misclassification: Strategy, Method and Application.* New York: Springer.

Yi, G. Y. and R. J. Cook (2002). Marginal methods for incomplete longitudinal data arising in clusters. *Journal of the American Statistical Association 97*, 1131–1147.

Yi, G. Y., R. J. Cook, and B. Chen (2010). Estimating functions for evaluating treatment effects in cluster-randomized longitudinal studies in the presence of drop-out and noncompliance. *Canadian Journal of Statistics 38*, 232–255.

Yi, G. Y., Y. Ma, and R. J. Carroll (2012). A functional generalized method of moments approach for longitudinal studies with missing responses and covariate measurement error. *Biometrika 99*, 151–165.

Yi, G. Y. and N. Reid (2010). A note on mis-specified estimating functions. *Statistica Sinica 20*, 1749–1769.

Yi, G. Y., Y. Yan, X. Liao, and D. Spiegelman (2018). Parametric regression analysis with covariate misclassification in main study/validation study designs. *International Journal of Biostatistics 14*, 20170002.

Yuan, K.-H. and R. I. Jennrich (1998). Asymptotics of estimating equations under natural conditions. *Journal of Multivariate Analysis 65*, 245–260.

Yuan, M. and Y. Lin (2006). Model selection and estimation in regression with grouped variables. *Journal of the Royal Statistical Society B 68*, 49–67.

Yuan, M. and Y. Lin (2007). Model selection and estimation in the gaussian Graphical model. *Biometrika 94*, 19–35.

Yule, G. U. (1926). Why do we sometimes get nonsense-correlations between time-series?–a study in sampling and the nature of time-series. *Journal of the Royal Statistical Society 89*, 1–63.

Yule, G. U. (1927). VII. On a method of investigating periodicities disturbed series, with special reference to Wolfer's sunspot numbers. *Philosophical Transactions of the Royal Society of London A 226*, 267–298.

Zeger, S. L. and M. R. Karim (1991). Generalized linear models with random effects; a Gibbs sampling approach. *Journal of the American Statistical Association 86*, 79–86.

Zeileis, A. (2006). Object-oriented computation of sandwich estimators. *Journal of Statistical Software 16*, 1–16.

Zeileis, A., S. Köll, and N. Graham (2020). Various versatile variances: An object-oriented implementation of clustered covariances in R. *Journal of Statistical Software 95*, 1–36.

Zeng, D., F. Gao, and D. Y. Lin (2017). Maximum likelihood estimation for semiparametric regression models with multivariate interval-censored data. *Biometrika 104*, 505–525.

Zhang, C.-H. (2010). Nearly unbiased variable selection under minimax concave penalty. *Annals of Statistics 38*, 894–942.

Zhang, C.-H. and S. S. Zhang (2014). Confidence intervals for low dimensional parameters in high dimensional linear models. *Journal of the Royal Statistical Society B 76*, 217–242.

Zhang, F. and C. Gao (2020). Convergence rates of variational posterior distributions. *Annals of Statistics 48*, 2180–2207.

Zhang, L. and B. A. Shaby (2022). Uniqueness and global optimality of the maximum likelihood estimator of the generalized extreme value distribution. *Biometrika 109*, 853–864.

Zhang, W., C. Leng, and C. Y. Tang (2015). A joint modelling approach for longitudinal studies. *Journal of the Royal Statistical Society B 77*, 219–238.

Zhang, Y., R. Li, and C.-L. Tsai (2010). Regularization parameter selections via generalized information criterion. *Journal of the American Statistical Association 105*, 312–323.

Zhao, P. and B. Yu (2006). On model selection consistency of Lasso. *Journal of Machine Learning Research 7*, 2541–2563.

Zhao, Q., P. Sur, and E. J. Candès (2022). The asymptotic distribution of the MLE in high-dimensional logistic models: Arbitrary covariance. *Bernoulli 28*, 1835–1861.

Zhu, Y. L. and N. Reid (1994). Information, ancillarity and efficiency in the presence of nuisance parameters. *Canadian Journal of Statistics 22*, 111–124.

Zou, H. (2006). The adaptive Lasso and its oracle properties. *Journal of the American Statistical Association 101*, 1418–1429.

Zou, H. and T. J. Hastie (2005). Regularization and variable selection via the elastic net. *Journal of the Royal Statistical Society B 67*, 301–320.

Zucchini, W. and I. L. MacDonald (2009). *Hidden Markov Models for Time Series: an Introduction using R*. Boca Raton, FL: Chapman and Hall/CRC.

List of Authors

Index

For Product Safety Concerns and Information please contact our EU representative GPSR@taylorandfrancis.com Taylor & Francis Verlag GmbH, Kaufingerstraße 24, 80331 München, Germany